Elucidation of Abiotic Stress Signaling in Plants

The above image represents a depiction of activation of different signaling pathways by diverse stimuli that converge to activate intricate signaling and interaction networks to counter stress (top panel). Since environmental stresses influence most significantly to the reduction in potential crop yield, progress is now largely anticipated through functional genomics studies in plants through the use of techniques such as large-scale analysis of gene expression pattern in response to stress and construction, analysis and use of plant protein interactome networks maps for effective engineering strategies to generate stress tolerant crops (top panel). The molecular aspects of these signaling pathways are extensively studied in model plant *Arabidopsis thaliana* and crop plant rice (*Oryza sativa*) (below).

Girdhar K. Pandey
Editor

Elucidation of Abiotic Stress Signaling in Plants

Functional Genomics Perspectives, Volume 1

Springer

Editor
Girdhar K. Pandey
Department of Plant Molecular Biology
University of Delhi South Campus
Dhaula Kuan, New Delhi, India

ISBN 978-1-4939-2210-9 ISBN 978-1-4939-2211-6 (eBook)
DOI 10.1007/978-1-4939-2211-6

Library of Congress Control Number: 2015935335

Springer New York Heidelberg Dordrecht London

Printed on acid-free paper

Springer Science+Business Media LLC New York is part of Springer Science+Business Media (www.springer.com)

Preface

Plants are considered the backbone of life on earth. The colorful life on this planet has emerged as a consequence of over 3.5 billion years of unceasing evolution. Life on earth cannot sustain without plants, as they harness solar energy to produce sugars and oxygen, the primary constituents for supporting life. Humans are primarily dependent on plants and have developed a systematic discipline called "agriculture" to cultivate or domesticate plants over a period of time for food, biofuel, and fodder. At present time, crop productivity faces a major challenge from rapidly growing population and diminishing fertile land due to excessive anthropogenic activities. In addition, expanding human population and climate changes due to increased exploitation of natural resources imposes several major unfavorable conditions that reduce the crop productivity. These unfavorable conditions are primarily categorized as physical (or *abiotic*) and biological (or *biotic*) variables hindering normal growth and development in plants. Interestingly, stress perceived by one plant species may not be a stress factor for another plant species due to different growth habits and adaptation acquired during the course of evolution. Because of domestication and cultivation of crop plants by humans over a period of 10,000 years, many of these wild traits responsible for adaptive responses were lost, increasing the vulnerability of crop plants to biotic and abiotic stresses. Under abiotic stresses, limitation of water (drought), extremes of temperature (both high and low temperatures), nutrient deficiency, and soil contaminated with salt and heavy metals or pollutants are the major environmental factors contributing to crop losses worldwide.

In the past, agriculture has relied on breeding approaches to develop high yielding crop varieties which can grow optimally under stress conditions without affecting crop yield and productivity. In an effort to find an alternative tool faster than the traditional breeding approach, the last two decades has seen the advent and development of genetic engineering. This technique involves the identification, transfer, and stable integration of desired genes into genomes of crop plants to generate transgenic plants, exhibiting improved trait for tolerance against one or other stress factors in contained experimental conditions such as green houses.

However, plants are constantly exposed to a multitude of stresses at any given time in the natural environment, and not much has been achieved till now to generate crop varieties that can tolerate these multiple stresses without yield penalty. In order to develop stress-tolerant crop varieties with the ability to withstand multiple stresses in their environmental growth condition, an in-depth and systematic understanding of stress sensing, signal transduction, and generation of response is required.

Evolutionarily, the major distinction between plants and animals in sensing and responding to a plethora of stresses is due to their sessile versus mobile nature, respectively. In the case of animals, the primary response against a particular stress is avoidance of stress, whereas in plants, due to their immobilization, development of stress tolerance is the only escape response. Moreover, plants lack a well-defined brain and nervous system unlike their animal counterpart, leading to development of higher degree of plasticity in their communication skills by numerically expanding their signal transduction machinery. Despite the variances amid plants and animals, many of the signal transduction components can be found to be conserved. These include receptors, second messengers, signal-transducing molecules like kinases, phosphatases, small and large G-protein, and others, which finally affect the activity of either transcription factors to regulate the gene expression or transporters/channels, metabolic enzymes, and cytoskeletal proteins to directly change the physiology of the cell. Additionally, analogous to networking in the nervous systems, the signaling pathways in plants also exhibit scale-free web of networks instead of linear or definite pathways. These scale-free networks constitute extremely connected points called *nodes* and *hubs*, which are responsible for efficient processing, channeling, and integration of multiple signaling pathways at a given time to generate specificity as well as cross talk in the signaling networks.

Plants primarily rely on the complex, intertwined, and dynamic signal transduction pathways for developing a higher order of networks. This involves sophisticated control circuits like the nervous system of animals, where they learn, generate memory, alter behavior, and develop intelligence, which make them ready for future challenges. In nutshell, the complex interplay of signal transduction networks and machinery in plants leads them to sense, process, and integrate the signals they confront in their environment. Plants also develop behavioral changes accordingly or develop cognition and storage of processed information to adapt in rapidly changing or variable environment.

Identification of the role of a single or set of genes involved in signal transduction pathway has enabled researchers to understand and develop linear or complex signaling pathways, or maps in response to particular stimuli. However, because of the complete genome sequencing of many plant species including crop plants, a drift towards understanding the stress-signaling pathways involved in single or multiple stresses using high-throughput approaches has emerged. In the post-genomic era, the development of *-omic*-based approaches such as transcriptomic, proteomic, metabolomic, interactomic, and phenomic in several model organisms have laid the foundation of functional genomics. This area of plant science deals with the understanding of large network of genes and proteins and integration of transcript data to proteins which then go to metabolite, and the complex and dynamic interaction develops a response or phenotype.

Elucidation of Abiotic stress signaling in Plants: *Functional Genomics Perspectives* comprises 30 chapters divided into two volumes (Volume I and II) in which some of the world's most well-known plant biologists have contributed in the field of stress signaling in plants with a special emphasis on functional genomics aspects. This book provides timely research in the field of stress-mediated signaling to develop a better and holistic understanding of stress perception and its transduction followed by the generation of response. In spite of the advent of different approaches to develop stress-tolerant crops towards multiple stress conditions in the field, the success in achieving this goal is still unsatisfactory. This is because stress tolerance is a very complex process involving plethora of components starting from stress sensing to generation of final adaptive response. As mentioned above, there are several factors, which act as nodes and hub in the signaling pathways, also serving as master-control switches in regulating a myriad of stress-signaling pathways by affecting diverse target genes or gene products to finally bring about a stress tolerance response. Therefore, in-depth understanding of these master-control switches and key components in signal transduction pathway will be highly beneficial for designing crop plants tolerant to multiple stresses in the field.

Towards achieving this goal, this book is divided into two volumes comprising five sections. Volume I consists of two sections with 14 chapters. The first section "Functional Genomics Approaches in Signal transduction" discusses three chapters on various approaches used to understand the signal transduction networks. These chapters will aware the readers on practical aspect of various "Omic"-based approaches such as transcriptomic, proteomic, phosphoproteomic, metabolomic, interactomic, and phenomic to understand the functions of genes and gene networks in signaling under stress.

The next section "Components of Signal Transduction" comprises 11 chapters discussing the different components of signal transduction pathways. The first three chapters focus on calcium signaling by describing the genes encoding for CAX (calcium-H^+-exchanger) involved in sequestration of calcium ions into vacuoles and maintenance of Ca^{2+} homeostasis. Chapters 5 and 6 discuss the role of Ca^{2+} signal decoding components like sensor and effector proteins. Here, CBLs, CIPKs, and CDPKs gene families have been extensively worked out in model plant Arabidopsis under abiotic stress condition and their role in other crop plant is being elucidated. Chapter 7 describes the role of ROS as redox signaling component in regulating multiple stress responses and in manipulation of ROS levels for imparting stress tolerance in crop plants. The role of MAP kinases as crucial signaling components in biotic as well as abiotic stresses has been discussed in Chapter 8. MAP kinases act as converging points for several signaling pathways, involving the phosphorylation-based relay of information to regulate a large number of targets such as transcription factors, other kinases, and cytoskeletal proteins in stress signaling. The functional role of small and large G-protein acting as molecular switches to regulate both biotic and abiotic stresses has been discussed in Chapter 9. Chapter 10 deals with the molecular analysis of ABA receptor and ABA signaling in both biotic and abiotic stresses and genetic engineering of ABA receptor for developing stress-tolerant crop varieties. Auxin has been very well known as a plant growth regulator for several decades, and its emerging role in regulating stress signaling and responses is covered extensively in Chapter 11. SA (salicylic acid) is majorly involved in regulating biotic stress, but its role is also appreciated well in abiotic stresses as described in Chapter 12. In Chapter 13, the newly emerging role of methyl glyoxal (MG), which is a cytotoxin generated from both enzymatic and nonenzymatic pathways of metabolic reaction, has been discussed during several abiotic stresses. Chapter 14 discusses the role of immunophilins in diverse biological processes including development and stress management.

Volume II is divided into three sections encompassing 16 chapters. The first section of volume II emphasizes the gene expression regulation of stress signaling, with four chapters discussing the role of transcription factors (mediator complex in Chapter 1 and transcription factors of legumes in Chapter 2) and non-coding and small RNA (Chapters 3 and 4) in regulating abiotic stress responses.

Section two of volume II, comprises ten chapters, discusses the functional genomics aspect of heat/high temperature (Chapter 5), cold/freezing (Chapter 6), drought and dehydration (Chapter 7), flooding and submergence (Chapter 8), salinity (Chapter 9), UV-light (Chapter 10), heavy metal (Chapter 11), nitrogen (Chapter 12), and aging/senescence (Chapter 13) stress signaling responses. In this section, a detailed emphasis has been given in elaborating the respective stress-signaling pathway with a goal of potential candidate genes, which could be used for development of tolerant crop varieties by genetic manipulation and molecular breeding approaches. Moreover, cross talk or overlap in execution of several common signaling components open the scope for taming multiple stresses in future biotechnological intervention.

In the last section of volume II, Chapters 14–16 focus on the development of stress-tolerant crops and sustainable agriculture by utilizing the genes of signal transduction pathways. With the in-depth understanding of several signal transduction components and signaling pathways, the ultimate goal is to utilize the mechanistic knowledge and translate into useful tools to generate the crop varieties by either genetic manipulation of these signaling components or utilization of this knowledge for molecular marker-assisted breeding, ultimately augmenting stress tolerance in crop plants without compromising crop productivity.

Despite rigorous attempts, not every aspect of signaling pathways and components could be discussed here. Nevertheless, I strongly believe that two volumes covering signal transduction machinery and their components in stress condition, with a special emphasis to functional genomics, will be enormously useful to students, teachers, and research scientists.

I am indebted to all the contributors of this work, which could not be possibly compiled without their significant contributions. At last, I would like to express my sincere thanks to Dr. M. C. Tyagi and Dr. Amita Pandey for critical reading and help in copy-editing of this book. I also express my thanks to Ms. Manisha Sharma for designing the theme page.

New Delhi, India

Girdhar K. Pandey, Ph.D.

Contents

Contributors

Hussain Ara National Institute of Plant Genome Research, New Delhi, India

Peter D. Bickerton Faculty of Life Sciences, University of Manchester, Manchester, UK

Maik Böhmer Institut für Biologie und Biotechnologie der Pflanzen, Westfälische Wilhelms Universität, Münster, Germany

Pratikshya Borah Interdisciplinary Centre for Plant Genomics, Department of Plant Molecular Biology, University of Delhi South Campus, New Delhi, Delhi, India

Danda Chapagai Department of Biological Sciences, East Tennessee State University, Johnson City, TN, USA

Viswanathan Chinnusamy Division of Plant Physiology, Indian Agricultural Research Institute, New Delhi, Delhi, India

Joo Hyuk Cho Department of Molecular Biology, PERI, Sejong University, Seoul, South Korea

Monika Dalal National Research Centre on Plant Biotechnology, New Delhi, Delhi, India

Mackenzie Davenport Department of Biological Sciences, East Tennessee State University, Johnson City, TN, USA

Phillip Dean Department of Biological Sciences, East Tennessee State University, Johnson City, TN, USA

Aigen Fu College of Life Sciences, Northwest University, Xi'an, Shaanxi, China

Mukesh Jain Functional and Applied Genomics Laboratory, National Institute of Plant Genome Research, New Delhi, Delhi, India

Arpana Katiyar Department of Botany, University of Delhi, New Delhi, India

Charanpreet Kaur International Centre for Genetic Engineering and Biotechnology, New Delhi, India

Nisha Khatri Department of Botany, University of Delhi, New Delhi, India

Jitendra P. Khurana Interdisciplinary Centre for Plant Genomics, Department of Plant Molecular Biology, University of Delhi South Campus, New Delhi, India

Kyung-Nam Kim Department of Molecular Biology, PERI, Sejong University, Seoul, South Korea

Dhirendra Kumar Department of Biological Sciences, East Tennessee State University, Johnson City, TN, USA

Yashwanti Mudgil Department of Botany, University of Delhi, New Delhi, India

Ramsong C. Nongpiur Stress Physiology and Molecular Biology Laboratory, School of Life Sciences, Jawaharlal Nehru University, New Delhi, India

Kamlesh Kant Nutan Stress Physiology and Molecular Biology Laboratory, School of Life Sciences, Jawaharlal Nehru University, New Delhi, India

Amita Pandey Department of Plant Molecular Biology, University of Delhi South Campus, New Delhi, India

Girdhar K. Pandey Department of Plant Molecular Biology, University of Delhi South Campus, New Delhi, India

Ashwani Pareek Stress Physiology and Molecular Biology Laboratory, School of Life Sciences, Jawaharlal Nehru University, New Delhi, India

Jon K. Pittman Faculty of Life Sciences, University of Manchester, Manchester, UK

Swatismita Dhar Ray Biotechnology and Bioresources Management Division, Darbari Seth Block, India Habitat Centre, The Energy and Resources Institute, New Delhi, India

Suchismita Roy Stress Physiology and Molecular Biology Laboratory, School of Life Sciences, Jawaharlal Nehru University, New Delhi, India

Eshan Sharma Interdisciplinary Centre for Plant Genomics, Department of Plant Molecular Biology, University of Delhi South Campus, New Delhi, India

Manisha Sharma Department of Plant Molecular Biology, University of Delhi South Campus, New Delhi, India

Raghvendra Sharma Functional and Applied Genomics Laboratory, National Institute of Plant Genome Research, New Delhi, India

Shweta Sharma International Centre for Genetic Engineering and Biotechnology, New Delhi, India

Swati Singh Department of Botany, University of Delhi, New Delhi, India

Sneh L. Singla-Pareek Department of Plant Molecular Biology, International Centre for Genetic Engineering and Biotechnology, New Delhi, India

Alok Krishna Sinha National Institute of Plant Genome Research, New Delhi, India

Neelam Soda Stress Physiology and Molecular Biology Laboratory, School of Life Sciences, Jawaharlal Nehru University, New Delhi, India

Praveen Soni Stress Physiology and Molecular Biology Laboratory, School of Life Sciences, Jawaharlal Nehru University, New Delhi, India

Sudhir Kumar Sopory International Centre for Genetic Engineering and Biotechnology, New Delhi, India

Jawaharlal Nehru University, New Delhi, India

Ashish Kumar Srivastava Nuclear Agriculture and Biotechnology Division, Bhabha Atomic Research Centre, Mumbai, India

Penna Suprasanna Nuclear Agriculture and Biotechnology Division, Bhabha Atomic Research Centre, Mumbai, India

About the Editor

Girdhar K. Pandey, Ph.D.

Girdhar K. Pandey born in Almora, Uttrakhand, India. He received his B.Sc. (Hon.) in Biochemistry from Delhi University in 1992 and M.Sc. in Biotechnology in year 1994 from Banaras Hindu University (BHU). Subsequently, he joined Ph.D. in the School of Life Sciences, Jawaharlal Nehru University (JNU) and worked in the field of calcium signal transduction under abiotic stresses in plants. He was awarded the Ph.D. degree in year 1999 and then pursued postdoctoral career at Department of Plant and Microbial Biology, University of California Berkeley in year 2000. There, he extended his work in the field of calcium-mediated signaling in Arabidopsis by studying CBL-CIPKs, phosphatases, channels/transporters, and transcription factors involved in abiotic stresses. He has been working as Associate Professor in the Department of Plant Molecular Biology, Delhi University South Campus since October 2007.

Pandey's research interests involve detail mechanistic interplay of signal transduction networks in plant under mineral nutrient deficiency (mostly potassium, calcium, and nitrate) and abiotic stresses such as drought, salinity, and oxidative stresses induced by heavy metals. His laboratory is working on the coding and decoding of mineral nutrient deficiency and abiotic stress signals by studying several signaling components such as phospholipases (PLA, PLC, and PLD), calcium sensors such as calcineurin B-like (CBL) and CBL-interacting protein kinases (CIPK), phosphatases (mainly PP2C and DSP), transcription factors (AP2-domain containing or ERF, WRKY), transporters and channels proteins (potassium and calcium channels/transporters), small GTPases, and Armadillo domain containing proteins in both Arabidopsis and rice. The long-term goal of his research group is to establish the mechanistic interplay and cross talk of mineral nutrient-deficient conditions and different abiotic stress signaling cascades in Arabidopsis and rice model system by using the advance tools of bioinformatics, genetics, cell biology, biochemistry, and physiology with greater emphasis on functional genomics approaches.

He has been awarded with Far Eastern Regional Research Organization (FERRO) fellowship to work at Beltsville Agricultural Research Center (BARC), United States Department of Agriculture, Beltsville, MD (1998). Later, he was awarded with Indian National Science Academy (INSA)-Deutsche Forschungsgemeinschaft (DFG) bilateral exchange visiting scientist fellowship in 2011. Also Department of Biotechnology (DBT), India, has awarded him with prestigious DBT-CREST Award (Cutting-edge Research Enhancement and Scientific Training) in 2011–2012. See Pandey's web page for further information about his lab and research work: https://sites.google.com/site/gkplab/home; http://www.dpmb.ac.in/index.php?page=girdhar-pandey.

Part I
Functional Genomics Approaches in Signal Transduction

Chapter 1
Towards Understanding Abiotic Stress Signaling in Plants: Convergence of Genomic, Transcriptomic, Proteomic, and Metabolomic Approaches

Praveen Soni, Kamlesh Kant Nutan, Neelam Soda, Ramsong C. Nongpiur, Suchismita Roy, Sneh L. Singla-Pareek, and Ashwani Pareek

Abstract All aspects of a plant's life—beginning with the seed germination and ending with the seed formation—are adversely affected by different abiotic stresses such as salinity, flood, drought, heat, cold, etc. Being sessile, plants have developed excellent mechanisms of stress perception and signal transduction. Multiple, complex, and dynamically intertwined interactions among nucleic acids, proteins, and metabolites determine the phenotype and final response of plants towards environmental stresses. In response to these stresses, a multitude of processes are activated which enable the plants to cope with these stresses up to a certain extent. These include alteration of expression of stress-responsive genes, production of stress proteins, alteration of ion transport, activation of various antioxidant systems, and compatible solute accumulation. Our knowledge of abiotic stress signaling has grown in leaps and bounds since the emergence and developments in the *omics* technologies. Genome-scale studies at transcript, protein, and metabolite levels provide information about dynamic changes taking place at these functional levels. For full understanding of signaling networks, it is essentially important to integrate all these aspects. This approach is of remarkable applicability when the aim is to understand how plants react to abiotic stresses. In order to understand molecular basis of stress tolerance along with signaling network under unfavorable environmental situations, recent progress on systematic use of *omics* technologies including genomics, transcriptomics, proteomics, and metabolomics has been summarized in this chapter.

P. Soni • K.K. Nutan • N. Soda • R.C. Nongpiur • S. Roy • A. Pareek (✉)
Stress Physiology and Molecular Biology Laboratory, School of Life Sciences, Jawaharlal Nehru University, New Delhi 110 067, India
e-mail: ashwanip@mail.jnu.ac.in

S.L. Singla-Pareek
Department of Plant Molecular Biology, International Centre for Genetic Engineering and Biotechnology, Aruna Asaf Ali Road, New Delhi 110 067, India

G.K. Pandey (ed.), *Elucidation of Abiotic Stress Signaling in Plants*,
DOI 10.1007/978-1-4939-2211-6_1

Furthermore, the integration of all these approaches, which provide systems biology method for understanding stress response in plants, is also discussed.

Keywords Abiotic stresses • Drought • Genomics • Metabolomics • Proteomics • Salinity • Transcriptomics

1.1 Introduction

Plants being sessile face various extreme environmental conditions throughout their life cycle and respond accordingly to maintain their vital metabolic homeostasis by regulating their gene activity. Thus, understanding the intricacies of these plant responses to environmental stress is the first step towards improving crop productivity under these unfavorable conditions. Further, it is also known that these abiotic stresses do not occur alone; rather their combinations contribute at unpredictable amounts to the overall stress perceived by the plant. The signaling pathways are very complex involving different molecules (Punjabi-Sabharwal et al. 2010). Consequently, it is reported that engineering genes related to protection and maintenance of cellular constituents can improve tolerance of plants towards stresses. The most important fact is to identify the candidate gene and characterize it in context of stress (Xianan and Baird 2003). In this regard, a host of genes have been identified and already reported to improve tolerance of plants towards stresses. Some of the genes are listed in Table 1.1. Similarly, analysis of transcriptomes can also pave way towards isolation of "candidate genes" which would be suitable for raising transgenic plants with improved tolerance. In recent past, several publications have reported that using the comparative transcriptomic approach between contrasting genotypes of rice (IR64 and Pokkali), several differentially regulated genes could be isolated, which then served as useful genes in functional genomic studies (Kumari et al. 2009; Karan et al. 2009; Mustafa et al. 2010; Kumar et al 2012; Soda et al. 2013). It is also true that the analysis of changes in proteome of crop plants in response to abiotic stresses can also be used as a starting point to fish out the genes, which may ultimately serve as the "candidate genes" (Ruan et al. 2011)

In this chapter, we have described the various "omics"-based approaches, which have been employed in recent years to understand the response of plants towards abiotic stresses. For brevity sake, we have restricted our discussion to the identification of signaling molecules, which have been studied by employing the particular "omics" approach. Technical details about the technique/approach have also been provided at appropriate places in the text.

Table 1.1 List of a few representative genes involved in plant abiotic stress responses

Name of the gene	Responsive to abiotic stress(es)	References
14.3.3 gene family	Salinity and drought	Chen et al. (2006)
Annexin	Salt	Lee et al. (2004)
ATAF	Drought, salinity, cold, and wounding	Christianson et al. (2010)
bZIP family	Drought, temperature, and salinity	Corrêa et al. (2008), Weltmeier et al. (2006), Nieva et al. (2005), Baena-González and Sheen (2008), Satoh et al. (2004), Alonso et al. (2009)
CBF/DREB families	Drought, cold, and salinity	Agarwal and Jha (2010), Trujillo et al. (2008)
Glycerol-3-phosphate acyltransferase gene	Cold	Yan et al. (2008)
H-ATPase	Cold	Hashimoto et al. (2009)
HSC 70	Cold	Folgado et al. (2013)
HVA1	Salinity and drought	Fu et al. (2007)
ICS	UV light	Catinot et al. (2008)
LOX	Drought and wounding	Yang et al. (2012), Andreou and Feussner (2009)
MAPK	Abiotic stresses	Pitzschke et al. (2009)
MEKK1 and ANP1	Oxidative and environmental stresses	Nakagami et al. (2006), Suarez-Rodriguez et al. (2007)
MPK3, MPK4, and MPK6	Abiotic stress and oxidative stress	Nakagami et al. (2006), Qiu et al. (2008)
MYB 4, 6, 7, and 44	Drought and salt	Yanhui et al. (2006)
NAC5	Cold, drought, and salt	Takasaki et al. (2010)
OsRMC	Salt	Guo and Song (2009)
SAMS	Salt	Pacheco et al. (2013)
SCF	Salt	Liu et al. (2013)
Vacuolar H^+-pyrophosphatase	Flooding	Komatsu et al. (2009b)
WRKY family	Salinity, temperature, drought, and oxidative stress	Qiu et al. (2009)

1.2 Genomic Approach

1.2.1 Advances in Plant Genomic Technologies

Functional genomic approaches with the help of high-throughput technology allow large-scale gene function analysis and interaction study of gene products at cellular and organism levels. The data collected from the completed sequencing genome projects provide valuable information about genes to be analyzed (Pérez-Clemente et al. 2013). The availability of this plant genome sequence information currently facilitates studying the function of genes on a genome-wide level (Feuillet et al. 2010; Chain et al. 2009). However, for many plants, genomes have not been sequenced, or sequencing has not been completed. In such cases, the lack of information is compensated, in part, by the availability of huge collection of cDNA sequences and expressed sequence tags (ESTs) (Marques et al. 2009). In functional genomic projects, various tools like cDNA libraries, microarray, serial analysis of gene expression (SAGE), and ESTs are widely used to analyze global gene expression profiles in plants. Identification of gene function by analyzing mutants generated through chemical and physical mutagenesis has become feasible for large-scale analysis due to knowledge of different markers (Lukowitz et al. 2000). Characterization of mutant is the best way to know the function of a given gene. In this way, large collections of mutants and their characterization can complement large-scale expression studies.

Abiotic stress tolerance is quantitative and complex trait controlled by multiple interacting genes in plants (Punjabi-Sabharwal et al. 2010). Advancement in molecular biology techniques provides the function of genes present in plants. It also dissected out the abiotic stress responses governed by one gene or by multiple genes linked to the particular trait called as quantitative trait loci (QTL). QTL mapping advancements lead to the emergence of better breeding approaches such as breeding by design and marker-assisted selection (Peleman and van der Voort 2003). However, understanding the adaptive processes and complexity of stress signaling is inadequate due to lack of complete knowledge of genes involved in plant stress responses.

1.2.2 Gene Expression and Regulation in Response to Abiotic Stresses

In response to environmental stresses, expression of a number of genes involved in the stress defensive mechanism gets activated. Studies of stress-responsive networks have been revolutionized by the use of latest technologies such as microarray and next-generation sequencing (NGS). Basically, regulation of gene expression in plants can be noticed at transcriptional, posttranscriptional, and posttranslational steps. Various elements and factors are involved in the regulation at each step.

Three major elements are involved in transcriptional regulation: chromatin and its remodeling and modification; *cis*-regulatory elements present in promoters and other regulatory sequences, such as enhancers, present downstream and upstream of the coding region; and transcription factor (TF) which binds on *cis*-regulatory elements. Cifre et al. (2005) reported chromatin remodeling and modification involved in response to plant abiotic stresses. The sensitization of stress responsiveness is called priming. Priming, which is preexposure of stress before the actual stress (Conrath et al. 2006; Zimmerli et al. 2009), activated the defensive mechanism and increases the stress tolerance ability of the plant. It was shown that in case of WRKY transcription factors, priming was associated with chromatin modification of promoter (Jaskiewicz et al. 2011).

Transcription factor (TF) plays a very crucial role on gene expression which is initiated by binding of transcription initiation factor IID (TFIID) on the regulatory region of promoter which subsequently forms the transcription initiation complex along with other components. Formation of transcription initiation complex initiates the transcription process by recruiting the RNA polymerase II (Juven-Gershon et al. 2008). Apart from transcription factors (TFs) involved in the normal process of transcription, many stress-responsive TFs such as members of basic leucine zipper (bZIP), zinc finger families, MYB, and dehydration-responsive element-binding (DREB) or C-repeat binding factor (CBF) have been reported to be involved in the regulation of gene expression in plants under stress responses. Most of these transcription factors (TFs) bind to the *cis*-acting element present on the promoter of their targeted stress-inducible gene and regulate its expression under stress (Hu et al. 2006). Recently, the role of WRKY transcription factors in plant salinity and drought stress responses were reported in plants (Chen et al. 2012; Golldack et al. 2011). Overexpression of various members of WRKY TFs were leads to increase in abiotic stress tolerance by regulating stress-related genes in rice (Song et al. 2010; Wu et al. 2009).

Plant-specific NAC TFs are reported to be involved in abiotic stress response such as drought and salinity in plants (Nakashima et al. 2012; Yamaguchi-Shinozaki et al. 1992). It was reported that a rice OsNAC6 TF expression, which shows higher homology with *Arabidopsis* abiotic stress-responsive NAC TFs (ANAC019, ANAC055, and ANAC072), is induced by drought, salinity, ABA, and cold (Ooka et al. 2003).

Members of the 14.3.3 gene family G-box factor 14-3-3b protein (GF14b) and G-box factor 14-3-3c protein (GF14c), induced by abiotic stresses such as salinity, drought, and ABA, have been reported in rice (Chen et al. 2006). Members of this family of protein are also regulated by stress-responsive TFs (Chen et al. 2006). Pulla et al. (2009) reported that *S*-adenosyl-L-methionine synthetase (SAMS) gene of *Panax ginseng* (*PgSAM*) expressed under various abiotic stresses in *Panax ginseng* and might be providing protection against environmental stresses.

Some of the drought-responsive genes like RD20, RD22, RD29B, COR47, ERD14, VSP2, and RHL41 are also responsive to other abiotic stresses, which show the role of one gene in multiple stresses (Debnath et al. 2011). In *Arabidopsis*, 67 genes were identified to be responsive for multiple abiotic stresses (Swindell 2006). Similarly, transcriptome analysis in rice under salinity, drought, cold, and ABA stress showed the induction of 73 genes. Among 73 stress-responsive genes, 57, 62, 36, and 43 were induced by salinity, drought, cold, and ABA, respectively (Rabbani et al. 2003). It was also observed that out of 73 stress-inducible genes identified in rice, 51 are common between rice and *Arabidopsis*. However, some of the genes are specific to rice only, suggesting the different stress responsiveness between rice and *Arabidopsis* (Rabbani et al. 2003).

Gene expression regulation occurs at transcription level but also at posttranscriptional level which includes mRNA processing (capping, splicing, and polyadenylation), mRNA nucleocytoplasmic trafficking, mRNA turnover and stability, and mRNA translation (Floris et al. 2009).

Posttranslational regulation is the third level of regulation which includes phosphorylation, sumoylation, and ubiquitination of proteins. Under abiotic stresses, posttranslational regulation like phosphorylation and dephosphorylation plays an important role in signaling which activates the defense mechanism of plants. Under drought and osmotic stress, various signal transduction cascades formed by SNF1-related protein kinases (SnRKs) and mitogen-activated protein kinases (MAPKs) activated the phosphorylation of specific molecules (Zhu 2002). In *Arabidopsis*, ABA-dependent responses to water deficit, like stomata closure, are known to be regulated by SnRK2 proteins (Yoshida et al. 2006).

1.2.3 Transgenic Approach to Understand Gene Functions

Elucidating the mechanisms of stress tolerance by raising transgenic plant by manipulating stress-specific genes through genetic engineering is gaining popularity. Nowadays, success has been achieved in genetic improvement of plants for better abiotic stress tolerance by manipulating the gene-encoding enzymes involved in the regulatory pathways (Kumar et al. 2012).

Abscisic acid (ABA), which is also called as stress hormone, is involved in the regulation of various adaptive mechanisms in plants under different environmental stresses (Arbona and Gómez-Cadenas 2008). Therefore, with an aim to increase tolerance against abiotic stresses, many transgenic plants have been raised by manipulating key enzymes involved in the ABA biosynthetic pathway (Ji et al. 2011).

Compatible solutes and chaperoning protect the plant by various abiotic stresses by protecting biomolecules and membranes (Zhang et al. 2008). Transgenic plants overexpressing the genes involved in the biosynthesis of these solutes enhance the drought and osmotic stress ability of plants (Zhang et al. 2008). Transgenic plants overexpressing many genes encoding stress-related biomolecules such as proline (Hmida-Sayari et al. 2005), LEA (Rohila et al. 2002), chloroplast glycerol-3-phosphate acyltransferase (Sui et al. 2007), etc., have shown higher tolerance to abiotic stresses.

Reactive oxygen species (ROS) production is a common factor among most stresses (Hirayama and Shinozaki 2010). ROS performs dual roles. ROS is not only toxic to cells, but it also plays an important role as a signaling molecule. Scavenging the highly active ROS is a very important strategy for plant defense in which a series of interlinked enzymes are involved. Overexpression of the ROS-scavenging enzymes such as superoxide dismutase, glutathione reductase, glutathione peroxidase, and ascorbate peroxidase helps the plants in stress tolerance (Tang et al. 2006).

But the main hurdle in plant transgenic technology is the lack of effective single-copy gene transfer in plants. Secondly, a well-standardized tissue culture protocol is lacking for many plant species. Although plant transformation without tissue culture is known in *Arabidopsis*, it can be used in other crops. Therefore, effective transformation methods and easy tissue culture protocol need to be discovered.

1.3 Transcriptomic Approach

The transcriptome is the sum total of entire RNA molecules (mRNA, tRNA, rRNA, miRNA, long noncoding RNA, and only recently circular RNA) present within one or a population of cells at a particular time point. Thus, at different conditions or at different time points, the transcriptome is subject to variation. Transcriptomics, in short, can be defined as that field of functional genomics which concerns the study of transcriptomes. Transcriptomics, however, mainly focuses on gene expression, i.e., mRNA transcripts, as differential gene expression has been known to alter phenotypes of cells or a population of cells or entire organisms. Ultimately, a cell's transcriptome determines its phenotype in terms of development, differentiation, and ability to respond to environmental stimuli.

1.3.1 Understanding Abiotic Stress Tolerance in Plants Through Transcriptomics: The General Approach

As has been stated, differential gene expression determines phenotype. In fact, difference in expression of even a single gene can lead to highly altered phenotype. For example, *Arabidopsis* mutants, in which the CBF2/DREB2C gene was disrupted, displayed higher capacity to tolerate freezing, salinity, and dehydration stress (Novillo et al. 2004). One can only imagine that if differential expression of one gene can alter multiple phenotypes, then differential expression of a set of genes would definitely lead to a higher degree of phenotypic variations. Furthermore, it is known that the response of plants to external stimuli involves the following sequence of events:

1. *Perception:* usually involves membrane-localized sensor or receptor proteins.
2. *Signaling*: upon perception of external stimuli, downstream signaling proceeds either through protein–protein interactions or via the application of secondary messengers.
3. *Altered gene expression*: usually involves gene expression regulation through transcription factors.

Plants need to constantly alter their transcriptome to adjust to any abiotic stress, and this ability to adjust forms the basis of stress tolerance. These aspects were highly considered when researchers first started to envisage the basis of abiotic stress tolerance in plants. Transcriptomic studies were carried out on a large scale with the intent to decipher the signaling components of the abiotic stress response of plants. The overall aim was to decipher how a stressful environment affects gene expression in plants. The general experimental plan involved:

1. The comparison of transcriptomes of contrasting genotypes of a particular organism or phenotypically contrasting organisms in terms of abiotic stress tolerance for the identification of transcripts responsible for stress tolerance

2. Comparisons of transcriptomes of untreated and stressed samples of the same genotype to identify stress-responsive genes

1.3.2 Transcriptomics: Tools and Technologies and Their Contributions

Various approaches have been used for plant abiotic stress-related transcriptomic studies. A few of them, which have been highly useful towards the understanding of abiotic stress response in plants, are briefly described below.

1.3.2.1 Suppression Subtractive Hybridization

This is a PCR-based technique, which involves the amplification of cDNA obtained from control (driver) and experimental (tester) transcriptomes. This technique, as the name suggests, involves the hybridization of driver and tester molecules, which eventually leads either to their amplification or elimination from amplification based on the hybrids formed. Further, there is an equalization of amplification where low differentially expressed target molecules are amplified exponentially, whereas the high differentially expressed transcripts are subjected to a PCR-suppression effect and hence exhibit suppressed amplification. Thus, suppression subtractive hybridization (SSH) is a powerful tool to identify differentially expressed transcripts as even the low differences in transcript abundance between control and test samples can be detected. Various stress-responsive genes have been identified using SSH. Using SSH, a total of 1,058 genes were identified to be differentially expressed from eight stress cDNA libraries of *Arabidopsis*, and out of which 55 % of the stress-induced transcripts were rarely expressed in unstressed plants, and 17 % of them were completely absent in *Arabidopsis* EST databases present at the time (Mahalingam et al. 2003). Using SSH, Gulyani and Khurana (2011) obtained 1920 clones representing 208 contigs and 151 singletons, which were drought regulated in two contrasting cultivars of mulberry. The greatest advantage of this technique is that it does have a prerequisite such as whole genome sequence of the organism. SSH can be performed for any organism subjected to any condition even one which has never been tested before. SSH is only one of the techniques, which involve large-scale Sanger sequencing of ESTs, but the experiments are low throughput. Other techniques that use Sanger sequencing for identification of genes include the tag-based methods such as massively parallel signature sequencing (MPSS) (Brenner et al. 2000), SAGE (Velculescu et al. 1995; Harbers and Carninci 2005), and cap analysis of gene expression (CAGE) (Kodzius et al. 2006; Shiraki et al. 2003). These tag-based methods are high throughput and have also contributed significantly to the understanding of the transcriptomic changes that occur in plants in response to stress.

1.3.2.2 Microarray

Microarray is probably the transcriptomic method, which has contributed the most towards understanding abiotic stress tolerance in plants. It is based on the hybridization of a nucleic acid sample (target) to an enormous set of oligonucleotide or, occasionally, full-length cDNA probes which are attached to a solid matrix surface. For gene expression profiling cDNA obtained from mRNA is used as the target sample. In microarray oligonucleotides, cDNA sequences, ESTs, or even genomic DNA segments are arrayed on a glass slide to a density of about 1,000 cm^{-2}, which are then hybridized with differently fluorescent-labeled cDNA obtained from control and test mRNA samples, respectively. The difference in the fluorescence intensity of the two fluors provides the parameter for measuring difference in gene expression between control and test samples. Differences in gene expression are represented usually as fold change between control and test samples or also by using a statistical method like ANOVA where the null hypothesis is kept that a gene is equally expressed in both control and test. Microarray technology was first demonstrated in 1995 where ESTs were used to analyze the differential expression of 48 *Arabidopsis* genes in roots and shoots (Schena et al. 1995). Since then the technology has grown in leaps and bounds to the scale of whole genome-scale transcriptome profiling. Stress-responsive genes from many species such as *Arabidopsis*, rice, maize, etc., have been identified using microarrays (Kreps et al. 2002; Seki et al. 2001; Rabbani et al. 2003; Kawaura et al. 2008). In fact, if looked at it from one perspective, microarray provides the platform to understand plants' abiotic stress response from a whole genome point of view. It provided the necessary knowledge to understand the various stress signaling pathways and how different stresses are connected with each other in terms of the response that they evoke.

Microarray has a lot of advantages over other transcriptomic techniques such as EST sequencing or SSH. Microarray allows for studies to be made on the whole genome level. It is a rapid, easy to use technique, which can be easily replicated. Initially, microarray technology had a lot of drawbacks with background noise especially with lowly expressed differentially regulated genes, but a lot of progress has been made on this front, and nowadays the data obtained is much more accurate and reliable. It must be said that microarrays, like all transcriptomic methods, form only the initial high-throughput screening of differential gene expression. The data obtained has to be verified through other, more accurate, methods such as qRT-PCR or Northern blots. Nevertheless, microarray does provide a researcher with a good starting point for his/her experiments. Despite the endless possibilities, microarray technology does have its drawbacks. First and foremost, oligoarrays of an organism's genomic sequence is a prerequisite for oligonucleotide probe design. On the other hand, cDNA arrays do not require genomic sequence to be known, but the cDNA probes do not represent the entire repertoire of genes present in the organism, and hence the data obtained can be incomplete. cDNA arrays are cheap, but they are inaccurate and cannot measure individual samples. Oligoarrays, in contrast, are much more accurate than cDNA arrays, but they are expensive and limited only to certain model species.

1.3.2.3 RNA Sequencing

In the last two decades, there has been tremendous progress in the field of nucleic acid sequencing. Whole genomes of many organisms have been sequenced, which provide a great deal of information towards understanding the abiotic stress response of plants at the molecular level. Apart from genome sequencing, RNA sequencing has also made a huge impact towards understanding transcriptomic networks in general. Already used for a wide taxonomic range from yeast to *Arabidopsis* to humans, this technology employs the sequencing or cDNA generated from the total RNA population of test samples. Perhaps the biggest advantage of RNA sequencing is that an organism's genome sequence or even an EST database is not a prerequisite for the data obtained to be interpreted and analyzed. Furthermore, data acquired from large-scale Sanger sequencing of ESTs are biased against low-abundance transcripts, time-consuming, and expensive (Filichkin et al. 2010). Moreover, de novo sequencing has enabled whole genome level transcriptomic studies to be carried out even on non-model species. Examples of the use of RNA (de novo) sequencing on non-model species include olive (Alagna et al. 2009), chickpea (Garg and Jain 2011), barley (Thiel et al. 2012), and garlic (Sun et al. 2012) besides many others. Although few studies connected with plant abiotic stress have been carried out using RNA sequencing, the potential of this technology in this regard is almost limitless. Through RNA sequencing, one can not only identify stress-responsive mRNA transcripts but also identify siRNAs, miRNAs, as well as long noncoding RNAs, all of which have been shown to regulate the stress response in the model plant *Arabidopsis* (Sunkar et al. 2007; Amor et al. 2009).

1.4 Proteomic Approach

1.4.1 The Rapid Rise of Proteomics in the Post-Genomic Era

Proteomics is the next step in the study of biological systems. It is more complicated than genomics because an organism's genome is comparably constant, whereas the proteome differs in both spatial and temporal manners. A certain set of genes are expressed in different cell types, which in turn lead to cell differentiation and myriad of cellular responses under different conditions. A decade before to specify cellular responses, transcriptome analysis was used to be a preferred technique, but cell transcriptome does not correlate with its protein content (Rogers et al. 2008). It is now known that the complete set of mRNA is not always translated into protein. Marc Wilkins and his colleagues in the early 1990s coined the word proteomics mirroring the word genomics. Proteomics is complimentary to genomics as it describes the whole compliment of a protein component encoded by its genome. It is the study of multiprotein system focusing on their interplay with multiple, distinct proteins in their roles as part of larger system or network.

Cellular physiology in an organism is underpinned by the composition and function of its proteome. The power of proteomics lies in their ability to quantitatively analyze large numbers of proteins, providing rich sets of information, which can explain the molecular mechanisms of cellular function. The plant proteomic literature is considered by two major clads: First is the descriptive cataloging of protein expression (Lee et al. 2011), organelle localization, or identification of protein complex (Remmerie et al. 2011). The second clad is made up of comparative studies that identify differences in protein profiles between contrasting genotypes (Pang et al. 2010) or between control and treated samples (Kosova et al. 2013). Signaling event in cells involves intricate networks encompassing feedback loops and cross talk with varied extracellular and intracellular signals integrating a large number of information pertaining to the internal state of the cell. Signal propagation, especially during abiotic stress condition, involves protein changes on three different levels, (1) regulated protein posttranslational modifications (PTMs); (2) protein–protein interactions (PPIs), often owing to PTMs; and (3) signal-induced protein expression changes. All these three levels are orchestrated in a highly dynamic and often spatially segregated manner, and they may themselves lead to changes in protein activity, localization, and association with small molecules such as phospholipids. A primary task of decoding the signaling cascade in stress condition is therefore the measurement of PTMs, protein interactions, and proteome dynamics. The principles of cell signaling have been worked out over decades using ingenious analytical approaches. Involving proteomic approach, researchers could deduce abundance of several stress- and defense-related proteins. Proteomics in the post-genomic era is contributing a lot in elucidating the signal transduction pathway when it comes to stress.

1.4.2 Proteomic Tools for Investigating the Proteome

The context of proteomics largely depends on systems biology, rather than the structural one. The chore of characterizing the proteome requires analytical methods to detect and quantify proteins in their modified and unmodified forms. Proteomics encompasses applications as mining, protein expression profiling, protein network mapping, and mapping of protein modifications. The field of proteomics largely relies on the basic instrumentation, affinity enrichment and depletion, quantification techniques, peptide and protein identification by mass spectrometry, structural remodeling, statistics, and data mining. Analytical study of proteomics involves the development and integration of tools leading to sensitive and specific means of identifying and characterizing proteins. In the following section, details of some of the most prevalent techniques for proteomic analysis are provided.

1.4.2.1 Polyacrylamide Gel Electrophoresis

This is the most widely used analytical technique for separation of proteins. In this technique, a protein sample is subjected to dissolve in a loading buffer having a thiol reductant. The basis of separation is binding of SDS to the protein, which imparts a negative charge to the protein when the gel is subjected to high voltage and the SDS protein complexes are allowed to migrate through a cross-linked polyacrylamide gel and resolved on the basis of their molecular mass. This is one of the classical methods to study protein expression, and a huge number of single proteins have been identified using this method. The degree of resolution of this kind of gel is modest, and the bands, which appear simply, may actually bear numerous other proteins as well.

1.4.2.2 Two-Dimensional Gel Electrophoresis

Two-dimensional gel electrophoresis (2DE) is one of the most popular techniques for studying multiple proteins simultaneously. In 2DE proteins are first separated on the basis of their isoelectric point (pI) by isoelectric focusing (IEF), and then these focused proteins are further resolved on the basis of mass by electrophoresis on a polyacrylamide gel. This technique provides a more comprehensive protein profiles, which helps in identification of differentially accumulated proteins in two contrasting samples. Using 2D gel electrophoresis the entire proteome of poplar under control and drought stress condition was chalked out, which provided clues regarding the way poplar copes with water deficit (Plomion et al. 2006). Komatsu et al. (2009) reported increase in Hsp70 abundance in root and cotyledons of soybean plants under flooding stress.

In proteomic studies, there are some reports on differential protein abundance of signaling related proteins. The main reason is due to the relatively low abundance of signaling proteins with respect to other cytosolic proteins making signaling proteins hardly detectable on two-dimensional gels. However, 2DE helps in the identification of a large number of proteins, but it still has some basic limitations, as its limited applicability to hydrophobic proteins and those of extreme molecular weight or isoelectric point.

1.4.2.3 Differential in Gel Electrophoresis

Two-dimensional electrophoresis is a powerful tool to explore the plant proteome and to unravel changes in protein expression between samples. However, gel to- gel variations are found in two-dimensional electrophoresis which always remains a big drawback. Nowadays, this inconvenience can be bypassed by the use of a technique known as "differential gel electrophoresis" or differential in gel electrophoresis (DIGE) introduced by Unlu et al. (1997). Basically, this technique uses separate pools of proteins that are covalently labeled with *N*-hydroxy succinimidyl derivatives of the fluorescent cyanine dyes (Cy2, Cy3, and Cy5), and these fluorescent labels modify the e-amino group of lysine residues in the proteins. By this means, approximately 3 % of the available protein and one single lysine per protein molecule get a labeled form. This method allows the analysis of up to three pools of protein samples at one go on a single 2D gel, thereby minimizing the problem of gel-to-gel variability. A standard dye (Cy2) is used to label an internal standard that consists of equal amounts of all the samples to be analyzed within the overall experiment. The inclusion of such an internal standard allows experimental errors to be corrected and therefore improves the quantitative comparison of protein expression (Alban et al. 2003). In globe artichoke, UV-C-modulated 119 proteins were identified, and most of the proteins differentially modulated by UV-C were located in the chloroplast, involved in photosynthesis, sugar metabolisms, protein folding, and abiotic stress (Falvo et al. 2012).

Advantages of 2D-polyacrylamide gel electrophoresis (PAGE)/DIGE are its capability of analyzing the large mass range and the whole proteome of cell at any one time. These techniques 2D-PAGE/DIGE are particularly good for looking at proteins within the mass range of 20–250 kDa and pI of 3–11. Using isoelectric focusing strips of specific range of pH and the specific percentage of second dimension PAGE gel, one can focus on proteins of interest. 2D-PAGE/DIGE has some associated disadvantages as well as it is a lower-throughput and time-consuming process involving many steps and requires a high level of expertise to obtain good results. This technique is also not a good technique for the analysis of extremely acidic, basic, or hydrophobic proteins such as membrane-bound proteins and also smaller proteins and peptides (<15 kDa).

1.4.2.4 Comparative Proteomics Using Gel-Free Techniques

To overcome the limited dynamic range drawback of 2DE/DIGE system, solution-based "shotgun" techniques based on two-dimensional liquid chromatography (2D-LC) are taken up as an alternative approach. In 2D-LC, low-pH reversed-phase (RP) LC is used as the second chromatographic dimension, coupled directly with tandem MS (MS/MS) detection. The gel-free technique has the added advantage of identifying low-abundance proteins, proteins with extreme molecular weights or pI values, and hydrophobic proteins that cannot be identified by using gel-based technique. The liquid chromatography–mass spectrometry (LC–MS)/MS technique is conceptually the opposite of gel 2DE approaches. Liquid chromatography is improving in terms of properties as high-pressure pumping systems to provide higher-resolution column packaging with smaller-diameter particles. Better separative mode of analytes from each other gives rise to more complete analyses with less interference. Improvements include faster separations to help efficiency and greater sensitivity by producing eluted molecules in a narrower time window increasing the eluted temporal concentration for enhanced signal. The use of LC–MS/MS often offers definitive identification of proteins from complex mixtures.

Multidimensional protein identification technology (MudPIT) incorporates separation of complex peptide mixtures on a strong cation exchange, reversed-phase column and MS/MS analysis (Wolters et al. 2001). MudPIT (online 2D LC–MS/MS system) has several advantages compared to gel-based methods by providing greater peak capacity, higher sensitivity, greater throughput, and higher degree of automation (Whitelegge 2002). MudPIT separates peptides in 2D liquid chromatography. This allows greater separation of peptides that can be directly interfaced with the ion source of a mass spectrometer, thus maximizing its sensitivity. One of the shortcomings of MudPIT is that it avoids band broadening associated with many chromatographic steps, which leads to decreased resolution. A combination of gel-based and gel-free proteomics has been used for identification of soybean plasma membrane proteins under flooding or osmotic stress (Komatsu et al. 2009), suggesting that these two methods are complementary to one another for protein identification. Similar quantitative label-free shotgun proteomic analysis has been carried out recently for analyzing long-distance drought signaling mechanism in rice grown in split root system.

Importantly, 2D-LC is highly compatible with a stable isotope label-based quantitative technique employing isobaric tags for relative and absolute quantitation (iTRAQ), allowing multiplexed (high-throughput) quantitative analysis across multiple biological samples. This intensity-based label-free approach does not require continually reidentifying (redundantly identifying) peptides in every sample under study because it decouples profiling from identification and subsequently links the profiling and identification databases in silico via accurate *m/z* and retention time. Unfortunately most proteins in a complex mixture fall into the category of limited identifications.

1.4.2.5 Labeled and Non-labeled Proteomic Approaches

An isotopic labeling strategy came into play, which can be used for comparing two samples, which have been labeled separately with a heavy or a light isotopic molecular tag, respectively. Mixtures of these samples were passed through the same chromatography and mass spectrometry channels leading to better comparison of the ratios of heavy- and light-labeled peptides. This method is known as isotope-coded affinity tags or ICAT. Subsequently, this approach has been amended with a revised method called iTRAQ (isobaric tag for relative and absolute quantification). Another labeling approach devised to feed a cell line or small organism in which isotope-labeled amino acids are used, "stable isotope labeling by amino acids in cell culture" or SILAC method (Schutz et al. 2011) is also gaining ground. The ICAT method of proteome comparisons offers some advantages over 2D/MALDI-TOF-based approach. Quantitative proteomic studies using ICAT or SILAC have also been taken up for studying endomembranous system as Golgi apparatus (Nikolovski et al. 2012). MS-based label-free quantification methodologies allow differential expression analyses of multiple samples. Such label-free quantification has been extensively used in unraveling response-related mechanism of soybean seedlings against flooding and osmotic stress condition. Abdallah et al. (2012) analyzed the *Medicago truncatula* membrane protein using iTRAQ.

1.4.3 Protein Identification by Mass Spectrometry

Mass spectrometry has become a firsthand tool for identification of the characteristic proteins. Mass spectrometers have essentially three parts, source, mass analyzer, and detector. Ionization is carried out by two methods: electrospray ionization (ESI) and matrix-assisted laser desorption and ionization (MALDI). For MALDI, the sample of interest is mixed with an excess of a small organic compound in solution; the solution is allowed to dry, and then a pulse of laser light irradiates the sample, releasing the charged analyte for mass analysis. Nowadays, developments in terms of dissociation or fragmentation for the purpose of peptide and protein sequence identifications and associated PTMs have been introduced as well.

1.4.4 Studying Protein–Protein Complexes

Densitometric electrophoretic gel analysis coupled with protein identification by mass spectrometry has enabled the researchers to identify differentially abundant protein spots. Validation of comparative proteomics should be done by protein functional analysis as well. Therefore, other approaches (e.g., PTMs, protein–protein interactions) come into utility. Cell signaling acts in a modular form, and few key experiments used for identification of putative interacting proteins include yeast two-hybrid technique and immunoprecipitation of protein of interest with associated protein (Co-IP).

The yeast two-hybrid system, pioneered by Stan Fields, has eased the transformed analysis of protein–protein interactions. The efficiency of any given PPI pair to couple a transcription activation domain to a DNA-binding domain and thereby their ability to drive the expression of sensitive reporter genes in yeast is the key to this technique. However, there are some well-appreciated caveats to this method as well, i.e., its tendency to generate a high level of false-positive and false-negative results. In an example, directed yeast two-hybrid analyses showed PPIs between MEKK1 and MKK1/MKK2, between MKK1/MKK2 and MPK4, and between MEKK1 and MPK4 (Ichimura et al. 1998). MAPK signaling pathway regulated under stress and development demonstrated the rapid and transient activation of MPK4 and MPK6 on treatment of low temperature, low humidity, and hyperosmolarity. Additionally, MKK1 is activated by salt, drought, and wounding and can phosphorylate MPK4, suggesting a role in abiotic stress signaling (Teige et al. 2004; Xing et al. 2008).

The other most commonly used technique for verification of PPIs is Co-IP. In a typical experiment, bait complexes are captured from a cell lysate using a specific antibody. The antibody is then immobilized using protein A or protein G covalently attached to sepharose beads. After washing of the beads, the antibody, the bait, and the proteins associated to the bait are eluted. The bound proteins are then identified by MS or by immunoblotting. The advantage of this approach is that endogenous protein complexes can be studied; hence, any artificial effects of affinity tags or overexpression can be easily avoided. The disadvantage of this technique is that highly specific antibodies are required. Proteomic approach did unleash a noble nuclear calcium-sensing and signaling pathway that is critical for salt stress tolerance in the reference plant *Arabidopsis* (Kader and Lindberg 2010). The nuclear-localized calcium-binding protein, RSA1 (SHORT ROOT IN SALT MEDIUM 1), which is required for salt tolerance, interacts with a bHLH transcription factor-RITF1. This interaction was identified by co-immunoprecipitation, and was further validated by BiFC (Guan et al. 2013). They have shown that RSA1 and RITF1 regulate the transcription of several genes involved in the detoxification of ROS generated by salt stress and that they also regulate the *SOS1* gene that encodes a plasma membrane Na^+/H^+ antiporter essential for salt tolerance.

If proteins interact in vivo, they are expected to be co-localized, or, at least, they will display overlapping distribution within the cell. Co-localization is defined as the presence of two or more different molecules residing at the same physical location in a cell. The intracellular localization of two (or more) proteins can be studied by bimolecular fluorescence complementation (BiFC) assay. BiFC assay is based on the structural complementation of two nonfluorescent N- and C-terminal fragments of a fluorescent protein when they are fused to a pair of interacting proteins. Proteins that are postulated to interact are fused to unfolded complementary fragments of a fluorescent reporter protein and expressed in live cells. Interaction of these proteins will bring the fluorescent fragments within proximity, allowing the reporter protein to reform in its native three-dimensional structure and emit its fluorescent signal. This fluorescent signal can be detected and located within the cell using confocal microscopy. BiFC has been successfully used for identification of several PPIs (Gokirmak et al. 2010; Robida and Kerppola 2009; Li et al. 2010; Riese et al. 2013).

1.4.5 Mapping the Proteome from Whole Plants to Subcellular Compartments

Organelle proteomics is the study of the protein complement of a specific subcellular compartment and involves subcellular fractionation followed by protein identification, typically by MS. Inside the eukaryotic cell, proteins are spatially organized according to their function. In addition, the enrichment of an organelle before proteomic analysis enables the identification of low-abundance proteins that would not be detected in unfractionated samples. The biggest challenge in organelle proteomics is the production of pure organelle fractions, which are important since the presence of contaminating proteins reduces the confidence level with which novel proteins can be assigned to a specific organelle. Two-dimensional PAGE has also been used to study the plant mitochondrial proteome (Kruft et al. 2001). In a recent study, Choudhary et al. (2009) analyzed nuclear proteome of evolutionary diverse organisms during dehydration stress. They have identified around 32 conserved proteins and other dehydration-regulated proteins (DRPs) by using 2DE analysis.

1.4.6 Signaling Proteins and Posttranslational Modifications

Protein and its PTMs of proteins greatly increase protein complexity and dynamics leading to co-coordinating the intricate regulation of signaling events. Advances in proteomic techniques currently accelerate the overall identification of PTMs in a signaling cascade. In a simple approach, modified proteins can be visualized in the gel by specific methods suitable for different PTMs (Patton 2002). Citing an example, in case of phosphorylated proteins, autoradiography of incorporated ^{32}P or ^{33}P and Western analysis with anti-phosphoamino acid antibodies are generally used. Similarly, radioactive or nonradioactive staining methods have been devised for the detection of glycoproteins, proteolytic modifications, nitrosylation, and methylation (Patton 2002).

Further aided by the sensitivity of mass spectrometers, analysis of in vivo protein phosphorylation at a proteome scale in plants has been possible as well (Kwon et al. 2006). Peptide modifications give rise to specific features of MS/MS spectra. In case of phosphorylation of serine and threonine residues, there is an elimination of phosphoric acid giving product ions of 49 and 98 units below doubly and singly charged precursor ions, respectively. Despite the recent rapid developments in the field of plant signal transduction and the increase in sensitivity of mass spectrometers, it has been technically challenging to identify tyrosine-phosphorylated residues by mass spectrometry.

1.5 Metabolomic Approach: Connecting Genes to Metabolites

1.5.1 Plant Biology Needs Genomics, Transcriptomics, Proteomics, and Beyond: Metabolomics

A systematic study about quantification and characterization of pools of metabolites is known as metabolomics. Metabolites represent the distinctive biochemical impressions of particular cellular processes occurring in an organism. Metabolites are the final product of integration of transcription, translation, protein modification and interaction, and biochemical reactions and regulatory mechanisms operative above them. Therefore, they are nearer to the phenotype than transcripts or translates alone. Metabolomic approaches have been effectively used to examine changes occurring at the metabolite level in plants in response to abiotic stress in order to find out signaling under stress. These studies have pointed out the primary metabolites, intermediates of respiration, and photosynthesis processes as indicators of metabolic malfunctioning. In contrast, changes in secondary metabolites are stress specific and further differ from species to species. Additionally, several secondary metabolites are common in abiotic and biotic stresses. It indicates their effective defense roles against both kinds of stresses. Furthermore, the presence/absence of definite metabolites and positive quantitative correlation between metabolic and transcript profiles provides precise markers (mQTL). These markers are used for identification of a candidate gene for crop improvement in breeding programs.

1.5.2 Methods of Analysis of Global Metabolite Pool: The "Metabolome"

A metabolome reflects the final phenotype of plants as an outcome of alterations in transcription and translation and changes in activity of different enzymes/proteins, imposed by various environmental stresses. In turn, metabolites too effect transcription, translation, and protein activity. Their studies provide a comprehensive investigation of essential constituents of the plant metabolic response to abiotic stresses. First, we describe methods commonly used for plant metabolome analysis.

There is no single ideal method currently available for identification and quantification of all metabolites found in a given plant or any other organism (Fernie 2007; Hall 2006; Saito et al. 2006). Therefore, for extensive analysis of metabolome, a combination of multiple techniques, for example, gas chromatography (GC), liquid chromatography (LC), capillary electrophoresis (CE) coupled with mass spectroscopy (MS), and NMR, should be used (Dettmer et al. 2007; Dunn 2008; Eisenreich and Bacher 2007).

In general, few milligrams of a plant tissue is sufficient for extraction of metabolites. Procedures currently used for extraction are biased for certain class of metabolites. For example, water/methanol is used for extraction of polar metabolites, while chloroform is used to extract nonpolar compounds (Tikunov et al. 2005). Before the analysis, metabolites are purified using liquid–liquid and solid-phase chromatographic methods. Gas chromatography–mass spectrometry (GC–MS) (Lisec et al. 2006), LC–MS (De Vos et al. 2007), and NMR (Ward et al. 2007) provide metabolic profile. Signals corresponding to different metabolites in a chromatogram are detected, quantified, and identified, and subsequently a matrix of data about metabolites and their intensities is generated. Finally, using various data mining techniques, the metabolic profile data are analyzed. In the text below, an overview of some of the methods being widely used is described briefly.

1.5.2.1 Gas Chromatography–Mass Spectrometry

It is the most common and widely used technique for plant metabolomics as it is able to identify and quantify primary metabolites such as sugars, sugar alcohols, polyamines, amino acids, and organic acids. In this technique, hydrophilic metabolites are made volatile by chemical derivatization. Subsequently, they are separated by gas chromatography, which separates compounds in complex mixtures with high resolution. These are ionized by electron impact ionization. Finally, fragmentation pattern of a metabolite is detected by mass spectrometry.

GC–MS can detect about 200 metabolites, out of which 50 % can be known by comparison with data profile of standard compounds (Kusano et al. 2007; Roessner et al. 2001; Schauer and Fernie 2006). This method has been extensively used in plant metabolomics, especially in studying the central pathways of primary metabolism (Fukushima et al. 2009; Kusano et al. 2007; Lisec et al. 2006; Roessner et al. 2001). Because of short running time and relatively low cost, this technique has been widely used. Therefore, standardized stable protocols starting from metabolite extraction to final data interpretation are available. Additionally, libraries of chromatogram and mass spectrum of a wide range of standard compounds are also available (Schauer and Fernie 2006). Many metabolite databases are also available for peak annotation (Kind et al. 2009; Kopka et al. 2005). GC–MS can only detect thermally stable volatile compounds; therefore, identification of high molecular weight compounds is relatively difficult by this technique.

1.5.2.2 Liquid Chromatography–Mass Spectrometry

LC–MS does not require sample pretreatment as required in GC–MS. It has been used in the study of plant secondary metabolites such as alkaloids and phenylpropanoids (Iijima et al. 2005; Matsuda et al. 2009, 2010). It provides more options for metabolite separation. Depending upon the aim of the study and metabolites of interest, ion exchange, reversed-phase, or hydrophobic interaction columns can be used for separation. That's why it is possible to analyze a wide diversity of metabolites in plants by LC–MS. Moreover, ultra-performance liquid chromatography (UPLC) has advanced the LC–MS technique because of its higher resolution and sensitivity (Rogachev and Aharoni 2012). In LC–MS, metabolites are ionized using ESI method, and ionized fragments are detected by MS. Combination of various MS techniques such as quadrupole TOF, ion trap, linear trap quadrupole-Orbitrap, and Fourier transform ion cyclotron resonance (FT-ICR)-MS has made LC–MS a more powerful technique. Therefore, metabolites of high molecular weight and less thermostability can also be detected by LC–MS. Moreover, plant hormones, phosphorylated metabolites, and membrane lipids can also be detected by advanced LC–MS techniques (Arrivault et al. 2009; Kanno et al. 2010; Okazaki et al. 2009). The combination of different MS techniques with LC causes difficulty in the identification of peaks in MS spectral libraries because of dependency of retention time and mass spectrum on the type of instrument used (Moco et al. 2006). It forces to create own LC–MS reference library for an individual research group. Due to incomplete knowledge of plant secondary metabolites and lack of standards, it is difficult to identify all metabolic derivatives (Bocker and Rasche 2008).

1.5.2.3 Capillary Electrophoresis-Mass Spectroscopy

CE-MS is the combination of separation process of capillary electrophoresis with mass spectrometry. For ionization, ESI method is generally used in CE-MS. Here, separation takes place based upon charge-to-mass ratio. Therefore, in addition to polar compounds such as organic and amino acids, ionic compounds such as nucleotides and phosphates can also be detected by CE-MS (Levandi et al. 2008; Monton and Soga 2007). It makes CE-MS a more potent technique than LC–MS (Ramautar et al. 2009, 2011). One of the distinctive benefits of CE-MS is that a very small amount (in nanoliters) of sample is required for analysis. Moreover, metabolic analysis of samples takes place within short time. Lack of reference libraries is also the drawback in CE-MS as in the case of LC–MS (Sugimoto et al. 2010).

1.5.3 Metabolic Tuning in Plants in Response to Abiotic Stresses

Plants have evolved extensive and highly effective mechanisms to sense, transduce, and respond to a variety of environmental stresses. Plants make adjustment at different levels of organization (molecular, cellular, tissue, organ, system, anatomical, morphological) to cope with stresses (Atkinson and Urwin 2012). Signaling and regulatory (transcription factors, phosphatases, kinases) mechanisms get activated under stress (Krasensky and Jonak 2012; Wang et al. 2009). Various signaling processes involving calcium (Pan et al. 2012; Reddy et al. 2011), ROS (Ahmad et al. 2010; Loiacono and De Tullio 2012), and kinase cascades (Baena-Gonzalez and Sheen 2008) are induced for signal transduction. At molecular level many changes take place such as alteration of expression of stress-related genes (Delano-Frier et al. 2011; Grativol et al. 2012; Shinozaki and Yamaguchi-Shinozaki 2007), activation of proteins involved in transportation, detoxification, protease and chaperonin activity, production of metabolites involved in osmo-protection, etc. These changes lead to stress-specific responses, which results into adaptation and tolerance towards abiotic stresses. It is necessary for plants to optimize cellular energy resources to survive under stress. Therefore, energetically expensive processes like translation, reproductive activities, and few biosynthetic pathways (photosynthesis, amino acid synthesis) are partially arrested, and energy resources are relocated to activate defensive mechanisms (Good and Zaplachinski 1994; Qu et al. 2011; Holcik and Sonenberg 2005).

One of the most common mechanisms initiated in response to stress is the production of compatible solutes. These include chemically diverse metabolites such as proline, asparagine, polyamines, glycine betaine, γ-amino-*N*-butyric acid (GABA), raffinose, trehalose, sucrose, and polyols (sorbitol, myo-inositol) (Krasensky and Jonak 2012; Banu et al. 2010). Proline has been found to accumulate under several stresses such as salinity, drought, extreme temperatures, heavy metals, etc. (Verslues et al. 2006; Verbruggen and Hermans 2008). Besides osmoprotectant, proline also acts as cryoprotectant, ROS scavenger, and signaling molecule. Synthesis of proline takes place in cytosol or chloroplast, while in mitochondria it is catabolized. Studies have shown that genes involved in proline metabolism are subjected to transcriptional regulation by stresses (Verslues and Sharma 2010). Drought stress upregulates proline biosynthesis genes, while it downregulates its catabolic genes. Moreover, Δ-1-pyrroline-5-carboxylate synthetase, the core enzyme for proline biosynthesis, is inhibited by feedback mechanism by proline level (Yoshiba et al. 1997; Hong et al. 2000). Reverse genetics studies (overexpression of biosynthesis genes or knockout of catabolic genes of proline) have also confirmed the role of proline in providing resistance towards various stresses in different plant species (Kishor et al. 1995; Yamada et al. 2005; Sawahel and Hassan 2002; Su and Wu 2004; Zhu et al. 1998; Vendruscolo et al. 2007; Gleeson et al. 2005; Parvanova et al. 2004; Nanjo et al. 1999; Deuschle et al. 2004; Borsani et al. 2005; Roosens et al. 2002; Wu et al. 2005).

Polyamines are positively charged nitrogenous molecules. Spermidine, spermine, and putrescine are common polyamines found in higher plants. They have been reported to reduce hydrogen peroxide level and lipid oxidation through increasing activities of peroxidase and catalase (Alcázar et al. 2010; Cuevas et al. 2009). It has been found that putrescine level, antioxidant enzyme activity, and carotenoid content are positively correlated to each other (Tun et al. 2006). Opening of stomatal aperture and transpiration rate are negatively correlated to putrescine level (Tun et al. 2006). Reduction in stomata opening as well as in transpiration rate has been found when putrescine level is high. Spermidine and spermine induce production of nitric oxide (NO), which is an important signaling molecule under abiotic stress (Verslues and Juenger 2011).

Carbohydrate metabolism is directly linked to photosynthesis, and both of which undergo readjustment under stress conditions. Starch and fructans are primarily used as energy source during the stress instead of glucose (Kaplan and Guy 2004). Under stress conditions, the amount of fructose and glucose increases (Iordachescu and Imai 2008). These soluble sugars function as osmolytes and help in maintaining turgor pressure of the plant cell (Sharp et al. 2004). Raffinose and stachyose also accumulate under different abiotic stresses in different plant species (Kaplan and Guy 2004). These are non-energetic oligosaccharides. They reduce oxidative damage through ROS scavenging (Molinari et al. 2004). Sugar alcohols (mannitol, sorbitol) and polyols also act as scavengers of hydroxyl radicals and provide stress tolerance (Szabados and Savouré 2010).

Secondary metabolites also provide advantages against stress in different plant species. Among them, carotenoids and phenolic compounds are important protectants against ultraviolet irradiation and excess light. Alkaloid glycosides and glucosinolates act as feeding deterrents against herbivores. Secondary metabolites show species specificity, and they also accumulate in specific cell, tissue, developmental stage, and stress (Arbona et al. 2009; Munns and Tester 2008; Karowe and Grubb 2011). Phenylalanine ammonia lyase, one of the important enzymes in the phenolic biosynthesis pathway, becomes more active during stresses (Wahid et al. 2007). Furthermore, higher concentration of phenylalanine has been found in tolerant genotypes in comparison to sensitive ones (Lugan et al. 2009).

1.5.4 Applications of Metabolomics in Understanding Plant Abiotic Stress Signaling

The metabolic alterations in plants due to stresses have been comprehensively investigated by metabolomic techniques. Furthermore, integration of other omics studies has considerably increased our knowledge about plant's responses to stresses. Metabolite profiling has been done in various stresses such as drought, extreme temperatures, and salinity for extensive analyses of different steps of signal transduction pathways.

ABA-dependent and ABA-independent signaling is initiated during drought (Yamaguchi-Shinozaki et al. 1992). Drought induces biosynthesis of ABA. Recognizing drought stress and activation of protective mechanisms is started in which ABA has an important role. It reduces water loss through transpiration via stomatal closure. It regulates transcription of many stress-responsive genes, which leads to drought stress tolerance in plants. Recently, metabolomic and transcriptomic approaches were integrated to analyze drought response of *Arabidopsis* using NCED3 (nine-*cis*-epoxycarotenoid dioxygenase) knockout mutant and the wild-type plant (Urano et al. 2009). NCED3 is required for the drought-inducible biosynthesis of ABA (Yamaguchi-Shinozaki et al. 1992). This combined study revealed a positive correlation between drought-inducible buildup of proline, saccharopine, branch-chain amino acids, and drought-inducible ABA-dependent expression of their key biosynthetic genes. This study identified ABA-dependent changes in metabolism and changes in transcription of corresponding genes under drought stress. Comparison of transcript and metabolite profiles of drought and salinity in grapevine revealed higher content of proline, glucose, and malate in drought-stressed plants in comparison to salinity-stressed plants (Cramer et al. 2007). A correlation was found between changes in the concentration of these metabolites and their corresponding transcripts abundance. During severe stresses such as drought coupled with heat stress, plant replaces proline with sucrose as the main osmoprotectant (Rizhsky et al. 2004).

Metabolomic studies have revealed that plants show complex metabolic responses depending upon severity and duration of salinity stress. In a time-course study of metabolite changes in cell cultures of *Arabidopsis thaliana* subjected to salinity stress, it was found that inositol and glycerol accumulate within 24 h of salinity stress, whereas sucrose and lactate accumulate late after 48 h. Biosynthesis of phenylpropanoid and glycine betaine was inducted during early response to salinity, whereas sucrose metabolism and glycolysis were reduced at late duration (Kim et al. 2007). In a different study, treatment of tobacco plants with different doses of salt revealed different metabolic responses (Zhang et al. 2011). 50 mM NaCl prompted sucrose accumulation after 1 day of treatment, while 500 mM NaCl treatment for additional 24 h caused accumulation of proline and more concentration of sucrose. Proline level remained high, and sucrose level decreased (although still higher than control) as continuation of stress up to 3–7 days. Concentrations of uracil, myo-inositol, tryptophan, valine, isoleucine, and asparagine increased significantly, while levels of fructose, glucose, fumarate, malate, formate, uridine hypoxanthine, and glutamine were reduced (Zhang et al. 2011). Metabolic responses in salinity stress also differ depending upon the tissue. Sucrose level increased in both roots and shoots of maize exposed to salinity stress, while other osmoprotectants showed discriminated pattern: malate, succinate, and GABA levels increased in roots, while glycine betaine, asparagine, and glutamate increased in shoots. Metabolic responses were more prominent in shoots than in roots (Gavaghan et al. 2011). These responses also vary with species to species.

GC–MS and microarray approaches were integrated to compare salinity stress responses of *Arabidopsis* and its closely related halophytic species, *Thellungiella halophila* (Gong et al. 2005). These two species showed drastically different metabolic profiles. *Thellungiella* upheld higher levels of metabolites in salinity stress and even in control condition in comparison to *Arabidopsis*. Various osmolytes, such as proline, sucrose, fructose, and malate, were found in higher amount in *Thellungiella* even in the absence of salinity stress. In agreement to these findings, transcriptomics revealed that *Thellungiella* maintained higher transcript abundance of that several stress-related genes even in the absence of stress. Outcomes of these integrated studies evidently demonstrate well preparedness of *Thellungiella* in anticipation of stress.

Metabolite profiling has been also done to understand plant responses to extreme temperatures (Cook et al. 2004; Maruyama et al. 2009; Kaplan et al. 2004; Wienkoop et al. 2008; Hannah et al. 2006). Comparison of metabolite profiles of plants subjected to cold and other stresses has identified some common metabolites. DREBs/CBFs (dehydration-responsive element-binding factors/C-repeats-binding factors) have been discovered as an important transcription factor regulating cold response (Cook et al. 2004; Maruyama et al. 2009). Metabolome of *Arabidopsis* transgenic plants overexpressing DREB1A/CBF3 exhibited a remarkable similarity with low-temperature-regulated metabolome, particularly sugars and sugar alcohols (Cook et al. 2004; Maruyama et al. 2009). Strikingly, accumulation of raffinose and galactinol induced under cold was correlated with the transcript level of the galactinol synthase 3 gene, which is the direct target of DREB1A/CBF3 (Cook et al. 2004; Maruyama et al. 2009). Similar outcomes were obtained by Maruyama et al. (2009) in case of DREB1A-overexpressing plants as these plants also contained increased amount of cold-inducible metabolites. Metabolic study also revealed that role of DREB2A is not significant in cold. Plants overexpressing this gene did not contain higher level of any low-temperature-regulated metabolites (Cook et al. 2004; Maruyama et al. 2009). In agreement, DREB2A overexpression plants are more tolerant to drought instead of cold (Yamaguchi-Shinozaki et al. 1992). These metabolic studies have highlighted the importance of sucrose, raffinose, galactinol, and myo-inositol in cold stress. Metabolite changes in *Arabidopsis* plants in response to heat shock and cold shock were analyzed and compared using GC–MS (Kaplan et al. 2004; Wienkoop et al. 2008). The majority of metabolites produced in response to these two stresses were found to be common. It indicates that some metabolites such as proline, glucose, fructose, galactinol, and raffinose functioning as compatible solutes play an important role in tolerance to extreme temperatures.

1.6 Signal Transduction Under Abiotic Stress

Using various omics technologies and their combination, attempts have been made to dissect out the signal transduction pathways operative under abiotic stresses. Here, we are summarizing the mechanism of signaling under abiotic stress identified, till now, by using various omics technologies.

Our knowledge is still very hazy about the receptors, which perceive abiotic stresses. Salinity stress is one of the major threats to agriculture. It inflicts "osmotic stress" arising from the lowered water potential in the soil and "ionic stress" triggered by intracellular increase in concentration of Na^+ and Cl^- ions, which disturbs the K^+–Na^+ ratio and hence ionic balance (Blumwald et al. 2000). A summary of signal transduction pathways operative in plants under various abiotic stresses is illustrated in Fig. 1.1.

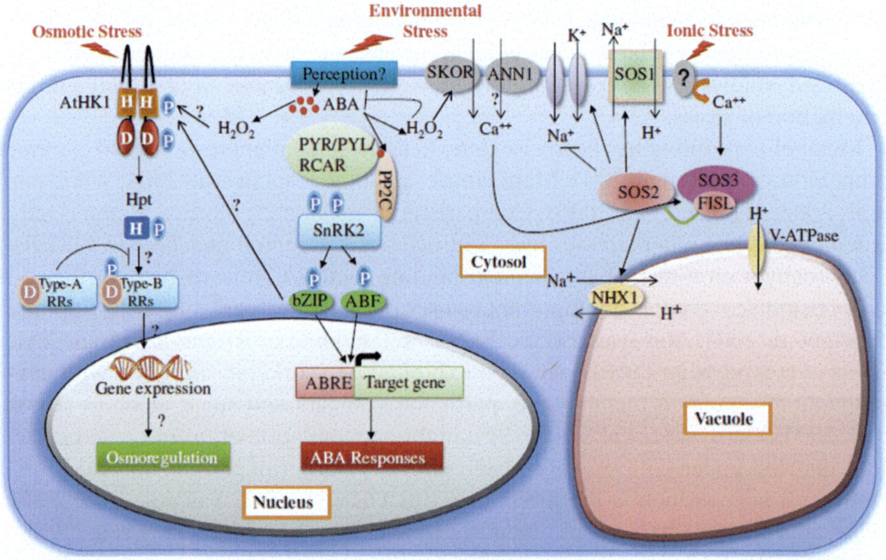

Fig. 1.1 Schematic representation of ABA signaling, osmosensing, and SOS pathway in a plant cell. ABA level rises under environmental stress condition, PYR/PYL/RCAR receptors bind ABA, and then receptor-ABA complex inhibits 2C-type protein phosphatases. These protein phosphatases act as co-receptors. Their inhibition triggers SNF1-type kinase cascade, which leads to ABA-dependent gene expression and consequently ABA responses. Membrane-bound receptors, histidine kinases, control a wide array of cellular mechanisms. Osmotic stress signal is perceived by one of the members of this large family of receptors, AtHK1, which transduces this signal to a histidine transferase (Hpt) through phosphorylation. Subsequently, Hpt interacts with response regulators (RRs). These passes signal to downstream components to activate or downregulate genes responsible for initiating osmo-protective responses. In plants, three members (SOS1, SOS2, and SOS3) play an important role in salt stress signaling. SOS1 transports sodium ions across the plasma membrane. SOS2 is a Ser/Thr protein kinase. SOS3 is a Ca^{++}-binding protein. Salt stress elicits Ca^{++} transients, which are perceived by SOS3. It interacts and activates SOS2. SOS2/SOS3 kinase complex phosphorylates several transporters localized on tonoplast and plasma membrane, including SOS1, which transports sodium ions across the plasma membrane. It restores the ionic homeostasis. H_2O_2, although through yet unknown mechanisms, plays an integral role in cross-linking various stress responses

In plants, relatively little information is available about osmosensing, but a number of plasma membrane-localized proteins have been thought to act as osmosensors (Grefen and Harter 2004). Homologues of yeast SLN1 (osmosensor in yeast) have been found in plants. In *Arabidopsis*, AtHK1 is supposed to play a role as putative osmosensor (Urao et al. 1999). Heterologous expression of AtHK1 cDNA in (*Sln1ΔSho1Δ*) double deletion mutant of yeast restores its viability on high-salinity medium, and it also triggers HOG1 pathway in yeast (Urao et al. 1999). *In planta*, AtHK1 is reported to be involved in water stress responses (Wohlbach et al. 2008). AtHK1 shows high expression in roots in comparison to other tissues, and its expression is induced by changes in external osmolarity, which strengthens the fact that it acts as an osmosensor. It is also noteworthy that mutant *Arabidopsis* plants of *AtHK1* gene are not able to survive under desiccation (Wohlbach et al. 2008). AtHK1 regulates not only water stress response but also desiccation procedures during maturation of seeds (Wohlbach et al. 2008). Interestingly, another histidine kinase in *Arabidopsis*—AtHK4, when expressed in yeast—shows regulation by fluctuations in turgidity in the presence of cytokinin (Reiser et al. 2003). In *Medicago sativa*, a salt stress-responsive histidine kinase has been identified (Coba de la Pena et al. 2008). In tobacco, a membrane-localized receptor-like kinase, NtC7, has been identified whose transcript level is upregulated by osmotic and salt stresses. Transgenic tobacco plants overexpressing this gene perform better than the wild-type plants under stress conditions (Tamura et al. 2003). Thus, quite a lot of candidates have been reported to function as osmosensors in plants.

The receptor/sensor, which perceives ionic stress signal in plants, has not been identified yet. Theoretically, it may be present at plasma membrane or within the cell. Several cytoplasmic enzymes are sensitive to Na^+, which might perform the function of Na^+ sensing (Zhu 2003). One of the probable candidates is a Na^+/H^+ antiporter SOS1 (Salt Overly Sensitive1) (Shi et al. 2000). It is plasma membrane localized and plays an important role in Na^+ efflux (Quintero et al. 2002). It has twelve hydrophobic transmembrane regions and a long hydrophilic C-terminal tail (Shi et al. 2000). It has been hypothesized that this cytoplasmic tail is important for Na^+ sensing (Zhu 2003). Probably SOS1 is a "transceptor" simultaneously performing dual function of Na^+ transport as well as Na^+ receptor, a phenomenon observed in case of other ions also such as NO_3^- (Conde et al. 2010). SOS1 and two more members—SOS2 and SOS3—were identified by screening of mutant population of *Arabidopsis* for altered tolerance towards salinity stress and identifying the responsible genetic alteration. As the name suggests, *sos* mutants of *Arabidopsis* are overly sensitive to salinity stress in comparison to wild-type plants. SOS2 is a serine/threonine kinase containing an N-terminal catalytic domain and a C-terminal auto-inhibitory domain. It shows strong similarity to the yeast SNF1 kinase. SOS3 is a Ca^{2+} sensor. It shows high similarity with members belonging to the neuronal Ca^{2+} sensor family (Zhu 2000; Silva and Gerós 2009). Salt stress elicits Ca^{2+} transients, which are perceived by SOS3. SOS3 relieves auto-inhibition of SOS2 and thereby activates it (Guo et al. 2001). SOS2/SOS3 kinase complex phosphorylates SOS1, which transports sodium ions across the plasma membrane. Thus these three components of SOS pathway constitute an operational module (Quintero et al. 2002).

ABA has an essential role in regulation of plant responses to drought and salinity (Cramer 2010; Hubbard et al. 2010; Kim et al. 2010; Chinnusamy et al. 2008). The downstream signaling of ABA involves transcriptional regulation of several stress-responsive genes and genes encoding for enzymes related to biosynthesis of many amino acids and sugars. It also controls stomatal aperture through regulation of ion and water transport in guard cells (Kim et al. 2010). The receptors of ABA hormone remained unidentified for a long time, but now, the receptors and the downstream signaling components have been discovered (Umezawa 2011; Ma et al. 2009; Park et al. 2009; Leung and Giraudat 1998), which involves receptors (PYR/PYL/RCAR), protein phosphatases (PP2C), and protein kinases (SnRK2/OST1) (Ma et al. 2009; Park et al. 2009). Various types of ABA receptors, soluble and membrane anchored, have been identified (Ma et al. 2009; Park et al. 2009). PYR/PYL/RCAR receptors, identified by 4 different groups, appear to be the most suitable one (Ma et al. 2009; Park et al. 2009; Santiago et al. 2009; Nishimura et al. 2010). ABI1 and ABI2 are 2C-type protein phosphatases, and they act as negative regulators of ABA signaling (Leung and Giraudat 1998). SNF1-related protein kinases belong to a multigene family of protein kinases and play a positive role in ABA signaling (Yoshida et al. 2002; Mustilli et al. 2002; Umezawa et al. 2010). PP2C protein phosphatase dephosphorylates SnRK2 at specific residues and represses its kinase activity. After binding ABA hormone, ABA+PYR1/PYLs/RCAR1 complex interacts with PP2C and impedes its phosphatase activity. This leads to activation of SnRK2, which in turn phosphorylates AREB/ABF bZIP transcription factors. These activated transcription factors induce expression of many stress-related genes, thereby ultimately resulting in ABA response.

1.7 Conclusions and Future Perspectives

Out of all the technologies mentioned in the text, transcriptomics remains one of the most popular ones in the field of plant stress physiology. Through the integration of information obtained from various transcriptomic approaches, one can design very elaborate, but accurate networks of gene expression regulation in response to abiotic stress. Stress response signaling pathways can be decoded but not without the aid from other fields of functional genomics. Furthermore, the first critical step towards stress response, i.e., stress perception, has nothing to do with gene regulation, but the first actual changes that occur are concerned with gene expression. Whether it is the production of miRNAs or the activation of genes, which are normally not expressed, the first changes on the major scale catalyzed by any environmental factor occur in the plant's transcriptomes. All other "omes" (proteome, metabolome, ionome, etc.) except for the genome are dependent on the transcriptomes. Thus, transcriptomics have not just enabled us to understand signaling under stress with respect to which genes are regulated but also provided the preliminary information for carrying out further detailed studies with respect to other "omics" approaches (Fig. 1.2).

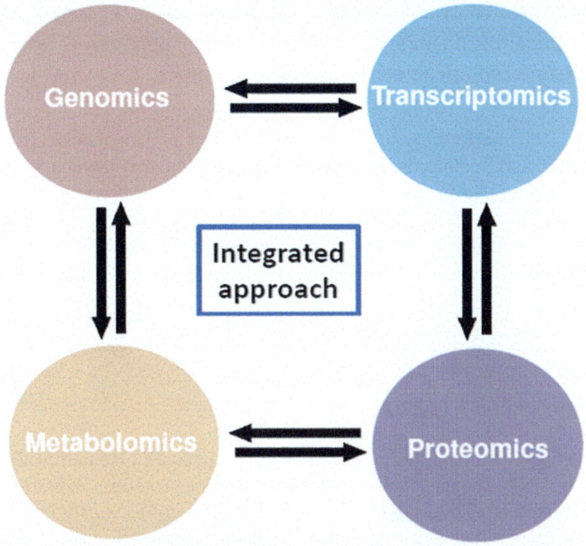

Fig. 1.2 A pictorial representation showing interrelation between "omics"

Though cells of an organism have a similar genome, each cell maintains a specific structure and plays a specific role under normal and stress conditions for survival. The key of understanding the complex nature of cellular responses lies in the study of transcriptomes and proteome of cells. The whole set of transcriptome does not represent the complete proteome of a cell. It is suggested that better understanding of the cell, its responses, and molecular mechanisms requires a complete and comprehensive protein reference map of the cell. The alterations at protein level have a vital effect on the structure and metabolism of a plant cell. The overall composition and function of a proteome is not only affected by quantitative changes in protein abundance but also by alterations in protein posttranslational modifications and protein–protein interactions. Therefore, for comprehensive analysis of proteome, these changes should also be surveyed. Moreover, the proteome study should be linked with other studies to unknot the intricacy of plant stress responses. This area has unlimited potential and recent advancements in the detection and isolation approaches, and improvements in mass spectroscopy instruments have made proteomic technologies more advanced (Kosova et al. 2013).

With the evolution of both techniques and technologies, there is a rapid growth of proteomics. MS is nowadays coming up as a "virtual imaging technique' in a way to directly analyze protein distributions in cells and tissue samples. Automation and robotics in as many steps as possible is being done to an increasing extent to enhance the speed and reliability of proteomic studies. Another new avenue for proteomics is the development of protein arrays similar to DNA microarrays. Though proteins lack the ability to hybridize with complimentary sequences, protein analysis by selective interaction of proteins or peptides with an array of different recognition elements is underway. An array of specific protein set will provide the platform to carry out highly parallel studies of protein–protein interaction and also show the effect of interaction when affected by signaling factors.

Acknowledgments The authors would like to thank the Department of Biotechnology, Department of Science and Technology (Government of India), and Jawaharlal Nehru University (through capacity building, CAS, and PURSE) for supporting research work in the laboratory. Award of research fellowship from the Council of Scientific and Industrial Research (CSIR) to S.R., R.N., and P.S. is thankfully acknowledged. Grants from DBT to K.K.N. and UGC to N.S. are also duly acknowledged.

References

Abdallah C, Sergeant K, Guillier C, Dumas-Gaudot E, Leclercq CC, Renaut J (2012) Optimization of iTRAQ labelling coupled to OFFGEL fractionation as a proteomic work flow to the analysis of microsomal proteins of *Medicago truncatula* roots. Proteome Sci 10(1):37

Agarwal PK, Jha B (2010) Transcription factors in plants and ABA dependent and independent abiotic stress signaling. Biol Planta 54(2):201–212

Ahmad P, Jaleel CA, Salem MA, Nabi G, Sharma S (2010) Roles of enzymatic and nonenzymatic antioxidants in plants during abiotic stress. Crit Rev Biotechnol 30(3):161–75

Alagna F, D'Agostino N, Torchia L, Servili M, Rao R, Pietrella M, Giuliano G, Chiusano ML, Baldoni L, Perrotta G (2009) Comparative 454 pyrosequencing of transcripts from two olive genotypes during fruit development. BMC Genomics 10:399

Alban A, David SO, Bjorkesten L, Andersson C, Sloge E, Lewis S, Currie I (2003) A novel experimental design for comparative two-dimensional gel analysis: two-dimensional difference gel electrophoresis incorporating a pooled internal standard. Proteomics 3(1):36–44

Alcázar R, Planas J, Saxena T, Zarza X, Bortolotti C, Cuevas J, Bitrián M et al (2010) Putrescine accumulation confers drought tolerance in transgenic *Arabidopsis* plants over-expressing the homologous arginine decarboxylase 2 gene. Plant Physiol Biochem 48:547–552

Alonso R, Oñate-Sánehez L, Weltmeier F, Ehlert A, Diaz I, Dietrich K et al (2009) A pivotal role of the basic leucine zipper transcription factor bZIP53 in the regulation of Arabidopsis seed maturation gene expression based on heterodimerization and protein complex formation. Plant Cell 21(6):1747–1761

Amor BB, Wirth S, Merchan F, Laporte P, d'Aubenton-Carafa Y, Hirsch J, Maizel A, Mallory A, Lucas A, Deragon JM, Vaucheret H, Thermes C, Crespi M (2009) Novel long non-protein coding RNAs involved in Arabidopsis differentiation and stress responses. Genome Res 19(1): 57–69

Andreou A, Feussner I (2009) Lipoxygenases-structure and reaction mechanism. Phytochemistry 70(13–14):1504–1510

Arbona V, Gómez-Cadenas A (2008) Hormonal modulation of citrus responses to flooding. J Plant Growth Regul 27(3):241–250

Arbona V, Iglesias DJ, Talón M, Gómez-Cadenas A (2009) Plant phenotype demarcation using nontargeted LC-MS and GC-MS metabolite profiling. J Agric Food Chem 57:7338–7347

Arrivault S, Guenther M, Ivakov A, Feil R, Vosloh D, van Dongen JT et al (2009) Use of reverse-phase liquid chromatography, linked to tandem mass spectrometry, to profile the Calvin cycle and other metabolic intermediates in Arabidopsis rosettes at different carbon dioxide concentrations. Plant J 59:826–839

Atkinson NJ, Urwin PE (2012) The interaction of plant biotic and abiotic stresses: from genes to the field. J Exp Bot 63(10):3523–43

Baena-González E, Sheen J (2008) Convergent energy and stress signaling. Trends Plant Sci 13(9):474–482

Banu MN, Hoque MA, Watanabe-Sugimoto M, Islam MM, Uraji M, Matsuoka K et al (2010) Proline and glycinebetaine ameliorated NaCl stress via scavenging of hydrogen peroxide and methylglyoxal but not superoxide or nitric oxide in tobacco cultured cells. Biosci Biotechnol Biochem 74(10):2043–9

Blumwald E, Aharon GS, Apse MP (2000) Sodium transport in plant cells. Biochim Biophys Acta 1465:140–151

Bocker S, Rasche F (2008) Towards de novo identification of metabolites by analysing tandem mass spectra. Bioinformatics 24:49–55

Borsani O, Zhu J, Verslues PE, Sunkar R, Zhu JK (2005) Endogenous siRNAs derived from a pair of natural cis-antisense transcripts regulate salt tolerance in Arabidopsis. Cell 123(7): 1279–91

Brenner S, Johnson M, Bridgham J, Golda G, Lloyd DH, Johnson D et al (2000) Gene expression analysis by massively parallel signature sequencing (MPSS) on microbead arrays. Nat Biotechnol 18(6):630–634

Catinot J, Buchala A, Abou-Mansour E, Métraux JP (2008) Salicylic acid production in response to biotic and abiotic stress depends on isochorismate in Nicotiana benthamiana. FEBS Lett 582(4):473–478

Chain PS, Grafham DV, Fulton RS, Fitzgerald MG, Hostetler J, Muzny D et al (2009) Genome project standards in a new era of sequencing. Science 326(5950):236–237

Chen F, Li Q, Sun L, He Z (2006) The rice 14-3-3 gene family and its involvement in responses to biotic and abiotic stress. DNA Res 13(2):53–63

Chen L, Song Y, Li S, Zhang L, Zou C, Yu D (2012) The role of WRKY transcription factors in plant abiotic stresses. Biochim Biophys Acta 1819(2):120–128

Chinnusamy V, Gong Z, Zhu JK (2008) Abscisic acid-mediated epigenetic processes in plant development and stress responses. J Integr Plant Biol 50(10):1187–1195

Choudhary MK, Basu D, Datta A, Chakraborty N, Chakraborty S (2009) Dehydration-responsive nuclear proteome of rice (Oryza sativa L.) illustrates protein network, novel regulators of cellular adaptation, and evolutionary perspective. Mol Cell Proteomics 8(7):1579–1598

Christianson JA, Dennis ES, Llewellyn DJ, Wilson WI (2010) ATAF NAC transcription factors: regulators of plant stress signaling. Plant Signal Behav 5(4):428–432

Cifre J, Bota J, Escalona JM, Medrano H, Flexas J (2005) Physiological tools for irrigation scheduling in grapevine (Vitis vinifera L.) An open gate to improve water-use efficiency? Agric Ecosysyst Environ 106(2–3):159–170

Coba de la Pena T, Carcamo CB, Almonacid L, Zaballos A, Lucas MM, Balomenos D et al (2008) A salt stress-responsive cytokinin receptor homologue isolated from Medicago sativa nodules. Planta 227:769–779

Conde A, Diallinas G, Chaumont F, Chaves M, Gerós H (2010) Transporters, channels, or simple diffusion? Dogmas, a typical roles and complexity in transport systems. Int J Biochem Cell Biol 42(6):857–868

Conrath U, Beckers GJM, Flors V, Garcia-Augustin P, Jakab G, Mauch F et al (2006) Priming: getting ready for battle. Mol Plant Microbe Interact 19(10):1062–1071

Cook D, Fowler S, Fiehn O, Thomashow MF (2004) A prominent role for the CBF cold response pathway in configuring the low temperature metabolome of Arabidopsis. Proc Natl Acad Sci U S A 101:15243–15248

Corrêa LG, Riaño-Pachón DM, Schrago CG, dos Santos RV, Mueller-Roeber B, Vincentz M (2008) The role of bZIP transcription factors in green plant evolution: adaptive features emerging from four founder genes. PLoS One 3(8):e2944

Cramer GR (2010) Abiotic stress & plant responses from the whole vine to the genes. Aust J Grape Wine Res 16:86–93

Cramer GR, Ergul A, Grimplet J, Tillett RL, Tattersall EA, Bohlman MC et al (2007) Water and salinity stress in grapevines: early and late changes in transcript and metabolite profiles. Funct Integr Genomics 7:111–134

Cuevas JC, López-Cobollo R, Alcázar R, Zarza X, Koncz C, Altabella T, Ferrando A et al (2009) Putrescine as a signal to modulate the indispensable ABA increase under cold stress. Plant Signal Behav 4:219–220

De Vos RC, Moco S, Lommen A, Keurentjes JJ, Bino RJ, Hall RD (2007) Untargeted large-scale plant metabolomics using liquid chromatography coupled to mass spectrometry. Nat Protoc 2(4):778–91

Debnath M, Pandey M, Bisen PS (2011) An omics approach to understand the plant abiotic stress. OMICS 15(11):739–762

Delano-Frier JP, Aviles-Arnaut H, Casarrubias-Castillo K, Casique-Arroyo G, Castrillon-Arbelaez PA, Herrera-Estrella L et al (2011) Transcriptomic analysis of grain amaranth (Amaranthus hypochondriacus) using 454 pyrosequencing: comparison with A. tuberculatus, expression profiling in stems and in response to biotic and abiotic stress. BMC Genomics 12:363

Dettmer K, Aronov PA, Hammock BD (2007) Mass spectrometry-based metabolomics. Mass Spectrom Rev 26:51–78

Deuschle K, Funck D, Forlani G, Stransky H, Biehl A, Leister D et al (2004) The role of (Delta)1-pyrroline-5-carboxylate dehydrogenase in proline degradation. Plant Cell 16(12):3413–25

Dunn WB (2008) Current trends and future requirements for the mass spectrometric investigation of microbial, mammalian and plant metabolomes. Phys Biol 5:11001

Eisenreich W, Bacher A (2007) Advances of high-resolution NMR techniques in the structural and metabolic analysis of plant biochemistry. Phytochemistry 68:2799–815

Falvo S, Di CM, Desiderio A, Benvenuto E, Moglia A, America T, Lanteri S, Acquadro A (2012) 2-D DIGE analysis of UV-C radiation-responsive proteins in globe artichoke leaves. Proteomics 12(3):448–460

Fernie AR (2007) The future of metabolic phytochemistry: larger numbers of metabolites, higher resolution, greater understanding. Phytochemistry 68:2861–80

Feuillet C, Leach JE, Rogers J, Schnable PS, Eversole K (2010) Crop genome sequencing: lessons and rationales. Trends Plant Sci 16(2):77–88

Filichkin SA, Priest HD, Givan SA, Shen R, Bryant DW, Fox SE et al (2010) Genome-wide mapping of alternative splicing in *Arabidopsis thaliana*. Genome Res 20(1):45–58

Floris M, Mahgoub H, Lanet E, Robaglia C, Menand B (2009) Post-transcriptional regulation of gene expression in plants during abiotic stress. Int J Mol Sci 10(7):3168–3185

Folgado R, Panis B, Sergeant K, Renaut J, Swennen R, Hausman JF (2013) Differential protein expression in response to abiotic stress in two potato species: *Solanum commersonii* Dun and *Solanum tuberosum* L. Int J Mol Sci 14(3):4912–4933

Fu D, Huang B, Xiao Y, Muthukrishnan S, Liang GH (2007) Overexpression of barley hva1 gene in creeping bentgrass for improving drought tolerance. Plant Cell Rep 26(4):467–477

Fukushima A, Kusano M, Redestig H, Arita M, Saito K (2009) Integrated omics approaches in plant systems biology. Curr Opin Chem Biol 13(5–6):532–8

Garg R, Jain M (2011) Pyrosequencing data reveals tissue-specific expression of lineage-specific transcripts in chickpea. Plant Signal Behav 6(11):1868–1870

Gavaghan CL, Li JV, Hadfield ST, Hole S, Nicholson JK, Wilson ID et al (2011) Application of NMR-based metabolomics to the investigation of salt stress in maize (Zea mays). Phytochem Anal 22(3):214–24

Gleeson D, Lelu-Walter M-A, Parkinson M (2005) Overproduction of proline in transgenic hybrid larch (*Larix* × *leptoeuropaea* (Dengler)) cultures renders them tolerant to cold, salt and frost. Mol Breeding 15(1):21–9

Gokirmak T, Paul AL, Ferl RJ (2010) Plant phosphopeptide-binding proteins as signaling mediators. Curr Opin Plant Biol 13(5):527–532

Golldack D, Lüking I, Yang O (2011) Plant tolerance to drought and salinity: stress regulating transcription factors and their functional significance in the cellular transcriptional network. Plant Cell Rep 30(8):1383–1391

Gong Q, Li P, Ma S, Indu Rupassara S, Bohnert HJ (2005) Salinity stress adaptation competence in the extremophile Thellungiella halophila in comparison with its relative Arabidopsis thaliana. Plant J 44:826–839

Good AG, Zaplachinski ST (1994) The effects of drought stress on free amino acid accumulation and protein synthesis in Brassica napus. Physiol Plant 90(1):9–14

Grativol C, Hemerly AS, Ferreira PC (2012) Genetic and epigenetic regulation of stress responses in natural plant populations. Biochim Biophys Acta 1819(2):176–85

Grefen C, Harter K (2004) Plant two-component systems: principles, functions, complexity and cross talk. Planta 219:733–742

Guan Q, Wu J, Yue X, Zhang Y, Zhu J (2013) A nuclear calcium-sensing pathway is critical for gene regulation and salt stress tolerance in Arabidopsis. PLoS Genet 9(8):e1003755

Gulyani V, Khurana P (2011) Identification and expression profiling of drought-regulated genes in mulberry (Morus sp.) by suppression subtractive hybridization of susceptible and tolerant cultivars. Tree Genet Genomes 7(4):725–738

Guo Y, Song Y (2009) Differential proteomic analysis of apoplastic proteins during initial phase of salt stress in rice. Plant Signal Behav 4(2):121–122

Guo Y, Halfter U, Ishitani M, Zhu JK (2001) Molecular characterization of functional domains in the protein kinase SOS2 that is required for plant salt tolerance. Plant Cell 13:1383–1400

Hall RD (2006) Plant metabolomics: from holistic hope, to hype, to hot topic. New Phytol 169: 453–68

Hannah MA, Wiese D, Freund S, Fiehn O, Heyer AG, Hincha DK (2006) Natural genetic variation of freezing tolerance in Arabidopsis. Plant Physiol 142:98–112

Harbers M, Carninci P (2005) Tag-based approaches for transcriptome research and genome annotation. Nat Methods 2(7):495–502

Hashimoto M, Toorchi M, Matsushita K, Iwasaki Y, Komatsu S (2009) Proteome analysis of rice root plasma membrane and detection of cold stress responsive proteins. Protein Pept Lett 16(6):685–697

Hirayama T, Shinozaki K (2010) Research on plant abiotic stress responses in the post-genomic era: past, present and future. Plant J 61(6):1041–1052

Hmida-Sayari A, Gargouri-Bouzid R, Bidani A, Jaoua L, Savouré A, Jaoua S (2005) Overexpression of D1-pyrroline-5-carboxylate synthetase increases proline production and confers salt tolerance in transgenic potato plants. Plant Sci 169(4):746–752

Holcik M, Sonenberg N (2005) Translational control in stress and apoptosis. Nat Rev Mol Cell Biol 6(4):318–27

Hong Z, Lakkineni K, Zhang Z, Verma DP (2000) Removal of feedback inhibition of delta (1)-pyrroline-5-carboxylate synthetase results in increased proline accumulation and protection of plants from osmotic stress. Plant Physiol 122(4):1129–36

Hu H, Dai M, Yao J, Xiao B, Li X, Zhang Q et al (2006) Overexpressing a NAM, ATAF, and CUC (NAC) transcription factor enhances drought resistance and salt tolerance in rice. Proc Natl Acad Sci U S A 103(35):12987–12992

Hubbard KE, Nishimura N, Hitomi K, Getzoff ED, Schroeder JI (2010) Early abscisic acid signal transduction mechanisms: newly discovered components and newly emerging questions. Genes Dev 24(16):1695–1708

Ichimura K, Mizoguchi T, Irie K, Morris P, Giraudat J, Matsumoto K, Shinozaki K (1998) Isolation of ATMEKK1 (a MAP kinase kinase kinase)-interacting proteins and analysis of a MAP kinase cascade in Arabidopsis. Biochem Biophys Res Commun 253:532–543

Iijima Y, Nakamura Y, Ogata Y, Tanaka K, Sakurai N, Suda K et al (2005) Metabolite annotations based on the integration of mass spectral information. Plant J 54(5):949–62

Iordachescu M, Imai R (2008) Trehalose biosynthesis in response to abiotic stresses. J Integr Plant Biol 50:1223–1229

Jaskiewicz M, Conrath U, Peterhänsel C (2011) Chromatin modification acts as a memory for systemic acquired resistance in the plant stress response. EMBO Rep 12(1):50–55

Ji X, Dong B, Shiran B, Talbot MJ, Edlington JE, Hughes T, White RG, Gubler F, Dolferus R (2011) Control of abscisic acid catabolism and abscisic acid homeostasis is important for reproductive stage stress tolerance in cereals. Plant Physiol 156:647–662

Juven-Gershon T, Hsu JY, Theisen JW, Kadonaga JT (2008) The RNA polymerase II core promoter-the gateway to transcription. Curr Opin Cell Biol 20(3):253–259

Kader MA, Lindberg S (2010) Cytosolic calcium and pH signaling in plants under salinity stress. Plant Signal Behav 5(3):233–238

Kanno Y, Jikumaru Y, Hanada A, Nambara E, Abrams SR, Kamiya Y, Seo M (2010) Comprehensive hormone profiling in developing Arabidopsis seeds: examination of the site of ABA biosynthesis, ABA transport and hormone interactions. Plant Cell Physiol 51:1988–2001

Kaplan F, Guy CL (2004) β-Amylase induction and the protective role of maltose during temperature shock. Plant Physiol 135:1674–1684

Kaplan F, Kopka J, Haskell DW, Zhao W, Schiller KC, Gatzke N et al (2004) Exploring the temperature-stress metabolome of Arabidopsis. Plant Physiol 136:4159–4168

Karan R, Singla-Pareek SL, Pareek A (2009) Histidine kinase and response regulator genes as they relate to salinity tolerance in rice. Funct Integr Genomics 9(3):411–417

Karowe DN, Grubb C (2011) Elevated CO_2 increases constitutive phenolics and trichomes, but decreases inducibility of phenolics in *Brassica rapa* (Brassicaceae). J Chem Ecol 37: 1332–1340

Kawaura K, Mochida K, Ogihara Y (2008) Genome wide analysis for identification of salt responsive genes in common wheat. Funct Integr Genomics 8(3):277–286

Kim JK, Bamba T, Harada K, Fukusaki E, Kobayashi A (2007) Time-course metabolic profiling in Arabidopsis thaliana cell cultures after salt stress treatment. J Exp Bot 58(3):415–24

Kim TH, Bohmer M, Hu H, Nishimura N, Schroeder JI (2010) Guard cell signal transduction network: advances in understanding abscisic acid, CO^2, and Ca^{2+} signaling. Annu Rev Plant Biol 61:561–591

Kind T, Scholz M, Fiehn O (2009) How large is the metabolome? A critical analysis of data exchange practices in chemistry. PLoS One 4(5):e5440

Kishor P, Hong Z, Miao GH, Hu C, Verma D (1995) Overexpression of (delta)-pyrroline-5-carboxylate synthetase increases proline production and confers osmotolerance in transgenic plants. Plant Physiol 108(4):1387–94

Kodzius R, Kojima M, Nishiyori H, Nakamura M, Fukuda S, Tagami M, Sasaki D, Imamura K, Kai C, Harbers M, Hayashizaki Y, Carninci P (2006) CAGE: cap analysis of gene expression. Nat Methods 3(3):211–222

Komatsu S, Wada T, Abalea Y, Nouri MZ, Nanjo Y, Nakayama N, Shimamura S, Yamamoto R, Nakamura T, Furukawa K (2009) Analysis of plasma membrane proteome in soybean and application to flooding stress response. J Proteome Res 8(10):4487–4499

Kopka J, Schauer N, Krueger S, Birkemeyer C, Usadel B, Bergmüller E et al (2005) GMD@CSB. DB: the Golm Metabolome Database. Bioinformatics 21(8):1635–8

Kosova K, Prail IT, Vitamvas P (2013) Protein contribution to plant salinity response and tolerance acquisition. Int J Mol Sci 14(4):6757–6789

Krasensky J, Jonak C (2012) Drought, salt, and temperature stress-induced metabolic rearrangements and regulatory networks. J Exp Bot 63(4):1593–608

Kreps JA, Wu Y, Chang HS, Zhu T, Wang X, Harper JF (2002) Transcriptome changes for Arabidopsis in response to salt, osmotic, and cold stress. Plant Physiol 130(4):2129–2141

Kruft V, Eubel H, Jansch L, Werhahn W, Braun HP (2001) Proteomic approach to identify novel mitochondrial proteins in Arabidopsis. Plant Physiol 127(4):1694–1710

Kumar G, Kushwaha HR, Panjabi-Sabharwal V, Kumari S, Joshi R, Karan R et al (2012) Clustered metallothionein genes are co-regulated in rice and ectopic expression of OsMT1e-P confers multiple abiotic stress tolerance in tobacco via ROS scavenging. BMC Plant Biol 12:107

Kumari S, Panjabi Nee Sabharwal V, Kushwaha HR, Sopory SK, Singla-Pareek SL et al (2009) Transcriptome map for seedling stage specific salinity stress response indicates a specific set of genes as candidate for saline tolerance in *Oryza sativa* L. Funct Integr Genomics 9:109–123

Kusano T, Yamaguchi K, Berberich T, Takahashi Y (2007) Advances in polyamine research in 2007. J Plant Res 120(3):345–50

Kwon SJ, Choi EY, Choi YJ, Ahn JH, Park OK (2006) Proteomics studies of post-translational modifications in plants. J Exp Bot 57(7):1547–1551

Lee S, Lee EJ, Yang EJ, Lee JE, Park AR, Song WH et al (2004) Proteomic identification of annexins, calcium-dependent membrane binding proteins that mediate osmotic stress and abscisic acid signal transduction in Arabidopsis. Plant Cell 16(6):1378–1391

Lee CP, Eubel H, O'Toole N, Millar AH (2011) Combining proteomics of root and shoot mitochondria and transcript analysis to define constitutive and variable components in plant mitochondria. Phytochemistry 72(10):1092–1108

Leung J, Giraudat J (1998) Abscisic acid signal transduction. Annu Rev Plant Physiol Plant Mol Biol 49:199–222

Levandi T, Leon C, Kaljurand M, Garcia-Canas V, Cifuentes A (2008) Capillary electrophoresis time of-flight mass spectrometry for comparative metabolomics of transgenic versus conventional maize. Anal Chem 80:6329–35

Li M, Berendzen KW, Schoffl F (2010) Promoter specificity and interactions between early and late Arabidopsis heat shock factors. Plant Mol Biol 73(4–5):559–567

Lisec J, Schauer N, Kopka J, Willmitzer L, Fernie AR (2006) Gas chromatography mass spectrometry-based metabolite profiling in plants. Nat Protoc 1(1):387–96

Liu D, Ford KL, Roessner U, Natera S, Cassin AM, Patterson JH et al (2013) Rice suspension cultured cells are evaluated as a model system to study salt responsive networks in plants using a combined proteomic and metabolomic profiling approach. Proteomics 13(12–13): 2046–2062

Loiacono FV, De Tullio MC (2012) Why we should stop inferring simple correlations between antioxidants and plant stress resistance: towards the antioxidomic era. OMICS 16(4):160–7

Lugan R, Niogret MF, Kervazo L, Larher FR, Kopka J, Bouchereau A (2009) Metabolome and water status phenotyping of *Arabidopsis* under abiotic stress cues reveals new insight into ESK1 function. Plant Cell Environ 32:95–108

Lukowitz W, Gillmor CS, Scheible WR (2000) Positional cloning in Arabidopsis. Why it feels good to have a genome initiative working for you. Plant Physiol 123(3):795–805

Ma Y, Szostkiewicz I, Korte A, Moes D, Yang Y, Christmann A et al (2009) Regulators of PP2C phosphatase activity function as abscisic acid sensors. Science 324(5930):1064–1068

Mahalingam R, Gomez-Buitrago A, Eckardt N, Shah N, Guevara-Garcia A, Day P et al (2003) Characterizing the stress/defense transcriptome of Arabidopsis. Genome Biol 4(3):20

Marques MC, Alonso-Cantabrana H, Forment J, Arribas R, Alamar S, Conejero V et al (2009) A new set of ESTs and cDNA clones from full-length and normalized libraries for gene discovery and functional characterization in citrus. BMC Genomics 10:428

Maruyama K, Takeda M, Kidokoro S, Yamada K, Sakuma Y, Urano K et al (2009) Metabolic pathways involved in cold acclimation identified by integrated analysis of metabolites and transcripts regulated by DREB1A and DREB2A. Plant Physiol 150(4):1972–80

Matsuda F, Shinbo Y, Oikawa A, Hirai MY, Fiehn O, Kanaya S et al (2009) Assessment of metabolome annotation quality: a method for evaluating the false discovery rate of elemental composition searches. PLoS One 4(10):e7490

Matsuda F, Hirai MY, Sasaki E, Akiyama K, Yonekura-Sakakibara K, Provart NJ et al (2010) AtMet Express development: a phytochemical atlas of Arabidopsis development. Plant Physiol 152:566–578

Moco S, Bino RJ, Vorst O, Verhoeven HA, de Groot J, van Beek TA et al (2006) A liquid chromatography–mass spectrometry-based metabolome database for tomato. Plant Physiol 141: 1205–1218

Molinari HBC, Marur CJ, Filho JCB, Kobayashi AK, Pileggi M, Júnior RPL et al (2004) Osmotic adjustment in transgenic citrus rootstock Carrizo citrange (*Citrus sinensis* Osb. × *Poncirus trifoliata* L. Raf.) overproducing proline. Plant Sci 167:1375–1381

Monton MR, Soga T (2007) Metabolome analysis by capillary electrophoresis–mass spectrometry. J Chromatogr A 1168:237–46

Munns R, Tester M (2008) Mechanisms of salinity tolerance. Annu Rev Plant Biol 59:651–81

Mustafiz A, Sahoo KK, Singla-Pareek SL, Sopory SK (2010) Metabolic engineering of glyoxalase pathway for enhancing stress tolerance in plants. Methods Mol Biol 639:95–118

Mustilli AC, Merlot S, Vavasseur A, Fenzi F, Giraudat J (2002) Arabidopsis OST1 protein kinase mediates the regulation of stomatal aperture by abscisic acid and acts upstream of reactive oxygen species production. Plant Cell 14(12):3089–3099

Nakagami H, Soukupová H, Schikora A, Žárský V, Hirt H (2006) A mitogen-activated protein kinase kinase kinase mediates reactive oxygen species homeostasis in Arabidopsis. J Biol Chem 281(50):38697–38704

Nakashima K, Takasaki H, Mizoi J, Shinozaki K, Yamaguchi-Shinozaki K (2012) NAC transcription factors in plant abiotic stress responses. Biochim Biophys Acta 1819(2):97–103

Nanjo T, Kobayashi M, Yoshiba Y, Sanada Y, Wada K, Tsukaya H et al (1999) Biological functions of proline in morphogenesis and osmotolerance revealed in antisense transgenic Arabidopsis thaliana. Plant J 18(2):185–93

Nieva C, Busk PK, Domínguez-Puigjaner E, Lumbreras V, Testillano PS, Risueño MC et al (2005) Isolation and functional characterisation of two new bZIP maize regulators of the ABA responsive gene rab28. Plant Mol Biol 58(6):899–914

Nikolovski N, Rubtsov D, Segura MP, Miles GP, Stevens TJ, Dunkley TP et al (2012) Putative glycosyltransferases and other plant Golgi apparatus proteins are revealed by LOPIT proteomics. Plant Physiol 160(2):1037–1051

Nishimura N, Sarkeshik A, Nito K, Park SY, Wang A, Carvalho PC et al (2010) PYR/PYL/RCAR family members are major in-vivo ABI1 protein phosphatase 2C-interacting proteins in Arabidopsis. Plant J 61(2):290–9

Novillo F, Alonso JM, Ecker JR, Salinas J (2004) CBF2/DREB1C is a negative regulator of CBF1/DREB1B and CBF3/DREB1A expression and plays a central role in stress tolerance in Arabidopsis. Proc Natl Acad Sci U S A 101(11):3985–90

Okazaki Y, Shimojima M, Sawada Y, Toyooka K, Narisawa T, Mochida K et al (2009) A chloroplastic UDP-glucose pyrophosphorylase from *Arabidopsis* is the committed enzyme for the first step of sulfolipid biosynthesis. Plant Cell 21:892–909

Ooka H, Satoh K, Doi K, Nagata T, Otomo Y, Murakami K et al (2003) Comprehensive analysis of NAC family genes in Oryza sativa and Arabidopsis thaliana. DNA Res 10(6):239, 247

Pacheco CM, Pestana-Calsa MC, Gozzo FC, Mansur Custodio Nogueira RJ, Menossi M, Calsa T Jr (2013) Differentially delayed root proteome responses to salt stress in sugar cane varieties. J Proteome Res 12(12):5681–5695

Pan Z, Zhao Y, Zheng Y, Liu J, Jiang X, Guo Y (2012) A high-throughput method for screening Arabidopsis mutants with disordered abiotic stress induced calcium signal. J Genet Genomics 39(5):225–35

Pang Q, Chen S, Dai S, Chen Y, Wang Y, Yan X (2010) Comparative proteomics of salt tolerance in *Arabidopsis thaliana* and *Thellungiella halophila*. J Proteome Res 9(5):2584–2599

Park SY, Fung P, Nishimura N, Jensen DR, Fujii H, Zhao Y et al (2009) Abscisic acid inhibits type 2C protein phosphatases via the PYR/PYL family of START proteins. Science 324(5930):1068–1071

Parvanova D, Ivanov S, Konstantinova T, Karanov E, Atanassov A, Tsvetkov T et al (2004) Transgenic tobacco plants accumulating osmolytes show reduced oxidative damage under freezing stress. Plant Physiol Biochem 42(1):57–63

Patton WF (2002) Detection technologies in proteome analysis. J Chromatogr B 771(1–2):3–31

Peleman JD, van der Voort JR (2003) Breeding by design. Trends Plant Sci 8(7):330–334

Pérez-Clemente RM, Vives V, Zandalinas SI, López-Climent MF, Muñoz V, Gómez-Cadenas A (2013) Biotechnological approaches to study plant responses to stress. Biomed Res Int 2013:654120

Pitzschke A, Schikora A, Hirt H (2009) MAPK cascade signaling networks in plant defence. Curr Opin Plant Biol 12(4):421–426

Plomion C, Lalanne C, Claverol S, Meddour H, Kohler A, Bogeat-Triboulot MB et al (2006) Mapping the proteome of poplar and application to the discovery of drought-stress responsive proteins. Proteomics 6(24):6509–6527

Pulla RK, Kim YJ, Parvin S, Shim JS, Lee JH, Kim YJ et al (2009) Isolation of S-adenosyl-L-methionine synthetase gene from Panax ginseng C.A. Meyer and analysis of its response to abiotic stresses. Physiol Mol Biol Plants 15(3):267–275

Punjabi-Sabharwal V, Karan R, Khan T, Pareek A (2010) Abiotic stress responses: complexities in gene expression. In: Pareek A, Sopory SK, Bonhert HJ, Govindjee (eds) Abiotic stress adaptation in plants: physiological, molecular and genomic foundation. Springer, Dordrecht, pp 177–198

Qiu JL, Zhou L, Yun BW, Nielsen HB, Fiil BK, Petersen K et al (2008) Arabidopsis mitogen-activated protein kinase kinases MKK1 and MKK2 have overlapping functions in defense signaling mediated by MEKK1, MPK4, and MKS1. Plant Physiol 148(1):212–222

Qiu D, Xiao J, Xie W, Cheng H, Li X, Wang S (2009) Exploring transcriptional signalling mediated by OsWRKY13, a potential regulator of multiple physiological processes in rice. BMC Plant Biol 9:74

Qu C, Liu C, Ze Y, Gong X, Hong M, Wang L et al (2011) Inhibition of nitrogen and photosynthetic carbon assimilation of maize seedlings by exposure to a combination of salt stress and potassium-deficient stress. Biol Trace Elem Res 144(1–3):1159–74

Quintero F, Ohta M, Shi H, Zhu J-K, Pardo J (2002) Reconstitution in yeast of the Arabidopsis SOS signaling pathway for Na^+ homeostasis. Proc Natl Acad Sci U S A 99(13):9061–9066

Rabbani MA, Maruyama K, Abe H, Khan MA, Katsura K, Ito Y et al (2003) Monitoring expression profiles of rice genes under cold, drought, and high-salinity stresses and abscisic acid application using cDNA microarray and RNA gel-blot analyses. Plant Physiol 133(4):1755–67

Ramautar R, Somsen GW, de Jong GJ (2009) CE-MS in metabolomics. Electrophoresis 30: 276–291

Ramautar R, Mayboroda OA, Somsen GW, de Jong GJ (2011) CE-MS for metabolomics: developments and applications in the period 2008–2010. Electrophoresis 32:52–65

Reddy AS, Ali GS, Celesnik H, Day IS (2011) Coping with stresses: roles of calcium- and calcium/calmodulin-regulated gene expression. Plant Cell 23(6):2010–32

Reiser V, Raitt D, Saito H (2003) Yeast osmosensor Slnl and plant cytokinin receptor Crel respond to changes in turgor pressure. J Cell Biol 161(6):1035–1040

Remmerie N, De VT, Valkenborg D, Laukens K, Smets K, Vreeken J et al (2011) Unraveling tobacco BY-2 protein complexes with BN PAGE/LC-MS/MS and clustering methods. J Proteomics 74(8):1201–1217

Riese F, Grinschgl S, Gersbacher MT, Russi N, Hock C, Nitsch RM et al (2013) Visualization and quantification of APP intracellular domain-mediated nuclear signaling by bimolecular fluorescence complementation. PLoS One 8(9):e76094

Rizhsky L, Liang H, Shuman J, Shulaev V, Davletova S, Mittler R (2004) When defense pathways collide. The response of Arabidopsis to a combination of drought and heat stress. Plant Physiol 134:1683–1696

Robida AM, Kerppola TK (2009) Bimolecular fluorescence complementation analysis of inducible protein interactions: effects of factors affecting protein folding on fluorescent protein fragment association. J Mol Biol 394(3):391–409

Roessner U, Luedemann A, Brust D, Fiehn O, Linke T, Willmitzer L et al (2001) Metabolic profiling allows comprehensive phenotyping of genetically or environmentally modified plant systems. Plant Cell 13(1):11–29

Rogachev I, Aharoni A (2012) UPLC-MS-based metabolite analysis in tomato. Methods Mol Biol 860:129–144

Rogers S, Girolami M, Kolch W, Waters KM, Liu T, Thrall B et al (2008) Investigating the correspondence between transcriptomic and proteomic expression profiles using coupled cluster models. Bioinformatics 24(24):2894–2900

Rohila JS, Jain RK, Wu R (2002) Genetic improvement of Basmati rice for salt and drought tolerance by regulated expression of a barley Hva1 cDNA. Plant Sci 163(3):525–532

Roosens NH, Bitar FA, Loenders K, Angenon G, Jacobs M (2002) Overexpression of ornithine-δ-aminotransferase increases proline biosynthesis and confers osmotolerance in transgenic plants. Mol Breed 9(2):73–80

Ruan SL, Ma HS, Wang SH, Fu YP, Xin Y, Liu WZ et al (2011) Proteomic identification of OsCYP2, a rice cyclophilin that confers salt tolerance in rice (Oryza sativa L.) seedlings when overexpressed. BMC Plant Biol 11(34):1–15

Saito K, Dixon RA, Willmitzer L (eds) (2006) Plant metabolomics, vol 57. Springer, Berlin

Santiago J, Rodrigues A, Saez A, Rubio S, Antoni R, Dupeux F, Park SY, Márquez JA, Cutler SR, Rodriguez PL (2009) Modulation of drought resistance by the abscisic acid receptor PYL5 through inhibition of clade A PP2Cs. Plant J 5:575–588

Satoh R, Fujita Y, Nakashima K, Shinozaki K, and Yamaguchi-Shinozaki K (2004) A novel subgroup of bZIP proteins functions as transcriptional activators in hypoosmolarity-responsive expression of the *ProDH* gene in *Arabidopsis*. Plant Cell Physiol 45:309–317

Sawahel WA, Hassan AH (2002) Generation of transgenic wheat plants producing high levels of the osmoprotectant proline. Biotechnol Lett 24(9):721–5

Schauer N, Fernie AR (2006) Plant metabolomics: towards biological function and mechanism. Trends Plant Sci 11(10):508–16

Schena M, Shalon D, Davis RW, Brown PO (1995) Quantitative monitoring of gene expression patterns with a complementary DNA microarray. Science 270(5235):467–70

Schutz W, Hausmann N, Krug K, Hampp R, Macek B (2011) Extending SILAC to proteomics of plant cell lines. Plant Cell 23(5):1701–1705

Seki M, Narusaka M, Abe H, Kasuga M, Yamaguchi-Shinozaki K, Carninci P et al (2001) Monitoring the expression pattern of 1300 Arabidopsis genes under drought and cold stresses by using a full-length cDNA microarray. Plant Cell 13(1):61–72

Sharp RE, Poroyko V, Hejlek LG, Spollen WG, Springer GK, Bohnert HJ et al (2004) Root growth maintenance during water deficits: physiology to functional genomics. J Exp Bot 55:2343–2351

Shi H, Ishitani M, Kim C, Zhu J-K (2000) The Arabidopsis thaliana salt tolerance gene SOS1 encodes a putative Na^+/H^+ antiporter. Proc Natl Acad Sci U S A 97(12):6896–6901

Shinozaki K, Yamaguchi-Shinozaki K (2007) Gene networks involved in drought stress response and tolerance. J Exp Bot 58(2):221–7

Shiraki T, Kondo S, Katayama S, Waki K, Kasukawa T, Kawaji H et al (2003) Cap analysis gene expression for high-throughput analysis of transcriptional starting point and identification of promoter usage. Proc Natl Acad Sci U S A 100(26):15776–15781

Silva P, Gerós H (2009) Regulation by salt of vacuolar H^+-ATPase and H^+ pyrophosphatase activities and Na^+/H^+ exchange. Plant Signal Behav 4:718–726

Soda N, Kushwaha HR, Soni P, Singla-Pareek SL, Pareek A (2013) A suite of new genes defining salinity stress tolerance in seedlings of contrasting rice genotypes. Funct Integr Genomics 13(3):351–365

Song Y, Chen LG, Zhang LP, Yu DQ (2010) Overexpression of OsWRKY72 gene interferes in the ABA signal and auxin transport pathway of Arabidopsis. J Biosci 35(3):459–471

Su J, Wu R (2004) Stress-inducible synthesis of proline in transgenic rice confers faster growth under stress conditions than that with constitutive synthesis. Plant Sci 166(4):941–8

Suarez-Rodriguez MC, Adams-Phillips L, Liu Y, Wang H, Su SH, Jester PJ et al (2007) MEKK1 is required for flg22-induced MPK4 activation in Arabidopsis plants. Plant Physiol 143(2): 661–669

Sugimoto M, Hirayama A, Robert M, Abe S, Soga T, Tomita M (2010) Prediction of metabolite identity from accurate mass, migration time prediction and isotopic pattern information in CE-TOFMS data. Electrophoresis 31:2311–2318

Sui N, Li M, Zhao SJ, Li F, Liang H, Meng QW (2007) Overexpression of glycerol-3-phosphate acyltransferase gene improves chilling tolerance in tomato. Planta 226(5):1097–1108

Sun X, Zhou S, Meng F, Liu S (2012) De novo assembly and characterization of the garlic (Allium sativum) bud transcriptome by Illumina sequencing. Plant Cell Rep 31(10):1823–1828

Sunkar R, Chinnusamy V, Zhu J, Zhu JK (2007) Small RNAs as big players in plant abiotic stress responses and nutrient deprivation. Trends Plant Sci 12(7):301–309

Swindell WR (2006) The association among gene expression responses to nine abiotic stress treatments in Arabidopsis thaliana. Genetics 174(4):1811–1824

Szabados L, Savouré A (2010) Proline: a multifunctional amino acid. Trends Plant Sci 15:89–97

Takasaki H, Maruyama K, Kidokoro S, Ito Y, Fujita Y, Shinozaki K et al (2010) The abiotic stress-responsive NAC-type transcription factor OsNAC5 regulates stress-inducible genes and stress tolerance in rice. Mol Genet Genomics 284(3):173–183

Tamura T, Hra K, Yamaguchi Y, Koizumi N, Sano H (2003) Osmotic stress tolerance of transgenic tobacco expressing a gene encoding a membrane-located receptor-like protein from tobacco plants. Plant Physiol 131(2):454–462

Tang L, Kwon SY, Kim SH, Kim JS, Choi JS, Cho KY et al (2006) Enhanced tolerance of transgenic potato plants expressing both superoxide dismutase and ascorbate peroxidase in chloroplasts against oxidative stress and high temperature. Plant Cell Rep 25(12):1380–1386

Teige M, Scheikl E, Eulgem T, Doczi R, Ichimura K, Shinozaki K et al (2004) The MKK2 pathway mediates cold and salt stress signaling in Arabidopsis. Mol Cell 15(1):141–152

Thiel J, Hollmann J, Rutten T, Weber H, Scholz U, Weschke W (2012) 454 Transcriptome sequencing suggests a role for two-component signalling in cellularization and differentiation of barley endosperm transfer cells. PLoS One 7(7):e41867

Tikunov Y, Lommen A, de Vos CH, Verhoeven HA, Bino RJ, Hall RD et al (2005) A novel approach for nontargeted data analysis for metabolomics. Large-scale profiling of tomato fruit volatiles. Plant Physiol 139(3):1125–37

Trujillo LE, Sotolongo M, Menéndez C, Ochogavía ME, Coll Y, Hernández I et al (2008) SodERF3, a novel sugarcane ethylene responsive factor (ERF), enhances salt and drought tolerance when overexpressed in tobacco plants. Plant Cell Physiol 49(4):512–525

Tun NN, Santa-Catarina C, Begum T, Silveira V, Handro W, Segal Floh EI et al (2006) Polyamines induce rapid biosynthesis of nitric oxide (NO) in *Arabidopsis thaliana* seedlings. Plant Cell Physiol 47:346–354

Umezawa T (2011) Systems biology approaches to abscisic acid signaling. J Plant Res 124(4): 539–548

Umezawa T, Nakashima K, Miyakawa T, Kuromori T, Tanokura M, Shinozaki K et al (2010) Molecular basis of the core regulatory network in ABA responses: sensing, signaling and transport. Plant Cell Physiol 51(11):1821–1839

Unlu M, Morgan ME, Minden JS (1997) Difference gel electrophoresis: a single gel method for detecting changes in protein extracts. Electrophoresis 18(11):2071–2077

Urano K, Maruyama K, Ogata Y, Morishita Y, Takeda M, Sakurai N et al (2009) Characterization of the ABA-regulated global responses to dehydration in Arabidopsis by metabolomics. Plant J 57:1065–1078

Urao T, Yakubov B, Satoh R, Yamaguchi-Shinozaki K, Seki M, Hirayama T et al (1999) A transmembrane hybrid-type histidine kinase in Arabidopsis functions as an osmosensor. Plant Cell 11(9):1743–1754

Velculescu VE, Zhang L, Vogelstein B, Kinzler KW (1995) Serial analysis of gene expression. Science 270(5235):484–487

Vendruscolo EC, Schuster I, Pileggi M, Scapim CA, Molinari HB, Marur CJ et al (2007) Stress-induced synthesis of proline confers tolerance to water deficit in transgenic wheat. J Plant Physiol 164(10):1367–76

Verbruggen N, Hermans C (2008) Proline accumulation in plants: a review. Amino Acids 35(4):753–9

Verslues PE, Juenger TE (2011) Drought, metabolites, and *Arabidopsis* natural variation: a promising combination for understanding adaptation to water-limited environments. Curr Opin Plant Biol 14:240–245

Verslues PE, Sharma S (2010) Proline metabolism and its implications for plant environment interaction. Arabidopsis Book 8:e0140

Verslues PE, Agarwal M, Katiyar-Agarwal S, Zhu J, Zhu JK (2006) Methods and concepts in quantifying resistance to drought, salt and freezing, abiotic stresses that affect plant water status. Plant J 45(4):523–39

Wahid A, Gelani S, Ashraf M, Foolad M (2007) Heat tolerance in plants: an overview. Environ Exp Bot 61:199–223

Wang XQ, Yang PF, Liu Z, Liu WZ, Hu Y, Chen H et al (2009) Exploring the mechanism of Physcomitrella patens desiccation tolerance through a proteomic strategy. Plant Physiol 149(4):1739–50

Ward JL, Baker JM, Beale MH (2007) Recent applications of NMR spectroscopy in plant metabolomics. FEBS J 274(5):1126–31

Weltmeier F, Ehlert A, Mayer CS, Dietrich K, Wang X, Schütze K et al (2006) Combinatorial control of Arabidopsis proline dehydrogenase transcription by specific heterodimerisation of bZIP transcription factors. EMBO J 25(13):3133–3143

Whitelegge JP (2002) Plant proteomics: BLASTing out of a MudPIT. Proc Natl Acad Sci U S A 99(18):11564–11566

Wienkoop S, Morgenthal K, Wolschin F, Scholz M, Selbig J, Weckwerth W (2008) Integration of metabolomic and proteomic phenotypes: analysis of data covariance dissects starch and RFO metabolism from low and high temperature compensation response in Arabidopsis thaliana. Mol Cell Proteomics 7:1725–1736

Wohlbach DJ, Quirino BF, Sussman MR (2008) Analysis of the *Arabidopsis* histidine kinase ATHK1 reveals a connection between vegetative osmotic stress sensing and seed maturation. Plant Cell 20:1101–1117

Wolters DA, Washburn MP, Yates JR III (2001) An automated multidimensional protein identification technology for shotgun proteomics. Anal Chem 73(23):5683–5690

Wu L, Fan Z, Guo L, Li Y, Chen ZL, Qu LJ (2005) Over-expression of the bacterial nhaA gene in rice enhances salt and drought tolerance. Plant Sci 168(2):297–302

Wu X, Shiroto Y, Kishitani S, Ito Y, Toriyama K (2009) Enhanced heat and drought tolerance in transgenic rice seedlings overexpressing OsWRKY11 under the control of HSP101 promoter. Plant Cell Rep 28(1):21–30

Xianan L, Baird WV (2003) Differential expression of genes regulated in response to drought or salinity stress in sunflower. Crop Sci 43:678–687

Xing Y, Jia W, Zhang J (2008) AtMKK1 mediates ABA-induced CAT1 expression and H_2O_2 production via AtMPK6-coupled signaling in Arabidopsis. Plant J 54(3):440–451

Yamada M, Morishita H, Urano K, Shiozaki N, Yamaguchi-Shinozaki K, Shinozaki K et al (2005) Effects of free proline accumulation in petunias under drought stress. J Exp Bot 56(417):1975–81

Yamaguchi-Shinozaki K, Koizumi M, Urao S, Shinozaki K (1992) Molecular cloning and characterization of 9 cDNAs for genes that are responsive to desiccation in Arabidopsis thaliana: sequence analysis of one cDNA clone that encodes a putative transmembrane channel protein. Plant Cell Physiol 33(3):217–224

Yan QC, Kuo MS, Li S, Bui HH, Peake DA, Sanders PE et al (2008) AGPAT6 is a novel microsomal glycerol-3-phosphate acyltransferase. J Biol Chem 283(15):10048–10057

Yang XY, Jiang WJ, Yu HJ (2012) The expression profiling of the lipoxygenase (LOX) family genes during fruit development, abiotic stress and hormonal treatments in cucumber (*Cucumis sativus* L.). Int J Mol Sci 13(2):2481–2500

Yanhui C, Xiaoyuan Y, Kun H, Meihua L, Jigang L, Zhaofeng G et al (2006) The MYB transcription factor superfamily of Arabidopsis: expression analysis and phylogenetic comparison with the rice MYB family. Plant Mol Biol 60(1):107–124

Yoshiba Y, Kiyosue T, Nakashima K, Yamaguchi-Shinozaki K, Shinozaki K (1997) Regulation of levels of proline as an osmolyte in plants under water stress. Plant Cell Physiol 38(10): 1095–102

Yoshida R, Hobo T, Ichimura K, Mizoguchi T, Takahashi F, Aronso J et al (2002) ABA-activated SnRK2 protein kinase is required for dehydration stress signaling in Arabidopsis. Plant Cell Physiol 43(12):1473–1483

Yoshida R, Umezawa T, Mizoguchi T, Takahashi S, Takahashi F, Shinozaki K (2006) The regulatory domain of SRK2E/OST1/SnRK2.6 interacts with ABI1 and integrates abscisic acid (ABA) and osmotic stress signals controlling stomatal closure in Arabidopsis. J Biol Chem 281(8): 5310–5318

Zhang YY, Li Y, Gao T et al (2008) Arabidopsis SDIR1 enhances drought tolerance in crop plants. Biosci Biotechnol Biochem 72(8):2251–2254

Zhang J, Zhang Y, Du Y, Chen S, Tang H (2011) Dynamic metabonomic responses of tobacco (Nicotiana tabacum) plants to salt stress. J Proteome Res 10:1904–14

Zhu JK (2000) Genetic analysis of plant salt tolerance using Arabidopsis. Plant Physiol 124(3):941–948

Zhu JK (2002) Salt and drought stress signal transduction in plants. Annu Rev Plant Biol 53:247–273

Zhu JK (2003) Regulation of ion homeostasis under salt stress. Curr Opin Plant Biol 6(5):441–445

Zhu B, Su J, Chang M, Verma DPS, Fan Y-L, Wu R (1998) Overexpression of a Δ1-pyrroline-5-carboxylate synthetase gene and analysis of tolerance to water- and salt-stress in transgenic rice. Plant Sci 139(1):41–8

Zimmerli L, Jakab G, Metraux JP, Mauch-Mani B (2009) Potentiation of pathogen specific defense mechanisms in Arabidopsis by beta-aminobutyric acid. Proc Natl Acad Sci U S A 97(23): 12920–12925

Chapter 2
Molecular Approaches in Deciphering Abiotic Stress Signaling Mechanisms in Plants

Swati Singh, Nisha Khatri, Arpana Katiyar, and Yashwanti Mudgil

Abstract There has been considerable interest in the area of abiotic stress research, especially in the direction of producing improved crop varieties that can encounter adversities of abiotic stresses such as heat, cold, drought, osmotic, and salinity effectively. These stresses can act alone or in combination cause greater damage to plants. Thus, in order to combat and survive in these extreme environmental conditions, plants have evolved tolerance mechanisms. These mechanisms include interconnected networks of signaling cascade, which involves role of a large number of genes and their products in tolerance mechanism. The clear-cut understanding of these stress tolerance/resistance mechanisms is critical in order to improve the performance of the plant. Deciphering such complex signaling cascades using traditional genomic approach has been difficult, and therefore, high-throughput functional genomics approaches need to be employed using tools like expression profiling, transcriptomics, proteomics, and metabolomics during tolerance response.

Keywords Abiotic stress • Legume • Transcription factor • WRKY family

2.1 Introduction

Dynamic climatic conditions altering the environment play a crucial role in plant performance. These altered environmental conditions act as an environmental stress to plants. Effect of abiotic stresses (temperature, drought, salt, etc.) on plants is drastic, both for vegetative and reproductive potential. These stress conditions lead to change in the resource availability of plants, which are of utmost importance for their optimum performance. One way to mitigate these conditions is environment-induced shift in the phenotype of the plants. The environmental cues that lead to change phenotype of plants are genetically controlled in a hierarchical fashion. A clear-cut understanding of the stress tolerance/resistance mechanisms is critical

---

S. Singh • N. Khatri • A. Katiyar • Y. Mudgil (✉)
Department of Botany, University of Delhi, New Delhi, Delhi 110 007, India
e-mail: ymudgil@gmail.com

G.K. Pandey (ed.), *Elucidation of Abiotic Stress Signaling in Plants*,
DOI 10.1007/978-1-4939-2211-6_2

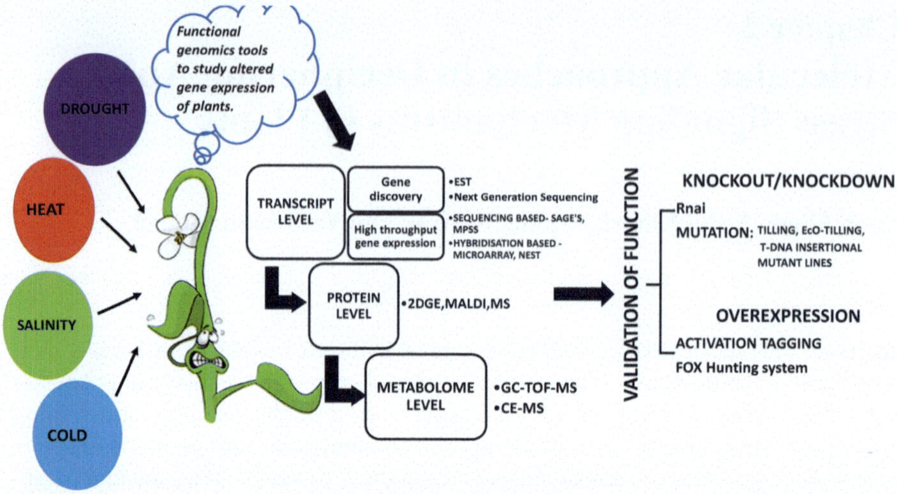

Fig. 2.1 Functional genomics strategies to study altered gene expression of plants under abiotic stress at different molecular levels. At transcript level—*EST* expressed sequence tag, *NGS* next-generation sequencing, *SAGE* serial analysis of gene expression, *MPSS* massively parallel signature sequencing, *microarray*, *NEST* nuclear expressed sequence tag. At protein level—*2DGE* two-dimensional gel electrophoresis, *MALDI* matrix-assisted laser desorption–ionization, *MS* mass spectrometry. At metabolome level—*GC-TOF MS* gas chromatography–time-of-flight mass spectrometry, *CE–MS* capillary electrophoresis–mass spectrometry. Genes identified through the above tools are further validated for their function through knockout/knockdown analysis (*RNAi* RNA interference; mutations—*TILLING* targeting induced local lesions in genomes; T-DNA insertional lines) and overexpression analysis (activation-tagging and FOX hunting system (full-length cDNA Over-eXpressing gene hunting system))

in order to improve plant performance. Thus, a holistic approach through ecological, molecular, and physiological perspectives to understand the behavior of plants is of great importance.

In the twentieth century, an integrated analysis of stress responses on whole plant using functional genomics approaches is being popularized over traditional approaches in understanding complicated plant behavior under stress conditions. The core behind the analyses of functional genomics is how the expression of genome is altered under the unpredictable environmental extremes, using multiple high-throughput screening of stress-treated seedlings at the genome level (next-generation sequencing), at transcriptome level (expressed sequence tag (EST), serial analysis of gene expression (SAGE), and microarray analysis), and at protein level (protein localization and protein–protein interaction). The functional genomics approach brings together physiological and phenotypic characteristics of plants by characterization of the diversity in gene functions under stressed conditions, through random/targeted mutagenesis, complementation analysis, and promoter-trapping strategies (reviewed by Sreenivasulu et al. 2007) (Fig. 2.1).

Although functional genomics approaches provide a gateway to extensive molecular analysis of genes and proteins involved in stress signaling, one of the major challenges faced by functional genomics biologists is to search the genes, which provide tolerance to plant without compromising with its growth and yield. In this chapter, we are trying to overview the abiotic stress signaling cascades for different stresses with functional genomics perspective.

2.2 Signaling Cascades in Abiotic Stress and Their Effect on Plants

2.2.1 Drought Stress

Drought stress involves multiple cellular and molecular pathways, which lead to change in many biochemical and physiological processes, which adversely affect plant growth, development, and ultimately productivity. Biosynthesis of phytohormone, abscisic acid (ABA), is the most pronounced and primary event, which occurs during drought stress. ABA plays a central role in drought stress by closure of stomata, which in turn reduces water loss and also induces stress-related gene expression (Kim et al. 2010; Pospíšilová 2003).

2.2.1.1 Effect on Plants

During drought stress, the normal bilayer structure of the membrane disrupts and becomes exceptionally porous. In water-deficit condition, due to disruption of membrane, attached proteins/enzymes may displace and lose their activity. Disruption of cellular compartmentalization and high concentration of cellular electrolytes also cause disruption of cellular metabolism. In addition to all this damage, enzymes located in cytosol may also exhibit reduced activity or can even undergo complete denaturation due to dehydration.

2.2.1.2 Signaling Cascade During Drought Stress

In *Arabidopsis*, upon exposure to drought stress, plasma membrane-located His kinase ATHK1 acts as a drought sensor (Wohlbach et al. 2008). GTG1 and GTG2 (G protein-coupled receptor [GPCR]-type G protein) are also hypothesized as ABA receptors (Pandey et al. 2009). ATHK1 senses modification in water status and upregulates transcription of ABA synthesis genes, resulting in an increase in intracellular ABA levels. Precursor of ABA is xanthoxin, which is derived from a direct cleavage of C_{40} carotenoids. This reaction seems to be a key step in ABA biosynthesis in plants. The enzyme, which cleaves C_{40} carotenoids, is 9-*cis*-epoxycarotenoid dioxygenase (NCED), which was originally identified from maize (Schwartz et al. 1997). NCED family has five members in *Arabidopsis*, and among all, AtNCED3 plays a crucial role in drought stress-inducible ABA biosynthesis. Drought stress induces expression of the gene encoding NCED (Iuchi et al. 2000; Xiong and Zhu 2003). An ATHK1-responsive ABA-independent pathway of stress-responsive gene expression also exists (Wohlbach et al. 2008). Increase in ABA level is perceived by soluble PYR/PYL/RCAR (pyrabactin resistance1/PYR1-like/regulatory component of ABA receptor) receptors, members of START superfamily of ligand-binding proteins (Ma et al. 2009; Park et al. 2009). After binding to its receptor, ABA-receptor complex then binds to type 2C protein phosphatase (PP2C) (Brandt et al. 2012; Fujii et al. 2009; Ma et al. 2009; Nishimura et al. 2009, 2010; Park et al. 2009; Santiago et al. 2009). PP2C is the inhibitor of Snf1-related protein kinase1. However, due to binding of PP2C with ABA-receptor complex, Snf1-related protein kinase1 now gets released from PP2C inhibition (Ma et al. 2009; Melcher et al. 2009). ABI1 and PP2CA are authentic SnRK1 phosphatases, which control SnRK1 activity and signaling during stress (Rodrigues et al. 2013). This activated Snf1-related protein kinase1 now phosphorylates downstream components like ion channels and transcription factors such as basic leucine zipper (bZIP), referred to as ABRE-binding proteins (AREB) or ABRE-binding factors (ABFs) that further regulate expression of concerned genes (Brandt et al. 2012; Fujita et al. 2013; Lopez-Molina et al. 2001). SnRK2 protein kinase (OST1/SRK2E) is activated by ABA and controls stomatal closure (Yoshida et al. 2002; Mustilli et al. 2002). SnRK2s are activated by drought, salinity, and ABA (Yoshida et al. 2002). The PYR/PYL/RCAR–PP2C–SnRK2 signaling module is conserved across land plants, and its key elements mediate ABA responses within the nonvascular plant *Physcomitrella patens* (Chater et al. 2011; Komatsu et al. 2009; Tougane et al. 2010).

Regulation of gene expression in drought stress is broadly classified into two categories, namely, ABA-independent and ABA-dependent regulatory systems (Yamaguchi-Shinozaki and Shinozaki 2005). Genes are dissected at transcriptional level for better understanding of the cascade; *cis*- and *trans*-acting factors have been analyzed critically. The promoter of drought-inducible genes contains two major *cis*-acting elements, ABRE (ABA-responsive element) and DRE (dehydration-responsive element)/CRT (C-repeat), both of which are involved in stress-inducible gene expression (Yamaguchi-Shinozaki and Shinozaki 1994, 2005). ABRE and DRE/CRT are *cis*-acting components that operate in ABA-dependent and ABA-independent gene expression, respectively, in response to abiotic stresses (Shinozaki and Yamaguchi-Shinozaki 2007). ABRE is the major *cis*-acting element involved in ABA-dependent pathways; other *cis* elements beside ABRE are MYC-like basic helix–loop–helix (bHLH) and MYB. MYC and MYB recognition sites are present in *RD22* promoter region and involved in drought- and ABA-induced gene expression of *RD22*. Transgenic plants that overproduce AtMYC2 and AtMYB2 have higher sensitivity towards ABA, thereby showing its role in drought stress (Abe et al. 2003). In ABA-independent gene expression, dehydration-responsive component (DRE) acts as a very important *cis*-acting component. The *RD29* gene induction with and without ABA suggests its involvement in both ABA-dependent and ABA-independent pathways. *RD29* promoter has 9-bp conserved sequence, TACCGACAT, dehydration-responsive element (DRE), and is responsible for regulating *RD29A* induction in the ABA-independent response to dehydration and cold (Yamaguchi-Shinozaki and Shinozaki 1994).

2.2.2 *Temperature Stress*

Temperature acts as a key physical parameter affecting growth and development of plants. Any fluctuation in the optimal level of temperature regime results in altered growth and productivity of plants. Mitigation of these changes in temperature extremes is brought through the evolution of complex mechanisms, which are illustrated by transcriptional changes observed in seedlings, leaves, roots, and pollen (reviewed by Zinn et al. 2010). Temperature stress results in multiple alterations of cellular mechanisms (Wang et al. 2003), and effectiveness of the stress relies upon multiple factors together with intensity, period, and rate of natural action (Wahid et al. 2007; Thakur et al. 2010). Broadly, temperature stress is classified as high-temperature (heat) and low-temperature (cold) stresses.

2.2.2.1 High-Temperature (Heat) Stress

Plants respond differently to varying temperature; regimes above an optimal level of induction result in development of heat stress on plants. Heat stress has been primarily divided into two major categories—the first one involves subjecting a plant growing in its natural temperature (~21 °C for *Arabidopsis thaliana*) to an abrupt rise in temperature (~42 to 45 °C for *Arabidopsis thaliana*) for a period of 0.5–1 h (Larkindale and Vierling 2008; Suzuki et al. 2008). The second type of heat stress is *priming*, during which plants are subjected to moderate heat stress (36–38 °C), for a few hours, and then again subjected to an abrupt episode of severe heat stress (46 °C) (Larkindale and Vierling 2008; Qin et al. 2008).

Effect on Plants

High temperature has an overall stressful effect on plant growth and development (Lobell et al. 2011); however, reproductive tissues are highly sensitive to heat waves. Slight increase in the optimal temperature regime bears drastic effects upon flowering time and total yield of plant (Mittler and Blumwald 2010; Zinn et al. 2010). At subcellular level, heat stress has been reported in fluidizing the membrane (by hampering their lipid content), instability of proteins and cytoskeletal structures, and efficiency of enzymes, thereby causing an overall metabolic imbalance within the cell (McClung and Davis 2010; Ruelland and Zachowski 2010; Suzuki et al. 2012).

Signaling Cascade During Heat Stress

Surviving such drastic effects of heat stress is achieved through initialization of a complex signaling cascade after sensing heat stress. Sensing occurs at plasma membrane of *Physcomitrella patens* (Saidi et al. 2009; Mittler et al. 2012). The signaling cascade, thereafter, results in opening of Ca^{2+} channel, which leads to rise in calcium concentrations within the cell. The increased Ca^{2+} influx thereafter can regulate many signaling pathways in plants (Zhang et al. 2009; Liu et al. 2008). In *Arabidopsis*, this increased calcium concentration could activate AtCaM3 (calmodulin) which may further downstream activate different transcription factors such as WRKY (Li et al. 2010) and HSF (heat-shock transcription factor) directly/indirectly (LIU et al. 2011). The indirect activation of HSF is a consequence of cascading activation of CBK (Ca^{2+}/calmodulin-binding protein kinase), which phosphorylates HSFA1a rendering it active (Liu et al. 2008). The activated HSFs further trigger increased expression of several stress-related transcripts, such as HSP (heat-shock proteins), of which few act as chaperons to stabilize protein (Wang et al. 2004), reviewed by Zinn et al. (2010). The advent of heat stress also has a potent role in increasing plasma membrane fluidity, which might further lead to the activation of phospholipase D (PLD) and phosphatidylinositol 4-phosphate 5-kinase (PIPK) and accumulation of PIP2 (phosphatidylinositol 4,5-bisphosphate) and IP_3 (inositol 1,4,5-trisphosphate)—a key member of lipid signaling (Mishkind et al. 2009). Increased lipid concentration within cell results in opening of channels and thereby influx of a calcium ion within cell, which eventually results in a heat-responsive signaling across cell. Although much have been discovered in downstream high-temperature signaling components, evidences for temperature-induced plasma membrane channel modulation and lipid signaling activation still remain unanswered in plants (Mittler et al. 2012).

2.2.2.2 Low-Temperature (Cold) Stress

Low-temperature stress is one of the most common stresses faced by plants growing in temperate regions of the world. Low-temperature stress has been broadly classified as chilling (0–15 °C) and freezing (<0 °C).

Effect on Plants

Of late, effects of both chilling and freezing stresses have been reported to be deleterious to overall plant growth and development. Symptoms shown in vegetative plant parts in response to chilling treatment include wilting, chlorosis, and reduced leaf expansion eventually leading to necrosis. Not only vegetative but reproductive potential of plants has also been reported to be severely affected leading to sterility in flowers in response to chilling (Wen et al. 2002). Freezing stress on the other hand induces membrane damage due to dehydration, thereby resulting in expansion-induced cell lyses and fracture lesions (Steponkus et al. 1993; Uemura and Steponkus 1997). Other major effects associated with cold stress include ice crystal formation in the apoplastic region of cell, which on further enlargement due to movement of cytosolic content acts as a mechanical strain on plasma membrane and cell wall leading to cell rupturing (McKersie and Bowley 1997).

Signaling Cascade During Cold Stress

The low-temperature signaling across plant cell is a complex phenomenon involving a myriad of transcriptional regulators, thereby affecting the downstream stress-induced gene expression. One of the critical components of this signaling includes a transcriptional activator family CBF (C-repeat-binding factors), comprising of CBF1–4 members, of which CBF1 and CBF3 are reported to be low-temperature-responsive factors (Novillo et al. 2004; Mahajan and Tuteja 2005). In response to cold, unknown sensors on the cell surface get activated, which further downstream activate a transcription factor ICE1 (inducer of cold expression). ICE1 belongs to a family of transcription factors ICE, which are constitutively expressed in all plant tissues, whose overexpression results in increased freezing tolerance in plants. The activation of ICE1 by phosphorylation of its MYC bHLH domain (Chinnusamy et al. 2003) results in binding of ICE1 to MYC recognition sequences present on the CBF3 promoter, thereby activating CBF3 transcription downstream (Chinnusamy et al. 2003). The nascent transcript of CBF3 thus formed binds to promoter region (bearing C-repeat conserved sequences CCGAC (CRT/DRE element)) of COR (cold-regulated) genes, thereby activating target gene and downstream components leading to induction of a low-temperature acclimations in *Arabidopsis*. The complexity of cold acclimatization is further augmented by upstream negative regulation of CBF through HOS1 (high expression of osmotically responsive gene) and a feedback inhibition through CBFs in CBF-dependent signaling cascade. The feedback inhibition of CBFs through knockout analysis of CBFs 1, 2, and 3 have shown a negative regulation of CBF1 and CBF3 by CBF2 and vice versa (Novillo et al. 2004).

2.2.2.3 Salinity Stress

Maintenance of an optimal saline environment within plant cell is an important aspect of growth and development of plant. Fluctuations in extracellular saline conditions result in hyper ionic and hyperosmotic conditions within cells, thereby altering the homeostasis of the cell. The major ions responsible for such fluctuations include Na^+, K^+, Mg^{2+}, and Cl^-, while maintenance of equilibrium during this ionic imbalance is maintained by Na^+, K^+, H^+, and Ca^{2+} signaling across cells (Mahajan and Tuteja 2005).

Effect on Plants

Increased Na^+ concentration within plant cell bears a toxic and inhibitory effect upon cell metabolism and functioning of several enzymes (Niu et al. 1995). The osmotic imbalance caused due to this increased Na^+ concentration also results in membrane disorganization and inhibition of cell division/expansion and growth. Overall, these imbalances result in the reduction of photosynthetic apparatus of plants with an increased formation of reactive oxygen species (ROS) (Mahajan and Tuteja 2005).

Signaling Cascade During Salinity Stress

In order to cope with the increased salt concentration, plants activate their SOS (salt overly sensitive) signaling pathway. The activation of the pathway is preceded by perceiving the altered ionic concentrations by unknown sensors present on the cell, which further downstream result in an increase in cytosolic Ca^{2+} influx (Knight et al. 1997) that acts as an elicitor for activation of SOS3—a myristoylated calcium-binding protein (Ishitani et al. 2000; Liu and Zhu 1998). The activated SOS3 further interacts with an active serine/threonine protein kinase SOS2 (Halfter et al. 2000; Liu et al. 2000) and regulates the expression of SOS1, a salt-tolerant effector gene encoding Na^+/H^+ antiporter (present on plasma membrane) (Halfter et al. 2000), which thereby maintains the sodium ion homeostasis within cell. Although, SOS signaling is an important aspect of maintaining the ionic balance within the cell, its negative regulation is also reported by AtHKT1, inhibiting the activity of SOS2–SOS3 complex (Rus et al. 2001; Uozumi et al. 2000). In addition to SOS1-mediated maintenance of osmotic balance within cell, SOS2-mediated activation of Na^+/H^+ exchange for vacuolar sequestration of increased Na^+ ions (Qiu et al. 2002) also results in maintaining the cellular homeostasis. Acclimation to salt stress at physiological level is achieved by accumulation of compatible solutes within cells, thereby preventing water loss and maintenance of turgor within cell (Bressan et al. 1998; Ford 1984). Accumulation of these osmolytes (sugars—fructose and sucrose, sugar alcohols, trehalose, fructans, glycine betaine, proline, and ectoin) does not configure any change on the metabolic functioning of the cell and helps in overall maintenance of osmotic balance within it (Delauney and Verma 1993; Louis and Galinski 1997; McCue and Hanson 1990).

2.2.2.4 ROS Stress

In order to cope with the stressful conditions, plants have evolved various regulatory pathways. Generation of ROS is one of them. When plants are subjected under environmental stress, several distinct types of ROS are generated within various intracellular compartments, either they can lead an oxidative damage or they might act as a signal.

Plants require light and oxygen for their growth and development. Under unfavorable conditions, there is an increase in the number of oxidizing compound compared with antioxidizing molecules (Shapiguzov et al. 2012). ROS are the reactive forms of molecular oxygen, which include hydroxyl radical (HO^-), superoxide (O_2^-), hydrogen peroxide (H_2O_2), and singlet oxygen ($^1O_2$) (Shapiguzov et al. 2012). ROS are accumulated into cells through various mechanisms like electron transport in the chloroplast and mitochondria or with the enzymatic mechanisms like cell wall peroxidase and NADPH oxidase (in plants known as respiratory burst oxidase homologues (RBOHs)) in extracellular space and peroxisome (Wrzaczek et al. 2013). Major contributors for ROS are the mitochondria in mammals, while in plants largest ROS producers are chloroplast and peroxisomes (Foyer et al. 2003;

Rhoads et al. 2006). In plants peroxisomes contribute 50 % of the total extracellular ROS production, and the rest part is contributed by RBOHs and from intracellular sources (O'Brien et al. 2012). In *Arabidopsis* a network of at least 152 genes are involved in managing the ROS level (Mittler et al. 2004). Silencing and knockdown studies have shown that peroxidase 33 (PRX33) and peroxidase 34 (PRX34) are the major contributors for ROS production in *Arabidopsis* (Bindschedler et al. 2006) and RBOHs constitute a multigene family comprised of ten genes (AtRBOHA–AtRBOHJ) (Baxter et al. 2013). During signaling mechanism of ROS, except H_2O_2, other forms of ROS are involved in localized signaling rather than the long distance (Jammes et al. 2009). H_2O_2 operates as a long-distance messenger due to its stability and ability to cross the membranes either through aquaporins or via diffusion mechanism (Bienert et al. 2007). Plant RBOHs are involved in different signaling pathways including root hair growth, stomatal closure, pollen stigma interaction, and plant defense and acclimation to different abiotic stresses (Torres and Dangl 2005; Jammes et al. 2009; Monshausen et al. 2007; Miller et al. 2009; McInnis et al. 2006; Suzuki et al. 2011).

Early signaling events during stress conditions involved increased flux of Ca^{2+} into the cytosol, protein phosphorylation, and activation of mitogen-activated protein kinases (MAPKs) (Benschop et al. 2007). Membrane permeable nature of H_2O_2 makes a key signaling molecule during different stress responses (Sagi et al. 2004). It is found that H_2O_2 accumulation in extracellular space causes initiation and self-propagation of rapid cell-to-cell signal in a RBOHD-dependent manner (Mittler et al. 2011). During signaling mechanism of ROS, it is found that any mechanical stimulus of plant tissue initiates an increase in the cytosolic Ca^{2+} from extracellular space; this increased Ca^{2+} then leads to ROS production through RBOHC-dependent manner, which at last regulates the root hair development (Monshausen et al. 2007; Takeda et al. 2008). Although ROS has been involved in various developmental processes, the specificity of signaling is not clear (Møller and Sweetlove 2010). There are various possibilities regarding the mode of action of ROS like the following: ROS may function as a general signal to activate cellular signaling network of cell to convey their specificity, and a good example of this signaling network is the activation of different MAPK cascades following ROS accumulation (Jammes et al. 2009; Teige et al. 2004), or it may contain any decoded messages, which have a specific pattern during its signaling mechanism (Mittler et al. 2011). The other possibility could be that each cellular compartment or individual cell has the specific ROS receptor, which conveys the message of specific ROS to other networks like Ca^{2+} or protein phosphorylation (Mittler et al. 2011).

Recently, a lot of information is accumulating about ROS signaling, but still there exist large gaps in ROS-mediated signaling mechanism such as whether ROS induce new signaling systems as second messengers or they are only involved in the regulation of already known signaling pathways. There is again no clear answer about the primary sensor of ROS and the primary gene response during ROS signaling. ROS sensors may be either transcription factors or protein kinases, which show activity change in response to various ROS (Fig. 2.2).

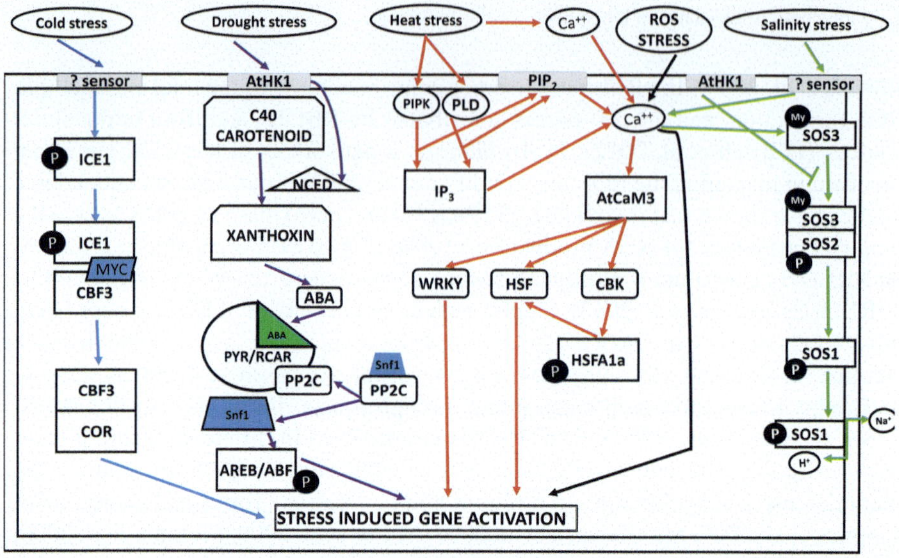

Fig. 2.2 Schematic illustration for abiotic stress signaling in plants. Advent of cold stress is perceived by an unknown sensor at plasma membrane, which further activates a signaling cascade downstream (*blue solid lines*). The activated unknown sensor phosphorylates ICE1 (inducer of cold expression) rendering it activated. This activated ICE1 binds to MYC domain of CBF3 (C-repeat-binding factor) promoter, thereby activating CBF3 transcription. CBF3 thus formed binds to conserved sequence in promoter region of COR (cold regulated) eventually activating target genes. Drought stress signaling (*purple solid lines*) is perceived through a plasma membrane-localized sensor ATHK1, which regulates expression of NCED (9-*cis*-epoxycarotenoid dioxygenase), a key enzyme in ABA (abscisic acid) biosynthetic pathway, regulating C40 carotenoid conversion to xanthoxin. ABA thus formed due to this regulation binds to its receptor PYR/PYL/RCAR (pyrabactin resistance1/PYR1-like/regulatory component of ABA receptor). This complex then binds to PP2C (type 2C protein phosphatase) thereby releasing and activating Snf1-related protein kinase1 (as PP2C acts as an inhibitor of Snf1). The activated Snf1 further activates ABRE-binding factors downstream, which eventually regulate expression of drought stress-inducible target genes in an ABA-dependent/ABA-independent manner. Heat stress signaling (*red solid lines*) is activated through modulation in cellular plasma membrane. This modulation acts as a signal for activating PIPK (phosphatidylinositol 4-phosphate 5-kinase) and PLD (phospholipase D). Both the molecules result in accumulation of second messenger IP_3 (inositol 1,4,5-trisphosphate) and PIP_2 (phosphatidylinositol 4,5-bisphosphate). This increased IP_3 concentration within cell results in an increase in cytosolic Ca^{2+} ion, thereby activating a heat-responsive signaling across cell. Heat stress also results in activation of Ca^{2+} ion channel thereby resulting in an increase in Ca^{2+} ion concentration within cell. The increased ion concentration further activates AtCaM3 (calmodulin) downstream, which in turn triggers activation of transcription factors such as WRKY and HSF (heat-shock protein). HSF is activated in response to heat stress both directly and indirectly, via CBK (Ca^{2+}/calmodulin-binding protein kinase), which phosphorylates Hsfa1, eventually cascading activation of HSF. The activated HSF and WRKY transcription factors finally help in activation of drought stress-induced genes. During salt stress signaling (*green solid lines*), an unknown salt sensor at plasma membrane leads to increase in cytosolic calcium level, which in turn binds to myristoylated SOS3 (salt overly sensitive) protein. This membrane-bound protein further binds to phosphorylated SOS2. This complex formation is however inhibited by AtHK1. The SOS3–SOS2 complex acts as a regulator of SOS1—a salt-tolerant effector gene encoding Na^+/H^+ antiporter, positioned at plasma membrane. The regulation of SOS1 through activation of SOS pathway results in maintaining cellular homeostasis. On the advent of abiotic stress, ROS (reactive oxygen species) accumulation and signaling (*black solid lines*) activates, both in cytosol and in apoplastic region. This increased ROS accumulation leads to an increased calcium ion concentration within the cell, which directly/indirectly helps in the activation of target genes responsive to abiotic stress

2.3 Functional Genomics of Abiotic Stress Signaling

Genetic variations lead to functional variation by changing the structure or function of the proteins. Functional genomics helps to explain functional differences in appearance or performance of altered gene. It enables to explore gene and protein functions and interactions on a global scale. The approach utilizes large-scale assays to study various genes or proteins in metabolic pathway under different environmental conditions. DNA's function is thus discovered by combining information from genes, transcripts, proteins, and metabolome using genomics, transcriptomics, proteomics, and metabolomics, respectively. Genomics involves study of structure, function, evolution, and mapping of genomes; on the other hand, transcriptomics includes structural and functional analyses of coding and noncoding ribonucleic acid. Proteomics deals with protein and posttranslational protein modification in conjunction with their regulatory pathway (Gupta et al. 2013). Posttranslational modification studies have revealed that the molecular mechanisms driving the responses of plants to environmental stresses include histone acetylation, methylation, ubiquitination, and phosphorylation. Combined effects of these modifications play an essential role in the regulation of stress-responsive gene expression (Yuan et al. 2013). Lastly, metabolomics also offers a robust tool to analyze varied metabolites in an integrated method that can act as a powerful tool in distinguishing the complicated network concerned in stress tolerance.

Plant usually modifies their "omics" profiles in order to cope with the environmental stresses. The main objective of the "omics" approaches is to find out the molecular interaction and their relationship with the signaling cascade and to process the information, which in turn connects specific signals with specific molecular responses (Gupta et al. 2013). As rightly said by Edwin Grant Conklin, "Life is not found in atoms or molecules or genes as such, but in organization; not in symbiosis but in synthesis."

2.3.1 Functional Genomics Approaches to Decode Abiotic Stress Signaling

In order to elucidate the role of intermediate components in the cascade of abiotic stress response and to trace out the entire stress response pathway, traditional approach of gene identification has been practiced in the past. Utilization of traditional breeding strategies, involving the use of genetic variation arising from varietal germplasm, interspecific or intergeneric hybridization, induced mutations, and somaclonal variation of cell and tissue cultures, has met with only limited success. These traditional practices have shown limited success in the generation of plant with improved stress resistance under field conditions due to the complex nature of stress tolerance traits, low genetic variance of yield under stress, and lack of economical selection techniques (Flowers and Yeo 1995; Frova et al. 1999; Ribaut et al. 1996, 1997).

In order to improve the existing technique for identification and introgression of genes and metabolite acting in the stress cascade, an updated approach of functional genomics has been implemented by the scientific community to determine the function of individual genes, pathways, and entire genomes. The basis of functional genomics underlies in the evaluation and study of the entire cell or organism as a system and in understanding how different biological processes occur within a system as a whole. The approach has shown quite an improvement and advancement over the last decade with involvement of high-throughput sequencing, proteomics, and metabolomics approaches. The basic outline under which functional genomics functions includes:

(a) Gene discovery
(b) High-throughput gene expression
(c) Validation of function and stress response mechanism

The role of these individual disciplines in crop improvement in response to stress tolerance has been discussed in detail in the following sections:

2.3.1.1 Gene Discovery

Mining of genes associated with abiotic stress mechanism in plants has been one of the primary steps in functional genomics, which leads to elucidation of an intercalating network of abiotic stress tolerance. Availability of genome-wide information in plants acts as a fundamental resource in fishing out candidate stress-related genes. Utilization of large-scale EST plant databases (such as National Center for Biotechnology Information (NCBI) UniGene, http://www.ncbi.nlm.nih.gov/; The Institute for Genomic Research (TIGR) Gene Indices, www.tigr.org; Sputnik, http://mips.gsf.de/proj/sputnik) (Vij and Tyagi 2007) acts as one of the most frequently used tools to explore gene catalogue of a species. ESTs are partial genetic sequences that are generated by single-pass sequencing of cDNA libraries (Bouchez and Höfte 1998), generated from various tissues, developmental stages, or treatments (Yamamoto and Sasaki 1997). Deciphering genes that are related to abiotic stress has been done through a comprehensive analysis of ESTs generated from different cDNA libraries of control and abiotic stress-treated tissues collected at various stages of development. Outcome of such analysis has led to the identification of gene number, gene content, and possible gene families involved in abiotic stress responses (Sreenivasulu et al. 2007). Further functional annotation of mined genes by comparative analysis of BLASTX and SwissProt database provides a possible outline of tolerance mechanism in plants, though ~20 to 30 % of these fished-out genes still remain unannotated.

Over the last decade, attempts have been made to analyze ESTs from salt-treated rice plants that resulted in the identification of genes responsible in cell rescue, transport, defense, and energy metabolism (Bohnert et al. 2001). The identification of abundantly expressed ESTs in libraries of a salt-treated halophyte *Thellungiella halophila* (Wang et al. 2004) as well as from monocots like barley, wheat, and maize

(Zhang et al. 2004) has also taken lead in the characterization of genes involved in stress mechanism. A recent comprehensive analysis of expressed sequence tags in common wheat has revealed that out of 0.68 million ESTs with 37,138 contigs and 215,199 singlets used, 55 % of the total singlets were attributed to stress-related libraries. Functional annotations of the contigs thereafter resulted in 21,125 sequences with molecular functions (redox, binding, catalytic activity, etc.), while 13,354 contig sets showed functions in biological processes such as transport, signaling, and metabolism. An overall annotation also showed that nearly 10.5 % of the gene still remains unannotated, while 3,500 new genes were identified through this approach (Manickavelu et al. 2012).

Thus, an overall analysis of different EST collections helps in the identification of stress-regulated genes, underlying regulatory and metabolic networks (Sreenivasulu et al. 2007). EST sequencing has been employed extensively in the absence of whole-genome sequences, notably in crops with large and repetitive genomes, although the entire transcriptome is unlikely to be resolved (Akpınar et al. 2013). Plant expressed sequence tags (ESTs) help in the well characterization of genes and clustering into nonredundant gene clusters (Vij and Tyagi 2007); however, the use of cDNA libraries as a source of these ESTs may underrepresent rare transcripts or transcripts that are not expressed under certain conditions.

2.3.1.2 High-Throughput Gene Expression

In order to have a broad view on global gene expression of a particular species in response to stress conditions, the use of high-throughput techniques for transcript profiling has been one of the successful approaches. The high-throughput approach utilized for transcript profiling has been broadly classified as hybridization-based, sequence-based, and proteomics-based tools and has been discussed in detail in this section.

Hybridization-Based Method

Evolution of microarray technology in the last decade has revolutionized the analysis of genome-wide gene expression. The technique employs arrays of cDNA libraries on a DNA chip, which are further hybridized to fluorescently labeled cDNA probe pairs prepared from RNA samples of different tissues and growth conditions (Seki et al. 2002). The first report of the use of microarray technology in expression profiling was done by Schena et al. (1995) in model organism *Arabidopsis*, wherein 48 genes were analyzed for differential expression in roots and shoots. Utilization of this technique to elucidate the role of genes of a particular species in response to abiotic stress treatment has been one of the popular approaches for genome profiling. One such attempt has been an extensive microarray expression data for *Arabidopsis* with nearly 7,000 independent full-length cDNA groups, showing a fivefold increase in 53 genes during cold stress, 277 genes during high temperature, and 194 genes during high-salinity stress, respectively (Seki et al. 2002). Detailed microarray expression data have also been constructed for rice as well as agronomical important plant species such as wheat (Ergen et al. 2009), barley (Close et al. 2004), maize (Luo et al. 2010), cotton (Ranjan et al. 2012), cassava (Utsumi et al. 2012), and tomato (Loukehaich et al. 2012) to unravel stress responses.

In order to analyze the complete repertoire of the drought-inducible genes in *Arabidopsis*, microarray has been performed. A 7,000 full-length complementary DNA microarray was used to spot 299 drought-inducible genes, 54 cold-inducible genes, 213 high-salinity-inducible genes, and 245 ABA-inducible genes (Shinozaki and Yamaguchi-Shinozaki 2007). Microarray analysis again broadly classified drought-inducible gene products into two categories. The first group is comprised of proteins like chaperones, late embryogenesis abundant (LEA) proteins, osmotin, antifreeze proteins, mRNA-binding proteins, key enzymes for osmolyte biosynthesis, water channel proteins, sugar and proline transporters, detoxification enzymes, and various proteases. The second group includes various transcription factors, protein kinases, protein phosphatases, enzymes involved in phospholipid metabolism, and other signaling molecules such as calmodulin-binding protein.

The use of a microarray to study global gene expression profiling in response to abiotic stress in rice was first reported by Kawasaki et al. (2001) who compared the gene expression profiles in salt-tolerant (var. Pokkali) and salt-sensitive (var. IR29) rice in response to salt stress. The analysis thus performed included 1,728 cDNA clones prepared from unstressed or salt-stressed roots of Pokkali, showing a delayed response in the expression pattern of IR29 as one of the major probable reasons of sensitiveness as compared to its tolerant variety (Pokkali) (Kawasaki et al. 2001). A rice genome array containing 48,564 japonica and 1,260 indica sequences has also been used to compare the transcriptome of salt-tolerant (FL478) and salt-sensitive (IR29) rice varieties. The response of the two varieties was strikingly different, with a much larger number of genes expressed in IR29 than in FL478 on exposure to salt stress. This difference in expression pattern is mainly attributed to the fact that FL478 is salt tolerant and maintains a low Na^+-to-K^+ ratio (Walia et al. 2005). In legumes, high-density cDNA-based arrays have been employed to study 27,513 unigene sets from various developmental stages and stress-exposed tissues of soybean. The result thus obtained identified a large number of genes involved during different biotic (herbicide and pathogen) and abiotic stresses (drought, heat, flooding) (Vodkin et al. 2004) reviewed by Kudapa et al. (2013). Other legumes such as chickpea have also been subjected to global gene expression analysis by various researchers, and in the year 2010 Mantri et al. (2010) identified differentially expressed genes in varying genotypes, in response to various abiotic stresses by comparing a 768 feature microarray of chickpea cDNA (559), grass pea cDNA (156), lentil resistance gene analogue cDNA (41), and control (12). Although microarray technique acts as a powerful tool in the evaluation of gene expression across species, its targeted mode of action acts as a hindrance in its utilization for species where the entire genome is not sequenced. For species where the entire genome sequence is unrevealed, a prior knowledge of transcript to be analyzed and thereby designing of probe remains out of question (Rensink and Buell 2005).

One of the other drawbacks of the microarray studies remains the posttranscriptional and posttranslational modification of transcript, which remains unsolved in this technique (Akpınar et al. 2013).

Another approach through which transcript profiling at genome level is performed is termed as nuclear expressed sequence tag (NEST) analysis. The technique employs a combination of fluorescence-assisted nucleus sorting and cDNA generation (based on the expression of nucleus-targeted green fluorescent protein [GFP], which is controlled by a cell-specific promoter) from the RNA of isolated nuclei (Macas et al. 1998). The isolated RNA in this technique overlays the posttranscriptional modifications and turnover of transcript in the cytosol, thereby accurately reflecting the total nuclear transcript abundance. The cell-specific cDNAs thus obtained are further characterized through differential-display reverse transcriptase-mediated PCR or by EST analysis. The use of NEST analysis in tobacco has revealed an approximate 25 % of salinity-induced transcripts that showed significant homology to functionally unknown genes (Cushman and Bohnert 2000).

Sequence-Based Approach

SAGE acts as one of the essential tools in the quantification of transcripts at a genome-wide scale. The techniques employs generation of unique short sequences tags of 9–17 base pairs (Velculescu et al. 1995; Saha et al. 2002) that are eventually concatenated and sequenced. Generation and quantification of transcript-specific tags help in estimating the expression level of the particular transcript eventually leading to the identification of novel expressed regions of the genome by providing an overall view of gene expression (Vij and Tyagi 2007). For the first time, Matsumura et al. (1999) successfully used this technique in rice seedlings, wherein 10,122 tags from 5,921 expressed genes from seedlings were analyzed and novel genes were identified. Of the many novel genes identified, 18 genes were found to be anaerobically induced and coding for prolamin, expansin, and glycine-rich cell wall protein, while six genes were found repressed under anaerobic conditions (Matsumura et al. 1999).

Over the last decade, reports of SAGE being utilized to analyze changes in gene expression in leaves of *Arabidopsis* in response to cold stress have come up (Jung et al. 2003). Revelation of SAGE analysis has shown that genes that are involved in anabolic pathway such as photosynthesis remain repressed under cold stress conditions, while expression of genes involved in cell rescue, defense, cell death, aging, protein synthesis, metabolism, and transport facilitation are upregulated. In order to identify genes responsible for the sensitiveness of *Arabidopsis* pollen towards cold stress, a comparison of control and stressed SAGE tags was made by Lee and Lee in the year 2003 (Lee and Lee 2003). The results thus obtained showed low accumulation of cold-responsive genes such as COR (cold responsive), beta amylase, and lipid transfer proteins in cold stress-treated pollens as compared to their control counterparts, suggesting it as one of the probable causes of cold susceptibility of *Arabidopsis* pollen (Chinnusamy et al. 2003).

Transcript profiling through SAGE is one of the popular methods in animal system; however, in plants, the technique still remains not widely applicable. As the SAGE tags are generally 9–17 bp in length, their applicability in tag-to-gene annotation remains limited to model organism where the entire genome sequence is known. In order to cope with these problems, variant of SAGE technology, SuperSAGE, has been of much use in plant species. The technique utilizes the same methodology as that of SAGE with only difference in tag length (26 bp), which remains longer than its precursor. The longer tag length has been a boon for transcript profiling in non-model organism as the tags themselves allow designing of PCR primer for cDNA synthesis of corresponding genes via RACE (Matsumura et al. 2003). The use of SuperSAGE technique in plant has been widely implemented for studying responses of genes under various biotic and abiotic stresses across different species, such as banana, tobacco, capsicum, chickpea, etc. Utilization of this technique for elucidating role of genes in abiotic stress response has helped in the elucidation of 3,000 stress-related transcripts from 360,000 transcripts of salt- and drought-related transcriptome of chickpea and lentil (Kudapa et al. 2013). The amenability of SuperSAGE with integrative transcriptome, SuperSAGE microarray, and next-generation sequencing has also led to an increase in its efficiency towards gene expression profiling (Matsumura et al. 2005). Application of NGS to gene expression analysis has led to the development of techniques like digital gene expression-TAG (DGE-TAG), DeepSAGE (Nielsen et al. 2006), RNA-seq (Marioni et al. 2008; Matsumura et al. 2010; Nagalakshmi et al. 2008), and eventually high-throughput SuperSAGE precisely known as deepSuperSAGE, which provide digital, highly accurate, easy-to-perform, and cost-effective solution to expression profiling over its hybridization counterpart microarray (Matsumura et al. 2010). Application of deepSuperSAGE in drought-treated and control chickpea roots has further identified 80,238 tags representing 17,493 unique transcripts (Molina et al. 2008, 2011).

An alternative approach to genome-wide transcript profiling through sequencing-based approach lies in massively parallel signature sequencing (MPSS). The technique follows similar tag-based approach to ascertain the expression pattern of mRNA within a species. These expressed mRNAs are further related to their corresponding ESTs or whole genome, thereby providing a larger perspective of expression profiling (Brenner et al. 2000; Pollock 2002). The basic idea behind the technique lies in the generation of longer sequence tags, which are ligated to microbeads and sequenced in parallel, thereby enabling analysis of millions of transcripts simultaneously (Brenner et al. 2000) with greater specificity and sensitivity and at a much faster speed. The use of MPSS has been widely popular and beneficial in species that lack a whole-genome sequence (Reinartz et al. 2002). In plants, MPSS has been employed in studying expression response of mRNA transcripts towards abiotic stress conditions. One of the earliest examples of the use of MPSS in gene expression profiling in response to abiotic stress conditions has been in studying ABA-responsive gene expression in ABA-insensitive mutants (*abi1-1*) and wild-type (WT) *Arabidopsis* plants. A study showed that a very few genes were affected by ABA in the ABA-insensitive mutant (*abi1-1*), whereas the ABA-responsive gene expression of the majority of genes was drastically impaired in the *abi1-1* mutant, thereby postulating existence of two ABA signaling pathways, only one of which is impaired in the *abi1-1* mutant (Hoth et al. 2002). Over the last decade, reports of MPSS analysis in expression profiling of small RNAs (Sunkar et al. 2007) having a distinct role in abiotic stress responses have also come into light. Although MPSS is a quick, reliable, and reproducible tool, the realization of its full potential in the global expression profiling towards abiotic stress response is yet to be achieved (Vij and Tyagi 2007) owing to its cost. As of now, plant MPSS database (http://mpss.udel.edu/) has been a repository and publicly available MPSS expression data of many plant species, some of which include *Arabidopsis*, rice, maize, soybean, and grapes. The resource for rice includes 20 MPSS libraries constructed from different tissues, of which three libraries belong to abiotic stress conditions, namely, cold, salt, and dehydration (Nakano et al. 2006).

2.3.1.3 Proteomics and Metabolomics Approach

Proteins act as one of the prime targets of abiotic stress response when looked up beyond their genetic counterpart. Identification of proteins as target of abiotic stress acts in nullifying the effect of posttranscriptional and translational modifications. Utilization of proteomic tools such as two-dimensional gel electrophoresis (2DGE), MALDI, and mass spectrometry (MS) in the identification of these target proteins has been a popular approach in *Arabidopsis* in response to cold stress. Analysis of *Arabidopsis* nuclear proteome thus revealed that out of 134 protein spots identified, nearly 30 % of the protein spots showed an altered response towards cold stress. Few of these proteins included heat-shock proteins, transcription factors (OBF2 and AtMYB2), DNA-binding proteins (Dr1 and DRT102), catalytic enzymes (phosphoglycerate kinase, serine acetyltransferase, and glyceraldehyde-3-phosphate dehydrogenase), and calmodulin, syntaxin, and germin-like proteins (Bae et al. 2003). A similar proteome analysis of rice plant using 2D gel electrophoresis showed that 42 out of 1,000 protein spots showed an altered response towards drought stress conditions (Salekdeh et al. 2002). Out of these 42 proteins with altered expression in stress conditions, 16 were identified as drought-responsive proteins using MS analysis. Detailed analysis of rice root proteome through 2DGE and MS analysis identified 54 proteins with affected expression in response to salt stress. Out of the 54 identified proteins, 6 proteins were identified and annotated as novel salt-responsive proteins, such as Cox6b-1, UGPase, GS root isozyme, α-NAC, putative splicing factor-like protein, and putative ABP (Yan et al. 2005). Extensive proteome analyses of rice seedling and anthers (Imin et al. 2004) in response to cold stress have also resulted in the identification of a number of cold stress-responsive genes like chaperones, proteases, and detoxifying enzymes (Cui et al. 2005). Other than *Arabidopsis* and rice, proteome approach to find proteins as putative targets of abiotic stress response has also been observed in tobacco apoplastic protein wherein

20 salt stress-responsive proteins were identified, which included chitinases, germin-like protein, lipid transfer proteins, and a few well-characterized stress-associated proteins (Dani et al. 2005).

Integrated metabolome and transcriptome analyses of model plants have also helped in the identification of metabolites in response to various stresses in plants. A comprehensive metabolite profiling of model plants in different abiotic stress (cold, drought, temperature, and salt) conditions also helps in fine-tuning the stress pathway of plants. Metabolite profiling analysis of *Arabidopsis* in response to dehydration stress using an integrative approach of GC-TOF MS, CE–MS, and DNA microarrays has revealed that accumulation of glucose, fructose, and branched-chain amino acid in response to dehydration stress is regulated by ABA accumulation during dehydration, thereby postulating an important role of ABA in stress response (Urano et al. 2010). Metabolic profiling has also revealed that sucrose replaces proline in plants as the major osmoprotectant during the more severe combined dehydration and heat-stress treatment (Rizhsky et al. 2004).

In response to cold stress, metabolite profiling of *Arabidopsis* has indicated that an increased tolerance to freezing stress in transgenic plants overexpressing DREB1A may be a consequence of the accumulation of low-temperature-regulated metabolites, especially raffinose, sucrose, myoinositol, and galactinol (Cook et al. 2004; Maruyama et al. 2009). In response to cold and heat shocks on *Arabidopsis*, using GC–MS (Kaplan et al. 2004) and GC-TOF MS (Wienkoop et al. 2008) has revealed that metabolites such as compatible solutes that are produced during extreme temperature overlap during both heat- and cold-shock treatments, thereby postulating that accumulation of compatible solutes bears an important role in temperature stress condition (Urano et al. 2010). Comparative metabolome analysis of a model halophyte *Thellungiella halophila* with *Arabidopsis* under salt stress conditions through GC–MS and microarrays (Gong et al. 2005) has shown that even in control conditions, *Thellungiella* contains higher levels of various osmolytes, such as fructose, sucrose, complex sugars, malate, and proline, and their respective biosynthetic genes compared with *Arabidopsis* (Gong et al. 2005). The results thus obtained suggest that a constant state of stress-anticipatory preparedness exists in *Thellungiella*.

2.4 Validation of Functions and Stress Response Mechanisms

In order to validate the function of genes mined from genome-wide analysis of a species, in response to abiotic stress mechanism, knockout analysis through RNA interference (RNAi), T-DNA insertional mutants, TILLING, and overexpression experiments has been carried out. These transgenic approaches have helped in validating the role of identified genes during different abiotic stresses faced by plants.

2.4.1 *Knockout/Knockdown Analysis*

2.4.1.1 RNA Interference in Gene Validation

RNA-mediated gene silencing, collectively termed as RNAi, has been employed as one of the promising tools in the validation of functional genes. The approach utilizes nearly 21–25-nt-long ds small RNAs, which are a product of small portion of desired gene's coding regions and act as a trigger for direct degradation of mRNA in homology-dependent posttranscriptional silencing (Grishok et al. 2001; Hamilton and Baulcombe 1999; Zamore et al. 2000) reviewed in McGinnis (2010). Not only posttranscriptional but also transcriptional regulation is achieved through RNAi-mediated silencing wherein such a construct is designed that the dsRNA thus formed bears homology to the promoter region of desired gene (Mette et al. 2000). siRNA-mediated DNA methylation and heterochromatin formation through epigenetic modulation are also one of the key approaches of transcriptional inhibition of the desired gene (Verdel et al. 2009) reviewed by McGinnis (2010). The technology finds its relevance in areas where antisense technologies fail to perform (Kudapa et al. 2013). The use of a single copy of transgene-induced RNAi, to efficiently block genes belonging to gene families or expressing during various time and places of developing tissues (Kudapa et al. 2013), helps in simplification of several phenotypic assays (such as several generation crosses for marker-assisted silencing) in a functional genomics project (McGinnis 2010).

However, early attempts of homologous recombination-based gene silencing in rice showed minimal success rate with very low efficiency rate (Hanin and Paszkowski 2003). Later attempts (Miki and Shimamoto 2004) showed 90 % silencing through RNAi vectors against 11 rice genes. In other crop plants such as legume, *Medicago* and soybean have shown the effectiveness of RNAi-induced gene silencing and validation of gene functions by efficient silencing of PIN (auxin efflux proteins) genes responsible for root nodule formation (Huo et al. 2006) and myoinositol-1-phosphate regulating cellular metabolism and growth, respectively (Nunes et al. 2006). Though RNAi-induced gene silencing has well been established in many plant species, application still remains limited due to its inefficient long-term establishment and inability to be utilized for large-scale functional analysis (Kudapa et al. 2013; Vij and Tyagi 2007). Other possible means for employing gene silencing in functional genomics has been the deployment of virus-induced gene silencing (VIGS) that provides long-term persistence during vegetative and in vitro propagation and results in generation of genotypically identical silenced genes (Kudapa et al. 2013; Vij and Tyagi 2007). In legumes, the approach has been well established in soybean and pea against genes responsible in biotic stress resistance (Gronlund et al. 2010; Kachroo and Ghabrial 2012). Implication of VIGS has been in both forward and reverse genetic approach to validate gene functions (Senthil-Kumar and Mysore 2011); however, its application is limited due to the unavailability of appropriate vector and vector delivery system in plants (Kudapa et al. 2013).

2.4.1.2 Gene Validation Through Mutations

There are two main complementary approaches developed for identifying target genes through mutations, namely, TILLING and T-DNA insertional mutant lines. Of the two approaches, T-DNA insertional mutant lines involve insertion of foreign DNA into candidate gene, for disrupting gene function and ascertaining their role through reverse genetics approach. In *Arabidopsis*, the technique involves the use of either transposable elements (Parinov et al. 1999) or T-DNA. The foreign DNA not only disrupts the expression of the gene into which it is inserted but also acts as a marker for subsequent identification of the mutation. Because *Arabidopsis* introns are small and there is very little intergenic material, any insertion of order ranging from 5 to 25 kb in length leads to a dramatic alteration in gene function (Radhamony et al. 2005). Although the approach has wide applications in crop improvement, its success has been limited to only few plant species, owing to their dependency on *Agrobacterium*-mediated T-DNA transmission or on endogenous transposon tagging system (McCallum et al. 2000). However, in a large population of T-DNA-transformed lines, one has a very good chance of finding a plant carrying a T-DNA insert within any gene of interest. Mutations that are homozygous lethal can also be maintained in the population in the form of heterozygous plants through this approach (Radhamony et al. 2005).

The second approach, TILLING (targeting induced local lesions in genomes), acts as an alternative approach to insertional mutagenesis, which was developed a decade ago by a graduate scholar Claire McCallum, who wanted to induce targeted mutations in chromomethylase gene of *Arabidopsis*. The method thus invented was the advancement over the existing reverse genetic tools to study genetic data of an organism. The main advantage of TILLING as a reverse genetics strategy envisages its application to any species plant (since it does not require transgenic or cell culture manipulations), regardless of its genome size and ploidy level. Chemical mutagens such as ethyl methanesulfonate (EMS) are generally used in this technique that provide a high frequency of point mutations distributed randomly in the genome. Initially, TILLING strategy was developed as a discovery platform for functional genomics, but it soon became a valuable tool in crop breeding as an alternative to the transgenic approach. The method combines a standard and efficient technique of mutagenesis in combination of a chemical mutagen with a sensitive DNA screening technique that identifies single-base mutations in a target gene. The approach relies primarily on the formation of DNA heteroduplexes. The general protocol for the formation of a tilling population includes the creation of mutated populations through chemical mutagenesis followed by the detection of mutations in target sequence through polymerase chain reaction, and subsequent phenotype thus obtained is eventually analyzed. One of the significant advantages of this technique includes production of an allelic series of mutations including hypomorphic alleles that are useful for genetic analysis. Success of TILLING approach over its T-DNA insertional counterpart has been demonstrated through its wide application across plant species. For a large number of agronomically important crops, including rice, barley, wheat, maize, sorghum, soybean, rapeseed, and tomato plants, TILLING has been utilized as a wonder technique for crop improvement (Kurowska et al. 2011).

By using these techniques, putative functions of abiotic stress-responsive genes were followed in *Arabidopsis*, rice, maize, and barley. In *Arabidopsis* (ecotype C24) 43,000 T-DNA insertional lines were generated (http://stress-genomics.org/stress.fls/tools/mutants.html), of which 30,000 lines were screened for stress-related gene regulation mutants. The mutants thus obtained, upon screening for salt-stressed conditions, showed hypersensitivity towards salt stress by demonstrating an impaired growth on increased salt conditions, thereby determining the role of knocked-out genes SOS1, SOS2, and SOS3 in maintaining and balancing ion transport across plant membrane (Kamei et al. 2005; Liu et al. 2008; Halfter et al. 2000).

Large-scale forward genetic screens have been used to identify abiotic stress response determinants in a T-DNA-mutagenized *Arabidopsis* population in the RD29a-LUC background. More than 200 mutants with altered stress/ABA response were identified from 250,000 independent insertional lines. These included mutations in genes coding for transcription factors, syntaxin, ABA biosynthetic enzyme, SUMO E3 ligase, and the sodium transporter HKT1 (Koiwa et al. 2006). T-DNA-tagged transgenic rice lines were evaluated for cold-responsive β-glucuronidase (GUS) expression, and detailed analysis led to the identification of two cold-responsive genes, namely, OsDMKT1 (putative demethyl menaquinone methyltransferase) and OsRLK1 (putative LRR-type receptor-like protein kinase) (Lee et al. 2004).

TILLING enables high-throughput analysis of a large number of mutants (McCallum et al. 2000). TILLING is applicable to virtually all genes in all species where mutations can be induced and have been reported in several crop species, including hexaploid wheat (Chen et al. 2012). TILLING mutants are reported in sorghum (Xin et al., 2008), maize (Till et al. 2004), barley (Caldwell et al. 2004), soybean (Cooper et al. 2008), rice (Cooper et al. 2013), and other crops, where further use of certain TILLING mutants in elucidation of stress responses has been demonstrated. In one such similar study, TILLING mutants for a specific kinase were used to assess salt stress response in legume species (de Lorenzo et al. 2009). Another variant of this strategy, referred to as EcoTILLING, has conjointly been developed to spot natural polymorphisms, similar to TILLING-assisted identification of evoked mutations. Polymorphisms demonstrating natural variation in germplasms are valuable tools in genetic mapping. Similar to TILLING, EcoTILLING is applicable to polyploid species, where it can be utilized to differentiate between alleles of homologous and paralogous genes (Comai et al. 2004). Recently, EcoTILLING not only provided allelic variants of a number of genes involved in salt stress response (Negrão et al. 2013), but transcription factors, diversifying stress responses, have also been targeted, to examine natural rice variants exposed to drought stress (Yu et al. 2012). The availability of comprehensive EST databases is central to the success of the above mentioned approaches to identify genes accurately and unambiguously.

2.4.2 Overexpression Analysis

To identify the function of any gene, insertional mutagenesis is also one of the effective techniques. Although loss-of-function mutations identified the function of genes in many organisms, there is a major limitation attached with it. After *Arabidopsis* genome sequencing, it has been discovered that most of the genes have closely related sequences in the genome that belongs to gene families (The *Arabidopsis* Genome Initiative 2000). So gene redundancy, where a gene family contains several genes that have a similar function, can compensate the loss-of-function phenotype, or there may exist alternative metabolic pathways and regulatory networks, which provide plants with the same gene product. This genetic redundancy may prevent elucidation of gene function by loss-of-function approaches. New advancement in this field came after 1992, named as gain-of-function approach, also known as activation-tagging technology. This was originally developed by Rick Walden and colleagues at the Max Planck Institute (Hayashi et al. 1992), who constructed a T-DNA vector with four copies of an enhancer element from the constitutively active promoter of the cauliflower mosaic virus (*CaMV*) *35S* gene (Odell et al. 1985). These enhancers can cause transcriptional activation of nearby genes, and as activated genes are associated with a T-DNA insertion, this approach came to be known as activation tagging. This technology has made the production of gain-of-function mutants possible. It has added advantage of screening of phenotypes in the T_1 generation, as all activation-tagged mutants are dominant. In the case of loss-of-function mutations, the phenotype is usually not seen until the T_2 generation. The T-DNA having tetramer of the *CaMV 35S* enhancer gets randomly inserted into the plant genome via *Agrobacterium* transformation. The inserted enhancer elements then activate the transcription of nearby genes. T-DNA insertion sites are then determined in the transformed mutants to identify candidate genes. Various techniques like plasmid rescue, inverse PCR, or adapter PCR methods can be used to determine the genomic fragment details near the T-DNA right and left border sequences. TAIL-PCR is also an efficient method to determine T-DNA insertion (Burke 2003; Yamamoto et al. 2003).

The activation-tagging system has been employed in identification of genes involved in biotic and abiotic stress responses. The *enhanced drought tolerance 1* (*edt1*) mutant is identified under drought condition using activation tagging. It has been shown that the enhanced drought tolerance is caused by T-DNA-tagged gene that encodes putative homeodomain-START transcription factor. Transgenic tobacco with overexpressed cDNA of this transcription factor also develops drought tolerance. So, activation tagging provides a key regulator that may be used to improve drought tolerance in plants (Yu et al. 2008). *CDR1-D* (constitutive disease resistance) and *FMO1–3D* (flavin-containing monooxygenase) mutants showed increased resistance during abiotic stress and are also identified through activation tagging (Koch et al. 2006; Xia et al. 2004).

However, just like two sides of a coin, activation-tagged mutagenesis does have some disadvantages. Activation of gene by a transcriptional enhancer is not restricted to one gene and in some cases may result into a complex phenotype caused by complex gene expression (Ichikawa et al. 2003). Therefore, it becomes difficult to identify, which gene corresponds to the mutant phenotype. To get rid of this problem, an alternative of gain-of-function technique is developed known as Full-length cDNA Over-eXpressing gene hunting system (FOX hunting system). This system can eliminate such complexity since the full-length cDNA is expressed in the correct orientation between the CaMV *35S* promoter and the *NOS* terminator. After identification, genomic PCR is performed to isolate the gene that results in mutant phenotype. During this approach, mutant lines were generated by ectopic expression of full-length cDNAs (Ichikawa et al. 2006). Novel salt stress tolerance genes were isolated from *Thellungiella halophila* using the FOX hunting system (Du et al. 2008).

Ectopic expression of some transcription factors may lead to lethality that makes analysis of gene difficult. In such cases, controlled gene expression using inducible system can work. Several ABA insensitivity and salt-tolerant mutants were isolated using this controlled system of induction by cloning an *Arabidopsis* cDNA library into the estradiol-inducible vector (Papdi et al. 2008).

Conclusion

The effective role of functional genomics in elucidation of complex regulatory pathways and respective signaling molecules has been enhanced by remarkable technical advancement in its sub-platforms, which includes genome, transcriptome, metabolome, fluxome, and ionome analyses, respectively. These analyses on the whole provide an overview of the entire cellular machinery. Implications of this advancement are immense and far reaching in many fields including medicine, agriculture, public health, defense, and more. In plants utilization of functional genomics approach has played a crucial role in understating the processes and molecular networks in response to stress conditions. The use of functional genomics to dissect out the role of individual gene, protein, and metabolite in an intriguing cascade of abiotic stress response provides a hint towards crop improvement against abiotic stress. With the advent of next-generation sequencing and RNA-seq techniques, cost-effective genome and transcriptome analysis has further strengthened the pillars of functional genomics. The use of high-throughput techniques has helped in mining greater details at each subcellular level, which upon integration through bioinformatics approach has revealed many of the signaling molecules intriguing in complex signaling cascades over the past few years. Thus, utilization of transgenic approach, molecular breeding, and map-based cloning together could further help in introgression of desired traits to make plant stress tolerant without compromising on its yield. Once identified, desirable QTLs can be utilized for extensive breeding to restore desirable traits along with the introgressed tolerance trait. Although the functional genomics approach is a notch higher in validating the role of genes in various molecular networks, it still bears challenges such as absence of a comprehensive bioinformatics database that can mine data from individual "omics" platform and provide a holistic view of cellular networks and machinery on the whole under any perturbed condition faced by plants. Thus, a comprehensive approach including functional, structural, and comparative genomics together is necessary to mitigate the various extremes of abiotic stress, through crop improvement program.

References

Abe H, Urao T, Ito T, Seki M, Shinozaki K, Yamaguchi-Shinozaki K (2003) Arabidopsis AtMYC2 (bHLH) and AtMYB2 (MYB) function as transcriptional activators in abscisic acid signaling. Plant Cell 15(1):63–78

Akpınar BA, Lucas SJ, Budak H (2013) Genomics approaches for crop improvement against abiotic stress. Scientific World Journal 2013

Bae MS, Cho EJ, Choi EY, Park OK (2003) Analysis of the Arabidopsis nuclear proteome and its response to cold stress. Plant J 36(5):652–663

Baxter A, Mittler R, Suzuki N (2014) ROS as key players in plant stress signalling. J Exp Bot 65(5):1229–1240

Benschop JJ, Mohammed S, O'Flaherty M, Heck AJ, Slijper M, Menke FL (2007) Quantitative phosphoproteomics of early elicitor signaling in Arabidopsis. Mol Cell Proteomics 6(7): 1198–1214

Bienert GP, Møller AL, Kristiansen KA, Schulz A, Møller IM, Schjoerring JK, Jahn TP (2007) Specific aquaporins facilitate the diffusion of hydrogen peroxide across membranes. J Biol Chem 282(2):1183–1192

Bindschedler LV, Dewdney J, Blee KA, Stone JM, Asai T, Plotnikov J, Denoux C, Hayes T, Gerrish C, Davies DR (2006) Peroxidase-dependent apoplastic oxidative burst in Arabidopsis required for pathogen resistance. Plant J 47(6):851–863

Bohnert HJ, Ayoubi P, Borchert C, Bressan RA, Burnap RL, Cushman JC, Cushman MA, Deyholos M, Fischer R, Galbraith DW (2001) A genomics approach towards salt stress tolerance. Plant Physiol Biochem 39(3):295–311

Bouchez D, Höfte H (1998) Functional genomics in plants. Plant Physiol 118(3):725–732

Brandt B, Brodsky DE, Xue S, Negi J, Iba K, Kangasjärvi J, Ghassemian M, Stephan AB, Hu H, Schroeder JI (2012) Reconstitution of abscisic acid activation of SLAC1 anion channel by CPK6 and OST1 kinases and branched ABI1 PP2C phosphatase action. Proc Natl Acad Sci 109(26):10593–10598

Brenner S, Johnson M, Bridgham J, Golda G, Lloyd DH, Johnson D, Luo S, McCurdy S, Foy M, Ewan M (2000) Gene expression analysis by massively parallel signature sequencing (MPSS) on microbead arrays. Nat Biotechnol 18(6):630–634

Bressan RA, Hasegawa PM, Pardo JM (1998) Plants use calcium to resolve salt stress. Trends Plant Sci 3(11):411–412

Burke E (2003) High-throughput TAIL-PCR as a tool to identify DNA flanking insertions. Plant Functional Genomics. Springer, In, pp 241–271

Caldwell DG, McCallum N, Shaw P, Muehlbauer GJ, Marshall DF, Waugh R (2004) A structured mutant population for forward and reverse genetics in Barley (Hordeum vulgare L.). Plant J 40(1):143–150

Chater C, Kamisugi Y, Movahedi M, Fleming A, Cuming AC, Gray JE, Beerling DJ (2011) Regulatory mechanism controlling stomatal behavior conserved across 400 million years of land plant evolution. Curr Biol 21(12):1025–1029

Chen L, Huang L, Min D, Phillips A, Wang S, Madgwick PJ, Parry MA, Hu Y-G (2012) Development and characterization of a new TILLING population of common bread wheat (Triticum aestivum L.). PLoS One 7(7):e41570

Chinnusamy V, Ohta M, Kanrar S, Lee BH, Hong X, Agarwal M, Zhu JK (2003) ICE1: a regulator of cold-induced transcriptome and freezing tolerance in Arabidopsis. Genes Dev 17(8): 1043–1054. doi:10.1101/gad.1077503

Close TJ, Wanamaker SI, Caldo RA, Turner SM, Ashlock DA, Dickerson JA, Wing RA, Muehlbauer GJ, Kleinhofs A, Wise RP (2004) A new resource for cereal genomics: 22K barley GeneChip comes of age. Plant Physiol 134(3):960–968

Comai L, Young K, Till BJ, Reynolds SH, Greene EA, Codomo CA, Enns LC, Johnson JE, Burtner C, Odden AR (2004) Efficient discovery of DNA polymorphisms in natural populations by ecotilling. Plant J 37(5):778–786

Cook D, Fowler S, Fiehn O, Thomashow MF (2004) A prominent role for the CBF cold response pathway in configuring the low-temperature metabolome of Arabidopsis. Proc Natl Acad Sci U S A 101(42):15243–15248

Cooper J, Till B, Laport R, Darlow M, Kleffner J, Jamai A, El-Mellouki T, Liu S, Ritchie R, Nielsen N (2008) TILLING to detect induced mutations in soybean. BMC Plant Biol 8(1):9

Cooper JL, Henikoff S, Comai L, Till BJ (2013) TILLING and ecotilling for rice. In: Rice protocols. Springer, pp 39–56

Cui S, Huang F, Wang J, Ma X, Cheng Y, Liu J (2005) A proteomic analysis of cold stress responses in rice seedlings. Proteomics 5(12):3162–3172

Cushman JC, Bohnert HJ (2000) Genomic approaches to plant stress tolerance. Curr Opin Plant Biol 3(2):117–124

Dani V, Simon WJ, Duranti M, Croy RR (2005) Changes in the tobacco leaf apoplast proteome in response to salt stress. Proteomics 5(3):737–745

de Lorenzo L, Merchan F, Laporte P, Thompson R, Clarke J, Sousa C, Crespi M (2009) A novel plant leucine-rich repeat receptor kinase regulates the response of Medicago truncatula roots to salt stress. Plant Cell 21(2):668–680

Delauney AJ, Verma DPS (1993) Proline biosynthesis and osmoregulation in plants. Plant J 4(2):215–223

Du J, Huang YP, Xi J, Cao MJ, Ni WS, Chen X, Zhu JK, Oliver DJ, Xiang CB (2008) Functional gene-mining for salt-tolerance genes with the power of Arabidopsis. Plant J 56(4):653–664

Ergen NZ, Thimmapuram J, Bohnert HJ, Budak H (2009) Transcriptome pathways unique to dehydration tolerant relatives of modern wheat. Funct Integr Genomics 9(3):377–396

Flowers T, Yeo A (1995) Breeding for salinity resistance in crop plants: where next? Funct Plant Biol 22(6):875–884

Ford CW (1984) Accumulation of low molecular weight solutes in water-stressed tropical legumes. Phytochemistry 23(5):1007–1015

Foyer CH, Parry M, Noctor G (2003) Markers and signals associated with nitrogen assimilation in higher plants. J Exp Bot 54(382):585–593

Frova C, Krajewski P, Di Fonzo N, Villa M, Sari-Gorla M (1999) Genetic analysis of drought tolerance in maize by molecular markers I. Yield components. Theor Appl Genet 99(1–2): 280–288

Fujii H, Chinnusamy V, Rodrigues A, Rubio S, Antoni R, Park S-Y, Cutler SR, Sheen J, Rodriguez PL, Zhu J-K (2009) In vitro reconstitution of an abscisic acid signalling pathway. Nature 462(7273):660–664

Fujita Y, Yoshida T, Yamaguchi-Shinozaki K (2013) Pivotal role of the AREB/ABF-SnRK2 pathway in ABRE-mediated transcription in response to osmotic stress in plants. Physiol Plant 147(1):15–27

Gong Q, Li P, Ma S, Indu Rupassara S, Bohnert HJ (2005) Salinity stress adaptation competence in the extremophile Thellungiella halophila in comparison with its relative Arabidopsis thaliana. Plant J 44(5):826–839

Grishok A, Pasquinelli AE, Conte D, Li N, Parrish S, Ha I, Baillie DL, Fire A, Ruvkun G, Mello CC (2001) Genes and mechanisms related to RNA interference regulate expression of the small temporal RNAs that control *C. elegans* developmental timing. Cell 106(1):23–34

Gronlund M, Olsen A, Johansen E, Jakobsen I (2010) Protocol: using virus-induced gene silencing to study the arbuscular mycorrhizal symbiosis in Pisum sativum. Plant Methods 6(1):28

Gupta B, Saha J, Sengupta A, Gupta K (2013) Plant abiotic stress: 'omics' approach. J Plant Biochem Physiol 1:e108

Halfter U, Ishitani M, Zhu JK (2000) The Arabidopsis SOS2 protein kinase physically interacts with and is activated by the calcium-binding protein SOS3. Proc Natl Acad Sci U S A 97(7):3735–3740. doi:10.1073/pnas.040577697

Hamilton AJ, Baulcombe DC (1999) A species of small antisense RNA in posttranscriptional gene silencing in plants. Science 286(5441):950–952

Hanin M, Paszkowski J (2003) Plant genome modification by homologous recombination. Curr Opin Plant Biol 6(2):157–162

Hayashi H, Czaja I, Lubenow H, Schell J, Walden R (1992) Activation of a plant gene by T-DNA tagging: auxin-independent growth in vitro. Science 258(5086):1350–1353

Hoth S, Morgante M, Sanchez J-P, Hanafey MK, Tingey SV, Chua N-H (2002) Genome-wide gene expression profiling in Arabidopsis thaliana reveals new targets of abscisic acid and largely impaired gene regulation in the abi1-1 mutant. J Cell Sci 115(24):4891–4900

Huo X, Schnabel E, Hughes K, Frugoli J (2006) RNAi phenotypes and the localization of a protein: GUS fusion imply a role for Medicago truncatula PIN genes in nodulation. J Plant Growth Regul 25(2):156–165

Ichikawa T, Nakazawa M, Kawashima M, Muto S, Gohda K, Suzuki K, Ishikawa A, Kobayashi H, Yoshizumi T, Tsumoto Y (2003) Sequence database of 1172 T-DNA insertion sites in Arabidopsis activation-tagging lines that showed phenotypes in T1 generation. Plant J 36(3): 421–429

Ichikawa T, Nakazawa M, Kawashima M, Iizumi H, Kuroda H, Kondou Y, Tsuhara Y, Suzuki K, Ishikawa A, Seki M (2006) The FOX hunting system: an alternative gain-of-function gene hunting technique. Plant J 48(6):974–985

Imin N, Kerim T, Rolfe BG, Weinman JJ (2004) Effect of early cold stress on the maturation of rice anthers. Proteomics 4(7):1873–1882

Initiative AG (2000) Analysis of the genome sequence of the flowering plant Arabidopsis thaliana. Nature 408(6814):796

Ishitani M, Liu J, Halfter U, Kim C-S, Shi W, Zhu J-K (2000) SOS3 function in plant salt tolerance requires N-myristoylation and calcium binding. Sci Signal 12(9):1667

Iuchi S, Kobayashi M, Yamaguchi-Shinozaki K, Shinozaki K (2000) A stress-inducible gene for 9-cis-epoxycarotenoid dioxygenase involved in abscisic acid biosynthesis under water stress in drought-tolerant cowpea. Plant Physiol 123(2):553–562

Jammes F, Song C, Shin D, Munemasa S, Takeda K, Gu D, Cho D, Lee S, Giordo R, Sritubtim S (2009) MAP kinases MPK9 and MPK12 are preferentially expressed in guard cells and positively regulate ROS-mediated ABA signaling. Proc Natl Acad Sci 106(48):20520–20525

Jung S-H, Lee J-Y, Lee D-H (2003) Use of SAGE technology to reveal changes in gene expression in Arabidopsis leaves undergoing cold stress. Plant Mol Biol 52(3):553–567

Kachroo A, Ghabrial S (2012) Virus-Induced Gene Silencing in Soybean. Antiviral Resistance in Plants. Springer, In, pp 287–297

Kamei A, Seki M, Umezawa T, Ishida J, Satou M, Akiyama K, ZHU JK, Shinozaki K (2005) Analysis of gene expression profiles in Arabidopsis salt overly sensitive mutants sos2–1 and sos3–1. Plant Cell Environ 28(10):1267–1275

Kaplan F, Kopka J, Haskell DW, Zhao W, Schiller KC, Gatzke N, Sung DY, Guy CL (2004) Exploring the temperature-stress metabolome of Arabidopsis. Plant Physiol 136(4): 4159–4168

Kawasaki S, Borchert C, Deyholos M, Wang H, Brazille S, Kawai K, Galbraith D, Bohnert HJ (2001) Gene expression profiles during the initial phase of salt stress in rice. Plant Cell 13(4):889–905

Kim T-H, Böhmer M, Hu H, Nishimura N, Schroeder JI (2010) Guard cell signal transduction network: advances in understanding abscisic acid, CO2, and Ca2+ signaling. Annu Rev Plant Biol 61:561

Knight H, Trewavas AJ, Knight MR (1997) Calcium signalling in Arabidopsis thaliana responding to drought and salinity. Plant J 12(5):1067–1078

Koch M, Vorwerk S, Masur C, Sharifi-Sirchi G, Olivieri N, Schlaich NL (2006) A role for a flavin-containing mono-oxygenase in resistance against microbial pathogens in Arabidopsis. Plant J 47(4):629–639

Koiwa H, Bressan RA, Hasegawa PM (2006) Identification of plant stress-responsive determinants in Arabidopsis by large-scale forward genetic screens. J Exp Bot 57(5):1119–1128

Komatsu K, Nishikawa Y, Ohtsuka T, Taji T, Quatrano RS, Tanaka S, Sakata Y (2009) Functional analyses of the ABI1-related protein phosphatase type 2C reveal evolutionarily conserved regulation of abscisic acid signaling between Arabidopsis and the moss Physcomitrella patens. Plant Mol Biol 70(3):327–340

Kudapa H, Ramalingam A, Nayakoti S, Chen X, Zhuang W-J, Liang X, Kahl G, Edwards D, Varshney RK (2013) Functional genomics to study stress responses in crop legumes: progress and prospects. Funct Plant Biol 40(12):1221–1233

Kurowska M, Daszkowska-Golec A, Gruszka D, Marzec M, Szurman M, Szarejko I, Maluszynski M (2011) TILLING-a shortcut in functional genomics. J Appl Genet 52(4):371–390

Larkindale J, Vierling E (2008) Core genome responses involved in acclimation to high temperature. Plant Physiol 146(2):748–761

Lee J-Y, Lee D-H (2003) Use of serial analysis of gene expression technology to reveal changes in gene expression in Arabidopsis pollen undergoing cold stress. Plant Physiol 132(2):517–529

Lee S-C, Kim J-Y, Kim S-H, Kim S-J, Lee K, Han S-K, Choi H-S, Jeong D-H, An G, Kim S-R (2004) Trapping and characterization of cold-responsive genes from T-DNA tagging lines in rice. Plant Science 166(1):69–79

Li S, Zhou X, Chen L, Huang W, Yu D (2010) Functional characterization of Arabidopsis thaliana WRKY39 in heat stress. Mol Cells 29(5):475–483

Liu J, Zhu J-K (1998) A calcium sensor homolog required for plant salt tolerance. Science 280(5371):1943–1945

Liu J, Ishitani M, Halfter U, Kim C-S, Zhu J-K (2000) The Arabidopsis thaliana SOS2 gene encodes a protein kinase that is required for salt tolerance. Proc Natl Acad Sci 97(7): 3730–3734

Liu HT, Gao F, Li GL, Han JL, Liu DL, Sun DY, Zhou RG (2008) The calmodulin-binding protein kinase 3 is part of heat-shock signal transduction in Arabidopsis thaliana. Plant J 55(5):760–773

LIU HC, LIAO HT, CHARNG YY (2011) The role of class A1 heat shock factors (HSFA1s) in response to heat and other stresses in Arabidopsis. Plant Cell Environ 34(5):738–751

Lobell DB, Schlenker W, Costa-Roberts J (2011) Climate trends and global crop production since 1980. Science 333(6042):616–620

Lopez-Molina L, Mongrand S, Chua N-H (2001) A postgermination developmental arrest checkpoint is mediated by abscisic acid and requires the ABI5 transcription factor in Arabidopsis. Proc Natl Acad Sci 98(8):4782–4787

Louis P, Galinski EA (1997) Characterization of genes for the biosynthesis of the compatible solute ectoine from Marinococcus halophilus and osmoregulated expression in Escherichia coli. Microbiology 143(4):1141–1149

Loukehaich R, Wang T, Ouyang B, Ziaf K, Li H, Zhang J, Lu Y, Ye Z (2012) SpUSP, an annexin-interacting universal stress protein, enhances drought tolerance in tomato. J Exp Bot 63(15): 5593–5606

Luo M, Liu J, Lee RD, Scully BT, Guo B (2010) Monitoring the expression of maize genes in developing kernels under drought stress using oligo-microarray. J Integr Plant Biol 52(12): 1059–1074

Ma Y, Szostkiewicz I, Korte A, Moes D, Yang Y, Christmann A, Grill E (2009) Regulators of PP2C phosphatase activity function as abscisic acid sensors. Science 324(5930):1064–1068

Macas J, Lambert GM, Dolezel D, Galbraith DW (1998) Nuclear expressed sequence tag(NEST) analysis: A novel means to study transcription through amplification of nuclear RNA. Cytometry 33(4):460–468

Mahajan S, Tuteja N (2005) Cold, salinity and drought stresses: an overview. Arch Biochem Biophys 444(2):139–158

Manickavelu A, Kawaura K, Oishi K, Shin T, Kohara Y, Yahiaoui N, Keller B, Abe R, Suzuki A, Nagayama T (2012) Comprehensive functional analyses of expressed sequence tags in common wheat (Triticum aestivum). DNA Res 19(2):165–177

Mantri NL, Ford R, Coram TE, Pang EC (2010) Evidence of unique and shared responses to major biotic and abiotic stresses in chickpea. Environ Exp Bot 69(3):286–292

Marioni JC, Mason CE, Mane SM, Stephens M, Gilad Y (2008) RNA-seq: an assessment of technical reproducibility and comparison with gene expression arrays. Genome Res 18(9):1509–1517

Maruyama K, Takeda M, Kidokoro S, Yamada K, Sakuma Y, Urano K, Fujita M, Yoshiwara K, Matsukura S, Morishita Y (2009) Metabolic pathways involved in cold acclimation identified by integrated analysis of metabolites and transcripts regulated by DREB1A and DREB2A. Plant Physiol 150(4):1972–1980

Matsumura H, Nirasawa S, Terauchi R (1999) Transcript profiling in rice (Oryza sativa L.) seedlings using serial analysis of gene expression (SAGE). Plant J 20(6):719–726

Matsumura H, Reich S, Ito A, Saitoh H, Kamoun S, Winter P, Kahl G, Reuter M, Krüger DH, Terauchi R (2003) Gene expression analysis of plant host–pathogen interactions by SuperSAGE. Proc Natl Acad Sci 100(26):15718–15723

Matsumura H, Ito A, Saitoh H, Winter P, Kahl G, Reuter M, Krüger DH, Terauchi R (2005) SuperSAGE. Cell Microbiol 7(1):11–18

Matsumura H, Yoshida K, Luo S, Kimura E, Fujibe T, Albertyn Z, Barrero RA, Krüger DH, Kahl G, Schroth GP (2010) High-throughput SuperSAGE for digital gene expression analysis of multiple samples using next generation sequencing. PLoS One 5(8):e12010

McCallum CM, Comai L, Greene EA, Henikoff S (2000) Targeting induced local lesions in genomes (TILLING) for plant functional genomics. Plant Physiol 123(2):439–442

McClung CR, Davis SJ (2010) Ambient thermometers in plants: from physiological outputs towards mechanisms of thermal sensing. Curr Biol 20(24):R1086–R1092

McCue KF, Hanson AD (1990) Drought and salt tolerance: towards understanding and application. Trends Biotechnol 8:358–362

McGinnis KM (2010) RNAi for functional genomics in plants. Brief Funct Genomics 9(2): 111–117

McInnis SM, Desikan R, Hancock JT, Hiscock SJ (2006) Production of reactive oxygen species and reactive nitrogen species by angiosperm stigmas and pollen: potential signalling crosstalk? New Phytol 172(2):221–228

McKersie BD, Bowley SR (1997) Active oxygen and freezing tolerance in transgenic plants. In: Li PH, Chen THH (eds) Plant cold hardiness. Plenum, New York, pp 203–214

Melcher K, Ng L-M, Zhou XE, Soon F-F, Xu Y, Suino-Powell KM, Park S-Y, Weiner JJ, Fujii H, Chinnusamy V (2009) A gate–latch–lock mechanism for hormone signalling by abscisic acid receptors. Nature 462(7273):602–608

Mette M, Aufsatz W, Van der Winden J, Matzke M, Matzke A (2000) Transcriptional silencing and promoter methylation triggered by double-stranded RNA. EMBO J 19(19):5194–5201

Miki D, Shimamoto K (2004) Simple RNAi vectors for stable and transient suppression of gene function in rice. Plant Cell Physiol 45(4):490–495

Miller G, Schlauch K, Tam R, Cortes D, Torres MA, Shulaev V, Dangl JL, Mittler R (2009) The plant NADPH oxidase RBOHD mediates rapid systemic signaling in response to diverse stimuli. Sci Signal 2(84):ra45

Mishkind M, Vermeer JE, Darwish E, Munnik T (2009) Heat stress activates phospholipase D and triggers PIP2 accumulation at the plasma membrane and nucleus. Plant J 60(1):10–21

Mittler R, Blumwald E (2010) Genetic engineering for modern agriculture: challenges and perspectives. Annu Rev Plant Biol 61:443–462

Mittler R, Vanderauwera S, Gollery M, Van Breusegem F (2004) Reactive oxygen gene network of plants. Trends Plant Sci 9(10):490–498

Mittler R, Vanderauwera S, Suzuki N, Miller G, Tognetti VB, Vandepoele K, Gollery M, Shulaev V, Van Breusegem F (2011) ROS signaling: the new wave? Trends Plant Sci 16(6):300–309

Mittler R, Finka A, Goloubinoff P (2012) How do plants feel the heat? Trends Biochem Sci 37(3):118–125

Molina C, Rotter B, Horres R, Udupa S, Besser B, Bellarmino L, Baum M, Matsumura H, Terauchi R, Kahl G, Winter P (2008) SuperSAGE: the drought stress-responsive transcriptome of chickpea roots. BMC Genomics 9(1):553

Molina C, Zaman-Allah M, Khan F, Fatnassi N, Horres R, Rotter B, Steinhauer D, Amenc L, Drevon J-J, Winter P, Kahl G (2011) The salt-responsive transcriptome of chickpea roots and nodules via deepSuperSAGE. BMC Plant Biol 11(1):31

Møller IM, Sweetlove LJ (2010) ROS signalling–specificity is required. Trends Plant Sci 15(7):370–374

Monshausen G, Bibikova T, Messerli M, Shi C, Gilroy S (2007) Oscillations in extracellular pH and reactive oxygen species modulate tip growth of Arabidopsis root hairs. Proc Natl Acad Sci 104(52):20996–21001

Mustilli A-C, Merlot S, Vavasseur A, Fenzi F, Giraudat J (2002) Arabidopsis OST1 protein kinase mediates the regulation of stomatal aperture by abscisic acid and acts upstream of reactive oxygen species production. Plant Cell 14(12):3089–3099

Nagalakshmi U, Wang Z, Waern K, Shou C, Raha D, Gerstein M, Snyder M (2008) The transcriptional landscape of the yeast genome defined by RNA sequencing. Science 320(5881): 1344–1349

Nakano M, Nobuta K, Vemaraju K, Tej SS, Skogen JW, Meyers BC (2006) Plant MPSS databases: signature-based transcriptional resources for analyses of mRNA and small RNA. Nucleic Acids Res 34(suppl 1):D731–D735

Negrão S, Cecília Almadanim M, Pires IS, Abreu IA, Maroco J, Courtois B, Gregorio GB, McNally KL, Margarida Oliveira M (2013) New allelic variants found in key rice salt-tolerance genes: an association study. Plant Biotechnol J 11(1):87–100

Nielsen KL, Høgh AL, Emmersen J (2006) DeepSAGE—digital transcriptomics with high sensitivity, simple experimental protocol and multiplexing of samples. Nucleic Acids Res 34(19):e133

Nishimura N, Hitomi K, Arvai AS, Rambo RP, Hitomi C, Cutler SR, Schroeder JI, Getzoff ED (2009) Structural mechanism of abscisic acid binding and signaling by dimeric PYR1. Science 326(5958):1373–1379

Nishimura N, Sarkeshik A, Nito K, Park SY, Wang A, Carvalho PC, Lee S, Caddell DF, Cutler SR, Chory J (2010) PYR/PYL/RCAR family members are major in vivo ABI1 protein phosphatase 2C-interacting proteins in Arabidopsis. Plant J 61(2):290–299

Niu X, Bressan RA, Hasegawa PM, Pardo JM (1995) Ion homeostasis in NaCl stress environments. Plant Physiol 109(3):735

Novillo F, Alonso JM, Ecker JR, Salinas J (2004) CBF2/DREB1C is a negative regulator of CBF1/DREB1B and CBF3/DREB1A expression and plays a central role in stress tolerance in Arabidopsis. Proc Natl Acad Sci U S A 101(11):3985–3990

Nunes AS, Vianna G, Cuneo F, Amaya-Farfán J, Capdeville G, Rech E, Aragão FL (2006) RNAi-mediated silencing of the myo-inositol-1-phosphate synthase gene (GmMIPS1) in transgenic soybean inhibited seed development and reduced phytate content. Planta 224(1):125–132. doi:10.1007/s00425-005-0201-0

O'Brien JA, Daudi A, Finch P, Butt VS, Whitelegge JP, Souda P, Ausubel FM, Bolwell GP (2012) A peroxidase-dependent apoplastic oxidative burst in cultured Arabidopsis cells functions in MAMP-elicited defense. Plant Physiol 158(4):2013–2027

Odell JT, Nagy F, Chua N-H (1985) Identification of DNA sequences required for activity of the cauliflower mosaic virus 35S promoter. Nature 313(6005):810–812

Pandey S, Nelson DC, Assmann SM (2009) Two novel GPCR-type g proteins are abscisic acid receptors in *Arabidopsis*. Cell 136(1):136–148

Papdi C, Ábrahám E, Joseph MP, Popescu C, Koncz C, Szabados L (2008) Functional identification of Arabidopsis stress regulatory genes using the controlled cDNA overexpression system. Plant Physiol 147(2):528–542

Parinov S, Sevugan M, Ye D, Yang W-C, Kumaran M, Sundaresan V (1999) Analysis of flanking sequences from dissociation insertion lines: a database for reverse genetics in Arabidopsis. Plant Cell 11(12):2263–2270

Park S-Y, Fung P, Nishimura N, Jensen DR, Fujii H, Zhao Y, Lumba S, Santiago J, Rodrigues A, Chow TF et al (2009) Abscisic acid inhibits type 2C protein phosphatases via the PYR/PYL family of START proteins. Science 324(5930):1068–1071

Pollock JD (2002) Gene expression profiling: methodological challenges, results, and prospects for addiction research. Chem Phys Lipids 121(1):241–256

Pospíšilová J (2003) Participation of phytohormones in the stomatal regulation of gas exchange during water stress. Biol Planta 46(4):491–506

Qin D, Wu H, Peng H, Yao Y, Ni Z, Li Z, Zhou C, Sun Q (2008) Heat stress-responsive transcriptome analysis in heat susceptible and tolerant wheat (Triticum aestivum L.) by using Wheat Genome Array. BMC Genomics 9(1):432

Qiu Q-S, Guo Y, Dietrich MA, Schumaker KS, Zhu J-K (2002) Regulation of SOS1, a plasma membrane Na+/H+ exchanger in Arabidopsis thaliana, by SOS2 and SOS3. Proc Natl Acad Sci 99(12):8436–8441

Radhamony RN, Mohan Prasad A, Srinivasan R (2005) T-DNA insertional mutagenesis in Arabidopsis: a tool for functional genomics. Electron J Biotechnol 8(1):82–106

Ranjan A, Pandey N, Lakhwani D, Dubey NK, Pathre UV, Sawant SV (2012) Comparative transcriptomic analysis of roots of contrasting Gossypium herbaceum genotypes revealing adaptation to drought. BMC Genomics 13(1):680

Reinartz J, Bruyns E, Lin J-Z, Burcham T, Brenner S, Bowen B, Kramer M, Woychik R (2002) Massively parallel signature sequencing (MPSS) as a tool for in-depth quantitative gene expression profiling in all organisms. Brief Funct Genomic Proteomic 1(1):95–104

Rensink WA, Buell CR (2005) Microarray expression profiling resources for plant genomics. Trends Plant Sci 10(12):603–609

Rhoads DM, Umbach AL, Subbaiah CC, Siedow JN (2006) Mitochondrial reactive oxygen species. Contribution to oxidative stress and interorganellar signaling. Plant Physiol 141(2): 357–366

Ribaut J-M, Hoisington D, Deutsch J, Jiang C, Gonzalez-de-Leon D (1996) Identification of quantitative trait loci under drought conditions in tropical maize. 1. Flowering parameters and the anthesis-silking interval. Theor Appl Genet 92(7):905–914

Ribaut J-M, Jiang C, Gonzalez-de-Leon D, Edmeades G, Hoisington D (1997) Identification of quantitative trait loci under drought conditions in tropical maize. 2. Yield components and marker-assisted selection strategies. Theor Appl Genet 94(6–7):887–896

Rizhsky L, Liang H, Shuman J, Shulaev V, Davletova S, Mittler R (2004) When defense pathways collide. The response of Arabidopsis to a combination of drought and heat stress. Plant Physiol 134(4):1683–1696

Rodrigues A, Adamo M, Crozet P, Margalha L, Confraria A, Martinho C, Elias A, Rabissi A, Lumbreras V, González-Guzmán M (2013) ABI1 and PP2CA phosphatases are negative regulators of Snf1-related protein kinase1 signaling in Arabidopsis. Plant Cell 25(10):3871–3884

Ruelland E, Zachowski A (2010) How plants sense temperature. Environ Exp Bot 69(3):225–232

Rus A, Yokoi S, Sharkhuu A, Reddy M, B-h L, Matsumoto TK, Koiwa H, Zhu J-K, Bressan RA, Hasegawa PM (2001) AtHKT1 is a salt tolerance determinant that controls Na+ entry into plant roots. Proc Natl Acad Sci 98(24):14150–14155

Sagi M, Davydov O, Orazova S, Yesbergenova Z, Ophir R, Stratmann JW, Fluhr R (2004) Plant respiratory burst oxidase homologs impinge on wound responsiveness and development in Lycopersicon esculentum. Plant Cell 16(3):616–628

Saha S, Sparks AB, Rago C, Akmaev V, Wang CJ, Vogelstein B, Kinzler KW, Velculescu VE (2002) Using the transcriptome to annotate the genome. Nat Biotechnol 20(5):508–512

Saidi Y, Finka A, Muriset M, Bromberg Z, Weiss YG, Maathuis FJ, Goloubinoff P (2009) The heat shock response in moss plants is regulated by specific calcium-permeable channels in the plasma membrane. Plant Cell 21(9):2829–2843

Salekdeh G, Siopongco J, Wade LJ, Ghareyazie B, Bennett J (2002) Proteomic analysis of rice leaves during drought stress and recovery. Proteomics 2(9):1131–1145

Santiago J, Dupeux F, Round A, Antoni R, Park S-Y, Jamin M, Cutler SR, Rodriguez PL, Márquez JA (2009) The abscisic acid receptor PYR1 in complex with abscisic acid. Nature 462(7273):665–668

Schena M, Shalon D, Davis RW, Brown PO (1995) Quantitative monitoring of gene expression patterns with a complementary DNA microarray. Science 270(5235):467–470

Schwartz SH, Tan BC, Gage DA, Zeevaart JA, McCarty DR (1997) Specific oxidative cleavage of carotenoids by VP14 of maize. Science 276(5320):1872–1874

Seki M, Narusaka M, Ishida J, Nanjo T, Fujita M, Oono Y, Kamiya A, Nakajima M, Enju A, Sakurai T (2002) Monitoring the expression profiles of 7000 Arabidopsis genes under drought, cold and high-salinity stresses using a full-length cDNA microarray. Plant J 31(3):279–292

Senthil-Kumar M, Mysore KS (2011) New dimensions for VIGS in plant functional genomics. Trends Plant Sci 16(12):656–665. doi:http://dx.doi.org/10.1016/j.tplants.2011.08.006

Shapiguzov A, Vainonen JP, Wrzaczek M, Kangasjärvi J (2012) ROS-talk–how the apoplast, the chloroplast, and the nucleus get the message through. Front Plant Sci 3

Shinozaki K, Yamaguchi-Shinozaki K (2007) Gene networks involved in drought stress response and tolerance. J Exp Bot 58(2):221–227

Sreenivasulu N, Sopory S, Kavi Kishor P (2007) Deciphering the regulatory mechanisms of abiotic stress tolerance in plants by genomic approaches. Gene 388(1):1–13

Steponkus PL, Uemura M, Webb MS (1993) A contrast of the cryostability of the plasma membrane of winter rye and spring oat-two species that widely differ in their freezing tolerance and plasma membrane lipid composition. In: Steponkus PL (ed) Advances in low-temperature biology, vol 2. JAI, London, pp 211–312

Sunkar R, Chinnusamy V, Zhu J, Zhu J-K (2007) Small RNAs as big players in plant abiotic stress responses and nutrient deprivation. Trends Plant Sci 12(7):301–309

Suzuki N, Bajad S, Shuman J, Shulaev V, Mittler R (2008) The transcriptional co-activator MBF1c is a key regulator of thermotolerance in Arabidopsis thaliana. J Biol Chem 283(14): 9269–9275

Suzuki N, Miller G, Morales J, Shulaev V, Torres MA, Mittler R (2011) Respiratory burst oxidases: the engines of ROS signaling. Curr Opin Plant Biol 14(6):691–699

Suzuki N, Koussevitzky S, Mittler R, Miller G (2012) ROS and redox signalling in the response of plants to abiotic stress. Plant Cell Environ 35(2):259–270

Takeda S, Gapper C, Kaya H, Bell E, Kuchitsu K, Dolan L (2008) Local positive feedback regulation determines cell shape in root hair cells. Science 319(5867):1241–1244

Teige M, Scheikl E, Eulgem T, Dóczi R, Ichimura K, Shinozaki K, Dangl JL, Hirt H (2004) The MKK2 pathway mediates cold and salt stress signaling in *Arabidopsis*. Mol Cell 15(1): 141–152

Thakur P, Kumar S, Malik JA, Berger JD, Nayyar H (2010) Cold stress effects on reproductive development in grain crops: an overview. Environ Exp Bot 67(3):429–443

Till BJ, Reynolds SH, Weil C, Springer N, Burtner C, Young K, Bowers E, Codomo CA, Enns LC, Odden AR (2004) Discovery of induced point mutations in maize genes by TILLING. BMC Plant Biol 4(1):12

Torres MA, Dangl JL (2005) Functions of the respiratory burst oxidase in biotic interactions, abiotic stress and development. Curr Opin Plant Biol 8(4):397–403

Tougane K, Komatsu K, Bhyan SB, Sakata Y, Ishizaki K, Yamato KT, Kohchi T, Takezawa D (2010) Evolutionarily conserved regulatory mechanisms of abscisic acid signaling in land plants: characterization of ABSCISIC ACID INSENSITIVE1-like type 2C protein phosphatase in the liverwort Marchantia polymorpha. Plant Physiol 152(3):1529–1543

Uemura M, Steponkus PL (1997) Effect of cold acclimation on membrane lipid composition and freeze-induced membrane destabilization. In: Li PH, Chen THH (eds) Plant cold hardiness. Plenum, New York, pp 171–179

Uozumi N, Kim EJ, Rubio F, Yamaguchi T, Muto S, Tsuboi A, Bakker EP, Nakamura T, Schroeder JI (2000) The Arabidopsis HKT1 gene homolog mediates inward Na+ currents in Xenopus laevis oocytes and Na+ uptake in Saccharomyces cerevisiae. Plant Physiol 122(4):1249–1260

Urano K, Kurihara Y, Seki M, Shinozaki K (2010) 'Omics' analyses of regulatory networks in plant abiotic stress responses. Curr Opin Plant Biol 13(2):132–138

Utsumi Y, Tanaka M, Morosawa T, Kurotani A, Yoshida T, Mochida K, Matsui A, Umemura Y, Ishitani M, Shinozaki K (2012) Transcriptome analysis using a high-density oligomicroarray under drought stress in various genotypes of cassava: an important tropical crop. DNA Res 19(4):335–345

Velculescu VE, Zhang L, Vogelstein B, Kinzler KW (1995) Serial analysis of gene expression. Science 270(5235):484–486

Verdel A, Vavasseur A, Le Gorrec M, Touat-Todeschini L (2009) Common themes in siRNA-mediated epigenetic silencing pathways. Int J Dev Biol 53(2):245–257

Vij S, Tyagi AK (2007) Emerging trends in the functional genomics of the abiotic stress response in crop plants. Plant Biotechnol J 5(3):361–380

Vodkin LO, Khanna A, Shealy R, Clough SJ, Gonzalez DO, Philip R, Zabala G, Thibaud-Nissen F, Sidarous M, Strömvik MV (2004) Microarrays for global expression constructed with a low redundancy set of 27,500 sequenced cDNAs representing an array of developmental stages and physiological conditions of the soybean plant. BMC Genomics 5(1):73

Wahid A, Gelani S, Ashraf M, Foolad M (2007) Heat tolerance in plants: an overview. Environ Exp Bot 61(3):199–223

Walia H, Wilson C, Condamine P, Liu X, Ismail AM, Zeng L, Wanamaker SI, Mandal J, Xu J, Cui X (2005) Comparative transcriptional profiling of two contrasting rice genotypes under salinity stress during the vegetative growth stage. Plant Physiol 139(2):822–835

Wang W, Vinocur B, Altman A (2003) Plant responses to drought, salinity and extreme temperatures: towards genetic engineering for stress tolerance. Planta 218(1):1–14

Wang W, Vinocur B, Shoseyov O, Altman A (2004) Role of plant heat-shock proteins and molecular chaperones in the abiotic stress response. Trends Plant Sci 9(5):244–252

Wen J-Q, Oono K, Imai R (2002) Two novel mitogen-activated protein signaling components, OsMEK1 and OsMAP1, are involved in a moderate low-temperature signaling pathway in rice. Plant Physiol 129(4):1880–1891

Wienkoop S, Morgenthal K, Wolschin F, Scholz M, Selbig J, Weckwerth W (2008) Integration of metabolomic and proteomic phenotypes analysis of data covariance dissects starch and RFO metabolism from low and high temperature compensation response in Arabidopsis thaliana. Mol Cell Proteomics 7(9):1725–1736

Wohlbach DJ, Quirino BF, Sussman MR (2008) Analysis of the Arabidopsis histidine kinase ATHK1 reveals a connection between vegetative osmotic stress sensing and seed maturation. Plant Cell 20(4):1101–1117

Wrzaczek M, Brosché M, Kangasjärvi J (2013) ROS signaling loops—production, perception, regulation. Curr Opin Plant Biol 16(5):575–582

Xia Y, Suzuki H, Borevitz J, Blount J, Guo Z, Patel K, Dixon RA, Lamb C (2004) An extracellular aspartic protease functions in Arabidopsis disease resistance signaling. EMBO J 23(4): 980–988

Xiong L, Zhu J-K (2003) Regulation of abscisic acid biosynthesis. Plant Physiol 133(1):29–36

Xin Z, Wang ML, Barkley NA, Burow G, Franks C, Pederson G, Burke J (2008) Applying genotyping (TILLING) and phenotyping analyses to elucidate gene function in a chemically induced sorghum mutant population. BMC Plant Biology 8 (1):103

Yamaguchi-Shinozaki K, Shinozaki K (1994) A novel cis-acting element in an Arabidopsis gene is involved in responsiveness to drought, low-temperature, or high-salt stress. Plant Cell 6(2):251–264

Yamaguchi-Shinozaki K, Shinozaki K (2005) Organization of *cis*-acting regulatory elements in osmotic-and cold-stress-responsive promoters. Trends Plant Sci 10(2):88–94

Yamamoto K, Sasaki T (1997) Large-scale EST sequencing in rice. Plant Mol Biol 35(1–2): 135–144

Yamamoto YY, Tsuhara Y, Gohda K, Suzuki K, Matsui M (2003) Gene trapping of the Arabidopsis genome with a firefly luciferase reporter. Plant J 35(2):273–283

Yan S, Tang Z, Su W, Sun W (2005) Proteomic analysis of salt stress-responsive proteins in rice root. Proteomics 5(1):235–244

Yoshida R, Hobo T, Ichimura K, Mizoguchi T, Takahashi F, Aronso J, Ecker JR, Shinozaki K (2002) ABA-activated SnRK2 protein kinase is required for dehydration stress signaling in Arabidopsis. Plant Cell Physiol 43(12):1473–1483

Yu H, Chen X, Hong Y-Y, Wang Y, Xu P, Ke S-D, Liu H-Y, Zhu J-K, Oliver DJ, Xiang C-B (2008) Activated expression of an Arabidopsis HD-START protein confers drought tolerance with improved root system and reduced stomatal density. Plant Cell 20(4):1134–1151

Yu S, Liao F, Wang F, Wen W, Li J, Mei H, Luo L (2012) Identification of rice transcription factors associated with drought tolerance using the ecotilling method. PLoS One 7(2):e30765

Yuan L, Liu X, Luo M, Yang S, Wu K (2013) Involvement of histone modifications in plant abiotic stress responses. J Integr Plant Biol 55(10):892–901

Zamore PD, Tuschl T, Sharp PA, Bartel DP (2000) RNAi: double-stranded RNA directs the ATP-dependent cleavage of mRNA at 21 to 23 nucleotide intervals. Cell 101(1):25–33

Zhang H, Sreenivasulu N, Weschke W, Stein N, Rudd S, Radchuk V, Potokina E, Scholz U, Schweizer P, Zierold U (2004) Large-scale analysis of the barley transcriptome based on expressed sequence tags. Plant J 40(2):276–290

Zhang W, Zhou R-G, Gao Y-J, Zheng S-Z, Xu P, Zhang S-Q, Sun D-Y (2009) Molecular and genetic evidence for the key role of AtCaM3 in heat-shock signal transduction in Arabidopsis. Plant Physiol 149(4):1773–1784

Zinn KE, Tunc-Ozdemir M, Harper JF (2010) Temperature stress and plant sexual reproduction: uncovering the weakest links. J Exp Bot 61(7):1959–1968

Chapter 3
Investigation of Plant Abiotic Stress Tolerance by Proteomics and Phosphoproteomics

Maik Böhmer

Abstract Abiotic stresses, including drought and salt, are among the most devastating threats for modern agriculture. Overcoming these threats by modern breeding technologies requires an intricate understanding of underlying signaling mechanisms in plants. This book chapter summarizes major achievements and novel technologies and approaches to elucidate plant abiotic stress responses using proteomics and phosphoproteomics. Proteomic and phosphoproteomic studies of drought and salt stress in model and major crop plants have seen a boost over recent years, as mass spectrometry-based techniques advanced and were more widely available to plant scientists. The key proteins and mechanisms identified in these studies present leads for successful breeding of abiotic stress tolerance in plants.

Keywords Protein phosphorylation • Phosphoproteomics • Proteomics • Quantitative mass spectrometry

3.1 Introduction

Increase in human population, a change in diet in developing countries toward meat and dairy products, and an increasing number of crops turned into biofuel require an enhancement in crop production over the next 40 years (Ray et al. 2013; Tester and Langridge 2010). Current growth rates suggest that the demand on crop production in 2050 cannot be achieved without a significant boost in productivity (Ray et al. 2013). Crop productivity can be improved in two ways, either by increasing the maximum crop yield or by preventing crop loss due to abiotic and biotic stresses, primarily drought, high salinity, and pathogens.

Most crops reach only about 30 % of the possible yield based on their genetic potential (Boyer 1982; Ciais et al. 2005). The main cause of yield loss is abiotic

M. Böhmer (✉)
Institut für Biologie und Biotechnologie der Pflanzen, Westfälische Wilhelms Universität, Schlossplatz 7, 48149 Münster, Germany
e-mail: m.boehmer@uni-muenster.de

G.K. Pandey (ed.), *Elucidation of Abiotic Stress Signaling in Plants*,
DOI 10.1007/978-1-4939-2211-6_3

stress, which affects over 90 % of rural land (Cramer et al. 2011). Nonoptimal temperatures, insufficient illumination, inadequate water supply, and scarce nutrient availability all lead to agricultural losses. Among the abiotic stress factors, drought stress is one of the most important causes of crop loss (Boyer 1982). Drought stress affects especially fields that are nonirrigated. Today, FAO data show that only 20 % of the world's croplands are irrigated, but those lands yield 40 % of the global harvest (Garces-Restrepo and Muñoz 2007). Salt stress affects more than 6 % of total land and about 20 % of irrigated land (Munns and Tester 2008). It was estimated that soil salinization may result in land loss of up to 50 % of all arable land by 2050 (Wang et al. 2003).

In order to improve crop productivity, breeders have to understand how plants respond to abiotic stress at the molecular level. In the last decade, gene expression analyses have provided researchers with a general understanding of plant responses to abiotic stress (Hirayama and Shinozaki 2010). Transcriptional analyses, however, do not consider the effect of translational and posttranslational modifications that can significantly change abundance as well as activity of proteins and that can only be detected by proteomic techniques. In addition, several studies have demonstrated that changes in the transcript level often do not correlate well to actual changes in protein abundance (Böhmer and Schroeder 2011; Piques et al. 2009).

3.2 Protein Expression Changes in Response to Abiotic Stress

3.2.1 2D Gel-Based Proteomics

Proteomic studies of plant responses to abiotic stress have primarily been studied in the model plants rice (*Oryza sativa*) and *Arabidopsis thaliana* but recently also in other crop plants. Studying crop plants is important since proteomic changes as a result of adaptation to abiotic stress in model organisms may not be widely applicable to other crop species (Moller and Tester 2007).

The most commonly used technique for quantitative proteomics over the last decade was 2D electrophoresis (2DE) of proteins followed by mass spectrometry. The great advantage of 2DE is that it uncouples quantification of protein changes from the identification of proteins. Quantitation of protein abundance is performed by comparing stained protein spot densities (Fig. 3.1). This reduces costs of subsequent mass spectrometric analyses as it only requires identifying differentially regulated protein spots.

The quality of 2DE greatly depends on sensitivity and linearity of the applied protein dye and uniformity of the 2D gels. To improve on the reproducibility and linearity of 2DE analyses, proteins from treated and control samples were stained with fluorescent dyes of different colors before they were run on the same 2D gel. The difference gel electrophoresis (DIGE) approach facilitates quantitative comparisons that were otherwise hindered by variations in spot migration (Unlu et al.

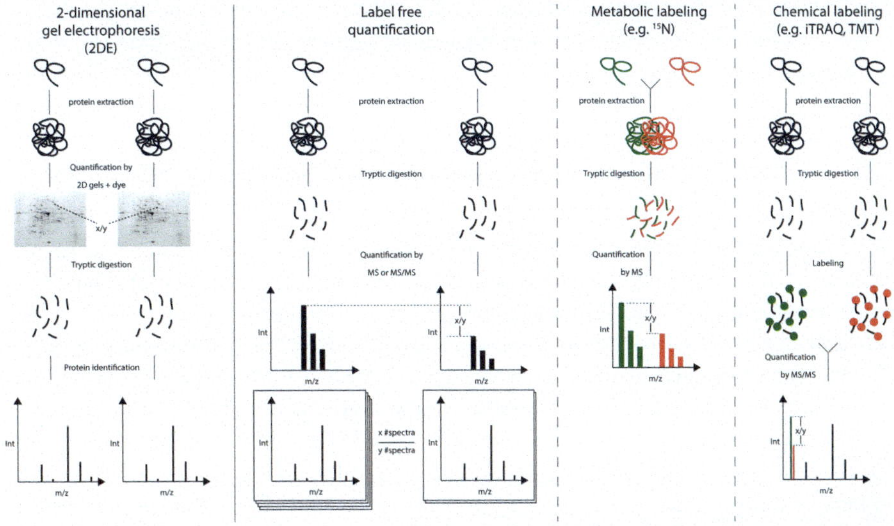

Fig. 3.1 Quantification techniques in plant proteomics. Quantification by two-dimensional gel electrophoresis (2DE) is based on comparison of stained protein spot densities. Proteins that are differentially expressed in different samples are then identified by mass spectrometry. In label-free quantification, proteins are extracted and enzymatically digested, and peptides are analyzed by mass spectrometry in parallel. Protein expression is quantified by precursor ion intensities or by the number of MS/MS spectra per protein. Metabolic labeling allows cell extracts to be mixed at an early stage, and protein extraction and peptide generation are performed together. Protein expression is quantified by precursor intensity. In chemical peptide labeling, proteins are extracted and peptides are generated and labeled in parallel. With isobaric tags, protein expression is quantified by the intensity of reporter ions in MS/MS spectra

1997). 2DE facilitates identification of proteins from non-sequenced or poorly annotated genomes since all peptides of a protein are localized in a single spot that can be analyzed by peptide mass fingerprinting (PMF) and/or by tandem mass spectrometry.

While transcriptomic techniques can detect and quantify millions of RNA molecules simultaneously, using these techniques requires knowledge of the genome sequence and gene models. Proteomic studies on the other hand are far more challenging due to different biochemical and physical properties of proteins, their differences in abundance, and no possible amplification mechanisms (Barkla et al. 2013a). Before next-generation sequencing (NGS) techniques were developed and genome sequences of most crop species were available, it was simpler to perform 2DE-based analyses of protein changes than to perform high-throughput transcriptomic analyses (Komatsu and Tanaka 2005).

Among the disadvantages of 2DE are limited resolution and dynamic range. Limited resolution can lead to the presence of multiple proteins per spot. In modern 2DE 10,000 protein spots representing 1,000 different proteins can be visualized on a single 2D gel (Gorg et al. 2009). Recent studies show that up to 60 % of resolved spots consist of multiple proteins (Lim et al. 2003; Campostrini et al. 2005; Yang et al. 2007). Therefore, quantifications by 2DE always need independent validation by western blot or ion counting. A further disadvantage is that membrane proteins are difficult to solubilize for 2DE.

3.2.2 Identification and Quantification by Shotgun Proteomics

Three developments have significantly altered quantitative proteomics: the higher availability of genome sequences for crop plants and other non-model plants based on recent genome projects, gel-free shotgun techniques to quantify proteins based on MS data, and computer-assisted identification of peptides and proteins from mass spectra, allowing the automated identification and quantification of thousands of peptides.

Genome sequences are now available for the model plants *Arabidopsis thaliana* (Arabidopsis Genome 2000) and rice (Sequencing Project International Rice 2005), but also for crop plant genomes, including maize (Schnable et al. 2009), potato (Potato Genome Sequencing Consortium 2011), tomato (Tomato Genome 2012), and others (summarized in Feuillet et al. 2011).

In addition to the genome sequence, proteomics requires a high-quality gene annotation (Vertommen et al. 2011; Carpentier et al. 2008). Often gene models are derived from EST libraries or are deferred from related, well-annotated genomes, like *Arabidopsis* or rice. In proteogenomics, high-quality peptide sequences from proteomic datasets are used to support annotation of new plant genomes (Castellana et al. 2008; Baerenfaller et al. 2008; Helmy et al. 2011; May et al. 2008). Also, the high degree of polyploidy in some crops, e.g., wheat with a hexaploid genome, increases the level of difficulty to identify individual protein isoforms. Novel technologies, like mass precursor alignment (MAPA), have been used to achieve isoform-specific protein identification in potato (Hoehenwarter et al. 2011; Ng et al. 2012; Hu et al. 2011). Due to these limitations, shotgun approaches make up only about 5 % of all proteomic studies on crop plants (Barkla et al. 2013b). Nevertheless, these approaches have been successfully applied to a number of crop species (Kosova et al. 2011; Zhang et al. 2012; Vanderschuren et al. 2013; Pechanova et al. 2013; Agrawal et al. 2013).

3.2.2.1 Targeted Analyses by Prefractionation

The majority of proteomic studies using shotgun proteomics or 2DE on whole seedlings predominantly identified highly abundant cytosolic proteins. This is in part due to the extraction methods but also due to limited dynamic range. Prefractionation on the tissue, cell type, or protein level allows for a more in-depth analysis of proteins, including transcription factors and kinases. Fractionation can be achieved on the organ, tissue, or cell level, by enriching organelles or subcellular compartments or by fractionation on the protein level. For protein prefractionation a number of techniques have been developed, including chromatography, fractional centrifugation, free-flow electrophoresis (FFE), and protein depletion. A recent study in

Arabidopsis thaliana identified 13,000 unique proteins by mass spectrometry, about one third of all annotated genes.

Organelles and individual cell types can respond very differently to abiotic stresses. Salinity response, for instance, in *A. thaliana* is regulated in a tissue- and cell-specific manner (Dinneny et al. 2008; Geng et al. 2013). A number of studies, therefore, specifically focused on subcellular regions or organelles. Initially, 2DE has been used to study organelle proteomes (Lilley and Dupree 2006; Bae et al. 2003) and is slowly replaced by shotgun techniques. Although for non-sequenced organisms 2DE is still in use, as for wheat mitochondria (Jacoby et al. 2010). Besides common stress-responsive proteins, studies on the nuclear proteomes of *Arabidopsis* and rice identified transcription factors that were regulated in response to cold and dehydration stress (Bae et al. 2003; Choudhary et al. 2009). Also cell suspension cells served as a single cell system to study, e.g., the effect of the drought hormone ABA (Böhmer and Schroeder 2011; Chen et al. 2010).

3.2.2.2 Quantitative Analyses in Shotgun Proteomics

Protein quantification on 2D gels is an absolute quantification approach. The intensity of a stained single protein spot corresponds to the abundance of the protein in the sample. Quantification by mass spectrometry is not quantitative due to great variability in chemical and physical properties of different tryptic peptides. This does not allow for a direct correlation between signal intensity and protein amount. Quantification by MS is therefore a relative quantification, where protein abundance before and after stimulus or between mutant and wild type is compared.

MS-based quantitation can be separated into two categories: labeling or label-free techniques (Schulze and Usadel 2010; Neilson et al. 2011). Techniques differ in their accuracy of quantification. Stable isotope labeling strategies are among the most accurate and provide relative standard deviations below 10 %. Label-free techniques based on peak intensities or extracted ion chromatograms provide relative standard deviation around 30 % and label-free techniques based on spectral counting can generate relative standard deviations up to 50 % (Schulze and Usadel 2010). The choice for a quantification technique depends on costs, accessibility of the desired sample to labeling, accuracy of quantification, the number of samples to compare, and the desired proteome coverage.

Label-free techniques are based on the quantification of ion intensities or spectral counts of identical peptides across various samples (Fig. 3.1). Statistical confidence can be gained by quantification of multiple peptides per protein. Label-free techniques are rather inexpensive, and sample complexity is not increased by combining multiple samples. Therefore, these techniques have a greater dynamic range and proteome coverage. The advantages come at the price of quantification accuracy (Schulze and Usadel 2010).

For label-free techniques, extracted ion chromatograms are generated for each peptide. The resulting chromatograms resemble the elution profile of the peptide from the LC column. The resulting peak volumes are a measure for the abundance of this protein in the sample. Due to different chemical and physical properties, only the intensities of the same peptides can be compared.

Without further bioinformatics analysis, only peptides that have been independently identified in multiple samples can be compared. As LC-MS/MS runs of the same sample generally results in only about a 60 % overlap in identified peptides, coverage drastically decreases when multiple samples are compared. As a solution, protein correlation profiling was developed that aligns total ion chromatograms of different samples. Therefore, peptides that have been identified in a single or few samples can be compared between multiple samples based on their chromatographic behavior (Schulze and Usadel 2010; Chen et al. 2010; Hoehenwarter et al. 2008, 2011).

Most widely used for quantitative proteomics in plants are ^{15}N metabolic labeling approaches (Engelsberger et al. 2006). Seedlings and cell culture are autotrophs and can be grown on ^{15}N-containing salts, resulting in quantitative labeling of the whole proteome (Fig. 3.1). Metabolic labeling has the advantage that treated and control samples can be mixed at an early stage of sample preparation thereby reducing variability. Quantification is performed by extracted ion chromatograms with the added advantage that no precursor alignment is needed since heavy and light forms of the peptide elute at the same time.

Chemical labeling techniques apply the label on the level of peptides or proteins (Fig. 3.1). Peptide labeling includes isobaric tags, commercially available as iTRAQ or TMT. In contrast to metabolic labeling and label-free techniques, quantification occurs at the MS2 level. Due to the isobaric tags, peptides from different samples have the same mass. Upon fragmentation, reporter ions appear in the low mass range, and ratios between reporter ions are a measure for relative quantification (Fig. 3.1). Quantification by isobaric tags has the disadvantage that only a single data point is measured per peptide. Also if the precursor mass window for fragmentation is too wide, additional peptide might become fragmented and add to the reporter ion counts. Isobaric tags therefore often underestimate actual changes in protein abundance.

3.2.3 Changes in Protein Abundance in Response to Abiotic Stress

A number of abiotic stresses were analyzed by proteomics, including cold, heat, waterlogging, drought, salinity, ozone treatment, hypoxia, anoxia, inadequate light, inadequate nutrition, exposure to heavy metal, and wounding (reviewed in Kosova et al. 2011). In the majority of studies, either proteomic compositions of stress-treated and control plants or different genotypes, ecotypes, or closely related plant species, e.g., *Arabidopsis thaliana* and the more salt-tolerant plant *Thellungiella halophila*, were studied (Barkla et al. 2013b).

A few proteins and signaling pathways were found to be commonly regulated among various stress stimuli, including metabolic proteins (carbohydrate, energy, nitrogen metabolism), compatible solutes, and stress-responsive proteins; protein synthesis, protein degradation, and protein turnover as well as reactive oxygen species formation and reactive oxygen species scavenging, including superoxide dismutase, ascorbate peroxidase, and glutathione reductase; and the phenylpropanoid pathway (Kosova et al. 2011; Suzuki et al. 2012; Böhmer and Schroeder 2011). In wheat, the first response to salt stress is a nonspecific response that becomes more specific over the course of weeks (Saqib et al. 2006). Also the initial responses to salt and water stress appear identical, and specificity develops over a period of time (Munns 2002).

Proteomic studies have analyzed a number of salt-responsive expression changes in various plant species mostly by 2DE, sometimes accompanied by DIGE. Protein changes were recorded in the model plants *Arabidopsis* (Lee et al. 2004; Jiang et al. 2007) and rice (Salekdeh et al. 2002; Yan et al. 2005; Chitteti and Peng 2007; Nohzadeh Malakshah et al. 2007; Cheng et al. 2009; Zhang et al. 2009; Li et al. 2010; Nam et al. 2012; Liu et al. 2012), but also in crop plants, including wheat (Peng et al. 2009; Wang et al. 2008; Ge et al. 2012), barley (Sugimoto and Takeda 2009; Witzel et al. 2009), maize (Zörb et al. 2004, 2010), and soybean (Aghaei et al. 2009; Sobhanian et al. 2010). Reviews summarized 905 proteins from 14 plant species (Zhao et al. 2013) or 560 unique salt-responsive proteins in 34 plant species (Zhang et al. 2012).

Proteins differentially expressed in response to salt stress are involved in salt perception and signal transduction, ROS formation or detoxification, calcium signaling, salt exclusion/compartmentalization, protein translation/turnover, cytoskeleton/cell wall, and carbohydrate and energy metabolism (Zhao et al. 2013; Zhang et al. 2012). Interestingly, besides the kinetics of protein changes in response to salt, a comparison of phylogenetically related but more tolerant or susceptible plant species, e.g., *Arabidopsis* and *Thellungiella halophila*, was done. *Thellungiella* showed much less changes in protein expression in response to salt stress than *Arabidopsis* (Pang et al. 2010).

3.3 Phosphorylation Changes in Response to Abiotic Stress

Proteins are not only regulated at the expression level but their activity can also be regulated by posttranslational modifications. Among the most prominent posttranslational modifications that affect protein activity are glycosylation, phosphorylation, and acetylation. When abiotic stress signals are perceived at the plasma membrane, signals often need to be transported to the plant nucleus to induce changes in the protein expression. Likewise, signals perceived in the cytosol may also have to be transported to channels in the plasma membrane. This is often achieved via phosphorylation cascades (Kersten et al. 2009). In plants, primarily serines and threonines are phosphorylated, while tyrosines account for only 2.4 % of all phosphorylations (Nakagami et al. 2010). What makes phosphoproteomic analyses particularly difficult is that phosphorylation stoichiometry can range from 90 to <10 %, all with potential biological effects (Schulze 2010).

3.3.1 *Quantification of Phosphorylation Changes by 2DE*

A number of studies have used 2D gels to quantify changes in phosphorylation in response to abiotic stress. Phosphorylation changes were detected by incorporation of isotope-labeled phosphate (Khan et al. 2005) or fluorescent, phosphoprotein-specific dyes (Zörb et al. 2010; Huang et al. 2011; Chitteti and Peng 2007; Ke et al. 2009; He and Li 2008; Guo et al. 2013). Intensity of the phosphoprotein-specific stain, often in comparison with a general protein stain, is an indicator for the phosphorylation status. Mass spectrometry is then used to identify the underlying protein. The actual phosphopeptides are generally not identified using this technique, and it is therefore difficult to confirm phosphorylation of the identified proteins, particularly if multiple proteins are present in the same spot.

3.3.2 *Shotgun Phosphoproteomics*

Tryptic digests of cell lysate generate a large number of peptides over a wide dynamic range. Phosphopeptides are strongly underrepresented, and, in addition, nonphosphorylated peptides suppress ionization of phosphorylated peptides. Therefore, in order to study phosphopeptides at the proteome level, enrichment is required.

3.3.2.1 Enrichment of Phosphopeptides

Enrichment via metal oxide affinity chromatography (MOAC), primarily via TiO_2, has established itself as the best method to enrich phosphopeptides although other techniques are in development, including electrostatic repulsion hydrophilic interaction chromatography (ERLIC) (Alpert 2008). TiO_2 is highly specific for phosphopeptide enrichments. For individual protein phosphopeptide enrichments, immobilized metal affinity chromatography (IMAC) is also used. Although less specific, it tends to enrich more phosphopeptides per protein. For proteome-wide analyses, phosphopeptides are further separated by strong cation exchange (SCX) chromatography.

Differences in enrichment techniques and equipment generate a bias in the subset of phosphoproteins that is analyzed in each study. This makes comparisons between studies very difficult. Previous large-scale studies have shown that in order to identify regulatory proteins, such as kinases and transcription factors, combinations of enrichment strategies with tissue-specific or subcellular targeting are necessary.

3.3.2.2 Mass Spectrometry of Phosphopeptides

Shotgun proteomics has in most cases replaced 2D gel-based phosphoproteomics. Shotgun proteomics can generate a vast amount of data that has to be analyzed for peptide identification as well as for phosphorylation site localization, which is not always unambiguously possible. Peptides are identified by database searches using database search algorithms, including MASCOT, SEQUEST, or the freeware tools X!Tandem and OMSSA.

The number of phosphorylation sites identified in individual studies has significantly increased over the years. In one of the earliest studies, a total of 283 phosphopeptides has been identified using SCX-IMAC enrichment (Nühse et al. 2004). Later studies increased the number of phosphorylation sites under similar conditions to 1,172 (Benschop et al. 2007), 3,029 (Reiland et al. 2009) and then to 5,143 phosphopeptides (Nakagami et al. 2010), using SCX-MOAC, respectively.

Search algorithms have been designed to identify peptides and not so much the actual phosphorylation site. In peptides with multiple serine, threonines, or tyrosines, a post database search validation step is necessary to statistically validate identified phosphorylation sites. A number of algorithms have been developed to facilitate this process. These include a-score (Beausoleil et al. 2006), PTMprophet (Shteynberg et al. 2012), PTM score (Olsen et al. 2006; Cox and Mann 2008), PhosphoRS (Taus et al. 2011), SloMo (Bailey et al. 2009), or the MASCOT delta score (Savitski et al. 2011). These algorithms calculate statistical probabilities for each potential phosphorylation site in a peptide.

Phosphopeptides identified in large-scale phosphoproteomic studies were deposited in plant phosphorylation databases. Among these are PhosPhAt (Heazlewood et al. 2008; Durek et al. 2010), P^{3}DB (Gao et al. 2009), and RIPP-DB (Nakagami et al. 2010). A combinatorial viewer that gathers information from all three portals is available as MASCP Gator (Joshi et al. 2011).

3.3.2.3 Quantification of Phosphorylation Changes

Quantification techniques that rely on MS data for quantification are well suited for the analysis of phosphorylation changes as they also confirm peptide phosphorylation. Quantification of phosphorylation changes has been performed with a number of techniques, including iTRAQ (Nühse et al. 2007), metabolic labeling (Benschop et al. 2007; Kline et al. 2010), and label-free techniques (Reiland et al. 2009, 2011; Chen et al. 2010; Chang et al. 2012).

Quantitative phosphoproteomic studies have been carried out in response to a stimulus (Chen et al. 2010; Kline et al. 2010) or comparing wild-type and mutant backgrounds (Wang et al. 2013; Umezawa et al. 2013). Most of the phosphoproteomic analyses of abiotic stress responses were carried out in *Arabidopsis*, but also samples from rice, corn, peanuts, and soybean have been analyzed (Rampitsch and Bykova 2012).

3.3.3 Kinase Substrate Identification

Currently, the annotation of the *Arabidopsis thaliana* genome accounts for around 1000 proteins harboring a protein kinase domain, twice as much as in mammals (Dissmeyer and Schnittger 2011). Besides kinases, *Arabidopsis* contains 112 phosphatase-coding genes (Kerk et al. 2002).

A number of kinase families are activated in response to abiotic stress or function in abiotic stress pathways, including CDPKs (Böhmer et al. 2006), CIPKs (Batistic and Kudla 2009; Luan 2009), SnRK2s, and MAPKs (Nakagami et al. 2005). Phosphorylation pathways are activated within minutes of a stimulus (Hsu et al. 2009). Kinases like SnRK2s and CDPKs are activated within 5 or 10 min after stress stimulus (Franz et al. 2013; Kim et al. 2011). The SOS pathway, for example, is a calcium-dependent protein kinase pathway that acts in salt response. SOS3/CBL4 is a calcineurin B-like calcium sensor that senses elevations in intracellular calcium and activates the Ser/Thr protein kinase SOS2/CIPK24 that activates SOS1, a plasma membrane-bound Na^+/H^+ antiporter (Qiu et al. 2002). A number of other protein kinases are differentially expressed in response to salt in various plant species (Peng et al. 2009; Chitteti and Peng 2007; Zörb et al. 2004). Recent phosphoproteomic studies also analyzed phosphorylation changes in response to salt stress (Chitteti and Peng 2007; Zhou et al. 2010; Hsu et al. 2009; Chang et al. 2012; Guo et al. 2013; Khan et al. 2005).

Phosphoproteomics offers a tool to identify kinase substrate pairs. Phosphorylation targets of the STN8 kinase were identified by a comparative phosphoproteomic approach (Reiland et al. 2011). In order to minimize plant adaptation in comparison with mutant and wild-type control plants, kinase-specific inhibitors have been used to specifically inhibit individual kinase isoforms prior to phosphoproteomic analysis (Böhmer and Romeis 2007).

3.4 Outlook and Perspective

The last decades of proteomic and phosphoproteomic studies went through two phases. In the first phase, peptides and phosphopeptides were discovered. In the second phase, relative changes in protein expression and posttranslational modifications were quantified using a number of techniques. In the future, specific quantification of individual marker proteins and phosphopeptides might complement unbiased discovery experiments. For this selected reaction monitoring (SRM) is a technique that can specifically be used to quantify known peptides for abundance or their phosphorylation status (Dubiella et al. 2013). SRM can also be used to measure the phosphorylation stoichiometry in dozens of peptides (Kline-Jonakin et al. 2011).

The information we have on protein phosphorylation in plants is far from complete. Only about 6 % of all phosphorylation sites have been found in at least two independent studies. 83 % of published phosphorylation sites have been identified only once (Durek et al. 2010). The more phosphorylation sites are identified, the better phosphorylation sites can be correlated with respective kinases and better predictions can be made for potential phosphorylation sites and for kinase substrates.

The accumulation of phosphorylation site information facilitates to build hypothesis of signaling events. A task that arises for the future is to develop techniques that identify the phosphorylation sites that have regulatory roles in vivo.

Acknowledgments Research in the author's laboratory is funded by the Deutsche Forschungsgemeinschaft (BO3155-3/1) and the European Space Agency (4000109583-ESA-CORA-GBF-2013-005-Böhmer).

References

Aghaei K, Ehsanpour AA, Shah AH, Komatsu S (2009) Proteome analysis of soybean hypocotyl and root under salt stress. Amino Acids 36(1):91–98

Agrawal GK, Jwa NS, Jung YH, Kim ST, Kim DW, Cho K, Shibato J, Rakwal R (2013) Rice proteomic analysis: sample preparation for protein identification. Methods Mol Biol 956: 151–184. doi:10.1007/978-1-62703-194-3_12

Alpert AJ (2008) Electrostatic repulsion hydrophilic interaction chromatography for isocratic separation of charged solutes and selective isolation of phosphopeptides. Anal Chem 80(1): 62–76. doi:10.1021/ac070997p

Arabidopsis Genome I (2000) Analysis of the genome sequence of the flowering plant Arabidopsis thaliana. Nature 408(6814):796–815. doi:10.1038/35048692

Bae MS, Cho EJ, Choi EY, Park OK (2003) Analysis of the Arabidopsis nuclear proteome and its response to cold stress. Plant J 36(5):652–663

Baerenfaller K, Grossmann J, Grobei MA, Hull R, Hirsch-Hoffmann M, Yalovsky S, Zimmermann P, Grossniklaus U, Gruissem W, Baginsky S (2008) Genome-scale proteomics reveals Arabidopsis thaliana gene models and proteome dynamics. Science 320(5878):938–941. doi:10.1126/science.1157956

Bailey CM, Sweet SM, Cunningham DL, Zeller M, Heath JK, Cooper HJ (2009) SLoMo: automated site localization of modifications from ETD/ECD mass spectra. J Proteome Res 8(4):1965–1971. doi:10.1021/pr800917p

Barkla BJ, Castellanos-Cervantes T, de Leon JL, Matros A, Mock HP, Perez-Alfocea F, Salekdeh GH, Witzel K, Zorb C (2013a) Elucidation of salt stress defense and tolerance mechanisms of crop plants using proteomics—current achievements and perspectives. Proteomics 13 (12–13):1885–1900. doi:10.1002/pmic.201200399

Barkla BJ, Vera-Estrella R, Pantoja O (2013b) Progress and challenges for abiotic stress proteomics of crop plants. Proteomics 13(12–13):1801–1815. doi:10.1002/pmic.201200401

Batistic O, Kudla J (2009) Plant calcineurin B-like proteins and their interacting protein kinases. Biochim Biophys Acta 1793(6):985–992. doi:10.1016/j.bbamcr.2008.10.006

Beausoleil SA, Villen J, Gerber SA, Rush J, Gygi SP (2006) A probability-based approach for high-throughput protein phosphorylation analysis and site localization. Nat Biotechnol 24(10):1285–1292. doi:10.1038/nbt1240

Benschop JJ, Mohammed S, O'Flaherty M, Heck AJ, Slijper M, Menke FL (2007) Quantitative phosphoproteomics of early elicitor signaling in Arabidopsis. Mol Cell Proteomics 6(7):1198–1214. doi:10.1074/mcp.M600429-MCP200

Böhmer M, Romeis T (2007) A chemical-genetic approach to elucidate protein kinase function in planta. Plant Mol Biol 65(6):817–827. doi:10.1007/s11103-007-9245-9

Böhmer M, Schroeder JI (2011) Quantitative transcriptomic analysis of abscisic acid-induced and reactive oxygen species-dependent expression changes and proteomic profiling in Arabidopsis suspension cells. Plant J 67(1):105–118. doi:10.1111/j.1365-313X.2011.04579.x

Böhmer M, Kurth J, Witte CP, Romeis T (2006) Function of plant calcium-dependent protein kinases in the activation of abiotic and pathogen-related stress responses and its potential application in the generation of stress-resistant plants. In: da Silva JA T (ed) Floriculture, ornamental and plant biotechnology: advances and topical issues , vol 3, 1st edn. Global science books, London, pp 367–372

Boyer JS (1982) Plant productivity and environment. Science 218(4571):443–448. doi:10.1126/science.218.4571.443

Campostrini N, Areces LB, Rappsilber J, Pietrogrande MC, Dondi F, Pastorino F, Ponzoni M, Righetti PG (2005) Spot overlapping in two-dimensional maps: a serious problem ignored for much too long. Proteomics 5(9):2385–2395. doi:10.1002/pmic.200401253

Carpentier SC, Panis B, Vertommen A, Swennen R, Sergeant K, Renaut J, Laukens K, Witters E, Samyn B, Devreese B (2008) Proteome analysis of non-model plants: a challenging but powerful approach. Mass Spectrom Rev 27(4):354–377. doi:10.1002/mas.20170

Castellana NE, Payne SH, Shen Z, Stanke M, Bafna V, Briggs SP (2008) Discovery and revision of Arabidopsis genes by proteogenomics. Proc Natl Acad Sci U S A 105(52):21034–21038. doi:10.1073/pnas.0811066106

Chang IF, Hsu JL, Hsu PH, Sheng WA, Lai SJ, Lee C, Chen CW, Hsu JC, Wang SY, Wang LY, Chen CC (2012) Comparative phosphoproteomic analysis of microsomal fractions of Arabidopsis thaliana and Oryza sativa subjected to high salinity. Plant Sci 185–186:131–142. doi:10.1016/j.plantsci.2011.09.009

Chen Y, Hoehenwarter W, Weckwerth W (2010) Comparative analysis of phytohormone-responsive phosphoproteins in Arabidopsis thaliana using TiO2-phosphopeptide enrichment and mass accuracy precursor alignment. Plant J 63(1):1–17. doi:10.1111/j.1365-313X.2010.04218.x

Cheng Y, Qi Y, Zhu Q, Chen X, Wang N, Zhao X, Chen H, Cui X, Xu L, Zhang W (2009) New changes in the plasma-membrane-associated proteome of rice roots under salt stress. Proteomics 9(11):3100–3114

Chitteti BR, Peng Z (2007) Proteome and phosphoproteome differential expression under salinity stress in rice (Oryza sativa) roots. J Proteome Res 6(5):1718–1727

Choudhary MK, Basu D, Datta A, Chakraborty N, Chakraborty S (2009) Dehydration-responsive nuclear proteome of rice (Oryza sativa L.) illustrates protein network, novel regulators of cellular adaptation, and evolutionary perspective. Mol Cell Proteomics 8(7):1579–1598. doi:10.1074/mcp. M800601-MCP200

Ciais P, Reichstein M, Viovy N, Granier A, Ogee J, Allard V, Aubinet M, Buchmann N, Bernhofer C, Carrara A, Chevallier F, De Noblet N, Friend AD, Friedlingstein P, Grunwald T, Heinesch B, Keronen P, Knohl A, Krinner G, Loustau D, Manca G, Matteucci G, Miglietta F, Ourcival JM, Papale D, Pilegaard K, Rambal S, Seufert G, Soussana JF, Sanz MJ, Schulze ED, Vesala T, Valentini R (2005) Europe-wide reduction in primary productivity caused by the heat and drought in 2003. Nature 437(7058):529–533. doi:10.1038/Nature03972

Cox J, Mann M (2008) MaxQuant enables high peptide identification rates, individualized p.p.b.-range mass accuracies and proteome-wide protein quantification. Nat Biotechnol 26(12): 1367–1372. doi:10.1038/nbt.1511

Cramer GR, Urano K, Delrot S, Pezzotti M, Shinozaki K (2011) Effects of abiotic stress on plants: a systems biology perspective. BMC Plant Biol 11:163. doi:10.1186/1471-2229-11-163

Dinneny JR, Long TA, Wang JY, Jung JW, Mace D, Pointer S, Barron C, Brady SM, Schiefelbein J, Benfey PN (2008) Cell identity mediates the response of Arabidopsis roots to abiotic stress. Science 320(5878):942–945. doi:10.1126/science.1153795

Dissmeyer N, Schnittger A (2011) The age of protein kinases. Methods Mol Biol 779:7–52. doi:10.1007/978-1-61779-264-9_2

Dubiella U, Seybold H, Durian G, Komander E, Lassig R, Witte CP, Schulze WX, Romeis T (2013) Calcium-dependent protein kinase/NADPH oxidase activation circuit is required for rapid defense signal propagation. Proc Natl Acad Sci U S A 110(21):8744–8749. doi:10.1073/pnas.1221294110

Durek P, Schmidt R, Heazlewood JL, Jones A, MacLean D, Nagel A, Kersten B, Schulze WX (2010) PhosPhAt: the Arabidopsis thaliana phosphorylation site database. An update. Nucleic Acids Res 38(Database issue):D828–834. doi:10.1093/nar/gkp810

Engelsberger WR, Erban A, Kopka J, Schulze WX (2006) Metabolic labeling of plant cell cultures with K(15)NO_3 as a tool for quantitative analysis of proteins and metabolites. Plant Methods 2:14. doi:10.1186/1746-4811-2-14

Feuillet C, Leach JE, Rogers J, Schnable PS, Eversole K (2011) Crop genome sequencing: lessons and rationales. Trends Plant Sci 16(2):77–88. doi:10.1016/j.tplants.2010.10.005

Franz S, Ehlert B, Liese A, Kurth J, Cazalé A-C, Romeis T (2013) Calcium-dependent protein kinase CPK21 functions in abiotic stress response in Arabidopsis thaliana. Mol Plant 6(270):rs8. doi:10.1093/mp/ssq064

Gao J, Agrawal GK, Thelen JJ, Xu D (2009) P3DB: a plant protein phosphorylation database. Nucleic Acids Res 37(Database issue):D960–962. doi:10.1093/nar/gkn733

Garces-Restrepo C, Muñoz G (2007) Irrigation management transfer, vol 32. Water reports. Food and Agriculture Organization of the United Nations

Ge P, Ma C, Wang S, Gao L, Li X, Guo G, Ma W, Yan Y (2012) Comparative proteomic analysis of grain development in two spring wheat varieties under drought stress. Anal Bioanal Chem 402(3):1297–1313. doi:10.1007/s00216-011-5532-z

Geng Y, Wu R, Wee CW, Xie F, Wei X, Chan PM, Tham C, Duan L, Dinneny JR (2013) A spatio-temporal understanding of growth regulation during the salt stress response in Arabidopsis. Plant Cell 25(6):2132–2154. doi:10.1105/tpc.113.112896

Gorg A, Drews O, Luck C, Weiland F, Weiss W (2009) 2-DE with IPGs. Electrophoresis 30(Suppl 1):S122–132. doi:10.1002/elps.200900051

Guo M, Gao W, Li L, Li H, Xu Y, Zhou C (2013) Proteomic and phosphoproteomic analyses of NaCl stress-responsive proteins in Arabidopsis roots. J Plant Interact 9(6):396–401. doi:10.1080/17429145.2013.845262

He H, Li J (2008) Proteomic analysis of phosphoproteins regulated by abscisic acid in rice leaves. Biochem Biophys Res Commun 371(4):883–888. doi:10.1016/j.bbrc.2008.05.001

Heazlewood JL, Durek P, Hummel J, Selbig J, Weckwerth W, Walther D, Schulze WX (2008) PhosPhAt: a database of phosphorylation sites in Arabidopsis thaliana and a plant-specific phosphorylation site predictor. Nucleic Acids Res 36(Database issue):D1015–1021. doi:10.1093/nar/gkm812

Helmy M, Tomita M, Ishihama Y (2011) OryzaPG-DB: rice proteome database based on shotgun proteogenomics. BMC Plant Biol 11(1):63. doi:10.1186/1471-2229-11-63

Hirayama T, Shinozaki K (2010) Research on plant abiotic stress responses in the post-genome era: past, present and future. Plant J 61(6):1041–1052. doi:10.1111/j.1365-313X.2010.04124.x

Hoehenwarter W, van Dongen JT, Wienkoop S, Steinfath M, Hummel J, Erban A, Sulpice R, Regierer B, Kopka J, Geigenberger P, Weckwerth W (2008) A rapid approach for phenotype-screening and database independent detection of cSNP/protein polymorphism using mass accuracy precursor alignment. Proteomics 8(20):4214–4225. doi:10.1002/pmic.200701047

Hoehenwarter W, Larhlimi A, Hummel J, Egelhofer V, Selbig J, van Dongen JT, Wienkoop S, Weckwerth W (2011) MAPA distinguishes genotype-specific variability of highly similar regulatory protein isoforms in potato tuber. J Proteome Res 10(7):2979–2991. doi:10.1021/Pr101109a

Hsu JL, Wang LY, Wang SY, Lin CH, Ho KC, Shi FK, Chang IF (2009) Functional phosphoproteomic profiling of phosphorylation sites in membrane fractions of salt-stressed Arabidopsis thaliana. Proteome Sci 7(1):42. doi:10.1186/1477-5956-7-42

Hu G, Houston NL, Pathak D, Schmidt L, Thelen JJ, Wendel JF (2011) Genomically biased accumulation of seed storage proteins in allopolyploid cotton. Genetics 189(3):1103–1115. doi:10.1534/genetics.111.132407

Huang C, Verrillo F, Renzone G, Arena S, Rocco M, Scaloni A, Marra M (2011) Response to biotic and oxidative stress in Arabidopsis thaliana: analysis of variably phosphorylated proteins. J Proteomics 74(10):1934–1949. doi:10.1016/j.jprot.2011.05.016

International Rice Genome Sequencing Project (2005) The map-based sequence of the rice genome. Nature 436(7052):793–800

Jacoby RP, Millar AH, Taylor NL (2010) Wheat mitochondrial proteomes provide new links between antioxidant defense and plant salinity tolerance. J Proteome Res 9(12):6595–6604. doi:10.1021/pr1007834

Jiang Y, Yang B, Harris NS, Deyholos MK (2007) Comparative proteomic analysis of NaCl stress-responsive proteins in Arabidopsis roots. J Exp Bot 58(13):3591–3607. doi:10.1093/jxb/erm207

Joshi HJ, Hirsch-Hoffmann M, Baerenfaller K, Gruissem W, Baginsky S, Schmidt R, Schulze WX, Sun Q, van Wijk KJ, Egelhofer V, Wienkoop S, Weckwerth W, Bruley C, Rolland N, Toyoda T, Nakagami H, Jones AM, Briggs SP, Castleden I, Tanz SK, Millar AH, Heazlewood JL (2011) MASCP gator: an aggregation portal for the visualization of Arabidopsis proteomics data. Plant Physiol 155(1):259–270. doi:10.1104/pp. 110.168195

Ke Y, Han G, He H, Li J (2009) Differential regulation of proteins and phosphoproteins in rice under drought stress. Biochem Biophys Res Commun 379(1):133–138. doi:10.1016/j.bbrc.2008.12.067

Kerk D, Bulgrien J, Smith DW, Barsam B, Veretnik S, Gribskov M (2002) The complement of protein phosphatase catalytic subunits encoded in the genome of Arabidopsis. Plant Physiol 129(2):908–925. doi:10.1104/pp. 004002

Kersten B, Agrawal GK, Durek P, Neigenfind J, Schulze W, Walther D, Rakwal R (2009) Plant phosphoproteomics: an update. Proteomics 9(4):964–988. doi:10.1002/pmic.200800548

Khan M, Takasaki H, Komatsu S (2005) Comprehensive phosphoproteome analysis in rice and identification of phosphoproteins responsive to different hormones/stresses. J Proteome Res 4(5):1592–1599. doi:10.1021/pr0501160

Kim T-H, Hauser F, Ha T, Xue S, Böhmer M, Nishimura N, Munemasa S, Hubbard K, Peine N, B-h L, Lee S, Robert N, Parker JE, Schroeder JI (2011) Chemical genetics reveals negative regulation of abscisic acid signaling by a plant immune response pathway. Curr Biol 21(11):990–997. doi:10.1016/j.cub.2011.04.045

Kline KG, Barrett-Wilt GA, Sussman MR (2010) In planta changes in protein phosphorylation induced by the plant hormone abscisic acid. Proc Natl Acad Sci U S A 107(36):15986–15991. doi:10.1073/pnas.1007879107

Kline-Jonakin KG, Barrett-Wilt GA, Sussman MR (2011) Quantitative plant phosphoproteomics. Curr Opinion Plant Biol 14(5):507–511. doi:10.1016/j.pbi.2011.06.008

Komatsu S, Tanaka N (2005) Rice proteome analysis: a step toward functional analysis of the rice genome. Proteomics 5(4):938–949. doi:10.1002/pmic.200401040

Kosova K, Vitamvas P, Prasil IT, Renaut J (2011) Plant proteome changes under abiotic stress—contribution of proteomics studies to understanding plant stress response. J Proteomics 74(8):1301–1322. doi:10.1016/j.jprot.2011.02.006

Lee S, Lee EJ, Yang EJ, Lee JE, Park AR, Song WH, Park OK (2004) Proteomic identification of annexins, calcium-dependent membrane binding proteins that mediate osmotic stress and abscisic acid signal transduction in Arabidopsis. Plant Cell 16(6):1378–1391. doi:10.1105/tpc.021683

Li XJ, Yang MF, Chen H, Qu LQ, Chen F, Shen SH (2010) Abscisic acid pretreatment enhances salt tolerance of rice seedlings: proteomic evidence. Biochim Biophys Acta 1804(4):929–940

Lilley KS, Dupree P (2006) Methods of quantitative proteomics and their application to plant organelle characterization. J Exp Bot 57(7):1493–1499. doi:10.1093/jxb/erj141

Lim H, Eng J, Yates JR 3rd, Tollaksen SL, Giometti CS, Holden JF, Adams MW, Reich CI, Olsen GJ, Hays LG (2003) Identification of 2D-gel proteins: a comparison of MALDI/TOF peptide mass mapping to mu LC-ESI tandem mass spectrometry. J Am Soc Mass Spectrom 14(9):957–970

Liu CW, Hsu YK, Cheng YH, Yen HC, Wu YP, Wang CS, Lai CC (2012) Proteomic analysis of salt-responsive ubiquitin-related proteins in rice roots. Rapid Commun Mass Spectrom 26(15):1649–1660

Luan S (2009) The CBL-CIPK network in plant calcium signaling. Trends Plant Sci 14(1):37–42

May P, Wienkoop S, Kempa S, Usadel B, Christian N, Rupprecht J, Weiss J, Recuenco-Munoz L, Ebenhoh O, Weckwerth W, Walther D (2008) Metabolomics- and proteomics-assisted genome annotation and analysis of the draft metabolic network of Chlamydomonas reinhardtii. Genetics 179(1):157–166. doi:10.1534/genetics.108.088336

Moller IS, Tester M (2007) Salinity tolerance of Arabidopsis: a good model for cereals? Trends Plant Sci 12(12):534–540. doi:10.1016/j.tplants.2007.09.009

Munns R (2002) Comparative physiology of salt and water stress. Plant Cell Environ 25(2):239–250

Munns R, Tester M (2008) Mechanisms of salinity tolerance. Annu Rev Plant Biol 59:651–681. doi:10.1146/annurev.arplant.59.032607.092911

Nakagami H, Pitzschke A, Hirt H (2005) Emerging MAP kinase pathways in plant stress signalling. Trends Plant Sci 10(7):339–346. doi:10.1016/j.tplants.2005.05.009

Nakagami H, Sugiyama N, Mochida K, Daudi A, Yoshida Y, Toyoda T, Tomita M, Ishihama Y, Shirasu K (2010) Large-scale comparative phosphoproteomics identifies conserved phosphorylation sites in plants. Plant Physiol 153(3):1161–1174. doi:10.1104/pp. 110.157347

Nam MH, Huh SM, Kim KM, Park WJ, Seo JB, Cho K, Kim DY, Kim BG, Yoon IS (2012) Comparative proteomic analysis of early salt stress-responsive proteins in roots of SnRK2 transgenic rice. Proteome Sci 10(1)

Neilson KA, Ali NA, Muralidharan S, Mirzaei M, Mariani M, Assadourian G, Lee A, van Sluyter SC, Haynes PA (2011) Less label, more free: approaches in label-free quantitative mass spectrometry. Proteomics 11(4):535–553. doi:10.1002/pmic.201000553

Ng DW, Zhang C, Miller M, Shen Z, Briggs SP, Chen ZJ (2012) Proteomic divergence in Arabidopsis autopolyploids and allopolyploids and their progenitors. Heredity 108(4):419–430. doi:10.1038/hdy.2011.92

Nohzadeh Malakshah S, Habibi Rezaei M, Heidari M, Salekdeh GH (2007) Proteomics reveals new salt responsive proteins associated with rice plasma membrane. Biosci Biotechnol Biochem 71(9):2144–2154

Nühse T, Stensballe A, Jensen O, Peck S (2004) Phosphoproteomics of the Arabidopsis plasma membrane and a new phosphorylation site database. Plant Cell 16(9):2394–2405

Nühse TS, Bottrill AR, Jones AME, Peck SC (2007) Quantitative phosphoproteomic analysis of plasma membrane proteins reveals regulatory mechanisms of plant innate immune responses. Plant J 51(5):931–9400. doi:10.1111/j.1365-313X.2007.03192.x

Olsen JV, Blagoev B, Gnad F, Macek B, Kumar C, Mortensen P, Mann M (2006) Global, in vivo, and site-specific phosphorylation dynamics in signaling networks. Cell 127(3):635–648. doi:10.1016/j.cell.2006.09.026

Pang Q, Chen S, Dai S, Chen Y, Wang Y, Yan X (2010) Comparative proteomics of salt tolerance in Arabidopsis thaliana and Thellungiella halophila. J Proteome Res 9(5):2584–2599. doi:10.1021/pr100034f

Pechanova O, Takac T, Samaj J, Pechan T (2013) Maize proteomics: an insight into the biology of an important cereal crop. Proteomics 13(3–4):637–662. doi:10.1002/pmic.201200275

Peng Z, Wang M, Li F, Lv H, Li C, Xia G (2009) A proteomic study of the response to salinity and drought stress in an introgression strain of bread wheat. Mol Cell Proteomics 8(12): 2676–2686

Piques M, Schulze WX, Hohne M, Usadel B, Gibon Y, Rohwer J, Stitt M (2009) Ribosome and transcript copy numbers, polysome occupancy and enzyme dynamics in Arabidopsis. Mol Syst Biol 5:314. doi:10.1038/msb.2009.68

Potato Genome Sequencing Consortium, Xu X, Pan S, Cheng S, Zhang B, Mu D, Ni P, Zhang G, Yang S, Li R, Wang J, Orjeda G, Guzman F, Torres M, Lozano R, Ponce O, Martinez D, De la Cruz G, Chakrabarti SK, Patil VU, Skryabin KG, Kuznetsov BB, Ravin NV, Kolganova TV, Beletsky AV, Mardanov AV, Di Genova A, Bolser DM, Martin DM, Li G, Yang Y, Kuang H, Hu Q, Xiong X, Bishop GJ, Sagredo B, Mejia N, Zagorski W, Gromadka R, Gawor J, Szczesny P, Huang S, Zhang Z, Liang C, He J, Li Y, He Y, Xu J, Zhang Y, Xie B, Du Y, Qu D, Bonierbale M, Ghislain M, Herrera Mdel R, Giuliano G, Pietrella M, Perrotta G, Facella P, O'Brien K, Feingold SE, Barreiro LE, Massa GA, Diambra L, Whitty BR, Vaillancourt B, Lin H, Massa AN, Geoffroy M, Lundback S, DellaPenna D, Buell CR, Sharma SK, Marshall DF, Waugh R, Bryan GJ, Destefanis M, Nagy I, Milbourne D, Thomson SJ, Fiers M, Jacobs JM, Nielsen KL, Sonderkaer M, Iovene M, Torres GA, Jiang J, Veilleux RE, Bachem CW, de Boer J, Borm T, Kloosterman B, van Eck H, Datema E, Hekkert B, Goverse A, van Ham RC, Visser RG (2011) Genome sequence and analysis of the tuber crop potato. Nature 475(7355):189–195. doi:10.1038/nature10158

Qiu QS, Guo Y, Dietrich MA, Schumaker KS, Zhu JK (2002) Regulation of SOS1, a plasma membrane Na+/H+ exchanger in Arabidopsis thaliana, by SOS2 and SOS3. Proc Natl Acad Sci U S A 99(12):8436–8441, Epub 2002 May 8428

Rampitsch C, Bykova NV (2012) The beginnings of crop phosphoproteomics: exploring early warning systems of stress. Front Plant Sci 3:144. doi:10.3389/fpls.2012.00144

Ray DK, Mueller ND, West PC, Foley JA (2013) Yield trends Are insufficient to double global crop production by 2050. PLoS One 8(6):e66428. doi:10.1371/journal.pone.0066428

Reiland S, Messerli G, Baerenfaller K, Gerrits B, Endler A, Grossmann J, Gruissem W, Baginsky S (2009) Large-scale Arabidopsis phosphoproteome profiling reveals novel chloroplast kinase substrates and phosphorylation networks. Plant Physiol 150(2):889–903. doi:10.1104/pp.109.138677

Reiland S, Finazzi G, Endler A, Willig A, Baerenfaller K, Grossmann J, Gerrits B, Rutishauser D, Gruissem W, Rochaix JD, Baginsky S (2011) Comparative phosphoproteome profiling reveals a function of the STN8 kinase in fine-tuning of cyclic electron flow (CEF). Proc Natl Acad Sci U S A 108(31):12955–12960. doi:10.1073/pnas.1104734108

Salekdeh GH, Siopongco J, Wade LJ, Ghareyazie B, Bennett J (2002) A proteomic approach to analyzing drought- and salt-responsiveness in rice. Field Crop Res 76(2–3):199–219

Saqib M, Zörb C, Schubert S (2006) Salt-resistant and salt-sensitive wheat genotypes show similar biochemical reaction at protein level in the first phase of salt stress. J Plant Nutr Soil Sci 169(4):542–548. doi:10.1002/jpln.200520557

Savitski MM, Lemeer S, Boesche M, Lang M, Mathieson T, Bantscheff M, Kuster B (2011) Confident phosphorylation site localization using the Mascot Delta Score. Mol Cell Proteomics 10(2):M110.003830. doi:10.1074/mcp.M110.003830

Schnable PS, Ware D, Fulton RS, Stein JC, Wei F, Pasternak S, Liang C, Zhang J, Fulton L, Graves TA, Minx P, Reily AD, Courtney L, Kruchowski SS, Tomlinson C, Strong C, Delehaunty K, Fronick C, Courtney B, Rock SM, Belter E, Du F, Kim K, Abbott RM, Cotton M, Levy A, Marchetto P, Ochoa K, Jackson SM, Gillam B, Chen W, Yan L, Higginbotham J, Cardenas M, Waligorski J, Applebaum E, Phelps L, Falcone J, Kanchi K, Thane T, Scimone A, Thane N, Henke J, Wang T, Ruppert J, Shah N, Rotter K, Hodges J, Ingenthron E, Cordes M, Kohlberg S, Sgro J, Delgado B, Mead K, Chinwalla A, Leonard S, Crouse K, Collura K, Kudrna D, Currie J, He R, Angelova A, Rajasekar S, Mueller T, Lomeli R, Scara G, Ko A, Delaney K, Wissotski M, Lopez G, Campos D, Braidotti M, Ashley E, Golser W, Kim H, Lee S, Lin J, Dujmic Z, Kim W, Talag J, Zuccolo A, Fan C, Sebastian A, Kramer M, Spiegel L, Nascimento L, Zutavern T, Miller B, Ambroise C, Muller S, Spooner W, Narechania A, Ren L, Wei S, Kumari S, Faga B, Levy MJ, McMahan L, Van Buren P, Vaughn MW, Ying K, Yeh C-T, Emrich SJ, Jia Y, Kalyanaraman A, Hsia A-P, Barbazuk WB, Baucom RS, Brutnell TP, Carpita NC, Chaparro C, Chia J-M, Deragon J-M, Estill JC, Fu Y, Jeddeloh JA, Han Y, Lee H, Li P, Lisch DR, Liu S, Liu Z, Nagel DH, McCann MC, SanMiguel P, Myers AM, Nettleton D, Nguyen J, Penning BW, Ponnala L, Schneider KL, Schwartz DC, Sharma A, Soderlund C, Springer NM, Sun Q, Wang H, Waterman M, Westerman R, Wolfgruber TK, Yang L, Yu Y, Zhang L, Zhou S, Zhu Q, Bennetzen JL, Dawe RK, Jiang J, Jiang N, Presting GG, Wessler SR, Aluru S, Martienssen RA, Clifton SW, McCombie WR, Wing RA, Wilson RK (2009) The B73 maize genome: complexity, diversity, and dynamics. Science 326(5956):1112–1115. doi:10.1126/science.1178534

Schulze WX (2010) Proteomics approaches to understand protein phosphorylation in pathway modulation. Curr Opin Plant Biol 13(3):280–287. doi:10.1016/j.pbi.2009.12.008

Schulze WX, Usadel B (2010) Quantitation in mass-spectrometry-based proteomics. Annu Rev Plant Biol 61:491–516. doi:10.1146/annurev-arplant-042809-112132

Shteynberg D, Deutsch E, Mendoza L, Slagel J, Lam HH, Nesvizhskii A, Moritz R (2012) PTMProphet: TPP software for validation of modified site locations on post-translationally modified peptides. Paper presented at the 60th ASMS Conference on Mass Spectrometry, Vancouver, BC, Canada, 20–24 May

Sobhanian H, Razavizadeh R, Nanjo Y, Ehsanpour AA, Jazii FR, Motamed N, Komatsu S (2010) Proteome analysis of soybean leaves, hypocotyls and roots under salt stress. Proteome Sci 8
Sugimoto M, Takeda K (2009) Proteomic analysis of specific proteins in the root of salt-tolerant barley. Biosci Biotechnol Biochem 73(12):2762–2765
Suzuki N, Koussevitzky S, Mittler R, Miller G (2012) ROS and redox signalling in the response of plants to abiotic stress. Plant Cell Environ 35(2):259–270. doi:10.1111/j.1365-3040.2011.02336.x
Taus T, Kocher T, Pichler P, Paschke C, Schmidt A, Henrich C, Mechtler K (2011) Universal and confident phosphorylation site localization using phosphoRS. J Proteome Res 10(12):5354–5362. doi:10.1021/pr200611n
Tester M, Langridge P (2010) Breeding technologies to increase crop production in a changing world. Science 327(5967):818–822. doi:10.1126/science.1183700
Tomato Genome C (2012) The tomato genome sequence provides insights into fleshy fruit evolution. Nature 485(7400):635–641. doi:10.1038/nature11119
Umezawa T, Sugiyama N, Takahashi F, Anderson JC, Ishihama Y, Peck SC, Shinozaki K (2013) Genetics and phosphoproteomics reveal a protein phosphorylation network in the abscisic acid signaling pathway in Arabidopsis thaliana. Sci Signal 6(270):rs8. doi:10.1126/scisignal.2003509
Unlu M, Morgan ME, Minden JS (1997) Difference gel electrophoresis: a single gel method for detecting changes in protein extracts. Electrophoresis 18(11):2071–2077. doi:10.1002/elps.1150181133
Vanderschuren H, Lentz E, Zainuddin I, Gruissem W (2013) Proteomics of model and crop plant species: status, current limitations and strategic advances for crop improvement. J Proteomics. doi:10.1016/j.jprot.2013.05.036
Vertommen A, Panis B, Swennen R, Carpentier SC (2011) Challenges and solutions for the identification of membrane proteins in non-model plants. J Proteomics 74(8):1165–1181. doi:10.1016/j.jprot.2011.02.016
Wang W, Vinocur B, Altman A (2003) Plant responses to drought, salinity and extreme temperatures: towards genetic engineering for stress tolerance. Planta 218(1):1–14. doi:10.1007/s00425-003-1105-5
Wang MC, Peng ZY, Li CL, Li F, Liu C, Xia GM (2008) Proteomic analysis on a high salt tolerance introgression strain of Triticum aestivum/Thinopyrum ponticum. Proteomics 8(7):1470–1489. doi:10.1002/pmic.200700569
Wang P, Xue L, Batelli G, Lee S, Hou Y-J, Van Oosten MJ, Zhang H, Tao WA, Zhu J-K (2013) Quantitative phosphoproteomics identifies SnRK2 protein kinase substrates and reveals the effectors of abscisic acid action. Proc Natl Acad Sci U S A 110(27):11205–11210. doi:10.1073/pnas.1308974110
Witzel K, Weidner A, Surabhi GK, Börner A, Mock HP (2009) Salt stress-induced alterations in the root proteome of barley genotypes with contrasting response towards salinity. J Exp Bot 60(12):3545–3557
Yan S, Tang Z, Su W, Sun W (2005) Proteomic analysis of salt stress-responsive proteins in rice root. Proteomics 5(1):235–244. doi:10.1002/pmic.200400853
Yang Y, Thannhauser TW, Li L, Zhang S (2007) Development of an integrated approach for evaluation of 2-D gel image analysis: impact of multiple proteins in single spots on comparative proteomics in conventional 2-D gel/MALDI workflow. Electrophoresis 28(12):2080–2094. doi:10.1002/elps.200600524
Zhang L, Tian LH, Zhao JF, Song Y, Zhang CJ, Guo Y (2009) Identification of an apoplastic protein involved in the initial phase of salt stress response in rice root by two-dimensional electrophoresis. Plant Physiol 149(2):916–928
Zhang H, Han B, Wang T, Chen S, Li H, Zhang Y, Dai S (2012) Mechanisms of plant salt response: insights from proteomics. J Proteome Res 11(1):49–67. doi:10.1021/pr200861w
Zhao Q, Zhang H, Wang T, Chen S, Dai S (2013) Proteomics-based investigation of salt-responsive mechanisms in plant roots. J Proteomics 82:230–253. doi:10.1016/j.jprot.2013.01.024

Zhou YJ, Gao F, Li XF, Zhang J, Zhang GF (2010) Alterations in phosphoproteome under salt stress in Thellungiella roots. Chin Sci Bull 55(32):3673–3679. doi:10.1007/s11434-010-4116-1

Zörb C, Schmitt S, Neeb A, Karl S, Linder M, Schubert S (2004) The biochemical reaction of maize (Zea mays L.) to salt stress is characterized by a mitigation of symptoms and not by a specific adaptation. Plant Sci 167(1):91–100

Zörb C, Schmitt S, Mühling KH (2010) Proteomic changes in maize roots after short-term adjustment to saline growth conditions. Proteomics 10(24):4441–4449

Part II
Components of Signal Transduction

Chapter 4
Role of Cation/Proton Exchangers in Abiotic Stress Signaling and Stress Tolerance in Plants

Peter D. Bickerton and Jon K. Pittman

Abstract The generation of cytosolic Ca^{2+} signals is an early response in plants to many environmental stresses. Ca^{2+} transporters play a critical role in the generation of these signals and the maintenance of cellular Ca^{2+} homeostasis. A major class of Ca^{2+} transporter is the Ca^{2+}/H^+ exchanger (CAX), which is almost ubiquitous throughout every domain of life but has been best characterised in higher plants. CAX transporters of *Arabidopsis thaliana* have been particularly well studied and have been shown to have diverse roles in vacuolar sequestration of Ca^{2+} and other cations, which provide an important mechanism for ion tolerance. Recent genetic studies of CAX transporters from various plant species indicate that these proteins may also play a role in the response to specific abiotic stresses. Altered sensitivity of CAX mutant plants to abiotic stresses such as salt stress and cold stress has been observed. The potential role of CAX-type cation transporters in environmental stress response is reviewed here. In particular, we discuss whether these high-capacity, low-affinity Ca^{2+} transporters play a central role in modulating Ca^{2+} signals in response to abiotic stresses.

Keywords Abiotic stress • Cation/proton exchanger • Calcium signalling • Metal transport • Vacuolar calcium

4.1 Introduction

Since the generation of the first living cells, there has been a requirement for all organisms to maintain strict calcium ion (Ca^{2+}) homeostasis (Case et al. 2007). As the fifth most abundant element on Earth, Ca^{2+} is readily available to cells, at high external concentrations. Ca^{2+} is incredibly useful to life, as a cell-signaling

P.D. Bickerton (✉) • J.K. Pittman
Faculty of Life Sciences, University of Manchester, Michael Smith Building, Oxford Road, Manchester M13 9PT, UK
e-mail: peter.bickerton@tgac.ac.uk; jon.pittman@manchester.ac.uk

G.K. Pandey (ed.), *Elucidation of Abiotic Stress Signaling in Plants*,
DOI 10.1007/978-1-4939-2211-6_4

molecule, as part of the binding matrix of cell walls, and as a vital component of teeth, bones and shells. However, the aspects of Ca^{2+}, which render it useful to cells, also mean that, at high concentrations, Ca^{2+} is significantly cytotoxic due to interaction with various cellular components. As a result of these conflicting effects of Ca^{2+} on cells, there exists a broad range of proteins, which control either influx or efflux of Ca^{2+}, plus those that buffer Ca^{2+} through direct binding. This tight regulation of cytosolic Ca^{2+} concentration means that Ca^{2+} is used throughout eukaryotic and prokaryotic organisms as a versatile secondary messenger of both biotic and abiotic stimuli. Of the Ca^{2+} efflux proteins, those that sequester Ca^{2+} into organelles are of particular interest when studying the role of Ca^{2+} in cell signaling, both as modulators of Ca^{2+} signals and as generators of organellar Ca^{2+} stores prior to stimulus-induced Ca^{2+} release (McAinsh and Pittman 2009).

Plant cells contain two major protein families, which actively remove Ca^{2+} from the cytosol. The first are Ca^{2+}-ATPases, which are high-affinity Ca^{2+} pumps present at various organellar membranes and the plasma membrane (Pittman et al. 2011). The second are ion-coupled Ca^{2+} exchangers, including Ca^{2+}/H^+ exchangers (CAX; also referred to as cation/H^+ exchangers; see Sect. 2.3), which are present mainly at the tonoplast and allow Ca^{2+} sequestration into the vacuole (Pittman 2011). CAXs are of interest when studying Ca^{2+} signaling due to their high-capacity, low-affinity transport of divalent cations including Ca^{2+} and because they are the major pathway for maintaining a large cellular store of Ca^{2+} in the vacuole, which can then potentially be released in order to generate cytosolic Ca^{2+} signals in response to external stimuli (Peiter 2011). Functional and genetic studies point towards varied roles for CAX in Ca^{2+} and transition metal homeostasis, development and abiotic stress responses in plants (Manohar et al. 2011), which will be the focus of this review.

4.2 Ca^{2+} Signaling in Plants

Transient increases in cytosolic Ca^{2+} concentration play an integral role in the signal transduction of external stimuli in plants (Dodd et al. 2010; Kudla et al. 2010; Bickerton and Pittman 2012). These Ca^{2+} signals take the form of transient spikes, extended elevations, oscillations, waves through the cytosol and are hypothesised to be specific for a given stimulus (McAinsh et al. 1992). Ca^{2+} signals occur in plants in response to a number of abiotic stimuli, such as salt stress, osmotic stress, cold stress and anoxia, and many of the downstream responses involving sensing and decoding of these signals have been relatively well defined (Lindberg et al. 2012). Figure 4.1 illustrates some simplified signal transduction pathways mediated by stimulus-induced cytosolic Ca^{2+} transients, often thought of as a 'Ca^{2+} signature' for that stimulus, followed by Ca^{2+} binding of sensors, activation of target proteins and eventually an appropriate plant response, such as transcriptional regulation of stress-responsive genetic elements. Many of the downstream elements in abiotic stress Ca^{2+} signaling pathways are well understood, and it is clear that Ca^{2+} transporters of different types are essential for the generation and shaping of a Ca^{2+} signal (McAinsh

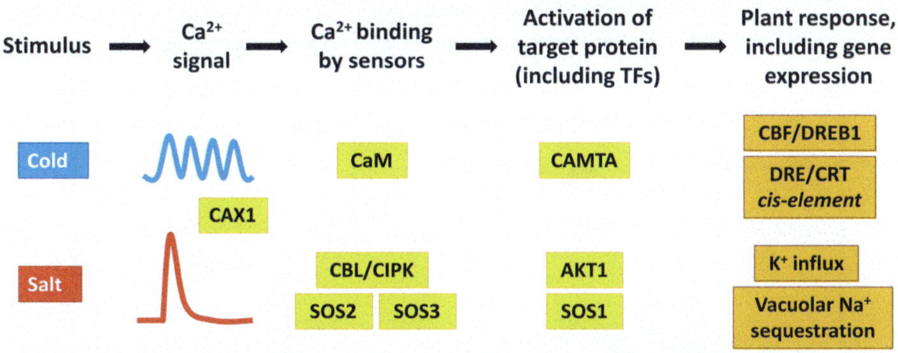

Fig. 4.1 Examples of Ca^{2+}-dependent abiotic stress response pathways. A stimulus, such as cold or salt, induces a specific Ca^{2+} signal, modulated by Ca^{2+} transporters, including Ca^{2+}/H^+ exchangers (CAX). The signal is decoded by sensors, which might include calmodulin (CaM) or a Ca^{2+}-binding protein-like (CBL) and CBL-interacting protein kinase (CIPK) complex (an example being the SOS2 and SOS3 complex). This leads to activation of target proteins, such as the Na^+/H^+ exchanger SOS1 or the K^+ channel AKT1, as well as transcription factors such as CaM-binding transcription activator (CAMTA), which leads to differential regulation of gene expression through induction of stress response genes, like those of the CBF/DREB1 family containing the DRE/CRT *cis*-elements

and Pittman 2009). However, it is less clear exactly how these Ca^{2+} transporters are regulated to allow the cell to generate a specific Ca^{2+} signature in response to a particular environmental stimulus.

One of the best studied and understood examples of an external stimulus-induced Ca^{2+} signaling response is that of guard cells stimulated with various abiotic components including the 'stress' hormone abscisic acid (ABA), CO_2, cold or H_2O_2 (Kim et al. 2010). For example, it is well established that stomatal closure induced by ABA is transduced by oscillations in cytosolic Ca^{2+} (McAinsh et al. 1992). Furthermore, the magnitude and pattern of the cytosolic Ca^{2+} oscillation, which can be manipulated by external application of Ca^{2+}, determines the degree of stomatal closure (McAinsh et al. 1995). Many of the molecular components that regulate the Ca^{2+}-dependent mechanisms of stomatal function are understood (Kim et al. 2010; Laanemets et al. 2013). For example, investigation of stomatal closure in the *Arabidopsis det3* mutant guard cells revealed interesting insights into the role of ion transporters. Guard cells treated with Ca^{2+}, ABA, cold and H_2O_2 were shown to exhibit differential Ca^{2+} signals in response to each stress, and each led to increases in stomatal closure (Allen et al. 2000). However, in *det3* mutants defective in the vacuolar H^+-ATPase (V-ATPase), Ca^{2+} oscillations in response to Ca^{2+} and H_2O_2 were abolished, and stomatal closure was inhibited, while ABA and cold were still able to elicit Ca^{2+} oscillations in *det3*, which led to stomatal closure akin to the wild type. The pH gradient generated by H^+ pumps such as the V-ATPase is required to energise H^+-coupled Ca^{2+} exchangers (Barkla et al. 2008). The V-ATPase-deficient *det3* has therefore been proposed to be defective in vacuolar Ca^{2+} sequestration, through inhibited CAX activity. The defective signaling of *det3* mutants in response to Ca^{2+} and H_2O_2, but not ABA and cold, highlights the complexity in plant Ca^{2+} signal-generating mechanisms and the importance of multiple Ca^{2+} transporters.

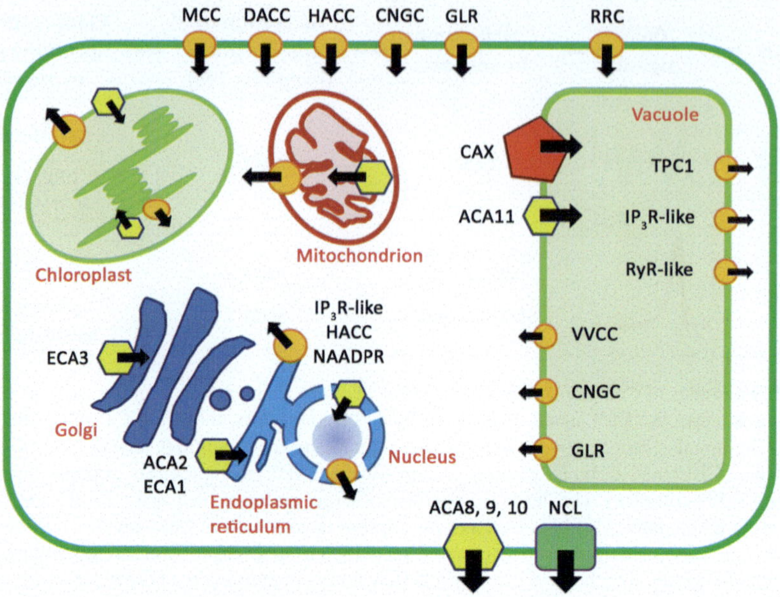

Fig. 4.2 Ca^{2+} transporters in a typical plant cell. Ca^{2+}-permeable influx channels (*orange circles*), Ca^{2+}-ATPases (*yellow hexagons*), Ca^{2+}/H^+ exchangers (CAX) (*red pentagon*) and Na^+/Ca^{2+} exchangers (NCL) (*green square*) at the plasma membrane and internal membranes perform cellular Ca^{2+} homeostasis including maintaining resting cytosolic Ca^{2+} concentration, shape Ca^{2+} signals and refill Ca^{2+} stores. *Black arrows* indicate the direction of Ca^{2+} transport. *ACA* autoinhibited Ca^{2+}-ATPase, *CNGC* cyclic nucleotide-gated channel, *DACC* depolarisation-activated Ca^{2+} channel, *ECA* ER-type Ca^{2+}-ATPase, *GLR* glutamate receptor, *HACC* hyperpolarisation-activated Ca^{2+} channel, *IP3R-like* inositol-1,4,5-trisphosphate receptor-like channel, *MCC* mechanosensitive Ca^{2+} channel, *NAADPR* nicotinic acid adenine dinucleotide phosphate receptor channel, *RRC* ROS-responsive channel, *RyR-like* ryanodine receptor-like, *TPC1* two-pore channel 1, *VVCC* vacuolar voltage-gated Ca^{2+} channel

4.2.1 Transporters in Plant Ca^{2+} Signaling

In order to shape Ca^{2+} signals, plant cells require a suite of transporters, which both bring Ca^{2+} into and remove Ca^{2+} from the cytosol (McAinsh and Pittman 2009; Spalding and Harper 2011) (Fig. 4.2). Ca^{2+}-permeable channels import Ca^{2+} into the cytosol from external and internal sources down a strong electrochemical gradient, while Ca^{2+} pumps and exchangers actively efflux Ca^{2+} from the cytosol either through sequestration into organelles or extrusion across the plasma membrane. These transporters are vital both in generating and modulating Ca^{2+} signals, as well as maintaining a resting cytosolic Ca^{2+} concentration of ~100 nM. Much research over the past few years has demonstrated the importance of Ca^{2+} influx pathways in plant Ca^{2+} signaling and transducing abiotic stress stimuli (Jammes et al. 2011; Spalding and Harper 2011; Swarbreck et al. 2013), but will not be the focus of this review.

4.2.2 Importance of Efflux Proteins in Shaping Ca^{2+} Signals

Considering the nature of the shape of Ca^{2+} signals, the Ca^{2+} efflux proteins are of particular interest, in both a capacity to mediate cytosolic Ca^{2+} signals and in maintenance of Ca^{2+} stores, both intracellular and apoplastic stores. The two most studied families of Ca^{2+} efflux proteins are Ca^{2+}-ATPases and CAXs (McAinsh and Pittman 2009; Pittman et al. 2011; Spalding and Harper 2011). Ca^{2+}-ATPases generally show high-affinity, low-capacity transport of Ca^{2+} out of the cytosol, while the opposite is true of CAXs. It has been proposed that the high affinity of Ca^{2+}-ATPases for Ca^{2+} may render them more important in maintaining Ca^{2+} homeostasis, while the high-capacity nature of CAX perhaps suggests an important role in mediating Ca^{2+} signals during transient large increases and oscillations (Hirschi 2001). Recent evidence from *Arabidopsis* also suggests the presence of a novel type of Na^+/Ca^{2+} exchanger at plant plasma membranes, which may play a role in mediating plant abiotic stress responses (see below).

The role of Ca^{2+}-ATPases in plant cellular Ca^{2+} homeostasis has long been established and is well understood, particularly due to the high conservation in structure and function of these Ca^{2+} pumps throughout eukaryotes (Baxter et al. 2003; Bonza and De Michelis 2011; Pittman et al. 2011). More recently, evidence for the importance of Ca^{2+}-ATPases in Ca^{2+} signaling and abiotic stress response has become stronger (Huda et al. 2013). In contrast, the roles of ion-coupled Ca^{2+} exchangers have been less clearly understood.

4.2.3 Ca^{2+}/H^+ Exchangers

Ca^{2+}/H^+ exchangers (CAXs) are a ubiquitous type of ion transporter, found throughout prokaryotes and eukaryotes, and related to other ion-coupled cation exchangers, including the Na^+/Ca^{2+} exchangers (NCXs), which are all members of the CaCA transporter superfamily (Emery et al. 2012). CAXs are widely present throughout plants, protists, bacteria, cyanobacteria, fungi, algae and some animals, although exceptions include mammals and insects (Manohar et al. 2011). Phylogenetic analysis reveals three types of CAX: type I, which includes plant, fungal, bacterial and protist, CAX type II found in fungi and some animals and bacterial type III CAX (Shigaki et al. 2006; Emery et al. 2012). Apart from in the yeast *Saccharomyces cerevisiae*, to date CAXs have been mostly studied from higher plants and mostly in *Arabidopsis* and rice (*Oryza sativa*). Plant CAXs were first characterised using yeast when *Arabidopsis* AtCAX1 and AtCAX2 were shown to rescue the Ca hypersensitivity phenotype of yeast mutants deficient in vacuolar Ca^{2+} sequestration (Hirschi et al. 1996). Similar suppression of yeast Ca sensitivity was shown with CAXs from other plants like mung bean and rice (Ueoka-Nakanishi et al. 2000; Kamiya et al. 2005). AtCAX1 was later shown only to rescue the Ca hypersensitivity phenotype upon truncation at the N-terminus, which revealed an N-terminal autoinhibitory domain that appears to be common to many plant and non-plant CAXs (Pittman and Hirschi 2001; Pittman et al. 2002b, 2009; Guttery et al. 2013). Subsequent genetic analysis of CAXs has confirmed their role as vacuolar Ca^{2+}/H^+ exchangers *in planta. Arabidopsis cax1* mutants showed a 40 % reduction in capacity of Ca^{2+}/H^+ transport at vacuolar membrane-enriched vesicles, while Ca^{2+}-ATPase activity increased compared to wild type (Cheng et al. 2003). It appears that there is a certain degree of CAX redundancy, as *cax1* mutants showed increased transcript abundance of *AtCAX3* and *AtCAX4* (Cheng et al. 2003; Conn et al. 2011). Similarly, in other CAX mutants, *CAX* gene induction has been observed (Connorton et al. 2012), highlighting the importance of a functional Ca^{2+}/H^+ exchanger in a plant cell, which is further demonstrated by the severe phenotype of mutant plants lacking multiple CAX isoforms (Cheng et al. 2005).

In addition to functioning as Ca^{2+}/H^+ exchangers, it is clear that many plant CAX proteins can transport a range of cations including Ca^{2+}, Cd^{2+}, Mn^{2+} and Zn^{2+}, as demonstrated through yeast expression (Shigaki et al. 2003) and from plant analysis (Koren'kov et al. 2007; Connorton et al. 2012), and can thus provide tolerance to metal stress through direct cation sequestration (Hirschi et al. 2000; Wu et al. 2011). Hence CAX can also be defined as cation/H^+ exchanger. It has also been shown that when *AtCAX1* and *AtCAX3* are co-expressed they can confer salt tolerance in yeast possibly due to Na^+/H^+ exchange activity (Zhao et al. 2009b). In contrast, some of the altered cation stress phenotypes observed in CAX knockout mutants might be due to indirect Ca^{2+} homeostasis effects. For example, increased Mg tolerance in *cax1* and *cax1/cax2* mutants does not appear to be due to direct Mg^{2+} transport but due to alterations in cellular Ca^{2+} (Connorton et al. 2012). In *Arabidopsis* there are 6 *CAX* genes, which cluster into two phylogenetic groups: type IA (*AtCAX1*, *AtCAX3* and *CAX4*) and type IB (*AtCAX2*, *AtCAX5* and *AtCAX6*). AtCAX1 and AtCAX3 appear to be important for Ca^{2+} homeostasis (Cheng et al. 2003; Conn et al. 2011), while AtCAX2 and AtCAX5 have a preference towards Mn^{2+} transport (Shigaki et al. 2003; Pittman et al. 2004; Edmond et al. 2009). However, there is no clear-cut distinction between the groups based on substrate specificity. AtCAX2 and AtCAX5 can also transport Ca^{2+}, while *AtCAX4* is transcriptionally induced by various metals (Cheng et al. 2002), and *cax4* knockdown mutants show sensitivity to metal stress (Mei et al. 2009). There may also be other physiological functions of CAX transporters. AtCAX1 and AtCAX3 play an important role in gas exchange and plant productivity, as well as for regulating apoplastic Ca concentration (Conn et al. 2011). In addition, these transporters have been implicated in the signaling of root-shoot phosphate transport in response to phosphate starvation (Liu et al. 2011), as well as regulation of apoplastic pH (Cho et al. 2012), which provides an interesting insight into the role for CAX in whole plant signaling. The diverse roles for CAX in the sequestration and homeostasis of Ca^{2+}, as well as providing tolerance to transition metal and salt stress, suggest CAXs are of great interest when elucidating plant responses to abiotic stress.

4.2.4 Na^+/Ca^{2+} Exchangers

Until recently, it was thought that only Ca^{2+}-ATPases and H^+-coupled Ca^{2+} exchangers were responsible for Ca^{2+} extrusion from the cytosol in plant cells, but that Na^+-coupled Ca^{2+} exchangers, which are prevalent in animal cells and encoded by NCX genes (Emery et al. 2012), were absent in higher plants. However, recent evidence suggests a role for Na^+/Ca^{2+} exchange (NCX) at the plasma membrane of *Arabidopsis* cells (Wang et al. 2012). The *Arabidopsis* NCX-like protein (AtNCL) shows structural and sequence similarity with CAX proteins, with regions of homology at the C-terminus and, like CAX, 10–11 transmembrane domains (Shigaki et al. 2006). AtNCL has two Ca^{2+}-binding EF-hands suggesting that the protein may be regulated by Ca^{2+} interaction. Recent phylogenetic analysis demonstrated that *Arabidopsis* has at least two NCL genes (also named "EFCAX") and showed that related genes are prevalent in the genomes of higher plants, moss and algae, and these genes are mostly related to higher plant CAX genes rather than any other exchanger families including the NCX family (Emery et al. 2012). When expressed in yeast, *AtNCL* was unable to suppress the yeast mutant Ca hypersensitive phenotype and did not appear to possess H^+/Ca^{2+} exchange activity but did show Na^+/Ca^{2+} exchange activity when expressed in mammalian cells (Wang et al. 2012). AtNCL–YFP fusion proteins localised to the *Arabidopsis* plasma membrane rather than the vacuole, and *AtNCL* expression was induced in response to a number of abiotic stimuli including salt, ABA, cold and heat. Furthermore, *atncl* knockout lines were less sensitive to salt stress compared to wild type and had reduced Na and increased Ca content compared to wild type under both normal conditions and salt stress, indicating that AtNCL functions in Ca^{2+} efflux at the plasma membrane (Wang et al. 2012). It appears that AtNCL represents a novel type of plant Ca^{2+} transporter, which may provide an interesting area for study particularly in terms of Ca^{2+} efflux in response to salt stress.

4.3 Functional Genomics Analysis of Plant CAX Genes

The determination of function of individual plant genes and gene families, such as the CAX genes, as well as interacting gene networks, has been clearly enhanced through the use of functional genomics techniques such as genome-scale transcriptomics (Maathuis et al. 2003; Bock et al. 2006; Narsai et al. 2010; Patel et al. 2012) and the systematic analysis of whole gene family function such as by gene knockout (O'Malley and Ecker 2010; Wang et al. 2013), cell-specific expression (Toufighi et al. 2005), subcellular localisation (Li et al. 2006; Winter et al. 2007) and protein interaction (Geisler-Lee et al. 2007; Lalonde et al. 2010; Ho et al. 2012). Furthermore, methods such as ionomics, to determine the genome-wide variation in element concentration upon changes in expression or knockout of each gene (Salt et al. 2008), are particularly useful for the study of ion transporters. Initially such approaches have focussed on *Arabidopsis* genes, although the more recent determination of genome sequences from multiple plant species has widened the scope of plant functional genomics, for example, the more detailed information now available from rice (Singh et al. 2014).

As described above, *Arabidopsis* has six CAX genes, although an earlier genome-wide phylogenetic analysis of putative cation transporters from *Arabidopsis* identified eleven CAX genes (Mäser et al. 2001). Subsequent analysis has determined that *AtCAX7* to *AtCAX11* are closer in sequence homology to another cation exchanger family called CCX (cation calcium exchanger) and were renamed *AtCCX1* to *AtCCX5* (Shigaki et al. 2006). Analysis of other plant genomes has observed that CAX genes are ubiquitous in higher plants although gene family size can vary from two identifiable CAXs in *Lotus japonicus* to fourteen in soybean, while many species, like rice, poplar and grape, have five or six (Emery et al. 2012). Of the six *Arabidopsis* CAX genes, *AtCAX1* to *AtCAX5* have been individually studied experimentally, especially through genetic analysis (see below in Sect. 4), but further information has been gained through functional genomics analysis. *AtCAX6* has not been successfully characterised to date, potentially due to low expression level (Edmond et al. 2009), but some information about this gene might be gained from transcriptomic data sets, although care should be taken in the interpretation of this data due to very high sequence identity between *AtCAX5* and *AtCAX6* and the possibility of cross-hybridisation. Likewise, there has been some experimental characterisation of five of the rice CAX genes (Kamiya et al. 2005), but no analysis has been done yet of *OsCAX4*. There are two members of the related EFCAX family in both *Arabidopsis* and rice (Emery et al. 2012), with experimental characterisation begun for one of the *Arabidopsis* genes (*AtNCL/AtEFCAX1*) (Wang et al. 2012). However, functional genomic information is starting to provide more insight into the roles of the rice CAX- and EFCAX-type Ca^{2+} exchangers (Singh et al. 2014).

Transcriptomic analysis has been particularly useful at indicating the potential roles of individual CAX proteins in specific abiotic stress responses. A summary of information for *Arabidopsis* and rice CAX and EFCAX genes determined from functional genomics analyses is provided in Table 4.1. One of the clear observations from this data is the wide range of abiotic stresses that appear to induce or repress expression of the CAX and EFCAX transporters. While some of the stress responses are shared between multiple CAX isoforms, there are clear distinctions and stress specificities, indicating that specific stress signals are modulated by specific transporters. For example, only *AtCAX2* appears to be strongly upregulated by hypoxia, while only *AtNCL* appears to be upregulated by heat stress (Table 4.1). Much of the genomic-scale data, particularly for *Arabidopsis* tissue-specific localisation and subcellular localisation, has been experimentally confirmed, but further experiments are required to validate many of the abiotic stress-induced expression changes for *Arabidopsis* and rice genes determined from microarray studies. Furthermore, analyses are needed to demonstrate whether these stress inductions have any functional relevance to plant abiotic stress tolerance and stress signaling, such as through reverse genetics studies, as described below.

Table 4.1 Summary of *Arabidopsis* (At) and rice (Os) CAX and EFCAX gene characteristics determined by functional genomics analyses

Gene name[a]	Gene ID	Tissue-specific expression	Cell-specific expression	Subcellular localisation	Abiotic stress increased expression[b]	Abiotic stress reduced expression[b]	Protein interactions[c]
AtCAX1	At2g38170	Leaves[d], flowers[d], seed[d], seedling[d]	Guard cell[e], mesophyll[e]	Vacuole[f]	Cold (1 h)[e], drought (1 h), Ca[e]	Osmotic (1 h), salt (6–12 h), heat (4 h), ABA (1 h), hypoxia, P starvation[e]	CAX3[g], CXIP4[g], GRXCP (CXIP1)[g], CIPK24 (SOS2)[g]
AtCAX2	At3g13320	Roots[d], leaves[d], flowers[d], seed, seedling	Guard cell, root epidermis[d]	Vacuole[f]	UV-B (1–6 h), wounding (1 h), cold (12 h), hypoxia, Zn[e]	Osmotic (1 h), salt (1 h), ABA (1–12 h), cold (1 h)	At2g41090 (EF-hand protein), CAM1 (TCH1), CAM7, NTL9
AtCAX3	At3g51860	Roots[d], flowers[d], seed[d], seedling[d], old leaves	Root cortex[d], guard cell (with ABA)[e]	Vacuole[f]	Osmotic (12–14 h), salt (6–24 h)[e], drought (12 h), UV-B (6–12 h), wounding (12 h), ABA (12 h), P starvation[e]	Mg[e]	CAX1[g]
AtCAX4	At5g01490	Roots[d], leaves[e], seed	Mesophyll, root cap[d], root epidermis[d]	Vacuole[f]	Cold (12–24 h), Mn[e], Ni[e]	Drought (0.25–0.5 h), oxidative (0.5 h), UV-B (0.25–0.5 h), wounding (0.5 h), ABA (12 h), heat (12 h)	None observed

(continued)

Table 4.1 (continued)

Gene name[a]	Gene ID	Tissue-specific expression	Cell-specific expression	Subcellular localisation	Abiotic stress increased expression[b]	Abiotic stress reduced expression[b]	Protein interactions[c]
AtCAX5	At1g55730	Roots[e], leaves[e], flowers[e], seed, seedling, old leaves	Guard cell, root epidermis and cortex	Plasma membrane? (vacuole in yeast[f])	Osmotic (6–24 h), salt (24 h), ABA (12 h), Mn[e]	Heat (1 h)	ACA10
AtCAX6	At1g55720	Flowers, dry seed, seedling, old leaves	Guard cell, root cortex	Plasma membrane?, mitochondria?	Osmotic (6–24 h), salt (24 h)	Heat (1 h)	ACA1
AtEFCAX1/AtNCL	At1g53210	Root[d], leaves[d], flowers, seed, seedling[d], young leaves[d]	Guard cell, mesophyll, root endodermis[d]	Plasma membrane[f], vacuole?	Cold (12–24 h)[e], salt (12–24 h)[e], drought (24 h), wounding (0.25–1 h), ABA (12 h)[e], heat (1–12 h)[e]	UV-B (1–6 h), ABA (1 h)	None observed
AtEFCAX2	At1g29020	Root, seedling root	Root cap, epidermis and cortex	Plasma membrane	None observed	Osmotic (3–24 h), salt (12 h), ABA (12 h)	None observed
OsCAX1a	Os01g37690	Leaves[d], inflorescence (P4, P5)[d], seedling root[e], flowers[d]	Guard cells[d], vascular bundles[d], trichomes[d]	Vacuole[f]	Ca[e]	Drought, salt[e], cold[e] (all 7-day seedlings), Mn[e], Ni[e], Mg[e], Cu[e]	None observed
OsCAX1b	Os05g51610	Young leaf, seedling root	ND	Vacuole	None observed	Drought (7-day seedling)	None observed
OsCAX1c	Os02g21009	Leaves[e]	ND	Vacuole	None observed	Drought, salt, cold (all 7-day seedlings)	None observed

OsCAX2	Os03g27960	Seedling root[e]	ND	Vacuole	Drought, salt (both 7-day seedlings)	None observed	None observed
OsCAX3	Os04g55940	Seedling root[e]	ND	Vacuole	None observed	Drought, salt (both 7-day seedlings)	None observed
OsCAX4	Os02g04630	Inflorescence (P6)	ND	Vacuole	Drought (7-day seedling)	None observed	None observed
OsEFCAX1	Os01g11414	Inflorescence (P4), seedling root	ND	Plasma membrane	Drought (7-day seedling)	None observed	None observed
OsEFCAX2	Os02g14980	Seed (S5)	ND	Plasma membrane	None observed	Salt, cold (both 7-day seedlings)	None observed

Gene expression information was obtained using the *Arabidopsis* and Rice eFP Browsers (Winter et al. 2007; Patel et al. 2012) except for metal-induced expression data collated from individual studies

Subcellular localisation was determined using the Cell eFP Browser (Winter et al. 2007) or from experimental data

Protein interactions were determined using the *Arabidopsis* and Rice Interactions Viewers (Geisler-Lee et al. 2007; Ho et al. 2012)

Genomics browsers at http://bar.utoronto.ca

ND, not determined

[a]OsCAX1a, OsCAX1b, OsCAX1c, OsCAX2, OsCAX3 and OsCAX4 are named as OsCAX1, OsCAX3, OsCAX4, OsCAX5, OsCAX6 and OsCAX2, respectively by Singh et al. (2014)

[b]For some stress conditions, the time point of expression (in hours or days) is given

[c]Only experimentally determined or high-confidence interactions are listed

[d]Confirmed experimentally using promoter-GUS analysis by Cheng et al. (2005), Kamiya et al. (2006), Mei et al. (2009), Pittman et al. (2004), Wang et al. (2012), Zhao et al. (2009b)

[e]Confirmed experimentally using Northern blotting or RT-PCR by Catalá et al. (2003), Cheng et al. (2002, 2005), Cho et al. (2012), Conn et al. (2011), Edmond et al. (2009), Hirschi (1999), Hirschi et al. (2000), Kamiya et al. (2005, 2006, 2012), Liu et al. (2011), Singh et al. (2014), Wang et al. (2012)

[f]Confirmed experimentally using fluorescent protein fusions or fractionation by Cheng et al. (2002, 2003, 2005), Edmond et al. (2009), Hirschi et al. (2000), Kamiya et al. (2006), Wang et al. (2012)

[g]Confirmed experimentally using yeast-2 hybrid, split ubiquitin analysis or biofluorescence complementation analysis by Cheng and Hirschi (2003), Cheng et al. (2004a, 2009b), Zhao et al. (2004a, b)

4.4 The Physiological Functions of Plant CAXs

Some of the proposed physiological roles of CAXs, determined principally from analysis of *Arabidopsis* and rice CAX genes, are summarised in Fig. 4.3. Genetic studies have highlighted the importance of CAX in maintaining Ca^{2+} homeostasis in plants. In tobacco plants over-expressing *AtCAX1*, Ca accumulation was 30 % higher in leaves compared with a vector control and was double in roots (Hirschi 1999). The transgenic plants showed a number of symptoms reminiscent of Ca^{2+} deficiency, such as sensitivity to K and Mg, and cold shock, while these were reversed upon application of Ca (Hirschi 1999). To complement this phenotype, it was later shown that *cax1* mutants could tolerate growth in serpentine solutions with a low Ca/Mg ratio, while mutants grew poorly in higher concentrations of Mg (Bradshaw 2005). A similar phenotype was observed when *AtCAX1* was over-expressed in tomato in order to increase total fruit Ca levels by vacuolar sequestration; however, the *AtCAX1*-over-expressing tomatoes showed an increased incidence of blossom-end rot, which is associated with localised cellular Ca^{2+} deficiency

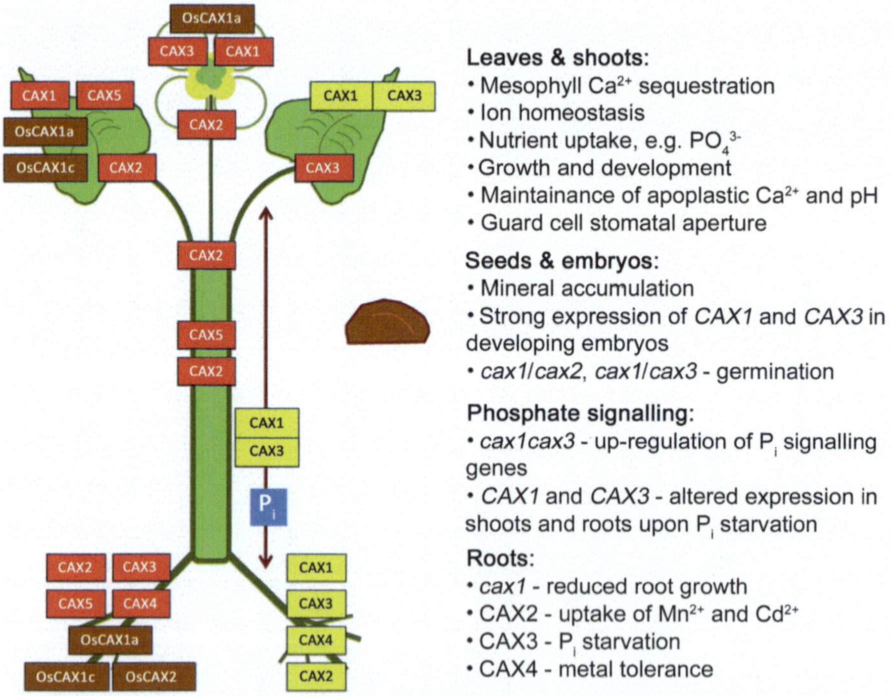

Fig. 4.3 Physiological roles of Ca^{2+}/H^+ exchangers (CAX) in plants. *Red* and *brown boxes* indicate the highest expression of various *Arabidopsis* and *Oryza sativa* (rice) CAX genes, respectively. *Yellow boxes* indicate where *Arabidopsis* CAX has been implicated in plant development and stress response, which are described in *bullet* points on the *right*, adjacent to their position on the plant

(Park et al. 2005). Due to the high level of Ca^{2+} sequestration performed by AtCAX1, attempts have been made to biofortify other crops, including carrots (Morris et al. 2008), as Ca deficiency remains one of the leading causes of malnutrition worldwide.

AtCAX1 is highly expressed in *Arabidopsis* leaves, while *AtCAX3* shows the highest expression in roots, but analysis of *cax1cax3* mutants indicates functional interaction between the two proteins in mediating ion homeostasis *in planta* (Cheng et al. 2005), which is explained both by genetic redundancy (Conn et al. 2011) and direct protein–protein interaction between both CAXs (Zhao et al. 2009a, b). As might also be expected, total Ca content is reduced slightly in *cax1* plants and substantially in *cax1cax3* plants (Catalá et al. 2003; Cheng et al. 2005; Conn et al. 2011). Furthermore, the Mg content of *cax1cax3* shoots was also lower than control, and, interestingly P, Mg and Zn levels were higher. It is therefore unclear whether the severe growth defects such as leaf-tip necrosis, stunted growth and small siliques exhibited by the *cax1cax3* plants are partly due to impaired Ca^{2+} homeostasis or alteration in concentration of other metals (Cheng et al. 2003). Further probing into the P accumulation phenotype revealed that *cax1cax3* mutants not only accumulate more P but also show a greater shoot/root P ratio under P starvation (Liu et al. 2011). *AtCAX1* expression levels were significantly reduced in shoots upon P starvation. Furthermore, the expression of phosphate starvation response genes was greatly altered in *cax1cax3* mutants under high P, which included the upregulation of several phosphate signaling genes, therefore implicating AtCAX1 and AtCAX3 in root-shoot signaling of P starvation in *Arabidopsis* (Liu et al. 2011). Another interesting gene upregulated in *cax1cax3* mutants is a glucose-6-phosphate/phosphate translocator *GPT2* (Liu et al. 2011), which is thought to have a role in dynamic acclimation of photosynthesis in response to environmental perturbations (Athanasiou et al. 2010), highlighting the variety of processes that are potentially mediated by CAX.

While *AtCAX1* and *AtCAX3* show differential tissue-specific expression in mature *Arabidopsis*, both genes are expressed in *Arabidopsis* seeds (Zhao et al. 2009b; Punshon et al. 2012). However, *AtCAX1* has the highest expression in the seed coat, in the cotyledon and at the heart stage of development, as well as in all stages of chalazal endosperm development, while *AtCAX3* showed mostly moderate expression, but was highly expressed in the linear cotyledon and mature green seeds. An analysis of the distribution of various elements in seeds found that *cax1* mutants had higher Ca, Cu and Mn content, while *AtCAX1*-over-expressing seeds showed a reduction in these elements, and *cax3* showed similar perturbances in Ca and Cu (Punshon et al. 2012). Using SXRF microtomography to further probe elemental distribution in seeds, it was found that Ca content was much higher in the seed coat of *cax1*, *cax3* and *cax1cax3* than wild type, while the opposite occurred in 35S::*AtCAX1* seeds (Punshon et al. 2012). Furthermore, Zn, K and Ca were higher in *cax1* and *cax3* seeds, but lower in those of 35S::*AtCAX1*. In endodermal cells of *cax1cax3* mutants, vacuolar storage of Ca, Cu, Fe, P, S and Zn was completely absent (Punshon et al. 2012). A significantly higher accumulation of Mn and Zn has also been measured in *cax2* seeds, while *cax2*, *cax3*, *cax1cax2* and *cax2cax3* mutants showed increased Fe, K and P (Connorton et al. 2012). The challenge is now to dissect the elemental changes that are due to direct ion transport by CAXs and those that are indirect effects. Perturbed seed metal homeostasis has been linked to significant reduction in germination in all mutants, although the strongest decrease was observed in *cax3* seeds, which showed only 25 % germination compared with over 80 % in wild-type plants (Connorton et al. 2012).

In mesophyll cells, AtCAX1 and AtCAX3 appear to be especially important in maintaining Ca^{2+} homeostasis, as well as physiological processes such as stomatal conductance. Ca content in mesophyll cells is significantly reduced in *cax1cax3* mutants, while the Ca^{2+}-ATPase mutant *aca4aca11* shows no change in Ca content, which suggests that CAX is the main route of Ca^{2+} sequestration into mesophyll vacuoles (Conn et al. 2011). Interestingly, *cax1cax3* mutants show a decrease in stomatal closure compared to wild-type plants (Conn et al. 2011), which was also observed to occur in *det3* mutants lacking V-ATPase activity (Allen et al. 2000). The control of CAX by V-ATPases has been well established (Krebs et al. 2010); therefore, AtCAX1 and AtCAX3 may regulate stomatal aperture via guard cell Ca^{2+} signaling. However, it has been argued that the 4-fold increase in apoplastic Ca in the *cax1cax3* mesophyll is the cause of stomatal aperture closure, leading to lower stomatal conductance, CO_2 limitation and thus reduced growth (Conn et al. 2011). However, it has also been suggested that a modulation of apoplastic pH in guard cells by AtCAX1 and AtCAX3, probably via plasma membrane H^+-ATPase activity, influences auxin transport activity and in turn regulates stomatal closure (Cho et al. 2012). Both scenarios are not mutually exclusive.

4.4.1 CAX in Abiotic Stress Signaling

A number of studies have implied a role for *Arabidopsis* CAX in mediating plant abiotic stress responses, particularly salt and cold stress. Expression of *AtCAX3* and *AtCAX4* was slightly upregulated in wild-type plants and markedly in *cax1* mutants under salt stress (Cheng et al. 2003). *AtCAX2* also showed slight induction in response to salt stress but with no difference in wild-type and *cax1* plants. Interestingly, the vacuolar Ca^{2+}-ATPase, *AtACA4*, was downregulated in response to high salt suggesting that CAXs are more important in vacuolar Ca^{2+} sequestration under salt stress conditions. *AtCAX1* and *AtCAX3* display functional interplay in a variety of processes, and their co-expression confers salt tolerance in yeast (Zhao et al. 2009b). However, some distinct roles have been observed for AtCAX3, particularly with regard to salt stress. In guard cells, for example, *AtCAX3* expression increases between 2.5- and 4.9-fold in response to ABA treatment (Leonhardt et al. 2004), while *cax3* plants show the most reduced growth on ABA (Connorton et al. 2012). An analysis of *cax3* and *cax1cax3* mutants showed that these plants have severely reduced root growth when treated with NaCl compared with both wild-type and *cax1* plants (Zhao et al. 2008). Furthermore, Ca^{2+}/H^+ exchange activity was also decreased in *cax3* but not in *cax1* mutants under salt stress. Therefore, the genetic evidence suggests that in *Arabidopsis* CAX activity is important in the salt stress response, with AtCAX3 appearing to be the most important isoform.

Slightly conflicting with this genetic data is evidence from yeast expression studies that implicates AtCAX1 as a component in a key Ca^{2+}-regulated salt stress response pathway. The CIPK-type protein kinase SOS2 (also known as CIPK24) is a central component in the salt overly sensitive (SOS) cell signaling response pathway to excess salt. The Ca^{2+}-binding CBL-type sensor SOS3 (CBL4) activates SOS2, and together this SOS2–SOS3 (CBL4/CIPK24) complex activates SOS1, a Na^+/H^+ exchanger at the plasma membrane (Ji et al. 2013). In a yeast assay, SOS2 was shown to activate AtCAX1 Ca^{2+}/H^+ exchange through direct protein interaction (Cheng et al. 2004b). It was not confirmed whether SOS2 activates AtCAX1 by phosphorylation, but when a candidate phosphorylated Ser residue (Ser-25) within the AtCAX1 N-terminus (Pittman et al. 2002a) is mutated (to S25A), SOS2 activation cannot occur. The regulation by SOS2 was also found to be independent of SOS3 (Cheng et al. 2004b). An interaction between these two proteins highlights a potentially important role for AtCAX1 in the signaling of salt stress, mediated in part by the SOS pathway, but interaction *in planta* has not been demonstrated.

Studies in other plant species such as rice, tobacco, soybean and the halophyte *Suaeda salsa* have further supported the role of CAX in salt stress. Expression of *SsCAX1* from *S. salsa* increases sensitivity to salt in transgenic *Arabidopsis* (Han et al. 2012). Likewise, expression of deregulated *AtCAX1* in tobacco increases salt sensitivity (Mei et al. 2007). This may indicate that alteration in net CAX activity causes a disruption in the plant's ability to mediate a normal salt stress response, possibly through disruption in modulation of the salt-induced Ca^{2+} signature. In contrast, high expression of a CAX gene was observed in a salt-tolerant rice variety under salt stress conditions, while rice CAX expression was reduced in salt-sensitive varieties (Senadheera et al. 2009). Moreover, in transgenic rice, over-expression of a V-ATPase subunit from *Spartina alterniflora*, a halophyte grass, increased salt tolerance (Baisakh et al. 2012). This coincided with an increase in Ca^{2+} sequestration and a substantial upregulation of several genes involved in the salt stress response, pointing towards a role for CAX in mediating salt tolerance in rice. Similarly, expression of a soybean CAX (*GmCAX1*) was induced by various treatments including salt and also ABA, PEG, Ca and Li, and when *GmCAX1* was over-expressed in *Arabidopsis*, greater tolerance to both Li and Na was conferred, while the plant accumulated less Na, K and Li (Luo et al. 2005). Interestingly, GmCAX1 was suggested to be localised to the plasma membrane rather than the tonoplast, indicating that GmCAX1 caused decreased salt stress sensitivity by cellular Na^+ efflux, potentially by this exchanger. The first clear-cut evidence of a CAX from a photosynthetic organism being able to directly transport Na^+ has come from the green alga *Chlamydomonas reinhardtii*. When heterologously expressed in yeast, CrCAX1 can transport Na^+ via H^+ exchange into yeast vacuoles (Pittman et al. 2009) (see below).

AtCAX1 has been implicated in controlling the expression of *CBF/DREB1* genes, which are induced under cold stress in *cax1* mutants (Catalá et al. 2003). *AtCAX1* is transiently inducible under cold stress in *Arabidopsis* leaves. However, expression of *AtCAX1* in response to cold was shown to be independent of ABA, as expression was equivalent in ABA-deficient and ABA-insensitive mutants and in wild-type plants (Catalá et al. 2003). More interestingly, *cax1* mutants showed increased tolerance to cold-acclimated freezing stress, coincident with induction of known cold-responsive genes by low temperature. CBF/DREB1 is a class of transcription factor that binds to specific *cis*-acting genetic elements (DRE, drought-responsive *cis*-element) in abiotic stress response genes (Agarwal et al. 2006). *CBF/DREB1* transcription factors are upregulated in response to a variety of abiotic stimuli, including dehydration, cold and salt (Narusaka et al. 2003). In *cax1* plants, under cold stress, the transcription factors *CBF1/DREB1B*, *CBF2/DREB1C* and *CBF3/DREB1A* were highly expressed than in wild-type plants (Catalá et al. 2003). AtCAX1 appears to be a negative regulator of the cold acclimation response, with the possible conclusion that Ca^{2+} signals modulated by AtCAX1 are required for proper generation of the cold response via transcription factor induction. In contrast, a CAX from cotton (GhCAX3), which can be upregulated by cold stress (as well as salt stress and ABA), positively confers a degree of cold tolerance when over-expressed in yeast and mature plants, although the *GhCAX3*-over-expression lines confer cold sensitivity on seed germination (Xu et al. 2013). Ca^{2+}/H^{+} exchangers therefore appear to play an important role in the signal transduction of cold stress, but there are differences between species, potentially due to the difference in expression pattern and species-specific signal transduction pathways.

Another potential signaling role for CAX is indicated by the tight relationship between CAX and pH, as pH modulation, like Ca^{2+} modulation, may mediate intracellular signaling processes (Felle 2001). There is increasing evidence that ion transporters like CAX that mediate H^{+} exchange play a role in pH homeostasis within cells (Pittman 2012). In addition, AtCAX1 and AtCAX2 activity can in turn be regulated by cytosolic pH (Pittman et al. 2005). Furthermore, there exist various examples of deregulation of V-ATPase activity in CAX mutants, while reduced activity of the plasma membrane H^{+}-ATPase has also been observed in *cax3* mutants (Barkla et al. 2008; Zhao et al. 2008). Recently there has been evidence of a link between AtCAX1/AtCAX3 activity, apoplastic pH in guard cells and IAA-mediated stomatal closure. *Arabidopsis cax1cax3* mutants failed to maintain apoplastic pH compared to wild type, which was observed alongside reduced inhibition of ABA-mediated signaling of stomatal closure by IAA (Cho et al. 2012). An analysis of gene expression in *Arabidopsis* roots showed upregulation of a number of genes involved in Ca^{2+} signaling in response to a lowering of pH, corresponding with over-expression of CaM motifs in the promoters of pH-responsive genes (Lager et al. 2010). Additionally, evidence suggests that pH and Ca^{2+} are modulated both in the cytosol and the apoplast in response to abiotic stresses, including salt and drought (Gao et al. 2004). Thus, the close relationship observed between CAX and pH indicates another link and potential mechanism in mediating plant abiotic stress responses.

The precise role of CAX in modulating plant Ca^{2+} signals has been hard to discern, potentially due to CAX gene redundancy in higher plants and the presence of other important Ca^{2+} efflux proteins, including Ca^{2+} pumps and NCX. This is in contrast to model organisms, such as *S. cerevisiae* or *Plasmodium*, which possess just one or two CAX genes and in which genetic analysis of CAXs have been very fruitful (Denis and Cyert 2002; Guttery et al. 2013). A photosynthetic model organism with great potential for further elucidating the role of CAX in cell signaling is *C. reinhardtii*. Compared to higher plants there is a lesser degree of gene redundancy in *Chlamydomonas*, which contains only two expressed CAX open reading frames compared with six in *Arabidopsis* and rice (Emery et al. 2012). Phylogenetic analysis of CAX proteins in *Chlamydomonas* shows that *CrCAX1* and *CrCAX1* cluster with type IC CAXs. Although distinct from the phylogenetic subgroups containing higher plant CAXs, these proteins share many features of the higher plant CAXs, including a similar post-translational regulatory mechanism, tonoplast localisation and Ca^{2+}/H^{+} exchange activity (Pittman et al. 2009). CrCAX1 is also interesting as it can transport a diverse range of cations including Na^{+}. The ability of *Chlamydomonas* CAX1 to transport Na^{+} provides an interesting target to further study the role of CAX in mediating salt stress, both in a direct Na^{+} homeostasis and Ca^{2+} signaling capacity, especially considering that when over-expressed in *Arabidopsis*, *CrCAX1* has been shown to increase plant salt tolerance (Pittman et al. 2009).

4.4.2 Role of CAX in Mediating Plant Responses to Toxic Metal Stress

Although AtCAX2 has been shown to transport Ca^{2+} into yeast vacuoles (Hirschi et al. 1996), *cax2* mutants of *Arabidopsis* do not show any significant alteration in vacuolar Ca^{2+}/H^{+} activity or any Ca sensitivity defects (Pittman et al. 2004). Vacuolar Mn^{2+}/H^{+} exchange is, however, reduced in *cax2* plants. *Arabidopsis cax2* and *cax2cax3* mutants were more recently shown to exhibit vastly reduced growth on high Mn (Connorton et al. 2012). AtCAX2 contains a three amino acid residue, which confers Mn^{2+} specificity, removal of which inhibits Mn^{2+} transport by AtCAX2 (Shigaki et al. 2003). Over-expression of *AtCAX2* in tobacco yielded increased accumulation of Ca, Cd and Mn, as well as tolerance to growth on high Mn, which points towards a role for AtCAX2 in sequestering heavy metals and ameliorating heavy metal stress (Hirschi et al. 2000). Expression of *AtCAX5*, a close homologue of *AtCAX2*, has been shown to increase under high Cd and Mn stress (Edmond et al. 2009). In yeast, AtCAX5 functions similarly to AtCAX2, whereby N-terminal truncation led to transport of Ca^{2+} and Mn^{2+} into yeast vacuoles. Additionally, the coincident expression of *AtCAX2* and *AtCAX5* in all plant tissues, including highest expression in the roots, suggests that these proteins play similar and potentially redundant roles in transporting transition metals such as Mn^{2+} and Cd^{2+} (Edmond et al. 2009). As yet the phenotype of *cax5* mutants has not been examined, and it is unknown whether Mn^{2+}/H^{+} exchange would be reduced in these plants. Furthermore, *cax2cax5* double mutants may indicate whether there is any functional interplay and/or redundancy between AtCAX2 and AtCAX5, as there is between AtCAX1 and AtCAX3.

AtCAX4 has high sequence identity with *AtCAX1*, yet *AtCAX4* was unable to suppress the Ca^{2+} hypersensitivity of the yeast mutant, and so direct evidence of AtCAX4 Ca^{2+} transport has been lacking (Cheng et al. 2002). Although *in planta* analysis indicates that AtCAX4 can mediate vacuolar Ca^{2+}/H^+ exchange (Mei et al. 2009), this CAX may be more important for transport of other metals. *AtCAX4* is expressed at low levels throughout *Arabidopsis* plants, although expression increases in response to Mn, Ni and Na conditions (Cheng et al. 2002). In *cax4* mutants, root length and lateral root number are reduced in response to high levels of Cd and Mn, while this phenotype is reversed upon co-expression of *AtCAX4* (Mei et al. 2009). Furthermore, tobacco plants over-expressing *AtCAX2* and *AtCAX4* show increased transport of Zn^{2+} and Cd^{2+} at the tonoplast, as well as increased root Cd content (Koren'kov et al. 2007). Furthermore, these plants show higher tolerance to growth on excess Zn, Mn and Cd.

Due to the observed effects of selected CAX proteins for providing tolerance to metal stress in plants through the vacuolar sequestration of toxic ions such as Mn^{2+}, Cd^{2+} and Zn^{2+}, there has been some interest in modifying plants in order to bioremediate toxic soils. To this end, a CAX mutant which confers high Cd^{2+} transport was over-expressed in petunia plants and conferred greater tolerance to growth on high levels of Cd (Wu et al. 2011). Transgenic plants were also shown to accumulate significantly greater levels of Cd, while plant development was unaffected until the flowering stage, which suggests that the over-expression of the CAX mutant may provide a useful route into remediating Cd-enriched soils. Bioremediation is not limited to soils but also can be performed in metal-polluted wastewaters. As *Chlamydomonas* CAX1 can transport Cd^{2+}, as well as Ca^{2+} and Na^+ (Pittman et al. 2009), further research into over-expression of *CrCAX1* in algae may lead to developments in the bioremediation of polluted ground waters, ponds and lakes.

Conclusions

Cation/H^+ exchangers clearly play an important role in allowing plants to respond to abiotic stresses and to tolerate these stresses to some degree. Part of this role appears to be in shaping stress-induced Ca^{2+} signals, and thus, these transporters are involved in allowing a plant cell to convert recognition of the stress into an appropriate response (Fig. 4.1). Genetic studies, principally from *Arabidopsis* knockout experiments, have provided clear evidence that specific CAX genes are required for providing tolerance to specific abiotic stresses rather than all stresses. However, future research is still required to understand how specific stresses activate Ca^{2+} transport of specific CAX isoforms. For example, why some CAX isoforms appear to be involved in cold stress pathways but other isoforms are not. This may partly be due to differences in tissue and cellular expression patterns and/or specific stress-induced transcriptional induction of *CAX* genes but, at the post-transcriptional level, may also be due to stress-induced protein activators that regulate specific CAXs through direct protein interaction, such as SOS2. Furthermore, strong evidence that

CAX-mediated Ca^{2+}/H^+ exchange does generate and/or modulate a Ca^{2+} signature is still lacking, partly due to the issue of genetic redundancy amongst the CAX gene family in most plant species. Another part of CAXs role in abiotic stress response is due to the transport of cations other than Ca^{2+}. It is clear that many CAX isoforms can transport a range of cations, including Mn^{2+}, Cd^{2+} and Zn^{2+}, which can cause plant toxicity when these metals are present in high abundance, and thus such CAXs can alleviate metal stress through vacuolar sequestration. There is also evidence that some CAX transporters can mediate Na^+ transport, including members of the newly discovered EFCAX family. Thus, in the cases where CAXs have been implicated in providing salt tolerance, careful analysis is required to determine whether this is due to direct Na^+ efflux from the cytosol or modulation of a salt-induced Ca^{2+} signature.

References

Agarwal M, Hao YJ, Kapoor A, Dong CH, Fujii H, Zheng XW, Zhu JK (2006) A R2R3 type MYB transcription factor is involved in the cold regulation of CBF genes and in acquired freezing tolerance. J Biol Chem 281:37636–37645

Allen GJ, Chu SP, Schumacher K, Shimazaki CT, Vafeados D, Kemper A, Hawke SD, Tallman G, Tsien RY, Harper JF, Chory J, Schroeder JI (2000) Alteration of stimulus-specific guard cell calcium oscillations and stomatal closing in Arabidopsis *det3* mutant. Science 289: 2338–2342

Athanasiou K, Dyson BC, Webster RE, Johnson GN (2010) Dynamic acclimation of photosynthesis increases plant fitness in changing environments. Plant Physiol 152:366–373

Baisakh N, RamanaRao MV, Rajasekaran K, Subudhi P, Janda J, Galbraith D, Vanier C, Pereira A (2012) Enhanced salt stress tolerance of rice plants expressing a vacuolar H^+-ATPase subunit c1 (SaVHAc1) gene from the halophyte grass *Spartina alterniflora* Loisel. Plant Biotechnol J 10:453–464

Barkla BJ, Hirschi KD, Pittman JK (2008) Exchangers man the pumps: Functional interplay between proton pumps and proton-coupled Ca^{2+} exchangers. Plant Signal Behav 3:354–356

Baxter I, Tchieu J, Sussman MR, Boutry M, Palmgren MG, Gribskov M, Harper JF, Axelsen KB (2003) Genomic comparison of P-type ATPase ion pumps in Arabidopsis and rice. Plant Physiol 132:618–628

Bickerton PD, Pittman JK (2012) Calcium signaling in plants. In: eLS. Wiley, New York

Bock KW, Honys D, Ward JM, Padmanaban S, Nawrocki EP, Hirschi KD, Twell D, Sze H (2006) Integrating membrane transport with male gametophyte development and function through transcriptomics. Plant Physiol 140:1151–1168

Bonza MC, De Michelis MI (2011) The plant Ca^{2+}-ATPase repertoire: biochemical features and physiological functions. Plant Biol (Stuttg) 13:421–430

Bradshaw HD (2005) Mutations in *CAX1* produce phenotypes characteristic of plants tolerant to serpentine soils. New Phytol 167:81–88

Case RM, Eisner D, Gurney A, Jones O, Muallem S, Verkhratsky A (2007) Evolution of calcium homeostasis: From birth of the first cell to an omnipresent signaling system. Cell Calcium 42:345–350

Catalá R, Santos E, Alonso JM, Ecker JR, Martínez-Zapater JM, Salinas J (2003) Mutations in the Ca^{2+}/H^+ transporter *CAX1* increase *CBF/DREB1* expression and the cold-acclimation response in Arabidopsis. Plant Cell 15:2940–2951

Cheng NH, Hirschi KD (2003) Cloning and characterization of CXIP1, a novel PICOT domain-containing *Arabidopsis* protein that associates with CAX1. J Biol Chem 278:6503–6509

Cheng NH, Pittman JK, Shigaki T, Hirschi KD (2002) Characterization of CAX4, an Arabidopsis H^+/cation antiporter. Plant Physiol 128:1245–1254
Cheng NH, Pittman JK, Barkla BJ, Shigaki T, Hirschi KD (2003) The Arabidopsis *cax1* mutant exhibits impaired ion homeostasis, development, and hormonal responses and reveals interplay among vacuolar transporters. Plant Cell 15:347–364
Cheng NH, Liu JZ, Nelson RS, Hirschi KD (2004a) Characterization of CXIP4, a novel *Arabidopsis* protein that activates the H^+/Ca^{2+} antiporter, CAX1. FEBS Lett 559:99–106
Cheng NH, Pittman JK, Zhu JK, Hirschi KD (2004b) The protein kinase SOS2 activates the *Arabidopsis* H^+/Ca^{2+} antiporter CAX1 to integrate calcium transport and salt tolerance. J Biol Chem 279:2922–2926
Cheng NH, Pittman JK, Shigaki T, Lachmansingh J, LeClere S, Lahner B, Salt DE, Hirschi KD (2005) Functional association of Arabidopsis CAX1 and CAX3 is required for normal growth and ion homeostasis. Plant Physiol 138:2048–2060
Cho D, Villiers F, Kroniewicz L, Lee S, Seo YJ, Hirschi KD, Leonhardt N, Kwak JM (2012) Vacuolar CAX1 and CAX3 influence auxin transport in guard cells via regulation of apoplastic pH. Plant Physiol 160:1293–1302
Conn SJ, Gilliham M, Athman A, Schreiber AW, Baumann U, Moller I, Cheng NH, Stancombe MA, Hirschi KD, Webb AAR, Burton R, Kaiser BN, Tyerman SD, Leigh RA (2011) Cell-specific vacuolar calcium storage mediated by CAX1 regulates apoplastic calcium concentration, gas exchange, and plant productivity in Arabidopsis. Plant Cell 23:240–257
Connorton JM, Webster RE, Cheng NH, Pittman JK (2012) Knockout of multiple Arabidopsis Cation/H^+ exchangers suggests isoform-specific roles in metal stress response, germination and seed mineral nutrition. PLoS One 7:e47455
Denis V, Cyert MS (2002) Internal Ca^{2+} release in yeast is triggered by hypertonic shock and mediated by a TRP channel homologue. J Cell Biol 156:29–34
Dodd AN, Kudla J, Sanders D (2010) The language of calcium signaling. Annu Rev Plant Biol 61:593–620
Edmond C, Shigaki T, Ewert S, Nelson M, Connorton J, Chalova V, Noordally Z, Pittman JK (2009) Comparative analysis of CAX2-like cation transporters indicates functional and regulatory diversity. Biochem J 418:145–154
Emery L, Whelan S, Hirschi KD, Pittman JK (2012) Protein phylogenetic analysis of Ca^{2+}/cation antiporters and insights into their evolution in plants. Front Plant Sci 3:1
Felle HH (2001) pH: Signal and messenger in plant cells. Plant Biol 3:577–591
Gao DJ, Knight MR, Trewavas AJ, Sattelmacher B, Plieth C (2004) Self-reporting Arabidopsis expressing pH and Ca^{2+} indicators unveil ion dynamics in the cytoplasm and in the apoplast under abiotic stress. Plant Physiol 134:898–908
Geisler-Lee J, O'Toole N, Ammar R, Provart NJ, Millar AH, Geisler M (2007) A predicted interactome for Arabidopsis. Plant Physiol 145:317–329
Guttery DS, Pittman JK, Frenal K, Poulin B, McFarlane LR, Slavic K, Wheatley SP, Soldati-Favre D, Krishna S, Tewari R, Staines HM (2013) The *Plasmodium berghei* Ca^{2+}/H^+ exchanger, PbCAX, is essential for tolerance to environmental Ca^{2+} during sexual development. PLoS Pathog 9:e1003191
Han N, Lan WJ, He X, Shao Q, Wang BS, Zhao XJ (2012) Expression of a *Suaeda salsa* vacuolar H^+/Ca^{2+} transporter gene in *Arabidopsis* contributes to physiological changes in salinity. Plant Mol Biol Rep 30:470–477
Hirschi KD (1999) Expression of Arabidopsis *CAX1* in tobacco: Altered calcium homeostasis and increased stress sensitivity. Plant Cell 11:2113–2122
Hirschi K (2001) Vacuolar H^+/Ca^{2+} transport: who's directing the traffic? Trends Plant Sci 6:100–104
Hirschi KD, Zhen RG, Cunningham KW, Rea PA, Fink GR (1996) CAX1, an H^+/Ca^{2+} antiporter from *Arabidopsis*. Proc Natl Acad Sci U S A 93:8782–8786
Hirschi KD, Korenkov VD, Wilganowski NL, Wagner GJ (2000) Expression of Arabidopsis CAX2 in tobacco. Altered metal accumulation and increased manganese tolerance. Plant Physiol 124:125–133

Ho C-L, Wu Y, H-B S, Provart N, Geisler M (2012) A predicted protein interactome for rice. Rice (N Y) 5:15

Huda KMK, Banu MSA, Tuteja R, Tuteja N (2013) Global calcium transducer P-type Ca^{2+}-ATPases open new avenues for agriculture by regulating stress signaling. J Exp Bot 64:3099–3109

Jammes F, Hu H-C, Villiers F, Bouten R, Kwak JM (2011) Calcium-permeable channels in plant cells. FEBS J 278:4262–4276

Ji HT, Pardo JM, Batelli G, Van Oosten MJ, Bressan RA, Li X (2013) The salt overly sensitive (SOS) pathway: established and emerging roles. Mol Plant 6:275–286

Kamiya T, Akahori T, Maeshima M (2005) Expression profile of the genes for rice cation/H^+ exchanger family and functional analysis in yeast. Plant Cell Physiol 46:1735–1740

Kamiya T, Akahori T, Ashikari M, Maeshima M (2006) Expression of the vacuolar Ca^{2+}/H^+ exchanger, OsCAX1a, in rice: Cell and age specificity of expression, and enhancement by Ca^{2+}. Plant Cell Physiol 47:96–106

Kamiya T, Yamagami M, Hirai MY, Fujiwara T (2012) Establishment of an in planta magnesium monitoring system using CAX3 promoter-luciferase in Arabidopsis. J Exp Bot 63:355–363

Kim TH, Bohmer M, Hu HH, Nishimura N, Schroeder JI (2010) Guard cell signal transduction network: Advances in understanding abscisic acid, CO_2, and Ca^{2+} signaling. Annu Rev Plant Biol 61:561–591

Koren'kov V, Park S, Cheng NH, Sreevidya C, Lachmansingh J, Morris J, Hirschi K, Wagner GJ (2007) Enhanced Cd^{2+}-selective root-tonoplast-transport in tobaccos expressing Arabidopsis cation exchangers. Planta 225:403–411

Krebs M, Beyhl D, Gorlich E, Al-Rasheid KAS, Marten I, Stierhof Y-D, Hedrich R, Schumacher K (2010) Arabidopsis V-ATPase activity at the tonoplast is required for efficient nutrient storage but not for sodium accumulation. Proc Natl Acad Sci U S A 107:3251–3256

Kudla J, Batistic O, Hashimoto K (2010) Calcium signals: the lead currency of plant information processing. Plant Cell 22:541–563

Laanemets K, Brandt B, Li J, Merilo E, Wang Y-F, Keshwani MM, Taylor SS, Kollist H, Schroeder JI (2013) Calcium-dependent and -independent stomatal signaling network and compensatory feedback control of stomatal opening via Ca^{2+} sensitivity priming. Plant Physiol 163:504–513

Lager I, Andreasson O, Dunbar TL, Andreasson E, Escobar MA, Rasmusson AG (2010) Changes in external pH rapidly alter plant gene expression and modulate auxin and elicitor responses. Plant Cell Environ 33:1513–1528

Lalonde S, Sero A, Pratelli R, Pilot G, Chen J, Sardi MI, Parsa SA, Kim D-Y, Acharya BR, Stein EV, Hu H-C, Villiers F, Takeda K, Yang Y, Han YS, Schwacke R, Chiang W, Kato N, Loque D, Assmann SM, Kwak JM, Schroeder JI, Rhee SY, Frommer WB (2010) A membrane protein/signaling protein interaction network for Arabidopsis version AMPv2. Front Physiol 1:24–24

Leonhardt N, Kwak JM, Robert N, Waner D, Leonhardt G, Schroeder JI (2004) Microarray expression analyses of Arabidopsis guard cells and isolation of a recessive abscisic acid hypersensitive protein phosphatase 2C mutant. Plant Cell 16:596–615

Li S, Ehrhardt DW, Rhee SY (2006) Systematic analysis of Arabidopsis organelles and a protein localization database for facilitating fluorescent tagging of full-length Arabidopsis proteins. Plant Physiol 141:527–539

Lindberg S, Kader MA, Yemelyanov V (2012) Calcium signaling in plant cells under environmental stress. In: Prasad MNV, Ahmad P (eds) Environmental adaptations and stress tolerance of plants in the era of climate change. Springer, New York, pp 325–360

Liu T-Y, Aung K, Tseng C-Y, Chang T-Y, Chen Y-S, Chiou T-J (2011) Vacuolar Ca^{2+}/H^+ transport activity is required for systemic phosphate homeostasis involving shoot-to-root signaling in Arabidopsis. Plant Physiol 156:1176–1189

Luo GZ, Wang HW, Huang J, Tian AG, Wang YJ, Zhang JS, Chen SY (2005) A putative plasma membrane cation/proton antiporter from soybean confers salt tolerance in *Arabidopsis*. Plant Mol Biol 59:809–820

Maathuis FJM, Filatov V, Herzyk P, Krijger GC, Axelsen KB, Chen SX, Green BJ, Li Y, Madagan KL, Sanchez-Fernandez R, Forde BG, Palmgren MG, Rea PA, Williams LE, Sanders D,

Amtmann A (2003) Transcriptome analysis of root transporters reveals participation of multiple gene families in the response to cation stress. Plant J 35:675–692

Manohar M, Shigaki T, Hirschi KD (2011) Plant cation/H^+ exchangers (CAXs): biological functions and genetic manipulations. Plant Biol 13:561–569

Mäser P, Thomine S, Schroeder JI, Ward JM, Hirschi K, Sze H, Talke IN, Amtmann A, Maathuis FJM, Sanders D, Harper JF, Tchieu J, Gribskov M, Persans MW, Salt DE, Kim SA, Guerinot ML (2001) Phylogenetic relationships within cation transporter families of Arabidopsis. Plant Physiol 126:1646–1667

McAinsh MR, Pittman JK (2009) Shaping the calcium signature. New Phytol 181:275–294

McAinsh MR, Brownlee C, Hetherington AM (1992) Visualising changes in cytosolic-free Ca^{2+} during the response of stomatal guard cells to abscisic acid. Plant Cell 4:1113–1122

McAinsh MR, Webb AAR, Taylor JE, Hetherington AM (1995) Stimulus-induced oscillations in guard cell cytosolic-free calcium. Plant Cell 7:1207–1219

Mei H, Zhao J, Pittman JK, Lachmansingh J, Park S, Hirschi KD (2007) *In planta* regulation of the *Arabidopsis* Ca^{2+}/H^+ antiporter CAX1. J Exp Bot 58:3419–3427

Mei H, Cheng NH, Zhao J, Park P, Escareno RA, Pittman JK, Hirschi KD (2009) Root development under metal stress in *Arabidopsis thaliana* requires the H^+/cation antiporter CAX4. New Phytol 183:95–105

Morris J, Hawthorne KM, Hotze T, Abrams SA, Hirschi KD (2008) Nutritional impact of elevated calcium transport activity in carrots. Proc Natl Acad Sci U S A 105:1431–1435

Narsai R, Castleden I, Whelan J (2010) Common and distinct organ and stress responsive transcriptomic patterns in *Oryza sativa* and *Arabidopsis thaliana*. BMC Plant Biol 10

Narusaka Y, Nakashima K, Shinwari ZK, Sakuma Y, Furihata T, Abe H, Narusaka M, Shinozaki K, Yamaguchi-Shinozaki K (2003) Interaction between two cis-acting elements, ABRE and DRE, in ABA-dependent expression of *Arabidopsis rd29A* gene in response to dehydration and high-salinity stresses. Plant J 34:137–148

O'Malley RC, Ecker JR (2010) Linking genotype to phenotype using the *Arabidopsis* unimutant collection. Plant J 61:928–940

Park S, Cheng NH, Pittman JK, Yoo KS, Park J, Smith RH, Hirschi KD (2005) Increased calcium levels and prolonged shelf life in tomatoes expressing Arabidopsis H^+/Ca^{2+} transporters. Plant Physiol 139:1194–1206

Patel RV, Nahal HK, Breit R, Provart NJ (2012) BAR expressolog identification: expression profile similarity ranking of homologous genes in plant species. Plant J 71:1038–1050

Peiter E (2011) The plant vacuole: Emitter and receiver of calcium signals. Cell Calcium 50: 120–128

Pittman JK (2011) Vacuolar Ca^{2+} uptake. Cell Calcium 50:139–146

Pittman JK (2012) Multiple transport pathways for mediating intracellular pH homeostasis: the contribution of H^+/ion exchangers. Front Plant Sci 3:11

Pittman JK, Hirschi KD (2001) Regulation of CAX1, an Arabidopsis Ca^{2+}/H^+ antiporter. Identification of an N-terminal autoinhibitory domain. Plant Physiol 127:1020–1029

Pittman JK, Shigaki T, Cheng NH, Hirschi KD (2002a) Mechanism of N-terminal autoinhibition in the *Arabidopsis* Ca^{2+}/H^+ antiporter CAX1. J Biol Chem 277:26452–26459

Pittman JK, Sreevidya CS, Shigaki T, Ueoka-Nakanishi H, Hirschi KD (2002b) Distinct N-terminal regulatory domains of Ca^{2+}/H^+ antiporters. Plant Physiol 130:1054–1062

Pittman JK, Shigaki T, Marshall JL, Morris JL, Cheng NH, Hirschi KD (2004) Functional and regulatory analysis of the *Arabidopsis thaliana* CAX2 cation transporter. Plant Mol Biol 56:959–971

Pittman JK, Shigaki T, Hirschi KD (2005) Evidence of differential pH regulation of the *Arabidopsis* vacuolar Ca^{2+}/H^+ antiporters CAX1 and CAX2. FEBS Lett 579:2648–2656

Pittman JK, Edmond C, Sunderland PA, Bray CM (2009) A cation-regulated and proton gradient-dependent cation transporter from *Chlamydomonas reinhardtii* has a role in calcium and sodium homeostasis. J Biol Chem 284:525–533

Pittman JK, Bonza MC, De Michelis MI (2011) Ca^{2+} pumps and Ca^{2+} antiporters in plant development. In: Geisler M, Venema K (eds) Transporters and pumps in plant signaling, signaling and communication in plants, vol 7. Springer, Heidelberg, pp 133–161

Punshon T, Hirschi K, Yang J, Lanzirotti A, Lai B, Guerinot ML (2012) The role of CAX1 and CAX3 in elemental distribution and abundance in Arabidopsis seed. Plant Physiol 158: 352–362

Salt DE, Baxter I, Lahner B (2008) Ionomics and the study of the plant ionome. Annu Rev Plant Biol 59:709–733

Senadheera P, Singh RK, Maathuis FJM (2009) Differentially expressed membrane transporters in rice roots may contribute to cultivar dependent salt tolerance. J Exp Bot 60:2553–2563

Shigaki T, Pittman JK, Hirschi KD (2003) Manganese specificity determinants in the *Arabidopsis* metal/H^+ antiporter CAX2. J Biol Chem 278:6610–6617

Shigaki T, Rees I, Nakhleh L, Hirschi KD (2006) Identification of three distinct phylogenetic groups of CAX cation/proton antiporters. J Mol Evol 63:815–825

Singh A, Kanwar P, Yadav AK, Mishra M, Jha SK, Baranwal V, Pandey A, Kapoor S, Tyagi AK, Pandey GK (2014) Genome-wide expressional and functional analysis of calcium transport elements during abiotic stress and development in rice. FEBS J 281:894–915

Spalding EP, Harper JF (2011) The ins and outs of cellular Ca^{2+} transport. Curr Opin Plant Biol 14:715–720

Swarbreck SM, Colaço R, Davies JM (2013) Plant calcium-permeable channels. Plant Physiol 163:514–522

Toufighi K, Brady SM, Austin R, Ly E, Provart NJ (2005) The botany array resource: e-Northerns, expression angling, and promoter analyses. Plant J 43:153–163

Ueoka-Nakanishi H, Tsuchiya T, Sasaki M, Nakanishi Y, Cunningham KW, Maeshima M (2000) Functional expression of mung bean Ca^{2+}/H^+ antiporter in yeast and its intracellular localization in the hypocotyl and tobacco cells. Eur J Biochem 267:3090–3098

Wang P, Li Z, Wei J, Zhao Z, Sun D, Cui S (2012) A Na^+/Ca^{2+} exchanger-like protein (AtNCL) involved in salt stress in *Arabidopsis*. J Biol Chem 287:44062–44070

Wang N, Long T, Yao W, Xiong L, Zhang Q, Wu C (2013) Mutant resources for the functional analysis of the rice genome. Mol Plant 6:596–604

Winter D, Vinegar B, Nahal H, Ammar R, Wilson GV, Provart NJ (2007) An "electronic fluorescent pictograph" browser for exploring and analyzing large-scale biological data sets. PLoS One 2

Wu Q, Shigaki T, Williams KA, Han JS, Kim CK, Hirschi KD, Park S (2011) Expression of an Arabidopsis Ca^{2+}/H^+ antiporter CAX1 variant in petunia enhances cadmium tolerance and accumulation. J Plant Physiol 168:167–173

Xu L, Zahid KR, He LR, Zhang WW, He X, Zhang XL, Yang XY, Zhu LF (2013) GhCAX3 gene, a novel Ca^{2+}/H^+ exchanger from cotton, confers regulation of cold response and ABA induced signal transduction. PLoS One 8:e66303

Zhao J, Barkla BJ, Marshall J, Pittman JK, Hirschi KD (2008) The *Arabidopsis cax3* mutants display altered salt tolerance, pH sensitivity and reduced plasma membrane H^+-ATPase activity. Planta 227:659–669

Zhao J, Connorton JM, Guo YQ, Li XK, Shigaki T, Hirschi KD, Pittman JK (2009a) Functional studies of split *Arabidopsis* Ca^{2+}/H^+ exchangers. J Biol Chem 284:34075–34083

Zhao J, Shigaki T, Mei H, Guo YQ, Cheng NH, Hirschi KD (2009b) Interaction between Arabidopsis Ca^{2+}/H^+ exchangers CAX1 and CAX3. J Biol Chem 284:4605–4615

Chapter 5
Decrypting Calcium Signaling in Plants: The Kinase Way

Swatismita Dhar Ray

Abstract Ca^{2+} is an indispensable messenger in cellular signaling for all eukaryotes. Plants respond to modulation in Ca^{2+} concentration, known as "Ca^{2+} signatures," which are induced in response to extrinsic and intrinsic cues. These Ca^{2+} signatures are perceived, decoded, and transduced to downstream response toolkit comprising of a complex group of Ca^{2+}-binding proteins that function as Ca^{2+} sensors. Ca^{2+}-dependent protein kinase (CPK), Ca^{2+} or Ca^{2+}/calmodulin-dependent protein kinase (CCaMK), calcineurin B-like protein (CBL), and their interacting kinases (CIPK) are directly or indirectly regulated by Ca^{2+}. These Ca^{2+}-regulated kinases are part of phosphorylation pathway that lead to regulation of ion channels, v-SNARE proteins, nitrate sensing, nodulation, and transcriptional factors for master regulation. Genome sequencing data of wide varieties of plant species along with high-throughput transcriptomic and functional genomic analysis has expedited revealing of multifaceted functions of these kinases in stress-signaling networks. Combining the transcriptomic and posttranscriptional proteomic regulatory mechanisms in CDPKs and CBL–CIPKs reveals an emerging evolutionary model. Subcellular proteomics and varying affinity for Ca^{2+} emerged as a crucial regulatory mechanism for transducing stress signal. Cross talk of isoforms and their interacting partners adds on to the humongous effect on increasing complexities among these signaling cascades. This chapter provides new insight about the colossal advancement in understanding of the regulatory mechanism and functionality involved in Ca^{2+} sensing by kinases in light of the information generated by genomic tools.

Keywords Abiotic stress • CBL proteins • CDPK proteins • CIPK proteins • Functional genomics • Genome-wide expression

---

S.D. Ray (✉)
Biotechnology and Bioresources Management Division, Darbari Seth Block, India Habitat Centre, The Energy and Resources Institute, Lodhi Road, New Delhi, Delhi 110 003, India
e-mail: swatismita.dhar@teri.res.in; swatidhar8@gmail.com

G.K. Pandey (ed.), *Elucidation of Abiotic Stress Signaling in Plants*,
DOI 10.1007/978-1-4939-2211-6_5

Abbreviations

ABA	Abscisic acid
ABF	ABRE-binding factor
ATP	Adenosine triphosphate
CCaMK	Calcium or calcium-/calmodulin-regulated kinases
CDPK/CPK	Calcium-dependent protein kinase
CLD	Calmodulin-like domain
CRK	CDPK-related kinase
GA	Gibberellins
HR	Hypersensitive response
HSP	Heat shock protein
IAA	Indole-3-acetic acid
JA	Jasmonic acid
MAPK	Mitogen-activated protein kinase
ROS	Reactive oxygen species
SA	Salicylic acid

5.1 Introduction

Plants being sessile organisms are constantly being challenged by several environmental cues. Stimuli generated by these cues regulate cellular and developmental programs. It is of immense importance to understand the underlying molecular mechanism of how plants perceive and respond to stress condition. Stress signal is perceived by the plant and transduced to downstream signaling cascade, which culminates in stress response. The stress-signaling pathway includes an array of proteins like kinases, transcription factors, cytoskeleton proteins, channel proteins, and chaperon proteins. Ca^{2+} is strongly established as secondary messenger working via "calcium decoders" like calmodulins and kinases in cellular signaling cascade (reviewed by Sanders et al. 1999; Knight and Knight 2001).

Ca^{2+} has long been recognized to be mediating plant's response to wide range of developmental and environmental cues. These "Ca^{2+} signatures" have been found to be generated due to application of different chemicals and release of caged inositol (1,4,5)-triphosphate, NAADP, IP_3, IP_6, sphingosine-1-phosphate, and cADPR (Clarkson et al. 1988; Gilroy et al. 1990; Franklin-Tong et al. 1996; Blatt et al. 1990; Allen et al. 2000). Even oscillatory changes of cellular Ca^{2+} concentrations were noted on applying ABA to guard cells of *Commelina* (McAinsh et al. 1995; Staxen et al. 1999). The concept of "Ca^{2+} signatures," i.e., precise regulation of cytosolic Ca^{2+} concentration by spatial, temporal, and concentration parameters by external or internal stimulus, was defined by A.M. Hetherington and his group, which could be monophasic and biphasic (Haley et al. 1995; Webb et al. 1996; Cessna et al. 1998; Plieth et al. 1999; Clayton et al. 1999; Kiegle et al. 2000; Toyota et al. 2008). The use of "cameleon" in plants opened new serge of precise data on Ca^{2+} signature depending on the stimulus (Allen et al. 1995; Iwano et al. 2009; Costa et al. 2010; Krebs et al. 2012). Modulation in cytosolic-free Ca^{2+} in response to stress is well reported; even recently a genome-wide study showed that 90 % of the 3,819 unigenes in tobacco, which differentially expressed on cryptogein elicitor by *Phytophthora cryptogea* treatment, have Ca^{2+}-dependent regulation (Sanders et al. 2002; Amelot et al. 2012). Over time, findings have substantiated on Ca^{2+} being secondary messenger in cellular signaling via "Ca^{2+} decoders" like calmodulins and kinases (Williamson and Ashley 1982; reviewed in Sanders et al. 1999; Knight 2000; Evans et al. 2001; Knight and Knight 2001; Batistič and Kudla 2012). Ca^{2+} signals are perceived by Ca^{2+} sensor proteins and relayed into downstream phosphorylation cascades for regulation of gene expression. The sensor proteins are categorized as "sensor relays" and sensor "responders" (Sanders et al. 2002). Calmodulins (CaM), calmodulin-like protein (CML), and calcineurin B-like protein (CBL) are "sensor relays" as they do not possess a catalytic domain and transmit the Ca^{2+} signal via protein–protein interaction (Fig. 5.1). Among these three proteins, CBLs interact with a family of protein known as CBL-interacting protein kinases (CIPKs). Hence, CBL–CIPK complexes are designated as bimolecular "sensor responders." CDPKs are also "sensor responder" through their kinase domain where Ca^{2+} signature is sensed by the calmodulin-like domain (Fig. 5.1; CLD; Harper et al. 1991; Harmon et al. 2001; Hrabak et al. 2003). CDPKs may have evolved as a fusion of two preexisting genes encoding for a Ca^{2+}-/calmodulin-dependent kinase with a calmodulin-like gene (Harper et al. 1991; Zhang and Choi 2001). CCaMKs are structurally similar to CDPKs, however, they consist of a dual-regulated kinase, a visinin-like domain for Ca^{2+}-binding and a Ca^{2+}/CaM-binding domain overlapping with the autoinhibitory domain. Interestingly, binding of Ca^{2+} to visinin-like domain increases affinity to Ca^{2+}/CaM and eventually binding of Ca^{2+}/CaM to CaM-binding domain-triggering displacement of autoinhibitory domain resulting to activation of catalytic domain (Patil et al. 1995; Sathyanarayanan et al. 2000; Sathyanarayanan and Poovaiah, 2004).

Ca^{2+}-mediated abiotic stress-signaling pathway is dominated by CDPK and CBL–CIPK proteins. The role of CCaMK is more pronounced in establishment of symbiosis as have been reported in *Lotus japonica, M. truncatula*, *Sesbania rostrata*, and *O. sativa* (Levy et al. 2004; Mitra et al. 2004; Godfroy et al. 2006; Tirichine et al. 2006; Chen et al. 2007; Capoen et al. 2009). A gain-of-function CCaMK mutant, *snf1,* in *L. japonica* developed spontaneous nodules in the absence of rhizobia (Tirichine et al. 2006). In addition, loss of function of MtDMI3/CCaMK in *M. truncatula* resulted in inability to form nodules (Gleason et al. 2006). With identification of downstream signaling components and the mechanism by which CCaMK decodes the microbial signals, it has been found that DMI3/CCaMK constitutes a branch point in symbiosis signaling, initiating both nodule organogenesis and rhizobial/AM fungal colonization signaling (Messinese et al. 2007; Chen et al. 2008; Yano et al. 2008; Hayashi et al. 2010; Shimoda et al. 2012; Takeda et al. 2012; Liao et al. 2012; Svistoonoff et al. 2013). Regarding involvement of CCaMK in abiotic stress, *PsCCaMK* in pea was reported to be induced in roots in response to low temperature and salinity (Pandey et al. 2002). Lately, OsDMI3 has been found to be involved in ABA signaling in rice, where ABA treatment induced the expression of *OsDMI3* and *OsMPK1* as well as the activities of

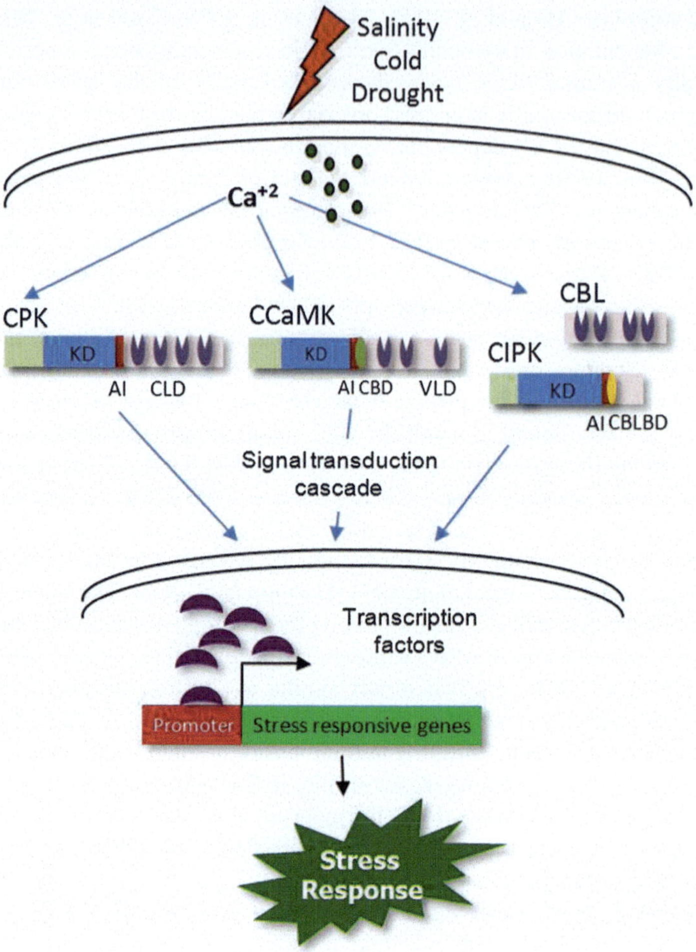

Fig. 5.1 Calcium-sensing kinase-mediated signal transduction process in plant cell. The signaling cascade is mediated by Ca^{2+}-dependent protein kinase (CPK), Ca^{2+}/calmodulin-dependent protein kinase (CCaMK), calcineurin B-like protein (CBL), and their interacting kinases (CIPK). *KD* kinase domain, *AI* autoinhibitory domain, *CLD* calmodulin-like domain, *CBD* calmodulin-binding domain, *VLD* visinin-like domain, *CBLBD* CBL-binding domain

OsDMI3 is increased. It was also found that OsDMI3 functions upstream to OsMPK1 in regulating the activities of antioxidant enzymes and the production of H_2O_2 in rice (Shi et al. 2012, 2014). In maize, *ZmCCaMK* was identified to be essential for ABA-induced antioxidant defense (Ma et al. 2012a). However, still there are not many reports on the role of CCaMK in abiotic stress; hence, in this article, I shall restrict the discussion to CDPK and CBL–CIPK proteins for their involvement in abiotic stress.

The focus of this review shall be on functional genomic aspects involving transcriptomic and proteomic regulation of CDPK and CBL–CIPK proteins. Emphasis will be on gene family wide transcriptomic studies in multiple plant species along with their functional analysis. Posttranslational proteomic regulatory mechanism of the said gene families will be discussed in details with an insight on cellular localization and its relevance with functionality. With more advent of information, the knowledge about the underlying complexities of cross-signaling and layered protein–protein interaction in plant signal transduction pathway is also increasing. Here we would also elucidate the role of Ca^{2+}-sensing kinases in the signaling cross talk on perceiving various internal as well as external cues.

5.2 Evolutionary Divergence of Ca^{2+} Kinase Gene Families

By virtue of technological advancement, we have access to completed large-scale genome sequencing data, contig data, and collection of EST data from large-scale EST sequencing projects. Availability of such enormous information has been instrumental in identifying large gene families and further functionally characterizing them for redundancy and divergence.

Arabidopsis codes for 34 CDPK genes, which are distributed among five chromosomes. Depending on sequence homology, they cluster into four distinct groups (Cheng et al. 2002; Hrabak et al. 2003). Similar attempts by using the rice genome sequence revealed 31 CDPK genes (Asano et al. 2005; Ray et al. 2007). Maize genome was found to harbor 40 CDPK genes (Kong et al. 2013; Ma et al. 2013). The size of maize genome (~2,300 Mb) is much larger than rice (~389 Mb) and *Arabidopsis* (125 Mb); however, the total number of CDPK genes was found to be similar across genus. Apple cultivar (*M.* × domestica), which has 881.3 Mb genome that underwent whole-genome duplication 50 million years ago, contains 30 CDPK genes, again a comparable number (Kanchiswamy et al. 2013). From *Populus*, 30 CDPK genes were identified (Zuo et al. 2013). *Vitis vinifera* has been identified to harbor 17 CDPK genes mapped onto 12 of 19 chromosomes (Chen et al. 2013a). *Triticum aestivum* harbors 20 CDPK genes where 14 CDPKs were identified from *Chlamydomonas reinhardtii,* a biflagellated green alga genome (Li et al. 2008; Liang and Pan 2013). A phylogenetic study of orthologous CDPKs including *A. thaliana*, *O. sativa*, apple, *P. trichocarpa,* and *Z. mays* also clustered into four clades as it was first observed in *Arabidopsis* (Cheng et al. 2002; Kanchiswamy et al. 2013). Even wheat CDPKs when compared with rice and *Arabidopsis* formed four clusters (Li et al. 2008). A phylogenetic analysis of rice, maize, and *Arabidopsis* CDPK superfamily revealed seven groups, where four clades belong to CDPKs and the others are CDPK-related kinases (CRKs), Ca^{2+}- and calmodulin-dependent protein kinases (CCaMKs), and phosphoenolpyruvate-related kinases (PEPRKs; Cheng et al. 2002; Asano et al. 2005; Kong et al. 2013). Since CDPK proteins belonging to three dicot (*Arabidopsis,* apple, and *Populus*) and three monocot (rice, maize, and wheat) species are represented in all four subgroups, it was implied that there are certain levels of divergence before bifurcation of monocot and dicot. However, major evolutionary change might not have happened in CDPK population after the bifurcation, and this could be indicative of their involvement in essential cell functions (Ray et al. 2007). It is suggested that after monocot and dicot divergence, these families might have expanded independently. In case of *Arabidopsis* and rice, it was observed that five *Arabidopsis* AtCPK genes (*AtCPK21*, *AtCPK22*, *AtCPK23*, *AtCPK27*, and *AtCPK31*) have arisen due to tandem duplication, whereas all nine pairs of rice CDPK genes (*OsCPK1/15*, *OsCPK2/14*, *OsCPK3/16*, *OsCPK5/13*, *OsCPK13/23*; *OsCPK11/17*, *OsCPK21/22*, *OsCPK25/26*, and *OsCPK24/28*) resulted only from segmental duplication event (Cheng et al. 2002; Ray et al. 2007). Similar observation was reported in maize genome where expansion of its CDPK genes was seen in group I members (maize-17; rice-11; *Arabidopsis*-10; Kong et al. 2013). Interestingly, apple CDPK genes were found to be more closely related to *Arabidopsis* and *Populus* CDPKs than to rice and maize CDPKs, emphasizing on the theory of dicot-specific origin (Kanchiswamy et al. 2013). The expansion of CDPK family due to genome expansion is best observed in *Populus*, whose genome has undergone at least three rounds of genome-wide duplication along with tandem, segmental duplication, and retroposition and replicative transposition (Tuskan et al. 2006). Of the 30 CDPK genes, 26 were found within the duplicated regions, and moreover, two paralogous gene pairs, *PtCDPK1/4* and *PtCDPK16/17*, underwent tandem duplication followed by segmental duplication (Zuo et al. 2013). On the other hand, *PtCDPK13* and *PtCDPK30* lacked their counterpart on the duplicated block, strongly implying that the dynamic rearrangement could be the reason for loss (Zuo et al. 2013). Fascinatingly, *Populus* genome showed higher rate (~80 %) of preferential retention of CDPK genes as well as other gene families (Zuo et al. 2013). The substitution rate ratio of nonsynonymous vs. synonymous was calculated for the paralogous 22 genes showing ~0.4 ratio, which signified good purifying selection pressure with less functional divergence during post-segmental duplication event (Zuo et al. 2013). In monocot species, wheat, CDPK genes showed high (often >80 %) homology with their rice counterpart where the average divergence time of rice–wheat CDPK orthologues was determined to be 50 MYA, which coincides with the time of these two species diverging from their last common ancestor (Li et al. 2008). Even divergence time of orthologous gene pairs *OsCPK8-TaCK12, OsCPK20-TaCPK7, OsCPK12-TaCPK18,* and *OsCPK19-TaCPK5* were similar to speciation time; however, divergence time of wheat homologues *TaCPK5-TaCPK18* was much earlier than speciation time, signifying that duplication of these occurred in much earlier ancestor (Li et al. 2008). To determine the positions of *Vitis vinifera* CDPK genes by phylogenetic analysis, CDPK genes ranging from green algae to flowering plants were compared, and the authors suggest of single origin of the VvCDPKs that can be dated back to the green algae, much before plants colonized the land. Moreover, they also opine that the split into four subfamilies may have happened in the ancestor of algae and land plants (Chen et al. 2013a, b, c). This latest work also supports the "multiple expansion" theory for CDPK gene family. Analysis also showed that group I predates the origin of seed plants, where three sub-subfamilies consist of an *Amborella* CDPK orthologue and two subfamilies include CDPKs from *Ginkgo* (Chen et al. 2013a, b, c). Hamel et al. (2013) studied the evolution of green plant CDPK genes using the genome sequence from early land plants and suggested that a consensus structure of the CDPK family is shared between all land plants; however, the green algae homologues form distinct clades (Hamel et al. 2013). Phylogenetic analysis of 48 putative CDPK genes from green alga belonging to eight *Chlorophyta* and *Streptophyta* species showed that two clades of the four groups have at least one representative member from each of the eight *Chlorophyta* species examined. Moreover on comparing the freshwater algae (*Chlamydomonas reinhardtii* and *Volvox carteri*) to the marine species (*M. pusilla* and *Ostreococcus* spp.), they found extensive gene loss due to reduction in genome size, which resulted in decrease in number of CDPK genes in these species in general (Hamel et al. 2013). The 35 novel CDPKs identified from bryophyte *Physcomitrella patens* and pteridophyte *Selaginella moellendorffii* were compared with rice, *Arabidopsis*, and poplar, where it was found that in angiosperms, CDPK genes evolved from the embryophytes by event of gene duplication. As evident with the angiosperms, the bryophytes also seem to have undergone extensive gene duplication events. With a relatively simple lifestyle, *P. patens* harbored 25 distinct CDPK members, which showed tight clustering of the isoforms in phylogenetic analysis indicating towards expansion due to recent duplications. However, only four pairs of CDPK could be defined as tandemly duplicated genes, which strongly implied that large-scale segmental duplications are largely responsible for the expansion of this gene family (Hamel et al. 2013). From tobacco, ice plant, chickpea, cucumber, cotton, ginseng, barley, alfalfa, potato, and *Petunia*, more than one CDPK homologues have also been characterized (Ray, 2012). CDPK isoforms show sequence homology ranging from 20 to 98 %, and determining functional redundancy or divergence among the homologues and orthologues remains a challenge.

If we look at the expression profile of the duplicated genes in rice, four pairs (*OsCPK1/15*, *OsCPK3/16*, *OsCPK2/14*, *OsCPK5/13*) have retained their expression, three pairs (*OsCPK13/23*, *OsCPK21/22*, *OsCPK24/28*) showed neo-functionalization, and only one pair (*OsCPK11/17*) resulted into pseudogenization. It has also been observed that divergence of expression has resulted due to modification in occurrence of *cis*-regulatory elements in the respective promoter regions (Ray et al. 2007). In maize genome, seven pairs of duplicated CDPK genes (*ZmCPK8/9*, *ZmCPK14/15*, *ZmCPK16/17*, *ZmCPK18/19*, *ZmCPK20/21*, *ZmCPK29/30*, and *ZmCPK39/40*) showed similar expression pattern in 50 developmental stages studied, i.e., retention of expression, whereas one pair (*ZmCPK23/24*) showed neo-functionalization. Among the dicots, in *Populus* three duplicated pairs (*PtCDPK2/3*, *PtCDPK16/17*, *PtCDPK23/24*) showed retention of expression (Zuo et al. 2013). Functional relatedness among CPK orthologues was studied by Li et al. (2008), and observed that wheat–rice CDPK orthologous gene pairs mostly have maintained similar expression pattern except for TaCPK8–OsCPK1 and TaCPK4–OsCPK24, implying of species-specific functional evolution of orthologues as have been noted in case of homologues. Even maize–rice orthologues (*ZmCPK11–OsCPK10*, *ZmCPK28–OsCPK19*, *ZmCPK29–OsCPK16*, *ZmCPK33–OsCPK8*) showed retention of expression profile (Kong et al. 2013). Hence, during evolution of CDPK genes, homologues and orthologues showed both retention and divergence of function, and for further investigation it could be interesting to see whether there is any underlying correlation of this phenomenon with the indispensability of the genes' function or balancing out the overall Ca^{2+}-signaling pathway.

To assess the functional redundancy or divergence of the CDPK gene family members in rice, Ray et al. (2007) clustered 31 CDPKs according to their expression profile in 17 stages of growth and development. Group 8 of the cluster consisted of six CDPKs (*OsCPK25/26*, *OsCPK6*, *OsCPK14*, *OsCPK2*, *OsCPK22*, and *OsCPK29*) preferentially expressing during the P6 stages of panicle development. Similar observation was reported in 2009 by Ye et al. who studied expression in 27 stages of rice development and found *OsCPK2*, *OsCPK22*, *OsCPK29*, and *OsCPK25*/26 to have stamen and panicle preferential expression. Interesting fact is that phylogenetically *OsCPK2*, *OsCPK14*, *OsCPK25*, and *OsCPK26* (group IIa) and *OsCPK22 and OsCPK29* (group IIIa) are related (Ray et al. 2007). Even maize CDPK genes *ZmCPK18, ZmCPK19, ZmCPK20, ZmCPK21*, and *ZmCPK32* belonging to group C of expression hierarchical clustering (of 50 stages) showed anther preferential expression and structurally cluster with groups IIa and IIIa of rice CDPKs mentioned above (Kong et al. 2013). PiCDPK1, AtCPK17 and AtCPK34, clustering in group IIa, along with PiCDPK2 and AtCPK24, clustering in group IIIa, also have been reported to have pollen-specific expression (Harper et al. 2004; Ray et al. 2007). Moreover, *PiCDPK1* and *PiCDPK2* are also known to be involved in pollen tube growth and polarity (Yoon et al. 2006). Li et al. in 2008 revealed that *TaCPK8* and *TaCPK13* clustering in group IIa have preferential expression in immature seeds and young spike, respectively. *PtCDPK23* and *PtCDPK24* belonging to the same group have also showed male catkin (cylindrical flower cluster) preferential expression (Zuo et al. 2013). These two clusters are excellent example of structural and functional relatedness that has been vividly noted in CDPK gene family across plant species. Similar structural and functional relatedness was also observed for rice, *Arabidopsis*, and maize abiotic stress-responsive CDPK genes in group I (Ray et al. 2007; Kong et al., 2013). *PtCDPK8* and *PtCDPK9,* induced by dehydration, were found to be orthologues of stress-responsive *AtCPK6* of group I. Even five genes (*VaCPK1, VaCPK2, VaCPK3, VaCPK20*, and *VaCPK26*) from *Vitis amurensis* differentially expressing under osmotic and temperature stresses belonged to group I (Dubrovina et al. 2013). Till now functions for all the homologues are not yet known so it will be improper to conclude that structural relatedness is absolute for functional colinearity. However, from the above findings, it can be implied that during evolution some degree of functional redundancy has been conserved.

CBLs as Ca^{2+} sensors are structurally conserved and have shown lesser expansion than the CDPKs. *Arabidopsis*, rice, and poplar harbor ten CBL genes each (Luan et al. 2002; Kolukisaoglu et al. 2004; Zhang et al. 2013); maize and grape have eight CBLs (Weinl and Kudla, 2009; Chen et al. 2011) where seven CBL genes have been identified from canola (Zhang et al. 2014). In *Arabidopsis*, *AtCBL1/9* and *AtCBL2/3* resulted due to segmental duplication, whereas *AtCBL3/7* was generated from tandem duplication (Kolukisaoglu et al. 2004). May be due to strong selection pressure for gene dosage effect, clustering of CBL genes on chromosomes was not found as was reported for NBS-LRR genes (Meyers et al. 2003). When expression analysis was performed for *AtCBL1* and *AtCBL9*, they revealed distinct expression profile under stress as well as interacting CIPK counterparts (Kolukisaoglu et al. 2004). Even further study showed *AtCBL1* to be part of glucose and GA signaling during germination and seedling development (Li et al. 2013), whereas *AtCBL9* acted as negative regulator of ABA-mediated signaling (Pandey et al. 2004). *AtCBL2/3* and *OsCBL2/3* duplication was species independent where *AtCBL2* and *AtCBL3* were involved in maintaining intracellular ion homeostasis via V-ATPase (Tang et al. 2012). In case of *OsCBL2,* it expressed in aleurone during germination and promoted vacuolation in aleurone cells. Even an orthologue, *HvCBL2*, from barley (*Hordeum vulgare*) with 91 % sequence identity, performed the same function, whereas *OsCBL3* was not detectable in aleurone tissue during germination but was upregulated by salt stress in seedlings (Hwang et al. 2005). These findings have just started to throw some light on the complexity of the signaling. Phylogenetic analysis of poplar, *Arabidopsis*, and rice CBL genes clustered them into 4 groups: A, B, C, and D. Parallel species-specific duplication event was noticed in groups B and C, namely, *PtCBL1/9, AtCBL1/9, OsCBL2/3, PtCBL2/5*, and *AtCBL2*/3. Even some branches were present in only one species, where one counterpart may have evolved or might have been lost in one species during evolution. Likewise, in the group D, rice CBLs clustered together (*OsCBL4/5/7/8*).

When the CIPK genes from rice (31), *Arabidopsis* (26), *Populus* (27), and maize (43) were compared, they clustered in two groups, intron-less and intron-rich (Chen et al. 2011). Species-specific evolution of CIPK genes was also widely noted as in *AtCIPK1/17, AtCIPK10/2, AtCIPK13/18, OSCIPK18/22, OsCIPK15/15, OsCIPK1/7, ZmCIPK1/17, ZmCIPK21/34, ZmCIPK16/20, ZmCIPK29/36, ZMCIPK6/27, PtCIPK6/20, PtCIPK5/10, PtCIPK15/16, PtCIPK7/8*, and *PtCIPK26/27*. During evolution, maize underwent several rounds of genome duplication resulting in an expanded genome (~2,400 Mb; Schnable et al. 2009). A branch of eight CIPK maize homologues (*ZmCIPK14, ZmCIPK15, ZmCIPK38, ZmCIPK39, ZmCIPK40, ZmCIPK41, ZmCIPK42*, and *ZmCIPK43*) evolved species specifically (Chen et al. 2011). Function of these genes will be interesting to know as they may have maize-specific activity. Even clear orthologous pairs were recognizable in this analysis, namely, *OsCIPK24/ZmCIPK24, OsCIPK3/ZmCIPK3, OsCIPK23/ZmCIPK23, OsCIPK7/ZmCIPK4, OsCIPk13/ZmCIPK13, OsCIPK25/ZmCIPK25, OsCIPK30/ZmCIPK30, OsCIPK26/ZmCIPK26, OsCIPK11/ZmCIPK11, OSCIPK10/ZmCIPK10, OsCIPK5/ZmCIPK5, AtCIPK14/PtCIPK21*, and *AtCIPK20/PtCIPK11* reflecting high evolutionary relationship between rice and maize CIPKS. From future studies, it will be interesting to observe how this structural and functional correlation takes shape.

5.3 Functional Genomic Relevance: Transcriptome Analysis of Ca^{2+}-Regulated Kinase Families in Abiotic Stress

Global transcriptomic analysis in plants under various abiotic stress conditions has been performed using technologies like microarray, q-PCR, large-scale northern analysis, RT-PCR, as well as deep transcriptome sequencing techniques, which provide us with a comprehensive spatial and temporal transcript regulation pattern of the Ca^{2+}-regulated kinase gene isoforms, which is indicative of its function (Ray et al. 2007, 2011; Wan et al. 2007; Venu et al. 2013). Transcriptome analysis of two drought-tolerant barley genotypes ("Martin" and "*Hordeum spontaneum* 41-1 (HS41-1)") and one drought-sensitive genotype ("Moroc9-75") using the 22K Affymetrix Barley 1 chip, during the reproductive stage under drought conditions, revealed that a CDPK gene is constitutively expressed in drought-tolerant genotypes, emphasizing on its involvement in drought tolerance (Guo et al. 2009). A genome-wide transcript abundance study of grape (*Vitis vinifera* L.) leaves under heat stress revealed upregulation of one CDPK (CF511469) and *CIPK25* gene (BQ797947; Liu et al. 2012). In another study, large-scale ESTs of abiotic stress-treated leaves, berries, and root of *Vitis vinifera* were generated and analyzed where transcripts encoding for *CIPK10* were overrepresented in both the stressed leaf and berries clusters (Tillett et al. 2011). Huang et al. (2012) performed microarray study of early Arsenic-induced gene expression in rice and showed that along with seven calmodulin genes, two CBL genes, one CIPK gene (*OsCIPK21*), and four CDPK genes (*OsCPK4, OsCPK13, OsCPK20*, and *OsCPK21*) were upregulated on As(V) treatment. De novo sequencing of an endangered desert shrub *Ammopiptanthus mongolicus* during cold acclimation revealed upregulation of three CDPK genes (Unigene567_All, Unigene13378_All, Unigene13204_All) and a CIPK gene (Pang et al. 2013). Venu et al. in 2013 performed both massively parallel signature sequencing (MPSS) and sequencing by synthesis (SBS) to identify differentially regulated genes from japonica rice under cold, drought, and salt stress. In the cold-stressed leaves, *OsCIPK15* and *OsCIPK24* were up- and down-regulated, respectively, in both the MPSS and SBS libraries. Under drought stress, *OsCIPK15* was upregulated in both the leaves and roots and was also induced by salt stress in root tissue. These studies were performed with an objective to identify global expression pattern, independent of specific gene families. However, with more studies identifying gene families with multiple isoforms, it has become a challenge to tease out individual function of all the isoforms and their extent of functional redundancy.

5.3.1 Global Transcriptome of CDPK Gene Family Across Plant Species

In rice, 15 CDPK genes have been found to be differentially regulated under abiotic stresses, viz., cold, salt, and dehydration as has been detected in microarray, large-scale EST data, and q-PCR analysis (Ray et al. 2007; Wan et al. 2007). Among the 40 CDPK genes from maize genome, transcript levels of *ZmCPK14* and *ZmCPK37* increased under salt stress. Under PEG treatment, *ZmCPK1, ZmCPK17, ZmCPK22*, and *ZmCPK28* were found to be >2-fold upregulated, whereas *ZmCPK11, ZmCPK14, ZmCPK31, ZmCPK37*, and *ZmCPK39* were

>1.2-fold upregulated. Six genes (*ZmCPK5, ZmCPK11, ZmCPK22, ZmCPK29, ZmCPK37*, and *ZmCPK39*) showed cold-induced expression on exposures to 4 °C. Ten CDPK genes (*ZmCPK1, ZmCPK14, ZmCPK17, ZmCPK28, ZmCPK29, ZmCPK31 ZmCPK5, ZmCPK22, ZmCPK37*, and *ZmCPK39*) were found to be induced by H_2O_2 treatment (Kong et al. 2013). Six of the maize CDPKs (*ZmCPK1, ZmCPK4, ZmCPK5, ZmCPK11, ZmCPK14*, and *ZmCPK17*) belonging to the phylogenetic group I showed differential regulation under cold, salt, desiccation stress, as well as ABA and H_2O_2 treatment by q-PCR analysis. In wheat, 14 CDPK genes were found to be induced by either/or cold (*TaCPK5, TaCPK15, TaCPK4, TaCPK6, TaCPK12, TaCPK3*, and *TaCPK7*), salt (*TaCPK10, TaCPK4, TaCPK19, TaCPK18*, and *TaCPK9*), drought (*TaCPK1, TaCPK9, TaCPK18*, and *TaCPK6*) stress, H_2O_2 (*TaCPK2, TaCPK9, TaCPK5, TaCPK7, TaCPK12, TaCPK1, TaCPK10,* and *TaCPK18*), and ABA (*TaCPK10, TaCPK5, TaCPK3, TaCPK4, TaCPK7, TaCPK6,* and *TaCPK9*) treatment. In the dicot species, *Vitis vinifera,* detailed analysis of publicly available Affymetrix microarray data revealed that nine VvCPKs (*VvCPK10, VvCPK12, VvCPK 7, VvCPK13, VvCPK2, VvCPK11, VvCPK3, VvCPK14,* and *VvCPK17*) are differentially regulated by cold, salt, and drought stress. *VvCPK17, VvCPK3,* and *VvCPK13* were significantly upregulated by drought stress. *VvCPK7* and *VvCPK17* were upregulated under short-term (24 h) but not long-term (24 h and 16 days) salt stress, whereas *VvCPK11*, *VvCPK12*, and *VvCPK14* showed reverse pattern of induction under salt stress. Interestingly, *VvCPK3* was upregulated both in the short and long time scale under salt stress. In *Vitis amurensis,* ten CDPK genes were detected by q-PCR to be actively expressing in cuttings under salt, desiccation (*VaCPK1, VaCPK2,VaCPK3, VaCPK9, VaCPK16, VaCPK21, VaCPK26*, and *VaCPK29*) and temperature stress (*VaCPK2, VaCPK3, VaCPK9, VaCPK13, VaCPK20, VaCPK21,VaCPK26,* and *VaCPK29*; Dubrovina et al. 2013). Microarray datasets under drought stress at root tips of two genotypes (Carpaccio and Soligo) of *Populus* were analyzed for CDPK gene expression. *PtCDPK27* and *PtCDPK28* were found upregulated, whereas *PtCDPK10* was downregulated in both the genotypes during early (36 h) and long-term (10 days) response to drought. Quite interestingly, *PtCDPK18* and *PtCDPK13* showed genotype-specific response where transcript abundance was noted under drought in Carpaccio. Q-PCR revealed transcript abundance of *PtCDPK9, PtCDPK12, PtCDPK18, PtCDPK19, PtCDPK21, PtCDPK22, PtCDPK27,* and *PtCDPK28* under drought stress and higher expression at 6 h except for *PtCDPK18,* which showed highest expression after 3 h of drought stress (Zuo et al. 2013). In apple cultivars (*Malus* × *domestica*) of the thirty CDPK genes analyzed, eleven were differential regulated in susceptible and resistant *Malus* × *domestica* apple genotypes following *Erwinia amylovora* infection (Kanchiswamy et al. 2013). *MdCPK19* and *MdCPK28* were significantly upregulated at 6 and 12 hpi (hours postinoculation) in the resistant genotype on pathogen challenge, whereas MdCPK26b was upregulated at 12 and 24 hpi. MdCPK8a was similarly upregulated at 6, 12, and 24 hpi in both resistant and susceptible cultivars (Kanchiswamy et al. 2013).

5.3.2 Global Transcriptome of CBL–CIPK Gene Family Across Plant Species

Global expression analysis of 6 CBL and 12 CIPK genes was done in canola seedlings under salt, cold, heat, oxidative, drought, low potassium stress, and ABA application (Zhang et al. 2014). *BnaCBL1, BnaCBL4,* and *BnaCBL9* were downregulated at the 6 h time point under salt stress. *BnaCBL1* and *BnaCBL10* transcript level showed a significant increase under different time points of cold treatment, whereas *BnaCBL2, BnaCBL3,* and *BnaCBL4* were downregulated. *BnaCBL1* and *BnaCBL9* were induced by heat treatment. Drought increased accumulation of *BnaCBL1* transcripts and downregulated *BnaCBL2* expression. *BnaCIPK3, BnaCIPK6, BnaCIPK9, BnaCIPK12, BnaCIPK15, BnaCIPK23,* and *BnaCIPK26* were induced by cold treatment. Salt treatment increased transcript accumulation of *BnaCIPK1, BnaCIPK3, BnaCIPK6, BnaCIPK7, BnaCIPK11, BnaCIPK15,* and *BnaCIPK24* at either 6 or 24 h time point. Oxidative stress upregulated the expression of *BnaCIPK6, BnaCIPK9, BnaCIPK15,* and *BnaCIPK26*. Expression of *BnaCIPK6* was upregulated by ABA treatment. Under drought stress *BnaCIPK6* was induced, whereas *BnaCIPK3* was repressed. In the same article, expression profiles of *Arabidopsis* CBL and CIPK genes under cold, salt, ABA, heat, oxidative, and low potassium stresses were also analyzed from publicly available microarray datasets by Goda et al. (2008) and Kilian et al. (2007). Among the *Arabidopsis* CBLs, *AtCBL2* and *AtCBL9* were found to be significantly regulated by ABA treatment. Under cold stress condition *AtCBL1, AtCBL4, AtCBL9,* and *AtCBL10* were differentially regulated, where *AtCBL1* was upregulated and the other three isoforms were downregulated. *AtCBL3* was upregulated by oxidative stress. However, under drought, heat, and salinity stress condition, none was found to be upregulated. *AtCBL4* and *AtCBL9* transcripts were downregulated under heat stress; *AtCBL2, AtCBL4,* and *AtCBL10* under salinity stress; and *AtCBL4* and *AtCBL9* by drought stress. Nine *Arabidopsis* CIPK genes were found to be differentially regulated (*AtCIPK3* and *AtCIPK15* repressed; *AtCIPK5, AtCIPK6, AtCIPK8, AtCIPK10, AtCIPK11, AtCIPK12, AtCIPK14, AtCIPK19,* and *AtCIPK25* upregulated) under ABA treatment. Cold stress significantly increased accumulation of *AtCIPK7, AtCIPK9, AtCIPK11,* and *AtCIPK12* transcripts while decreased that of *AtCIPK17*. Under drought stress condition, *AtCIPK6, AtCIPK9, AtCIPK11,* and *AtCIPK23* were significantly upregulated, whereas *AtCIPK15* and *AtCIPK17* were downregulated. Unlike CBL genes, CIPK isoforms were found to be induced under heat (*AtCIPK1, AtCIPK11, AtCIPK15,* and *AtCIPK24*), salinity (*AtCIPK6, AtCIPK7, AtCIPK9, AtCIPK11,* and *AtCIPK12*), oxidative (*AtCIPK23*) stress, and low-K^+ (*AtCIPK9*) condition (Zhang et al. 2014). For global expression analysis of CBL and CIPK gene family in rice under varied stress conditions (salt, cold, heat, drought, and low potassium), the rice 57 K microarray datasets from Jain et al. (2007), Ma et al. (2012a, b), and Walia et al. (2005) were used as reference. Among the 34 OsCIPK genes, they found that 94 % (32/34) of them were differentially regulated under either of the stress conditions. Temperature stress (heat and cold) regulated expression of a significant number of CIPK genes. Under cold stress,

OsCIPK2, OsCIPK7, OsCIPK9, OsCIPK14/15, OsCIPK19, OsCIPK24, OsCIPK26, and *OsCIPK27* were upregulated, whereas *OsCIPK1, OsCIPK4, OsCIPK5, OsCIPK6, OsCIPK8, OsCIPK11, OsCIPK12, OsCIPK18, OsCIPK22, OsCIPK23,* and *OsCIPK32*/33 were downregulated. On the other hand, heat stress significantly induced the expression of *OsCIPK6, OsCIPK8, OsCIPK14/15, OsCIPK24, OsCIPK25,* and *OsCIPK29* and decreased the transcript accumulation of *OsCIPK1, OsCIPK2, OsCIPK4, OsCIPK5, OsCIPK7, OsCIPK9, OsCIPK10, OsCIPK12, OsCIPK16, OsCIPK17, OsCIPK19, OsCIPK21, OsCIPK22, OsCIPK23, OsCIPK27,* and *OsCIPK32*/33 (Zhang et al. 2014). Xiang et al. in 2007 also reported stress-induced expression profiles of rice CIPK genes. Fifteen genes (*OsCIPK1, OsCIPK2, OsCIPK5, OsCIPK9, OsCIPK11, OsCIPK12, OsCIPK15, OsCIPK17, OsCIPK20, OsCIPK21, OsCIPK22, OsCIPK23, OsCIPK24, OsCIPK29,* and *OsCIPK30*) were induced by drought, 12 genes were induced by salinity stress (*OsCIPK7, OsCIPK8, OsCIPK9, OsCIPK10, OsCIPK11, OsCIPK15, OsCIPK16, OsCIPK17, OsCIPK21, OsCIPK22, OsCIPK29,* and *OsCIPK30*), PEG treatment (*OsCIPK1, OsCIPK3, OsCIPK9, OsCIPK12, OsCIPK15, OsCIPK16, OsCIPK17, OsCIPK21, OsCIPK22, OsCIPK23, OsCIPK24,* and *OsCIPK29*), and ABA-induced expression of 16 (*OsCIPK1, OsCIPK2, OsCIPK3, OsCIPK5, OsCIPK7, OsCIPK9, OsCIPK11, OsCIPK12, OsCIPK15, OsCIPK16, OsCIPK17, OsCIPK20, OsCIPK22, OsCIPK24, OsCIPK29,* and *OsCIPK30*) genes. However, only three genes (*OsCIPK1*, *OsCIPK3*, and *OsCIPK9*) were induced by cold stress (Xiang et al. 2007). Similar differential inducibility of rice CIPK genes during salt, dehydration, and cold stress as well as ABA treatment was also reported by Gu et al. in 2008. As observed among CDPK orthologues, interspecies comparison of CBL and CIPK orthologues revealed both diversification and retention of expression pattern reflecting on their functional attribute. *OsCBL1* and *BnaCBL1* orthologous were both induced by cold stress. However, orthologous *OsCBL10* were induced and *BnaCBL10* repressed by drought stress treatment. *OsCIPK6* and *OsCIPK9* were induced by drought and cold treatment, respectively, as found in case of their orthologous gene *BnaCIPK6* and *BnaCIPK9*, respectively. However, *OsCIPK5* and *OsCIPK23* were downregulated by cold or heat stress, unlike their respective orthologues *BnaCIPK6* and *BnaCIPK23* (Zhang et al. 2014).

In maize, global expression pattern of 37 CIPK genes was assessed where 24, 31, 20, and 19 ZmCIPK isoforms were, respectively, upregulated by salt, PEG, cold, and heat (Chen et al. 2011). Only seven of them (*ZmCIPK2, ZmCIPK4, ZmCIPK5, ZmCIPK23, ZmCIPK24, ZmCIPK32,* and *ZmCIPK36*) were upregulated by all of these four stresses. On comparing the expression pattern ZmCIPKs in two inbred lines (cold tolerant and sensitive) of maize, 13 genes (*ZmCIPK10, ZmCIPK12, ZmCIPK14, ZmCIPK16, ZmCIPK19, ZmCIPK22, ZmCIPK24, ZmCIPK26, ZmCIPK27, ZmCIPK28, ZmCIPK30, ZmCIPK32,* and *ZmCIPK36*) were differentially responsive to cold stress between the two inbred lines, and seven genes (*ZmCIPK14, ZmCIPK16, ZmCIPK22, ZmCIPK24, ZmCIPK27, ZmCIPK32,* and *ZmCIPK36*) were significantly upregulated in cold-tolerant line (Chen et al. 2011). More studies are being done with multiple plant species and specially varying genotypes that have not been sequenced yet. These analyses would complement the existing evidence of involvement of these Ca^{2+}-regulated kinases in abiotic stress-signaling cross talk.

5.4 Posttranslational Proteomic Regulation of Ca^{2+}-Regulated Kinases

Structurally CDPKs consist of four domains (Cheng et al. 2002). The variable N-terminal domain contains myristoylation/palmitoylation sites for subcellular targeting (Cheng et al. 2002; Boudsocq and Sheen 2013). The catalytic kinase domain is followed by an autoinhibitory domain (Harmon et al. 1994) and the CLD (calmodulin-like domain) that contains EF hands (Cheng et al. 2002). A relatively short variable C-terminal domain follows the CLD. Ca^{2+} ion binds to the CLD triggering a conformational change resulting in release of the catalytic domain. The Ca^{2+}-mediated activation model has been demonstrated by removing the CLD, which resulted in an inactive kinase, and the activity was rescued by addition of exogenous CLD protein in the presence of Ca^{2+}. Moreover, truncated CDPK protein, without both the CLD and autoinhibitory domain, produced a constitutively active kinase (Harper et al. 1994; Harmon et al. 1994; Huang et al. 1996; Yoo and Harmon 1996). It has been found that CLD is composed of two globular EF structural domains (N-lobe, C-lobe), each containing a pair of Ca^{2+}-binding site. Of the two globular EF structural domains (N-lobe, C-lobe), at a lower cytosolic Ca^{2+} level, the C-lobe interacts with the junction and eventually with an increase in Ca^{2+} level; the ion binds to N-lobe and triggers the conformational change and activation of the enzyme (Christodoulou et al. 2004). Sensitivity to Ca^{2+} of LeCPK2, expressing in flowers and responding to heat/cold stress, mechanical wounding, and phytohormones (ethylene, MJ, and SA), has been reported to be dependent on the 161 residues of CLD (Chang et al. 2009, 2011). Even differential activity of AtCPK21 was found to be regulated by EF motifs, where preferentially the N-terminal EF hand pair controlled the specificity of AtCPK21 function (Franz et al. 2011). In chickpea, differential Ca^{2+} sensitivity of homologous CDPKs was found to confer functional specificity (Syam and Chelliah 2006a, b).

Proteomics proved to be a powerful tool in exploring biological mechanisms of proteins at the system level. Phosphoproteome analysis of *Arabidopsis* showed significantly altered phosphorylation of 50 different phosphopeptides on ABA treatment for 5–30 min. Among the multiple calcium-related proteins, phosphorylation of AtCPK5/6 decreased on ABA treatment (Kline et al. 2010). On collating expression profiling data and phosphoproteomic data in Physcomitrella patens performed in laboratories worldwide, it brought much deeper insight in the abiotic stress-responsive mechanisms including dehydration, salt, cold, and ABA treatment, where a lot of proteins were found to be regulated coordinately. Among them a CDPK protein was found differentially regulated during dehydration and salinity but not under cold stress (Wang et al. 2012b). In another study, functional phosphoproteomic profiling of phosphorylation sites in membrane fractions of salt-stressed *Arabidopsis* using Zr^{4+}-IMAC magnetic beads to enrich phosphopeptides from the membrane fraction of 3-days salt-stressed *Arabidopsis* was done, where an aquaporin, AtPIP2;4, was identified to be phosphorylated, which is also known to be a substrate of a CDPK protein (Hsu et al., 2009). In more targeted studies, *in vitro* autophosphorylation has been reported for eight CDPKs from *Arabidopsis* (AtCPK1, AtCPK4, AtCPK5, AtCPK10, AtCPK11, AtCPK16, AtCPK28, and AtCPK34) and PfCPK1 from *P. falciparum* (Hegeman et al. 2006). Thirty-five phosphorylation sites have been detected and 15 sites were found to cluster into five conserved groups, which were distributed in kinase, CLD, and N-terminal variable domains. Largely, the frequency of sites was found to be higher in the variable N-terminal domain. Interestingly, junction domain was devoid of any phosphorylation site (Hegeman et al. 2006). Autophosphorylation sites have also been mapped for tomato (CDPK1; Rutschmann et al. 2002; Chang et al. 2011), tobacco (NtCPK2; Glinski et al. 2003), potato (StCDPK3; Grandellis et al. 2012), ice plant (McCPK1; Chehab et al. 2004), and rice (OsCPK10; Fu and Yu 2013). For quite some time the function of variable N-terminal domain in regulation of activity as well as substrate specificity of CDPKs was elusive. In NtCDPK2 and NtCDPK3, stress-inducible phosphorylation site was identified exclusively in N-terminal domain where phosphorylation was found to be differential despite >90 % overall sequence identity. Domain swap and mutation studies of the respective N-terminal domains clearly established their role in regulation (Witte et al. 2010). NtCDPK1 also showed N-terminal domain-mediated substrate specificity where it was found that phosphorylation of NtCDPK1 at Ser-114 of RSG (repression of shoot growth) mediated its binding to 14-3-3 proteins. This resulted in sequestering of RSG in the cytoplasm and inhibited GA biosynthesis (Fukazawa et al. 2000, 2010; Igarashi et al. 2001; Ishida et al. 2004, 2008). Mutation (R10A) in the N-terminal domain of NtCDPK1 reduced RSG recognition. More crucially, chimeric AtCPK9 with N-terminal domain of NtCDPK1 was found to phosphorylate RSG, although native AtCPK9 neither binds nor phosphorylates RSG (Ito et al. 2010). A plastid glutamine synthetase of *M. truncatula* (MtGS2) on phosphorylation (Ser-97) by CDPK facilitated the formation of the GS2-14-3-3 complex. This complex was found to be further recognized by a plant protease (Lima et al. 2006a). Interestingly, in a subsequent article, it was also reported that GS1, an isoenzyme of GS2, is also phosphorylated by CDPK. However, the phosphorylated GS1 does not interact with 14-3-3 proteins indicating towards the complexity involved in posttranscriptional regulation of isoenzymes mediated by phosphorylation (Lima et al. 2006b). SOS1 (plasma membrane sodium/proton exchanger salt-overly-sensitive 1) was also reported to be relieved from autoinhibition upon phosphorylation of the autoinhibitory domain by SOS2-SOS3 (calcium-dependent protein kinase complex; Quintero et al. 2011). Another CDPK gene from *Arabidopsis*, AtCPK28, was found to be phosphorylated at several sites *in vivo*, and site-specific mutation at these phosphorylation sites resulted in reduced *in vitro* activity again indicating towards the posttranslational regulation (Matschi et al. 2013). Unlike CDPKs, which along with calcium-binding domains harbor a catalytic domain in a single protein, CBLs contain only calcium-binding domains. Therefore, CBLs transit the Ca^{2+} signals through interaction with target proteins, CBL-interacting protein kinases (CIPKs; Batistic and Kudla 2004). CBL proteins have sequence similarity with the regulatory B subunit of calcineurin (CNB) and neuronal calcium sensor 1 (NCS-1) found in animals. The number of EF hands varies in CBLs as well as several EF hands are found with amino acid substitutions that impair Ca^{2+} binding (Kolukisaoglu et al. 2004). These structural variations could be responsible for differential activation of CBL–CIPK complexes depending on stimulus-specific Ca^{2+} spike. Structural analyses of CBLs have revealed that NCS-1 possesses a wide hydrophobic crevice at their surface situated between the two EF hand-containing lobes (Strahl et al. 2003). The hydrophobic crevice is highly conserved, and in the non-CIPK-bound form, the C-terminal tail of the CBL proteins binds to the hydrophobic crevice and helps prevent nonspecific protein interactions (Batistic and Kudla 2009). CIPKs contain a C-terminal regulatory domain and an N-terminal protein kinase domain. The regulatory C-terminal domain harbors an NAF/FISL motif required for interaction with CBLs. Binding of CBLs to the NAF motif removes the autoinhibitory domain from the kinase domain, conferring activation of the kinase (Albrecht et al. 2001). Autophosphorylation and transphosphorylation have been found to enhance activation process of CIPKs (Gong et al. 2002; Guo et al. 2001). Crystallization study of *Arabidopsis* CBL proteins revealed that in spite of an uncommon Ca^{2+}-binding loop in the first EF hand, it binds to Ca^{2+} (Nagae et al. 2003). Even binding of Ca^{2+} to CBL is regulated by its state of interaction with CIPK proteins. Non-CIPK-interacting form of AtCBL2 was found to bind only two Ca^{2+} ions on first and last EF hands (Nagae et al. 2003). However, when in complex with AtCIPK14, all four EF hands of AtCBL2 were bound to Ca^{2+} (Akaboshi et al. 2008). It was also found that AtCBL4 binds to four Ca^{2+} even in CIPK-unbound form unlike AtCBL2 (Sanchez-Barrena et al. 2005). Remarkably, when in complex with AtCIPK24, AtCBL4 binds to Ca^{2+} in the first and last EF hands, as happens to AtCBL2 in CIPK-unbound form (Sánchez-Barrena et al. 2007).

Phosphorylation of CBL proteins acts as regulatory mechanism in the same way as it does in CDPK proteins. It was reported in *Arabidopsis* that phosphorylation at the conserved PFPF motif at the C-terminus of CBL proteins by CIPK stabilizes the interaction between them (Du et al. 2011). Phosphorylation of CBL/SCaBP1 by CIPK/PKS5 neutralized the inhibitory effect of PKS5 on plasma membrane H^+-ATPase activity (Du et al. 2011). The AtCBL1–CIPK23 and AtCBL9–CIPK23 complexes as well as individual CBL proteins compete with AtCIPK23 and phosphorylate and regulate AKT1 to protect plants from low potassium stress (Xu et al. 2006; Ren et al. 2013). CIPKs and CBLs have also been found to be regulated by protein phosphatase. Lan et al. (2011) showed that CIPKs interacted with PP2Cs via protein phosphatase interaction (PPI) motif as well as kinase domain. Furthermore, specific CBL proteins were found to physically interact and inactivate PP2C phosphatases and recover the CIPK-dependent AKT1 channel activity (Lee et al. 2007; Lan et al. 2011). It has also been shown that CIPK interacts with protein phosphatase ABA-insensitive 1 (ABI1) or ABA-insensitive 2 (ABI2) proteins and may be regulating the activity by either phosphorylation/dephosphorylation or physiological interaction (Nakashima and Yamaguchi-Shinozaki, 2013). Interestingly Ca^{2+} was found to have inhibitory effect on the activity of VfCBL1-VfCIPK1, in *Vicia faba* (Tominaga et al. 2010). These findings suggest that the complexity of CBL–CIPK posttranslation regulation is rather disarming as it is regulated by Ca^{2+}, phosphorylation of CBL by CIPKs, and thereafter interaction with downstream proteins, even interaction of CBL directly with downstream protein, as well as interaction with phosphatases.

5.5 Subcellular Proteomics and Functional Relevance

Cellular positioning of proteins is an important aspect for specific information processing. Subcellular localization facilitates access and availability of interacting partners that are determinants for transduction of signals (Sachs and Engelman 2006). CDPK proteins harbor a highly variable N-terminal domain, which contains *N*-myristoylation and palmitoylation sites that are responsible for subcellular targeting (Cheng et al. 2002). *N*-myristoylation and palmitoylation signal is abundant among proteins those are involved in signal transduction (Taniguchi 1999; Iwanaga et al. 2009). *N*-myristoylation and palmitoylation are posttranslational attachment of myristic acid to the glycine and palmitic acid to cysteine residue, respectively, preferably in the N-terminal domain, which facilitates anchoring to membrane (Taniguchi 1999; Iwanaga et al. 2009). Out of 34 CDPKs from *Arabidopsis*, 24 have been predicted to contain myristoylation site (Cheng et al. 2002); similarly in rice, out of 31 CDPKs, 18 proteins bear this site (Asano et al. 2005; Ray et al. 2007). In case of wheat 17 of the 40 CDPK isoforms have been predicted to have myristoylation site (Li et al. 2009; Kong et al. 2013). Fourteen of the 30 CDPK genes from *Populus* are predicted to have myristoylation site, and 28 of them bear palmitoylation site (Zuo et al. 2013). In apple all 30 CDPKs have palmitoylation site (Kanchiswamy et al. 2013). In agreement to the structural diversity, CDPKs are located on plasma membrane (Yoon et al. 1999; Dammann et al. 2003), cytosol (Dammann et al. 2003; Ray et al., unpublished data), nucleus (Patharkar and Cushman 2000; Dammann et al. 2003), peroxisome (Dammann et al. 2003), ER membrane (Lu and Hrabak 2002), cytoskeleton (Putnam-Evans et al. 1989), mitochondria (Pical et al. 1993), as well as endosperm storage vesicles (Anil et al. 2003). In a subcellular proteomic localization study, eight *Arabidopsis* CDPK proteins (AtCPK1, AtCPK3, AtCPK4, AtCPK7, AtCPK8, AtCPK9, AtCPK16, AtCPK21, and AtCPK28) were analyzed by generating transgenic lines expressing AtCPK–GFP fusion construct (Dammann et al. 2003). AtCPK3 and AtCPK4 localized in nucleus as well as cytosol, suggesting them to be soluble proteins with potential to target nucleus; six CDPK proteins (AtCPK7, AtCPK8, AtCPK9, AtCPK16, AtCPK21, and AtCPK28) localized on nuclear membrane (Dammann et al. 2003). Unlikely, AtCPK1 localized with peroxisomal bodies (Dammann et al. 2003). In a subsequent study AtCPK1 was found to regulate plant innate immunity via SA-dependent signaling pathway and showed dual localization in lipid bodies and peroxisomes (Coca and San Segundo 2010). Fatty acids were transferred from oil bodies to peroxisomes for fatty acid β-oxidation, and it was observed that the lipid bodies containing AtCPK1 are located near peroxisomes, where it may be facilitating defense response (Coca and San Segundo 2010). In 2008, Benetka et al. reported that AtCPK2, AtCPK9, and AtCPK13 localized on plasma membrane where AtCPK2 localized also on some distinct spots in the cytosol (not nucleus), whereas AtCPK6 localized on membrane and nucleus. They also demonstrated that disruption in myristoylation site led to loss of target location and reallocating these proteins preferably to cytosol (Benetka et al. 2008). The determinant role of myristoylation and palmitoylation in localization has been reported for AtCPK16, which localizes predominantly on the plasma membrane. Mutation at myristoylation site (replacing glycine on position 2 with alanine) relocated AtCPK16 to chloroplasts. Conversely, the mutant protein, which can be myristoylated but not palmitoylated, re-localized to the nucleus (not to chloroplast). The protein with impaired myristoylation as well as palmitoylation localized back to chloroplasts, implying that myristoylation alone inhibits chloroplast localization, while the presence of both the signals determines membrane localization (Stael et al. 2011). Native StCDPK1, LeCPK1, and NtCPK5 proteins were found to be membrane bound. However, mutation at myristoylation and palmitoylation site targeted protein to cytoplasm (Rutschmann et al. 2002; Raices et al. 2003; Wang et al. 2005).

Colinearity of cellular compartmentalization and function has been demonstrated in multiple CDPK proteins. In potato, StCDPK5 was observed to be dominantly localized to the plasma membrane and also activating the plasma membrane NADPH oxidase (RBOH; for respiratory burst oxidase homologue) StRBOHB by direct phosphorylation of the N-terminal region. Mutations of N-terminal myristoylation and palmitoylation sites resulted in loss of plasma membrane localization and capacity to activate StRBOHB *in vivo*. A tomato SlCDPK2, which also contains myristoylation and palmitoylation sites in its N-terminus, phosphorylated StRBOHB *in vitro* but not *in vivo*. Variable domain swapping of StCDPK5 with that of SlCDPK2 abolished the activation and phosphorylation abilities of StRBOHB *in vivo* and re-localized to the trans-Golgi network, as observed for SlCDPK2. Conversely, chimeric SlCDPK2 localized to the plasma membrane and activated StRBOHB, conferring the role of variable domains in substrate specificity and proper subcellular localization of CDPK proteins (Asai et al. 2013). In a series of studies published over time, AtCPK4, AtCPK5, AtCPK6, and AtCPK11 showed dual functionalities and their cellular localization seems to aid the same. AtCPK11 and its interacting partner, AtDi19, a zinc-finger protein, were found to colocate in nucleus, which is consistent with their *in vitro* interaction (Milla et al. 2006a). AtCPK11 was also found to be localized in cytoplasm. Functionally AtCPK11- and AtDi19-related genes are reported to be stimulated by drought and salt stresses (Milla et al. 2006b). Moreover, AtCPK11 along with AtCPK4 has been found to also phosphorylate transcription factors ABF1 and ABF4 while mediating ABA-signaling stress response (Zhu et al. 2007). Furthermore, AtCPK11, AtCPK4, AtCPK5, and AtCPK6 have also been reported to have role in innate immunity. They have been shown to regulate ROS production by directly phosphorylating the NADPH oxidase as well as regulate early target genes, which are also modulated by flg22 (within 30–60 min; Boudsocq et al. 2010).

Association of AtCPK5 with the plasma membrane could be removed by converting glycine at the proposed site of myristate attachment to alanine (G2A). In transgenic plants, the G2A mutation completely abolished AtCPK5 membrane association. Even the authors demonstrated importance of the first sixteen amino acids of AtCPK5 for plasma membrane localization (Lu and Hrabak 2002). Therefore, dual location of AtCPK4, AtCPK5, AtCPK6, and AtCPK11 could be transducing a quick response on Ca^{2+} sensing by phosphorylating downstream proteins, while phosphorylation of the transcription factors by the nuclear-localized ones would regulate gene expression. Similar event was noted for AtCPK32, which interacted with ABF4 and localized in the nucleus (Choi et al. 2005). Again AtCPK23 along with AtCPK21 was involved in drought response via ABI1-mediated regulation of guard cell anion channel SLAC1, where membrane localization of both the CDPK proteins is in tune with their functional role in guard cell signaling (Geiger et al. 2010; Franz et al. 2011). On studying localization using nanodomain marker remorin, ABI1 and AtCPK21 were found to co-localize with the marker remorin, and AtCPK21 was found to interact and activate SLAH3 (Demir et al. 2013). Dual localization of CDPK proteins as found in *Arabidopsis* has also been reported in rice where OsCPK13 localized in cytoplasm and nucleus and it functionally confers salt/drought tolerance, but the interacting proteins and signaling pathways are yet to be identified for this enzyme; hence, the relevance of dual subcellular localization remains unclear (Saijo et al. 2000; Ray et al. unpublished data). AhCPK2 from *Arachis hypogaea* was found to be located on membrane and soluble fractions under normal and stressed condition, respectively, but not in the nuclear fraction. But then again, it was clearly detectable in the nuclear fraction of the cells when subjected to 0.4 M sucrose for 4 days. It has also been demonstrated that this differential nuclear localization is specific to stress response, since following auxin treatment to *Arachis* cells, the subcellular localization remained unchanged (Raichaudhuri et al. 2006). On the same note, McCPK1 also showed stress-dependent differential localization in cellular compartments (Patharkar and Cushman 2000; Chehab et al. 2004). Yeast-two hybrid analysis revealed McCSP1 (two-component pseudoresponse regulator class of transcription factor; Patharkar and Cushman 2000), McCAP1 (*M. crystallinum* CPK1 adapter protein 1 having a coiled-coil structure; Patharkar and Cushman 2006), and McCAP2 (*M. crystallinum* CPK1 adapter protein 2; Chehab et al. 2007) as interacting proteins of McCPK1. Under salt stress condition McCPK1 was found to phosphorylate McCSP1 in a calcium-dependent manner and co-localize in the nucleus. However, in control condition, when McCPK1 remained associated with the plasma membrane, McCSP1 exclusively localized to the nucleus (Patharkar and Cushman 2000). McCAP1 was also identified to be a substrate of McCPK1. Further, McCPK1 and McCAP1 were found to co-localize in nucleus and cytoplasmic strands of plant cell on exposure to low-humidity condition. Taken together, McCAP1 might be anchoring McCPK1 to cytoskeleton at the time of stress condition (Fig. 5.2; Patharkar and Cushman 2006). On the other hand, another CAP protein, McCAP2, was not phosphorylated by McCPK1 but co-localized with McCPK1 in vesicular and actin microfilament structures as well as ER under low-humidity (40 %) condition, whereas under high relative humidity (80 %), McCPK1 relocated to the plasma membrane, but McCAP2 remained to the vesicle-like structures. McCAP2 also co-localized with AtVTI1a, a v-SNARE protein known to localize to the trans-Golgi network (TGN) and prevacuolar compartments (PVCs; Zheng et al. 1999). Evidently, McCPK1 does not phosphorylate McCAP2, but it might be an adapter protein in the protein complex facilitating vesicle transport (Chehab et al. 2007), and McCPK1 is involved in maintaining water balance of ice plant to changing relative humidity as well as high salinity. CDPK from *Zingiber officinale*, ZoCDPK1–GFP fusion protein preferably

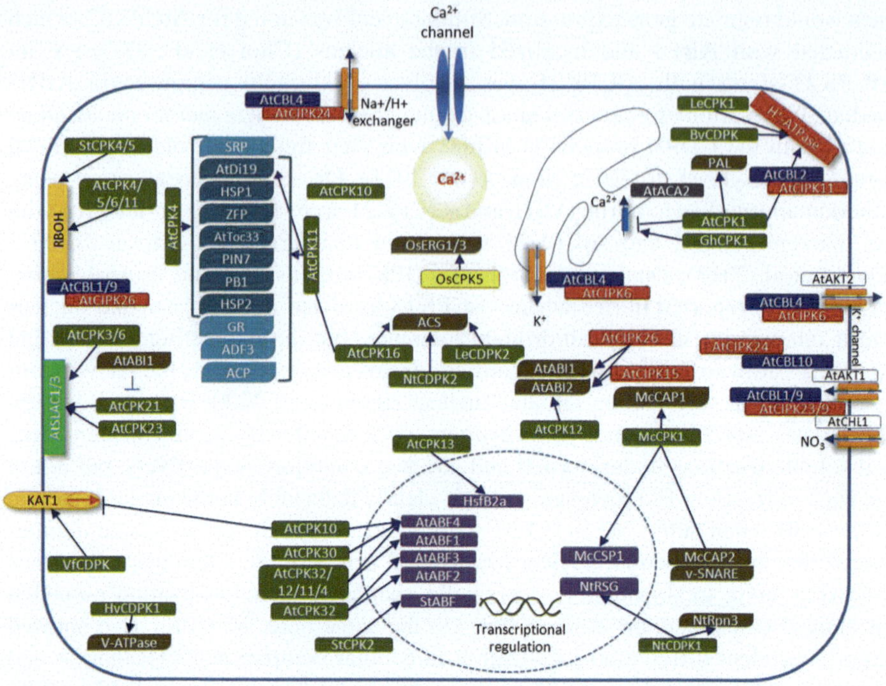

Fig. 5.2 A schematic model of CDPK and CBL–CIPK protein and their targets in plant cell. This model is not exhaustive, all orthologues are not represented, and examples across plant taxa have been included. CDPK and CBL–CIPK proteins interact with target proteins at multiple subcellular locations. The physiological functions of the sensors are mentioned in text. *SRP* serine-rich protein, *ZFP* zinc-finger protein, *SLAC1* slow anion channel associated 1, *KAT1* potassium channel in *Arabidopsis thaliana* 1, *ABF* ABRE-binding protein/factor, *GR* geranylgeranyl reductase, *HSP1* heat shock protein 1, *PIN7* pin-formed 7, *CSP1* CDPK substrate protein 1, *CAP1* CDPK adapter protein 1, *ADF3* actin depolarizing factor 3, *ACP* acyl-carrier protein, *ABI* ABA insensitive, *ACS* 1-aminocyclopropane-1-carboxylic acid synthase, *PAL* phenylalanine lyase, *RSG* repression of shoot growth, *Toc33* translocon at the outer envelope membrane of chloroplast protein, *Rpn3* regulatory subunit of 26S proteosome, *PB1* Phox and Bem1 domain protein, *RBOH* respiratory burst oxidase or NADPH oxidases, *ERG1* elicitor-responsive gene 1, *CHL1* chlorate-resistant mutant 1

localized in the nucleus and to a reasonable level in cytosol. Under salinity and drought stress conditions, additional GFP signals were detected in plasma membrane but not in cytosol, fairly indicative of re-localization of ZoCDPK1 in response to stress conditions (Vivek et al. 2013). OsCPK4 and OsCPK18 were also located in plasma membrane and on mutation of the myristoylation site, relocated to cytoplasm. These CDPK proteins were found to be inducible by AM fungus *Glomus intraradices;* therefore, their membrane location might be instrumental in recognizing the symbiotic signal from the fungi (Campos-Soriano et al. 2011). Interestingly, PiCDPK1 localizes preferably at the periphery of the pollen tube consistent with the plasma membrane, unlike PiCDPK2, which localized to internal membrane compartments (Yoon et al. 2006). Truncated PiCDPK1 (ΔN-PiCDPK1), on losing its target signal, localized in the cytosol rather than the plasma membrane and, interestingly, resulted in normal pollen tube growth in contrast to the loss of polarization in pollen tubes, which was exhibited on expression of constitutively active protein. Subcellular localization of several other CDPK proteins is reported; however, they are yet to be substantially correlated to their function, namely, from rice (Morello et al. 1993; Abo-el Saad and Wu 1995), OsCPK2 (Martín and Busconi 2000), OsCPK24 (Zhang et al. 2005), OsCPK21 (Asano et al. 2011, Ray et al. unpublished data), capsicum CaCDPK3 (Chung et al. 2004), chickpea CaCPK1 and CaCPK2 (Syam and Chelliah, 2006a), *Chlamydomonas* CrCDPK3 (Liang and Pan, 2013), *and PeCPK10* (Chen et al. 2013a, b, c). These findings emphasize on assessment of subcellular localization of CDPKs as an important parameter for determining the cellular functionality and their role in signaling cascade.

CBL proteins ranging from algae, fern, moss, *Arabidopsis* (dicot), and rice (monocot) harbor an N-terminal myristoylation and *S*-acylation motif (Kolukisaoglu et al. 2004; Batistic et al. 2008). Interestingly, myristoylated isoforms clustered in two neighboring branches indicating loss of lipid modification in a single early event during evolution (Kolukisaoglu et al. 2004). High-throughput cellular localization studies of all ten *Arabidopsis* CBL proteins revealed that their variable N-terminal extensions determine their subcellular targeting. The CBL proteins AtCBL1, AtCBL4, AtCBL5, and AtCBL9 harbor N-terminal myristoylation as well as *S*-acylation sites, which mediate plasma membrane targeting (D'Angelo et al. 2006; Cheong et al. 2007; Batistic et al. 2008, 2010; Held et al. 2011). Ringlike distribution of AtCBL1:GFP at the plasma membrane in protoplasts was observed. Myristoylation mutant (AtCBL1G2A:GFP and AtCBL1G2AC3S:GFP) proteins showed cytoplasmic and nucleoplasmic distribution. AtCBL1C3S:GFP fluorescence was also detected at the nuclear envelope. It was also demonstrated that acylation stabilizes the plasma membrane binding of AtCBL1 as well as crucial for endoplasmic reticulum (ER)-to-plasma membrane trafficking (Batistic et al. 2008). Co-expressing wild-type AtCBL1 with AtCIPK1 complemented temperature sensitivity in *cdc25* yeast mutant; however, none of the mutated AtCBL1 proteins (G2A, C3S, or G2AC3S) were able to complement (Batistic et al. 2008). This clearly emphasizes on the importance of both myristoylation and *S*-acylation for targeting of the AtCBL1/CIPK1 complex to the plasma membrane in yeast. The localization of AtCBL4–GFP and AtCBL5–GFP fusion proteins was initially detected throughout the plant cells, including cytoplasm and the nucleus. Further, co-localization analysis of AtCBL4::GFP with AtCBL1n::OFP (plasma membrane marker protein created by the fusion of the AtCBL1 N-terminus to an OFP) confirmed the plasma membrane localization of AtCBL4 along with nucleus and cytoplasm. Fractionation experiments detected AtCPK5 in both soluble and membrane fractions. Hence, AtCBL1 and AtCBL9 preferentially receive Ca^{2+} signal at membrane, whereas AtCBL4 and AtCBL5 also function in cytoplasm and nucleus (Batistic et al. 2008, 2010). AtCBL2 localized on tonoplast and AtCBL10 localized to the endosomal and tonoplast compartments (Kim et al. 2007; Batistic et al. 2008). Batistic et al. (2010) showed that N-terminal extension of AtCBL2, AtCBL3,

AtCBL6, and AtCBL10 varied from the myristoylated CBL proteins and formed a separate phylogenetic group. AtCBL2::GFP co-localized at the tonoplast with AtCBL2::OFP, AtCBL3::OFP, AtCBL6::OFP, and OFP:: AtCBL10, respectively, indicating that all four CBL proteins are targeted to the same compartment (Batistic et al. 2010). AtCBL7 and AtCBL3 arose by tandem duplication, but there is significant variation in the N-terminal domains (Kolukisaoglu et al. 2004). GFP fused with AtCBL7 was found to be distributed in cytoplasm and nucleus. Co-localization studies with marker proteins for ER, Golgi, and tonoplast did not indicate any association of AtCBL7 with these compartments. AtCBL8 localized as AtCBL7 but it was also found on plasma membrane. AtCBL8 is phylogenetically most closely related to AtCBL4, but unlike AtCBL4 lacks the myristoylation motif (Batistic et al. 2010). Batistic et al. (2008) revealed that 12 aa of the N-terminus of AtCBL1 are sufficient for dual lipid modification and the plasma membrane targeting. The same group in 2010 made GFP fusion construct with N-termini of AtCBL3 (22 aa), AtCBL4 (18 aa), AtCBL5 (12 aa), AtCBL7 (23 aa), AtCBL9 (12 aa), AtCBL8 (16 aa), and AtCBL10 (40 aa) and observed localization in leaf epidermal cells and protoplasts. AtCBL3n::GFP localized at the tonoplast and AtCBL7n::GFP in cytoplasm and nuclei in accordance to their respective full-length proteins. Similarly, AtCBL10n::GFP localized in the endosomal and tonoplast compartments as per the full-length AtCBL10 protein (Kim et al. 2007). On the other hand, N-terminally truncated AtCBL10Δn::GFP (CBL10 36–246 aa-GFP) protein and AtCBL10ΔTMD::GFP accumulated exclusively in the cytoplasm and nucleus, indicating that the N-terminal 40 amino acids of AtCBL10 are required for proper targeting. In contrast to AtCBL5, full-length protein (cytoplasm and nucleus) AtCBL5n::GFP localized exclusively to plasma membrane. Similarly they observed an apparent plasma membrane localization of AtCBL4n::GFP and AtCBL8n::GFP, respectively. The role of N-terminal domain for plasma membrane localization was confirmed in AtCBL9n::GFP. However, the differences in the localization of the AtCBL4, AtCBL5, and AtCBL8 proteins suggest of a more complex regulation where the N-terminal domain is important but not exclusively determining the targeting (Batistic et al. 2010). In rice *OsCBL1, OsCBL2, OsCBL3,* and *OsCBL4* fused to GFP were transiently expressed in barley aleurone protoplasts for proteomic localization. OsCBL2 and OsCBL3 are targeted to the tonoplast, whereas OsCBL4 is to the plasma membrane. OsCBL4 has a possible myristoylation site, which is in tune with its plasma membrane localization. However, specific localization was not apparent for OsCBL1 (Hwang et al. 2005). In maize, ZmCIPK16 was found to be distributed in the nucleus, plasma membrane, and cytoplasm; however, its interacting partners ZmCBL3, ZmCBL4, and ZmCBL5 localized in the plasma membrane (Zhao et al. 2009); in canola BnaCBL1 protein localized in nuclei and plasma membrane of leaf epidermal cells. BnaCBL9 and BnaCBL3 localized in the cytoplasm and nuclei, whereas BnaCBL10 was found to be located on tonoplast along with cytoplasm and nucleus (Zhang et al. 2014).

An intriguing aspect of CBL–CIPK is their regulation of specific activity. Activity of the complex is regulated by differential affinity of CBL proteins for Ca^{2+}, alternative complex formation with multiple CIPKs as well as spatial property. AtCIPK24 and its interaction with AtCBL4 and AtCBL10 demonstrated correlation between activity and cellular location. The SOS (salt-overly-sensitive) pathway was delineated by identifying mutants (SOS3/CBL4, SOS2/CIPK24, and SOS1/Na^+–H^+ antiporter) in a forward genetic screen of *Arabidopsis* plants, which were sensitive to high salt conditions (Zhu et al. 1998). The calcium sensor SOS3/AtCBL4 interacted with SOS2/AtCIPK24 to the plasma membrane where they activated the SOS1/Na^+–H^+ exchanger leading to extrusion of toxic sodium ions from the cell of roots (Halfter et al. 2000; Qiu et al. 2002; Quintero et al. 2002). In leaves, AtCIPK24 interacted with AtCBL10 localized on vacuolar membrane and consequently sequestered cytosolic sodium into vacuole (Kim et al. 2007; Waadt et al. 2008). Bimolecular fluorescence complementation (BiFC) analysis was done for interaction of AtCIPK14 with AtCBL2, AtCBL3, and AtCBL8. AtCIPK14 and the respective CBL proteins were cloned to N- and C-terminus of the yellow fluorescent protein (YFP), respectively, and co-infiltrated into *N. benthamiana* leaves with plasma membrane and tonoplast markers tagged to OFP. AtCBL2 and AtCBL3 were found to interact with AtCIPK14 at the tonoplast (Batistic et al. 2010; Tang et al. 2012). AtCBL8–CIPK14 complex formed exclusively at the plasma membrane, as localization was detected when N-terminus of AtCBL8 was fused to GFP, whereas native AtCBL8 localized in cytoplasm and nucleus (Batistic et al. 2010). The data also indicate of dual functioning mode of AtCIPK14 depending on the interacting CBL protein. AtCBL3 was also found to be fully sufficient for dynamic translocation of AtCIPK5 between the cytoplasm and the tonoplast in a dose-dependent manner (Schlücking et al. 2013).

CIPK proteins from *Arabidopsis* and rice did not reveal any apparent lipid modification or targeting motif on sequence analysis (Kolukisaoglu et al. 2004). GFP fusion proteins of AtCIPK1, AtCIPK23, and AtCIPK24 were found to localize throughout the plant cell, including the cytoplasmic and nuclear compartments (D'Angelo et al. 2006; Cheong et al. 2007; Kim et al. 2007). Batistic et al. (2010) extended the analysis of *Arabidopsis* CIPK localization by expressing 12 GFP fusion proteins (AtCIPK1, AtCIPK2, AtCIPK3, AtCIPK4, AtCIPK7, AtCIPK8, AtCIPK10, AtCIPK14, AtCIPK17, AtCIPK23, and AtCIPK24) in *N. benthamiana* leaves and found that they have similar localization pattern, i.e., in cytoplasm and nucleus.

Taken together cellular localization could be a mechanism for conferring specificity to spatial and temporal deciphering of Ca^{2+} signals. However, this mode of regulation will have better understanding as and when functions of more CBL–CIPK complexes are elucidated in future studies.

5.6 Interaction and Complexities of Ca^{2+}-Regulated Kinase in Abiotic Stress Network

CDPKs have been found to be intricately involved in abiotic stress regulation by transducing the stress signal to other signaling molecules (i.e., glutamine synthetase) or directly to transcriptional activators like RSG (Table 5.1; Ludwig et al. 2004; Lima et al. 2006a, 2006b; Ito et al. 2010; Ray 2012). A significant cross talk at the level of CDPK-mediated signal transduction prevails between cold, drought, and salinity stress, where ABA is found to be regulating cellular homeostasis (Table 5.1).

Table 5.1 CDPK genes characterized to be involved in abiotic stress across plant species

Species	Gene name	Characterization	Reference
Arabidopsis thaliana	*AtCPK1*	Enhances NADPH activity on overexpression; mediates pathogen resistance; involved in secondary metabolism via the activation of ROS production	Xing et al. (2001), Coca and San Segundo (2010), Shkryl et al. (2011), Bulgakov et al. (2011)
	AtCPK3	Functions in guard cell ion channel regulation; salt stress acclimation	Mori et al. (2006), Mehlmer et al. (2010)
	AtCPK4	Overexpression increased ABA sensitivity and salt hypersensitivity in seedling growth and affects stomatal regulation; involved in primary responses in innate immune signaling	Zhu et al. (2007), Boudsocq et al. (2010)
	AtCPK5	Involved in primary responses in innate immune signaling; along with NADPH oxidase facilitates rapid signal propagation for defense-response activation at distal sites within the plant	Boudsocq et al. (2010), Dubiella et al. (2013)
	AtCPK6	Functions in guard cell ion channel regulation; salt stress acclimation; convergent point of stomatal closure in response to abiotic and biotic stress	Mori et al. (2006), Boudsocq et al. (2010), Xu et al. (2010), Munemasa et al. (2011), Ye et al. (2013)
	AtCPK10	Abscisic acid and Ca^{2+}-mediated stomatal regulation in response to drought stress; activated stress and ABA-inducible promoter	Urao et al. (1994), Sheen (1996), Zou et al. (2010)
	AtCPK11	Overexpression increased ABA sensitivity and salt hypersensitivity in seedling growth and affected stomatal regulation; involved in primary responses in innate immune signaling; responsive to drought and salt stress; interacts with AtDi19 protein	Urao et al. (1994), Milla et al. (2006b), Zhu et al. (2007), Boudsocq et al. (2010)
	AtCPK12	Negatively regulates ABA signaling	Zhao et al. (2011)
	AtCPK16	Involved in root gravitropism by phosphorylating AtACS7	Huang et al. (2013a, b)
	AtCPK21	Regulates guard cell anion channel SLAC1; responsive to abiotic stress; involved in ABA signaling via inhibition of ABI1 and activation of the SLAC1	Geiger et al. (2010), Franz et al. (2011), Demir et al. (2013)

	AtCPK23	Responsive to drought and salt stresses via stomatal closure; regulates guard cell anion channel SLAC1	Ma and Wu (2007), Geiger et al. (2010)
	AtCPK30	Activated stress and ABA-inducible promoter	Sheen (1996)
	AtCPK32	Induced by ABA, salt stress, touch, wounding, and darkness	Choi et al. (2005), Chotikacharoensuk et al. (2006)
Arachis hypogaea	*AhCPK2*	Responsive to drought stress	Raichaudhuri et al. (2006)
Cicer arietinum	*CaCPK1*	Induced by biotic stress; expressed abundantly in roots	Syam and Chelliah (2006a)
	CaCPK2	Induced by dehydration stress and GA; expressed abundantly in roots	Syam and Chelliah (2006a)
Capsicum annuum	*CaCDPK3*	Induced by ABA, salicylic acid, jasmonic acid, and ethephon	Chung et al. (2004)
Chlamydomonas reinhardtii	*CrCDPK3*	Involved in flagellar biogenesis	Liang and Pan (2013)
Cucumis sativus	*CsCDPK5*	Induced by cytokinin, IAA, ABA, GA	Kumar et al. (2004)
Fragaria x *ananassa*	*FaCDPK1*	Induced by low-temperature stress and expressed in developing fruit	Llop-Tous et al. (2002)
Hevea brasiliensis	*HbCDPK1*	Induced by mechanical wounding, jasmonic acid, and ethephon; preferential transcript accumulation in latex	Zhu et al. (2010)
Hordeum vulgare	*HvCDPK3*	Constitutive active expression showed compromised penetration of pathogen and resistance to powdery mildew	Freymark et al. (2007)
	HvCDPK4	Constitutive active expression showed compromised penetration of pathogen and resistance to powdery mildew	Freymark et al. (2007)
Isatis indigotica	*IiCPK2*	Induced by salinity, cold stress, and GA	Lu et al. (2006)
Medicago sativa	*MsCK1*	Induced by cold stress	Monroy and Dhindsa (1995)
	MsCK2	Induced by cold stress	Monroy and Dhindsa (1995)
	MsCPK3	Induced by heat stress and 2,4-D	Davletova et al. (2001)
Mesembryanthemum crystallinum	*McCPK1*	Responsive to drought and salt stress; subcellular localization affected by salt and water deficit condition; interaction with v-SNARE family protein and McCAP1, a novel coiled-coil protein	Patharkar and Cushman (2000), Chehab et al. (2004), Patharkar and Cushman (2006)

(continued)

Table 5.1 (continued)

Species	Gene name	Characterization	Reference
Nicotiana tabacum	*NtCDPK1*	Induced by Ca^{2+}, GA, ABA, cytokinin, methyl jasmonate, wounding, fungal elicitors, chitosan, and salt stress; interacts with NtRpn3 regulatory subunit of 26S proteosome and regulates cell division, differentiation, and cell death; phosphorylates RSG and mediates GA biosynthesis	Yoon et al. (1999), Lee et al. (2003), Ishida et al. (2008)
	NtCDPK2	Induced by fungal elicitor and osmotic stress	Romeis et al. (2001), Ludwig et al. (2005)
	NtCDPK3	Induced by fungal elicitor and osmotic stress	Romeis et al. (2001)
	NtCPK4	Induced by salt stress and GA; transcript accumulates on stigma surface and in anther during its early developmental stage	Zhang et al. (2005)
Oryza sativa	*OsCPK1*	Specifically activated by sucrose starvation; involved in mechanism to prevent drought stress injury during germination by negatively regulating the expression of GA biosynthesis enzymes and activate expression of a 14-3-3 protein, GF14c	Ho et al. (2013)
	OsCPK4	Induced by arsenate [As(V)] treatment	Huang et al. (2012)
	OsCPK5	Mediates response to calcium ions or fungal elicitor via phosphorylating OsERG1 and OsERG3	Kang et al. (2013)
	OsCPK7	Induced by cold, GA, and JA; overexpression conferred cold tolerance and induction of CRTintP1 and calreticulin, which also conferred cold tolerance to rice	Breviario et al. (1995), Frattini et al. (1999), Yang et al. (2003), Akimoto-Tomiyama et al. (2003), Abbasi et al. (2004), Komatsu et al. (2007)
	OsCPK9	Induced by *Magnaporthe grisea*	Asano et al. (2005)
	OsCPK10	Strongly induced by *Magnaporthe grisea* elicitor; overexpression of constitutively active OsCPK10 in *Arabidopsis* and rice-enhanced resistance to *Pseudomonas syringae* pv. Tomato and *M. grisea*, respectively; activated expression of SA- and JA-related defense genes in both the transgenics	Fu and Yu (2013)
	OsCPK12	Overexpression conferred tolerance to salt stress and reduced resistance to blast disease	Asano et al. (2011)

	OsCPK13	Responsive to cold, salt, and dehydration; express predominantly in vascular tissue of crown and roots, vascular bundle, and central cylinder; induced by fungal elicitor; overexpression in *Sorghum* gives lesion-mimic phenotype and upregulation of pathogen-related proteins; induced by ther arsenate [As(V)] treatment	Saijo et al. (2000), Saijo et al. (2001), Akimoto-Tomiyama et al. (2003), Mall et al. (2011), Huang et al. (2012)
	OsCPK15	Induced by fungal elicitor	Akimoto-Tomiyama et al. (2003)
	OsCPK20	Induced by fungal elicitor; induced by ther arsenate [As(V)] treatment	Akimoto-Tomiyama et al. (2003), Huang et al. (2012)
	OsCPK21	Overexpression confers salt tolerance; induced by ther arsenate [As(V)] treatment	Asano et al. (2011), Huang et al. (2012)
	OsCPK24	Induced by fungal elicitor	Akimoto-Tomiyama et al. (2003), Zhang et al. (2005)
Phalaenopsis amabilis	*PaCDPK1*	Induced under cold stress, wounding, and pathogen attack	Tsai et al. (2007)
Panax ginseng	*PgCDPK1c*	Induced by salt stress	Kiselev et al. (2009)
	PgCDPK2c	Induced by salt stress	Kiselev et al. (2009)
	PgCDPK4a	Induced by salt stress	Kiselev et al. (2009)
	PgCDPK1b	Repressed by salt stress	Kiselev et al. (2009)
	PgCDPK3a	Repressed by salt stress	Kiselev et al. (2009)
Populus euphratica	*PeCPK10*	Induced by salt, cold, and drought stresses; overexpression in *Arabidopsis* conferred tolerance to cold and drought stress	Chen et al. (2013a, b, c)
Solanum tuberosum	*StCDPK4*	Responsive to fungal elicitor and phosphorylates downstream St RBOHB (respiratory burst oxidase homologue) resulting in ROS production	Kobayashi et al. (2007)
	StCDPK5	Responsive to fungal elicitor and phosphorylates downstream St RBOHB (respiratory burst oxidase homologue) resulting in ROS production	Kobayashi et al. (2007)

(continued)

Table 5.1 (continued)

Species	Gene name	Characterization	Reference
	RiCDPK2	*A. solani*-stimulated activity of *RiCDPK2* in the host suppresses hypersensitive cell death; mediate SpA-induced signaling during interaction with A. solani	Hassan et al. (2012, 2013)
Solanum lycopersicum	*LeCDPK1*	Induced by fungal elicitor, H_2O_2, wounding	Chico et al. (2002)
	LeCPK2	Induced by wounding and phytohormone treatment; preferentially expressed during flower development	Chang et al. (2009)
Tritichum aestivum	*TaCPK2*	Associated with multiple disease resistance in grass by modulating WRKY45-1 and associated defense-response genes	Geng et al. (2013)
Vicia faba	*VfCPK1*	Induced by drought, ABA, and $CaCl_2$	Liu et al. (2006)
Vigna radiata	*VrCPK1*	Induced by IAA treatment, mechanical strain, and salt stress	Botella et al. (1996)
Zea mays	*ZmCPK1*	Induced by cold stress	Berberich and Kusano (1997)
	ZmCPK10	Induced by fungal infection as well as fungal elicitor treatment	Murillo et al. (2001)
	ZmCPK11	Responsive to wounding and control vacuolar function; component of touch- and wound-induced pathway(s)	Szczegielniak et al. (2005, 2012)
	ZmCPK4	Positively regulated abscisic acid signaling and enhanced drought stress tolerance	Jiang et al. (2013)
Zingiber officinale	*ZoCDPK1*	Induced by high salt, drought, and JA; overexpression in tobacco-conferred tolerance to salinity and drought stress	Vivek et al. (2013)

The table has been sorted according to alphabetical order of species and then by gene name

The already complex signaling network turned out to be much more complicated when Uno et al. (2009) identified the interacting partners of AtCPK4 and AtCPK11 (share 95 % similarity) by high-throughput yeast-two hybrid interaction. AtCPK4 interacted with 14 redundant proteins (five most redundant interacting proteins are AtDi19, HSP1, serine-rich protein, zinc-finger protein, and AtToc33), and 16 proteins have yielded single hit. On the other hand, AtCPK11 interacts with 24 different proteins, in which 13 were redundant preys. Moreover, of the redundant ones, both the CDPKs interact with AtDi19, HSP1, HSP2, zinc-finger protein, AtToc33, Pin7, and PB1 domain-containing protein. Hence, to highlight the above, AtCPK4 and AtCPK11 interact with proteins involved in wide cellular processes like hormone signaling (Pin7), translocation of nuclear-encoded preprotein into chloroplast (AtToc33), and stress-response factors (HSP1, HSP2, AtDi19). Strikingly, even after having 95 % protein identity, they showed specific interactions suggesting that CDPK proteins have precise posttranslational regulation. Even these interacting proteins are localized in different compartments in the cell like chloroplast, nucleus (Uno et al. 2009). Substrate specificity of CDPK proteins is an immerging area, which is instrumental in understanding the downstream signaling machinery. In an attempt to identify substrate of AtCPK1, AtCPK10, AtCPK16, and AtCPK34 by SPA assay, a large subset of 103 target peptides were identified to be phosphorylated. Among them the largest subset (33) was found to be phosphorylated by all three AtCPK1, AtCPK10, and AtCPK34. However, each of these CDPK proteins also phosphorylated a distinct subset of peptides, suggesting that each isoform has unique substrate specificity too. AtCPK34 had the largest number of isoform-specific substrates which is 31 (e.g., syntaxin, metal ion-binding protein, protein kinase, transporter, etc.), whereas AtCPK10 and AtCPK1 had only three (e.g., phospholipase) and two (e.g., nitrate reductase) substrates, respectively. Moreover, kinetic differences between AtCPK16, AtCPK1, AtCPK10, and AtCPK34 were also observed (Curran et al. 2011). Conclusively, CDPK proteins percept a wide range of Ca^{2+} signals and integrate at different levels of signaling cascade, modulating the appropriate downstream responses.

When the complexity of interaction was analyzed at genic level, some functional clarity was available; however, new dimensions of regulatory aspect also immerged. *AtCPK4* and *AtCPK11* were also found to be induced by ABA application, and their loss-of-function mutant (*cpk4* and *cpk11*) and double mutant (*cpk4cpk11*) showed ABA-insensitive phenotypes during seed germination and stomatal movement, which decreased salt stress tolerance in seedlings (Fig. 5.2; Zhu et al. 2007). Conversely, as expected the overexpressing *AtCPK4* and *AtCPK11* lines were more sensitive to ABA during seedling growth and stomatal movement. Interestingly, ABA-responsive genes, namely, *ABF1*, *ABF2*, *AGF4*, *ABI4*, *ABI5*, *RD29A*, *RAB18*, *KIN1*, *KIN2*, and *ERD10*, were found to be downregulated in mutant lines and upregulated in overexpression lines (Zhu et al. 2007). Cytoplasmic and nuclear localization of these two CDPKs could be facilitating interaction with nuclear-localized transcription factors for delayed ABA response as well as early response, by phosphorylating signaling molecules in cytosol (Dammann et al. 2003; Zhu et al. 2007). Association of ABA in stimulating cytosolic concentration of Ca^{2+} for stomatal closure is well documented (Song et al. 2008; Kim et al. 2011). Hence, existence of Ca^{2+}-sensing signal transducers like CDPK proteins is obvious in regulation of stomatal conductance (Hubbard et al. 2011). *AtCPK3* and *AtCPK6* isolated from guard cell-enriched cDNA library were found to express in guard cell as well as mesophyll cells (Kwak et al. 2002). *cpk3cpk6* double-mutant plants showed partial impairment of ABA-induced and Ca^{2+}-reactive stomatal closure, whereas long-term Ca^{2+}-programmed stomatal closure was not. This differential regulation of R (rapid)- and S (slow)-type anion channels is found to be regulated by parallel signal transduction mechanism (Mori et al. 2006). MJ (methyl jasmonate) and ABA signaling have been found to have partial overlap in regulation of guard cell signaling (Munemasa et al. 2007; Saito et al. 2008). On screening *AtCPK3, AtCPK6, AtCPK4,* and *AtCPK11* disruption mutants for stomatal phenotype, *AtCPK6* disruption (*cpk6*) mutants showed interference in MJ-mediated activation of I_{Ca} channels and *S*-type anion channels. Further AtCPK6 was found to be a positive regulator of MJ signaling in *Arabidopsis* guard cells by a feedback loop mechanism (Munemasa et al. 2011). An *AtCPK10* mutant (*cpk10*), which was found to be sensitive to drought and complemented lines, showed recovery of phenotype and also showed ABA- and Ca^{2+}-mediated inhibition of the inward K^+ currents (Zou et al. 2010). The same was observed in *hsp1* mutants, where HSP1 (heat shock protein 1) protein is the interacting partner of AtCPK10. Pulled together, AtCPK10 and HSP1 function in the regulation of stomatal movements via ABA- and Ca^{2+}-signaling pathways during drought stress (Zou et al. 2010). In an earlier report, AtCPK10 and AtCPK11 have also been found to be inducible under cold, salinity, and drought stress by Urao et al. (1994). AtCPK23 has been found to be involved in regulating activity of guard cell anion channel and K^+ uptake (Ma and Wu 2007; Geiger et al. 2010). On expressing *SLAC1* together with *AtCPK3*, *AtCPK6*, *AtCPK21*, *AtCPK23*, *AtCPK31*, *ABI1,* and *HAB1* in *X. laevis* oocytes, it was found that on reception of ABA in stomata, ABI1 is inactivated by the ABA receptors, which relieves inhibition of AtOST1, AtCPK23, and AtCPK21, and finally SLAC1, a guard cell anion channel. In turn SLAC1 is activated by phosphorylation resulting in concomitant water loss from guard cells and closure of stomatal pore (Geiger et al. 2010). Ma and Wu in 2007 found that *cpk23* mutant exhibited enhanced tolerance to drought and salt stresses, while the *AtCPK23* overexpression lines were more sensitive to them, and the complementary lines displayed recovery of phenotype comparable to wild-type plants (Ma and Wu 2007). They correlated this phenomenon to the reduced stomatal aperture in mutant lines and further found that AtCPK23 mediated salt tolerance by regulating K^+ uptake (Ma and Wu 2007). In an earlier report another K^+ channel, KAT1, has also been reported to be phosphorylated by CDPK from *Vicia faba* guard cells (Li et al. 1998). In *Arabidopsis*, negative regulation of ABA by CDPK has also been observed where AtCPK12 phosphorylated a protein phosphatase, ABI2, and also phosphorylated ABF1 and ABF4 *in vitro*. Consequently, during seed germination and post-germination growth, AtCPK12 modulated ABA signaling in a loop regulation (Zhao et al. 2011). An interesting report by Sheen (1996) in protoplast transient expression system demonstrated that *AtCPK10* and *AtCPK30* were activated in response to ABA. AtCPK32 was found to phosphorylate ABF4, a transcriptional regulator in

ABA-dependent signaling cascade at its C2–C3 conserved region. Overexpression of *AtCPK32* also resulted in induction of ABA-responsive genes. The cross talk in ABA signaling is more eminent as not only AtCPK32 interacted with multiple members of ABF family (i.e., ABF1, ABF2, and ABF3) but also ABF4 interacted with AtCPK10 and AtCPK30 of CDPK gene family (Choi et al. 2005). ABA-mediated signaling of CDPK genes was also demonstrated in multiple plant species. *ACPK1*, expressed in the mesocarp of grape berries, could be positively involved in ABA-signaling pathway and promoting plasmalemma H^+-ATPase-powered activity in grape berry (Yu et al. 2007). Further heterologous expression of *ACPK1* in *Arabidopsis* implicated it to be a positive regulator in ABA signaling (Yu et al. 2007). *VfCPK1* isolated from epidermal peels of broad bean (*Vicia faba* L.) leaves showed differential accumulation of mRNA and protein in leaves treated with ABA and drought stress (Liu et al. 2006). In potato StCDPK2 phosphorylated StABF1 *in vitro* (Muñiz García et al. 2011). All these reports emphasized that the pivotal role of CDPK proteins plays an important role in ABA-mediated stress-signaling pathway across species. Characterizing *OsCPK13* (*OsCDPK7*) to be a nodal player in cold- and salt-/drought-responsive pathways was a pioneering study in the field of CDPK genes (Saijo et al. 2000). Working on this hypothesis, *OsCPK13* was expressed in *Sorghum* sp. However, the transgenic lines did not show enhanced tolerance to cold/salt stress, but showed a lesion mimic phenotype and 2-D gel analysis revealed accumulation of multiple PR-10 proteins, implying induction of response to biotic stress (Mall et al. 2011). Noticeably, OsCPK13 may be the junction point between two distinct pathways of cold and salt/drought stress tolerance as well as biotic stress under specific circumstances. OsCPK7 (OsCDPK13) protein was found induced and phosphorylated by cold stress as well as GA treatment (Yang et al. 2003; Abbasi et al. 2004). Proteomic analysis of lines overexpressing *OsCPK7* and its related proteins, calreticulin and CRTintP1 (calreticulin-interacting protein 1), revealed that OsCPK7 might be involved in sugar-sensing pathway and damage repair caused by low-temperature stress (Komatsu et al. 2007). Hence, divergent involvement of OsCPK7 in conferring cold stress tolerance via sugar-sensing pathway, involvement in seed development, and induction by GA but suppression in response to ABA and brassinolide is evident (Kawasaki et al. 1993); Abbasi et al. 2004; Komatsu et al. 2007). Another CDPK gene, OsCDPK1, was found to be specifically activated by sucrose starvation and during post-germination seedling growth as well as involved in drought tolerance and negatively regulating the expression of enzymes essential for GA biosynthesis; however, it activated the expression of a 14-3-3 protein, GF14c. Indicating that OsCDPK1 could be central to multiple signaling pathways (Ho et al. 2013).

CBL–CIPK complex has been well reported to be involved in salt, drought, flooding, cold, and wounding stress as well as low-K^+ stress (Table 5.2; reviewed by Pandey 2008; Batistic and Kudla 2009; Das and Pandey, 2010; Bailey-Serres and Voesenek 2010). It was first demonstrated in *Arabidopsis* by high-throughput yeast-2 hybrid analyses that different CBL proteins interact with diverse CIPK proteins and the specificity of this interaction determines the network outcome. *Arabidopsis* AtCBL1 and AtCBL9 were found to interact with AtCIPK1, AtCIPK5,

Table 5.2 CBL–CIPK genes characterized to be involved in abiotic stress across plant species

Species	Gene name	Characterization	Reference
Arabidopsis thaliana	*AtCBL1*	Expresses in stems and roots; induced by wounding, drought, salt, cold stress, and ABA treatment; overexpression shows enhanced tolerance to salt and drought but reduced tolerance to freezing	Kudla et al. (1999), Cheong et al. (2007), Pandey et al. (2004), Kolukisaoglu et al. (2004)
	AtCBL2	Responsive to light; serves as molecular links between calcium signaling and V-ATPase for ion homeostasis	Nozawa et al. (2001), Tang et al. (2012)
	AtCBL3	Serves as molecular links between calcium signaling and V-ATPase for ion homeostasis	Tang et al. (2012)
	AtCBL4/SOS3	Mediate salt stress adaptation by regulating the Na+/H+ antiporter	Zhu et al. (1998), Halfter et al. (2000), Qiu et al. (2002), Quintero et al. (2002)
	AtCBL5	Expresses significantly in green tissues but not in roots; overexpression enhances tolerance to high salt or drought responses in transgenic *Arabidopsis*	Cheong et al. (2010)
	AtCBL9	Highly inducible by cold, drought, salinity, and ABA; functions as a negative regulator of calcium-induced ABA-signaling and ABA biosynthesis pathway	Pandey et al. (2004)
	AtCBL10	Induced by salt stress; mediates salt tolerance by regulating ion homeostasis in *Arabidopsis*	Kim et al. (2007)
	AtCIPK1	Induced by osmotic stress response; by alternative binding to CBL1 or CBL9, CIPK1 acts as a convergence point for ABA-dependent and ABA-independent stress responses	D'Angelo et al. (2006)
	AtCIPK3	Induced by cold, drought, salt, wounding stress, and ABA; interacts with AtCBL10 and regulates ABA response during seed germination	Kim et al. (2003), Pandey et al. (2008)
	AtCIPK6	Induced by osmotic, salt, and drought stresses in shoots and roots; overexpression in *Arabidopsis* increased plant tolerance to salt stress, but sensitivity to ABA; involved in salt stress signaling and reduces shoot-to-root auxin transport	Tripathi et al. (2009), Tsou et al. (2012), Chen et al. 2013a, b, c

	AtCIPK8	Regulates the low-affinity phase of the primary nitrate response and also induces expression of primary nitrate-response genes	Hu et al. (2009)
	AtCIPK9	Induced by cold, salt, osmotic stress, and wounding; interacts with CBL3 and regulates K^+ uptake in *Arabidopsis* roots and maintains K+ homeostasis	Pandey et al. (2007), Liu et al. (2013)
	AtCIPK15	Interacts with ABI2 hence might be an integral part of ABA-signaling pathway	Guo et al. (2002)
	AtCIPK23	Regulated by K+ uptake and NO3-uptake	Xu et al. (2006), Cheong et al. (2007), Hashimoto and Kudla (2011)
	AtCIPK24/SOS2	Involved in SOS pathway for conferring salt tolerance	Halfter et al. (2000), Qiu et al. (2002), Quintero et al. (2002)
	AtCIPK26	Overexpression confers hypersensitivity to ABA; interacts with ABI1 and ABI2, two negative regulators of ABA signaling; specifically interacts with the N-terminal domain of RBOHF and phosphorylates it; involved in ROS production by RBOHF in HEK293T cells along with either CBL1 or CBL9	Drerup et al. (2013), Lyzenga et al. (2013)
Brassica napus	*BnCBL1*	Induced by salt, osmotic stress, and ABA in root and low-phosphate stress in leaves; overexpression conferred salt and low-phosphate tolerance in transgenic *Arabidopsis*	Chen et al. (2012)
	BnCIPK6	Induced by high salt, osmotic stress, low phosphate, and ABA in root; overexpression in *Arabidopsis cipk6* mutant rescued the low-phosphate-sensitive and ABA-insensitive phenotypes	Chen et al. (2012)
	BnCIPK24	Expression in *Arabidopsis sos2* mutant complemented the function of AtCIPK24/sos2	Zhang et al. (2014)
Brassica juncea	*BjSOS3*	Structural and functional orthologue of *Arabidopsis sos3*	Khushwaha et al. (2011)
Cicer arietinum	*CaCIPK6*	Overexpression conferred salt tolerance in transgenic tobacco	Tripathi et al. (2009)

(continued)

Table 5.2 (continued)

Species	Gene name	Characterization	Reference
Glycine max	*GmCBL1*	Induced by abiotic stress and plant hormone; overexpression enhanced tolerance to salt, drought stress, and hypocotyl elongation in transgenic *Arabidopsis*	Li et al. (2012)
Gossypium hirsutum	*GhCIPK6*	Induced by salt, drought, and ABA treatments; overexpression conferred tolerance against salt, drought, and ABA stresses in *Arabidopsis*	He et al. (2013)
Hordeum brevisubulatum	*HbCIPK2*	Induced by salt, drought stress, and ABA treatment; it had a role in preventing K(+) loss in root and accumulation of less Na(+) in shoot in turn maintaining K(+)/Na(+) homeostasis	Li et al. (2012)
Malus domestica	*MdCIPK6L*	Induced by salt stress; overexpression enhances tolerance to salt, osmotic, drought, and chilling stress in transgenic *Arabidopsis*, tomato, and apple	Wang et al. (2012a, b)
	MdSOS2	Showed highest similarity to AtCIPK24/SOS2 and functionally complemented the *Arabidopsis* sos2 mutant	Hu et al. (2012)
Oryza sativa	*OsCBL1*	Induced by salt and cold stress	Hwang et al. (2005), Zhang et al. 2014)
	OsCBL2	Upregulated by GA in aleurone layer; highly expressed in roots of seedlings and tillering plants, panicle, seed, and aleurone layer of mature grain; promotes vacuolation of barley aleurone cells in the presence of GA	Hwang et al. (2005)
	OsCBL3	Induced by salt stress in rice seedling	Hwang et al. (2005)
	OsCBL4	Orthologue of AtCBL4/SOS3 and involved in SOS pathway	Martínez-Atienza et al. (2007)
	OsCBL8	Overexpression in rice conferred tolerance to salt stress	Gu et al. (2008)

	OsCIPK3	Induced by cold stress and to a lesser extent by ABA and PEG treatment; overexpression enhances tolerance of transgenic plants to cold shock; negatively regulates salt stress tolerance in rice	Xiang et al. (2007)
	OsCIPK12	Induced by drought stress, ABA, and PEG treatment; overexpression increased drought tolerance in vegetative stage of transgenic rice	Xiang et al. (2007)
	OsCIPK14	Nodal role in MAMP-induced defense signaling	Kurusu et al. (2010)
	OsCIPK15	Induced by drought, salt stress, ABA, and PEG treatment; overexpression conferred salt tolerance in rice; involved in MAMP-induced defense signaling	Xiang et al. (2007), Kurusu et al. (2010)
	OsCIPK23	Induced by drought stress and PEG treatment; expresses in pistil during pollination; overexpression in rice conferred drought tolerance	Xiang et al. (2007), Yang et al. (2008)
	OsCIPK24	Induced by drought stress, ABA, and PEG treatment	Xiang et al. (2007)
	OsCIPK31	Induced by light, high salt, osmotic stresses, and ABA responses during the seed germination and seedling stage	Piao et al. (2010)
Pisum sativum	*PsCBL*	Induced by cold, salt, wounding stress, and salicylic acid treatment	Mahajan et al. (2006), Tuteja and Mahajan, (2007)
	PsCIPK	Induced by cold, salt, wounding stress, and salicylic acid treatment	Mahajan et al. (2006), Tuteja and Mahajan, (2007)
Populus euphratica	*PeCIPK26/PeSOS2*	Interact with PeCBL1, PeCBL4/PeSOS3, PeCBL9, and PeCBL10; overexpression conferred salt tolerance *Arabidopsis cipk24* mutant	Zhang et al. (2013)
Populus trichocarpa	*PtSOS2*	Heterologous expression of PtSOS2 rescued salt-sensitive phenotype of *Arabidopsis sos2* mutants	Tang et al. (2010)
	PtSOS3	Heterologous expression of PtSOS3 functionally complemented *Arabidopsis sos3* mutants	Tang et al. (2010)
Solanum lycopersicum	*SlCBL10*	Interacted with SlCIPK6; overexpression in *N. benthamiana* leaves causes accumulation of ROS	de la Torre et al. (2013)

(continued)

Table 5.2 (continued)

Species	Gene name	Characterization	Reference
Triticum aestivum	*TaCIPK14*	Induced under cold, salt, PEG, ABA, ethylene, and H2O2; overexpression conferred salinity and cold tolerance in transgenic tobacco	Deng et al. (2013a)
	TaCIPK29	Induced by salt stress, cations, and ROS; overexpression helped in maintaining cation content in transgenic tobacco	Deng et al. (2013b)
Zea mays	*ZmCBL4*	Induced by salt stress, LiCl, ABA, and PEG treatment; *ZmCBL4* complemented the salt hypersensitivity in *Arabidopsis sos*3 mutant	Wang et al. (2007)
	ZmCIPK16	Induced by salt, heat, osmotic, drought stress, PEG, and ABA treatment; overexpression conferred salt tolerance in *sos2* mutant of *Arabidopsis*	Zhao et al. (2009), Chen et al. (2011)

The table has been sorted according to alphabetical order of species and then by gene name

AtCIPK8, AtCIPK11, AtCIPK12, AtCIPK17, AtCIPK18, AtCIPK23, AtCIPK24, and AtCIPK26 (Kolukisaoglu et al. 2004). AtCBL3 was found to interact strongly with AtCIPK1, AtCIPK2, AtCIPK4, AtCIPK6, AtCIPK7, AtCIPK11, and AtCIPK13 and weakly with AtCIPK9, AtCIPK12, and AtCIPK14 (Albrecht et al. 2001). In canola, six *BnaCBLs* (BnaCBL1, BnaCBL2, BnaCBL3, BnaCBL4, BnaCBL9, BnaCBL10) and 17 *BnaCIPKs* were studied for interaction. BnaCBL1 interacted significantly with a subset of 13 CIPKs, BnaCBL2 interacted with 12 BnaCIPKs, BnaCBL3 interacted with only eight CIPKs, BnaCBL4 interacted with 14 BnaCIPKs, BnaCBL9 exhibited a significant interaction with 12 CIPKs, and BnaCBL10 was found to interact with only seven BnaCIPKs. BnaCIPK5, BnaCIPK10, and BnaCIPK26 showed interactions with all of the six BnaCBL. On comparing the interaction pattern between canola and *Arabidopsis* orthologues, they observed that predicting gene function based on orthologues is not accurate; hence, individual functional analysis is required. On characterizing these proteins functionally, AtCBL1 has been found to be induced by drought and cold but not by ABA (Kudla et al. 1999; Albrecht et al. 2003). Even glucose and gibberellin (GA) signals have been found to induce *AtCBL1* during germination and seedling development (Liang and Pan 2013b) indicating towards involvement in cross response. *AtCBL9* is homologous to *AtCBL1*; however, *AtCBL9* was inducible by multiple stress signals as well as ABA in young seedlings (Pandey et al. 2004). *AtCBL9* mutant showed drastic alteration in its response to ABA, and further characterization showed that AtCBL9 acts as a negative regulator in ABA signaling and is also involved in ABA biosynthesis under stress (Pandey et al. 2004). AtCIPK1 have been found to strongly interact with both AtCBL1 and AtCBL9 (Albrecht et al. 2001; Kolukisaoglu et al. 2004). Interestingly, the AtCBL1–AtCIPK1 complex was found to be involved in ABA-dependent stress responses, while the AtCBL9–AtCIPK1 complex modulated ABA-independent stress responses (D'Angelo et al. 2006). On the other hand AtCBL9 interacted with AtCIPK3 and negatively regulated ABA response during seed germination (Pandey et al. 2008). AtCBL1 and AtCBL9 were also found to interact with AtCIPK23 and act as upstream regulator of AtCIPK23 (Xu et al. 2006; Cheong et al. 2007). *Arabidopsis* double-mutant *cbl1cbl9* exhibited sensitivity to low K^+, whereas overexpression of *AtCBL1* or *AtCBL9* increased K^+ content in root and shoots even under low-K^+ condition (Xu et al. 2006). Taken together AtCBL1 or AtCBL9-CIPK23 complex phosphorylated the K^+ channel AKT1 as well as phosphorylation of AtCBL1 by AtCIPK23 and PP2C phosphatases mediated K^+ uptake under limited K^+ conditions. Moreover, AtCBL1/CBL9-CIPK23 complex is also found to regulate nitrate sensing and uptake in plants (Fig. 5.2; Xu et al. 2006; Li et al. 2006; Cheong et al. 2007; Hashimoto and Kudla 2011; Lan et al. 2011; Hashimoto et al. 2012). AtCIPK9 was also found to interact with AtCBL3 and maintain K^+ homeostasis under low-K^+ stress in *Arabidopsis* (Pandey et al. 2008; Liu et al. 2013). *cipk9* and *cbl3* mutant was found to be hypersensitive to low-K^+ conditions, and overexpressing lines for both the genes resulted in a low-K^+-sensitive phenotype (Pandey et al. 2008; Liu et al. 2013). Orthologue of *Arabidopsis* AKT1 channel was identified from grapevine (*VvK1*.1), which was found to have role in K^+ uptake from the soil. Heterologous expression of AtCBL1-AtCIPK23 in *Xenopus oocytes* with *VvK1*.1 demonstrated shared functionality with *Arabidopsis* AKT1 (Cuéllar et al. 2010). Another K^+ channel-encoding gene, *VvK1.2* when co-expressed with VvCIPK04-VvCBL01 in *Xenopus oocytes*, gave rise to K^+ currents, which are orthologues of AtCIPK23 and AtCBL1, respectively. When VvCIPK03–VvCBL02 were co-expressed with *VvK1.2,* much smaller K^+ currents were generated. VvCBL01 and VvCIPK04 were also found to control the activity of AKT1 in *Arabidopsis* roots (Cuéllar et al. 2013). Recently, it was shown that co-expression of either *AtCBL1* or *AtCBL9* with *AtCIPK26* strongly enhances ROS production by phosphorylating NADPH oxidase RBOHF (Sagi and Fluhr 2006; Kimura et al. 2013; Drerup et al. 2013). AtCBL1-AtCIPK7 is induced by cold, and concurrently this complex has been predicted to be having role in regulation of cold response (Kolukisaoglu et al. 2004; Huang et al. 2011). Heterologous expression of *PeCBL1,* which is an orthologue of AtCBL1 from *Populus euphratica*, in the corresponding *Arabidopsis cbl1cbl9* double mutant displayed increase in the resistance of sensitive phenotypes to low-K^+ stress. PeCBL1 was also found to interact with PeCIPK24, PeCIPK25, and PeCIPK26. Hence, taken together PeCBL1 in complex with PeCIPK24, PeCIPK25, and PeCIPK26 has been confirmed to be involved in regulation of Na^+/K^+ homeostasis (Zhang et al. 2010, 2013). Lyzenga et al. (2013) have demonstrated that overexpression of *AtCIPK26* conferred hypersensitivity to ABA, and AtCIPK26 was found to interact with ABI1 and ABI2, two negative regulators of ABA signaling, confirming its role in ABA signal transduction. They also showed that AtCIPK26 is degraded by KEG through ubiquitin–proteasomal degradation (Lyzenga et al. 2013). ABI2 was also found to interact with AtCIPK15, which might also be an integral part of ABA-signaling pathway (Guo et al. 2002). Another CIPK gene from *Arabidopsis*, *AtCIPK8*, was found to be involved in early nitrate-signaling pathway. Nitrate-induced *AtCIPK8* expression and complementation of knockout mutant line showed that AtCIPK8 positively regulated nitrate transporter genes and required for nitrate assimilation (Hu et al. 2009). CIPK23 is also involved in the primary nitrate response by phosphorylation of CHL1 (NRT1.1) (Ho et al. 2009).

Transcript accumulation of *AtCIPK6* increased under salt, drought stress, and ABA treatment. *Arabidopsis* overexpressing *AtCIPK6* gene showed increase in salt tolerance but decrease in sensitivity to ABA (Tsou et al. 2012; Chen et al. 2013a, b, c). In cotton, *GhCIPK6* gene, which is an orthologue of *AtCIPK6,* was also induced by salt, drought, and ABA treatments. Overexpression of *GhCIPK6* in *Arabidopsis* conferred tolerance to salt, drought stress, and ABA treatment (He et al. 2013). In *Brassica napus*, interaction of BnCIPK6 and BnCBL1 was demonstrated by BiFC in plant cells. Both the genes were inducible by salt and osmotic stresses, phosphorous starvation, and ABA treatment. Transgene expression of *BnCIPK6* and *BnCBL1* enhanced high salinity and low phosphate tolerance; moreover, *BnCIPK6* complemented *Arabidopsis cipk6* mutant emphasizing their role in high salinity, phosphorous deficiency, and ABA signaling (Chen et al. 2012). Overexpression of a constitutively active mutant of *CaCIPK6* from *Cicer arietinum*, homologue of *AtCIPK6*, promoted salt tolerance in transgenic tobacco (Tripathi et al. 2009). In pea, orthologue of *AtCBL2* and *AtCIPK18* was identified as *PsCBL* and *PsCIPK*, respectively, which were interacting partners and also were coordinately upregulated in response to salinity, cold stress, wounding, salicylic acid, and Ca^{2+} but surprisingly not to ABA and dehydration (Tuteja and Mahajan, 2007; Mahajan et al. 2006).

Through forward genetic screens, SOS pathway comprising of SOS3/CBL4 and SOS2/CIPK24 was identified that mediates salt stress adaptation by regulating the Na^+–H^+ antiporter/SOS1 at the plasma membrane (Zhu et al. 1998; Halfter et al. 2000; Qiu et al. 2002; Quintero et al. 2002). CIPK24/SOS2 as well as AtCBL10 directly interacted with AtCBL10 and conferred salt stress tolerance to *Arabidopsis* (Fig. 5.2; Kim et al. 2007; Waadt et al. 2008; Ren et al. 2013). Conservation of the SOS pathway has been reported in rice, *Brassica juncea*, maize, tomato, *Populus trichocarpa*, *Populus euphratica*, and apple (Martínez-Atienza et al. 2007; Wang et al. 2007, 2012a, b; Kumar et al. 2009; Zhao et al. 2009; Kushwaha et al. 2011; Hu et al. 2012; Tang et al. 2010; Lv et al. 2014; Oh et al. 2008). From monocots like rice and wheat, using reverse functional genetic tool such as gain/loss of function of OsCIPK3, OsCIPK12, OsCIPK15, OsCIPK23, OsCIPK31, and TaCIPK14 has been demonstrated to be intricately involved in diverse abiotic stress pathways, namely, cold, salt, PEG, ABA, ethylene, and H_2O_2 (Xiang et al. 2007; Yang et al. 2008; Piao et al. 2010; Deng et al. 2013a). With more studies being reported from multiple plant species, the nitty-gritty of Ca^{2+} signaling network and role of CDPK and CBL–CIPK proteins will be further elucidated.

5.7 Conclusion and Perspectives

The role of Ca^{2+}-regulated kinases in abiotic stress signaling are well established by now. In the recent year, we have witnessed significant progress in our understanding of the underlying Ca^{2+} decoding process, which specially includes the structural determinants like delineation of essential amino acids required for specific activity. A major achievement was knowledge about the interacting partners, which reflects on the multiple layers that are involved in stress-signaling regulation. It has been established that not only transcriptional regulation but also posttranslational regulations like affinity for binding Ca^{2+}, concentration gradient, cellular proteomic localization, and varying affinity of interacting proteins are strong determinants of regulation. However, it is still not clear to what degree the affinity for Ca^{2+} is influencing the signal perception and transduction. We are yet to understand the mechanism by which the Ca^{2+} signatures are differentiated by the sensing molecules and the factors that determine specificity to the transducers for well-defined stress cross-talk response. More data on phosphorylating downstream partners and the complexes formed *in vivo* shall explain more regarding the signaling network functioning in response to multiple stress condition. It will be interesting to know why nature has parallel Ca^{2+}-sensing systems comprising of CaM, CDPK, and CBL proteins when each family has multiple gene members. Even within the families, redundancy in function among isoforms is evident and that stances a question that, what could be the underlying complementarity of these parallel systems? Not only within abiotic stress but convergence of the pathways is noted for biotic stress as well as developmental processes, which impart complexity in the functional role of signaling pathways. The challenge still remains in understanding the intricacies of molecular and biochemical processes that are nodal to these physiological events.

Acknowledgment I would also like to thank Dr. Akhilesh K. Tyagi for providing valuable suggestion in the Ca^{2+} signaling work performed.

References

Abbasi F, Onodera H, Toki S, Tanaka H, Komatsu S (2004) *OsCDPK13*, a calcium-dependent protein kinase gene from rice, is induced by cold and gibberellin in rice leaf sheath. Plant Mol Biol 55:541–552

Abo-el Saad M, Wu R (1995) A rice calcium-dependent protein kinase is induced by gibberellin. Plant Physiol 108:787–793

Akaboshi M, Hashimoto H, Ishida H, Saijo S, Koizumi N, Sato M, Shimizu T (2008) The crystal structure of plant-specific calcium-binding protein AtCBL2 in complex with the regulatory domain of AtCIPK14. J Mol Biol 377:246–257

Akimoto-Tomiyama C, Sakata K, Yazaki J, Nakamura K, Fujii F, Shimbo K, Yamamoto K, Sasaki T, Kishimoto N, Kikuchi S, Shibuya N, Minami E (2003) Rice gene expression in response to N-acetylchitooligosaccharide elicitor: Comprehensive analysis by DNA microarray with randomly selected ESTs. Plant Mol Biol 52:537–551

Albrecht V, Ritz O, Linder S, Harter K, Kudla J (2001) The NAF domain defines a novel protein-protein interaction module conserved in Ca^{2+}-regulated kinases. EMBO J 20(5):1051–1063

Albrecht V, Weinl S, Blazevic D, D'Angelo C, Batistic O, Kolukisaoglu U, Bock R, Schulz B, Harter K, Kudla J (2003) The calcium sensor CBL1 integrates plant responses to abiotic stresses. Plant J 36(4):457–470

Allen GJ, Muir SR, Sanders D (1995) Release of Ca^{2+} from individual plant vacuoles by both InsP3 and cyclic ADP-ribose. Science 268:735–737

Allen GJ, Chu SP, Schumacher K, Shimazaki CT, Vafeados D, Kemper A, Hawke SD, Tallman G, Tsien RY, Harper JF et al (2000) Alteration of stimulus-specific guard cell calcium oscillations and stomatal closing in *Arabidopsis* det3 mutant. Science 289:2338–2342

Amelot N, Dorlhac de Borne F, San Clemente H, Mazars C, Grima-Pettenati J, Brière C (2012) Transcriptome analysis of tobacco BY-2 cells elicited by cryptogein reveals new potential actors of calcium-dependent and calcium-independent plant defense pathways. Cell Calcium 51:117–130

Anil VS, Harmon AC, Rao KS (2003) Temporal association of Ca^{2+}-dependent protein kinase with oil bodies during seed development in *Santalum album* L.: its biochemical characterization and significance. Plant Cell Physiol 44(4):367–376

Asai S, Ichikawa T, Nomura H, Kobayashi M, Kamiyoshihara Y, Mori H, Kadota Y, Zipfel C, Jones JD, Yoshioka H (2013) The variable domain of a plant calcium-dependent protein kinase (CDPK) confers subcellular localization and substrate recognition for NADPH oxidase. J Biol Chem 288(20):14332–14340

Asano T, Tanaka N, Yang G, Hayashi N, Komatsu S (2005) Genome-wide identification of the rice calcium-dependent protein kinase and its closely related kinase gene families: Comprehensive analysis of the CDPKs gene family in rice. Plant Cell Physiol 46:356–366

Asano T, Hakata M, Nakamura H, Aoki N, Komatsu S, Ichikawa H, Hirochika H, Ohsugi R (2011) Functional characterisation of OsCPK21, a calcium-dependent protein kinase that confers salt tolerance in rice. Plant Mol Biol 75(1–2):179–191

Bailey-Serres J, Voesenek LA (2010) Life in the balance: a signaling network controlling survival of flooding. Curr Opin Plant Biol 13(5):489–494
Batistic O, Kudla J (2004) Integration and channeling of calcium signaling through the CBL calcium sensor/CIPK protein kinase network. Planta 219(6):915–24
Batistic O, Kudla J (2009) Plant calcineurin B-like proteins and their interacting protein kinases. Biochim Biophys Acta 1793(6):985–992
Batistič O, Kudla J (2012) Analysis of calcium signaling pathways in plants. Biochim Biophys Acta 1820(8):1283–1293
Batistic O, Sorek N, Schültke S, Yalovsky S, Kudla J (2008) Dual fatty acyl modification determines the localization and plasma membrane targeting of CBL/CIPK Ca^{2+} signaling complexes in *Arabidopsis*. Plant Cell 20(5):1346–1362
Batistic O, Waadt R, Steinhorst L, Held K, Kudla J (2010) CBL-mediated targeting of CIPKs facilitates the decoding of calcium signals emanating from distinct cellular stores. Plant J 61:211–222
Benetka W, Mehlmer N, Maurer-stroh S, Sammer M, Koranda M, Betschinger J, Knoblich JA, Teige M, Eisenhaber F (2008) Experimental testing of predicted myristoylation targets involved in asymmetric cell division and calcium-dependent signaling. Cell Cycle 7(23):3709–3719
Berberich T, Kusano T (1997) Cycloheximide induces a subset of low temperature-inducible genes in maize. Mol Gen Genet 254:275–283
Blatt MR, Thiel G, Dr T (1990) Reversible inactivation of K+ channels in Vicia stomatal guard cells following the photolysis of caged inositol 1,4,5-triphosphate. Nature 346:766–769
Botella JR, Arteca JM, Somodevilla M, Arteca RN (1996) Calcium-dependent protein kinase gene expression in response to physical and chemical stimuli in mungbean (*Vigna radiata*). Plant Mol Biol 30:1129–1137
Boudsocq M, Sheen J (2013) CDPKs in immune and stress signaling. Trends Plant Sci 18(1): 30–40
Boudsocq M, Willmann MR, McCormack M, Lee H, Shan L, He P, Bush J, Cheng S, Sheen J (2010) Differential innate immune signalling via Ca^{2+} sensor protein kinases. Nature 464(7287):418–422
Breviario D, Morello L, Giani S (1995) Molecular cloning of two novel rice cDNA sequences encoding putative calcium-dependent protein kinases. Plant Mol Biol 27(5):953–967
Bulgakov VP, Gorpenchenko TY, Shkryl YN, Veremeichik GN, Mischenko NP, Avramenko TV, Fedoreyev SA, Zhuravlev YN (2011) CDPK-driven changes in the intracellular ROS level and plant secondary metabolism. Bioeng Bugs 2(6):327–330
Campos-Soriano L, Gómez-Ariza J, Bonfante P, San Segundo B (2011) A rice calcium-dependent protein kinase is expressed in cortical root cells during the presymbiotic phase of the arbuscular mycorrhizal symbiosis. BMC Plant Biol 19:11–90
Capoen W, Den Herder J, Sun J, Verplancke C, De Keyser A, De Rycke R, Goormachtig S, Oldroyd G, Holsters M (2009) Calcium spiking patterns and the role of the calcium/calmodulin-dependent kinase CCaMK in lateral root base nodulation of *Sesbania rostrata*. Plant Cell 21:1526–1540
Cessna SG, Chandra S, Low PS (1998) Hypo-osmotic shock of tobacco cells stimulates Ca^{2+} fluxes deriving first from external and then internal Ca^{2+} stores. J Biol Chem 273:27286–27291
Chang WJ, Su HS, Li WJ, Zhang ZL (2009) Expression profiling of a novel calcium-dependent protein kinase gene, *LeCPK2*, from tomato (*Solanum lycopersicum*) under heat and pathogen-related hormones. Biosci Biotechnol Biochem 73(11):2427–2431
Chang W, Fu G, Chen X, Zhu J, Zhang Z (2011) Biochemical characterization of a calcium-sensitive protein kinase LeCPK2 from tomato. Indian J Biochem Biophys 48(3):148–153
Chehab EW, Patharkar OR, Hegeman AD, Taybi T, Cushman JC (2004) Autophosphorylation and subcellular localization dynamics of a salt- and water deficit-induced calcium-dependent protein kinase from ice plant. Plant Physiol 135(3):1430–1446
Chehab EW, Patharkar OR, Cushman JC (2007) Isolation and characterization of a novel v-SNARE family protein that interacts with a calcium-dependent protein kinase from the common ice plant, *Mesembryanthemum crystallinum*. Planta 225(4):783–799

Chen C, Gao M, Liu J, Zhu H (2007) Fungal symbiosis in rice requires an ortholog of a legume common symbiosis gene encoding a Ca^{2+}/calmodulin-dependent protein kinase. Plant Physiol 145:1619–1628
Chen C, Ane JM, Zhu H (2008) OsIPD3, an ortholog of the *Medicago truncatula* DMI3 interacting protein IPD3, is required for mycorrhizal symbiosis in rice. New Phytol 180:311–315
Chen X, Gu Z, Xin D, Hao L, Liu C, Huang J, Ma B, Zhang H (2011) Identification and characterization of putative CIPK genes in maize. J Genet Genomics 38(2):77–87
Chen L, Ren F, Zhou L, Wang QQ, Zhong H, Li XB (2012) The *Brassica napus* calcineurin B-Like 1/CBL-interacting protein kinase 6 (CBL1/CIPK6) component is involved in the plant response to abiotic stress and ABA signalling. J Exp Bot 63(17):6211–6222
Chen F, Fasoli M, Tornielli GB, Dal Santo S, Pezzotti M, Zhang L, Cai B, Cheng ZM (2013a) The evolutionary history and diverse physiological roles of the grapevine calcium-dependent protein kinase gene family. PLoS One 8(12):e80818
Chen J, Xue B, Xia X, Yin W (2013b) A novel calcium-dependent protein kinase gene from *Populus euphratica*, confers both drought and cold stress tolerance. Biochem Biophys Res Commun 441(3):630–663
Chen L, Wang QQ, Zhou L, Ren F, Li DD, Li XB (2013c) *Arabidopsis* CBL-interacting protein kinase (CIPK6) is involved in plant response to salt/osmotic stress and ABA. Mol Biol Rep 40(8):4759–4767
Cheng S, Willmann MR, Chen H, Sheen J (2002) Update on calcium signaling calcium signaling through protein kinases. The *Arabidopsis* calcium-dependent protein kinase gene family. Plant Physiol 129:469–485
Cheong YH, Pandey GK, Grant JJ, Batistic O, Li L, Kim BG, Lee SC, Kudla J, Luan S (2007) Two calcineurin B-like calcium sensors, interacting with protein kinase CIPK23, regulate leaf transpiration and root potassium uptake in *Arabidopsis*. Plant J 52(2):223–239
Cheong YH1, Sung SJ, Kim BG, Pandey GK, Cho JS, Kim KN, Luan S (2010) Constitutive overexpression of the calcium sensor CBL5 confers osmotic or drought stress tolerance in Arabidopsis. Mol Cells. 28;29(2):159–165
Chico MJ, Raíces M, Téllez-Iñón TM, Ullao MR (2002) A calcium-dependent protein kinase is systemically induced upon wounding in tomato plants. Plant Physiol 128:256–270
Choi HI, Park HJ, Park JH, Kim S, Im MY, Seo HH, Kim YW, Hwang I, Kim SY (2005) *Arabidopsis* calcium-dependent protein kinase AtCPK32 interacts with ABF4, a transcriptional regulator of abscisic acid-responsive gene expression, and modulates its activity. Plant Physiol 139:1750–1761
Chotikacharoensuk T, Arteca RN, Arteca JM (2006) Use of differential display for the identification of touch-induced genes from an ethylene-insensitive *Arabidopsis* mutant and partial characterization of these genes. J Plant Physiol 163(12):1305–1320
Christodoulou J, Malmendal A, Harper JF, Chazin WJ (2004) Evidence for differing roles for each lobe of the calmodulin-like domain in a calcium-dependent protein kinase. J Biol Chem 279(28):29092–29100
Chung E, Park JM, Oh SK, Joung YH, Lee S, Choi D (2004) Molecular and biochemical characterization of the *Capsicum annuum* calcium-dependent protein kinase 3 (*CaCDPK3*) gene induced by abiotic and biotic stresses. Planta 220(2):286–95
Clarkson DT, Brownlee C, Ayling SM (1988) Cytoplasmic calcium measurements in intact higher plants cells: results from fluorescence ratio imaging of fura-2. J Cell Sci 91:71–80
Clayton H, Knight MR, Knight H, McAinsh MR, Hetherington AM (1999) Dissection of the ozone-induced calcium signature. Plant J 17:575–579
Coca M, San Segundo B (2010) AtCPK1 calcium-dependent protein kinase mediates pathogen resistance in *Arabidopsis*. Plant J 63(3):526–540
Costa A, Drago I, Behera S, Zottini M, Pizzo P, Schroeder JI, Pozzan T, Lo Schiavo F (2010) H_2O_2 in plant peroxisomes: an *in vivo* analysis uncovers a Ca^{2+}-dependent scavenging system. Plant J 62:760–772
Cuéllar T, Pascaud F, Verdeil JL, Torregrosa L, Adam-Blondon AF, Thibaud JB, Sentenac H, Gaillard I (2010) A grapevine Shaker inward K(+) channel activated by the calcineurin B-like

calcium sensor 1-protein kinase CIPK23 network is expressed in grape berries under drought stress conditions. Plant J 61(1):58–69

Cuéllar T, Azeem F, Andrianteranagna M, Pascaud F, Verdeil JL, Sentenac H, Zimmermann S, Gaillard I (2013) Potassium transport in developing fleshy fruits: the grapevine inward K(+) channel VvK1.2 is activated by CIPK-CBL complexes and induced in ripening berry flesh cells. Plant J 73(6):1006–1018

Curran A, Chang IF, Chang CL, Garg S, Miguel RM, Barron YD, Li Y, Romanowsky S, Cushman JC, Gribskov M, Harmon AC, Harper JF (2011) Calcium-dependent protein kinases from *Arabidopsis* show substrate specificity differences in an analysis of 103 substrates. Front Plant Sci 2:36

Dammann C, Ichida A, Hong B, Romanowsky SM, Hrabak EM, Harmon AC, Pickard BG, Harper JF (2003) Subcellular targeting of nine calcium-dependent protein kinase isoforms from *Arabidopsis*. Plant Physiol 132:1840–1848

D'Angelo C, Weinl S, Batistic O, Pandey GK, Cheong YH, Schültke S, Albrecht V, Ehlert B, Schulz B, Harter K, Luan S, Bock R, Kudla J (2006) Alternative complex formation of the Ca-regulated protein kinase CIPK1 controls abscisic acid-dependent and independent stress responses in *Arabidopsis*. Plant J 48(6):857–872

Das R, Pandey GK (2010) Expressional analysis and role of calcium regulated kinases in abiotic stress signaling. Curr Genomics 11(1):2–13

Davletova S, Mesyaros T, Miskolczi P, Oberschall A, Torok K, Magyar Z, Dudits S, Deak M (2001) Auxin and heat shock activation of a novel member of the calmodulin-like *domain protein kinase gene family in cultured alfalfa cells*. J Exp Bot 52:215–221

de la Torre F, Gutiérrez-Beltrán E, Pareja-Jaime Y, Chakravarthy S, Martin GB, del Pozo O (2013) The tomato calcium sensor Cbl10 and its interacting protein kinase *Cipk6* define a signaling pathway in plant immunity. Plant Cell 25(7):2748–64

Demir F, Horntrich C, Blachutzik JO, Scherzer S, Reinders Y, Kierszniowska S, Schulze WX, Harms GS, Hedrich R, Geiger D, Kreuzer I (2013) *Arabidopsis* nanodomain-delimited ABA signaling pathway regulates the anion channel SLAH3. Proc Natl Acad Sci U S A 110(20):8296–8301

Deng X, Hu W, Wei S, Zhou S, Zhang F, Han J, Chen L, Li Y, Feng J, Fang B, Luo Q, Li S, Liu Y, Yang G, He G (2013a) *TaCIPK29*, a CBL-interacting protein kinase gene from wheat, confers salt stress tolerance in transgenic tobacco. PLoS One 8(7):e69881

Deng X, Zhou S, Hu W, Feng J, Zhang F, Chen L, Huang C, Luo Q, He Y, Yang G, He G (2013b) Ectopic expression of wheat *TaCIPK14*, encoding a calcineurin B-like protein-interacting protein kinase, confers salinity and cold tolerance in tobacco. Physiol Plant 149(3):367–377

Drerup MM, Schlücking K, Hashimoto K, Manishankar P, Steinhorst L, Kuchitsu K, Kudla J (2013) The Calcineurin B-like calcium sensors CBL1 and CBL9 together with their interacting protein kinase CIPK26 regulate the *Arabidopsis* NADPH oxidase RBOHF. Mol Plant 6(2):559–569

Du W, Lin H, Chen S, Wu Y, Zhang J, Fuglsang AT, Palmgren MG, Wu W, Guo Y (2011) Phosphorylation of SOS3-like calcium-binding proteins by their interacting SOS2-like protein kinases is a common regulatory mechanism in *Arabidopsis*. Plant Physiol 156(4):2235–2243

Dubiella U, Seybold H, Durian G, Komander E, Lassig R, Witte CP, Schulze WX, Romeis T (2013) Calcium-dependent protein kinase/NADPH oxidase activation circuit is required for rapid defense signal propagation. Proc Natl Acad Sci U S A 110(21):8744–8749

Dubrovina AS, Kiselev KV, Khristenko VS (2013) Expression of calcium-dependent protein kinase (CDPK) genes under abiotic stress conditions in wild-growing grapevine *Vitis amurensis*. J Plant Physiol 170(17):1491–500

Evans NH, McAinsh MR, Hetherington AM (2001) Calcium oscillations in higher plants. Curr Opin Plant Biol 4:415–420

Franklin-Tong VE, Drobak BK, Allan AC, Watkins P, Trewavas AJ (1996) Growth of pollen tubes of *Papaver rhoeas* is regulated by a slow-moving calcium wave propagated by inositol 1,4,5-trisphosphate. Plant Cell 8:1305–1321

Franz S, Ehlert B, Liese A, Kurth J, Cazalé AC, Romeis T (2011) Calcium-dependent protein kinase CPK21 functions in abiotic stress response in *Arabidopsis thaliana*. Mol Plant 4(1):83–96

Frattini M, Morello L, Breviario D (1999) Rice calcium-dependent protein kinase isoforms OsCDPK2 and OsCDPK11 show different responses to light and different expression patterns during seed development. Plant Mol Biol 41(6):753–764

Freymark G, Diehl T, Miklis M, Romeis T, Panstruga R (2007) Antagonistic control of powdery mildew host cell entry by barley calcium-dependent protein kinases (CDPKs). Mol Plant Microbe Interact 20(10):1213–1221

Fu L, Yu X (2013) An C Overexpression of constitutively active OsCPK10 increases *Arabidopsis* resistance against *Pseudomonas syringae* pv. tomato and rice resistance against *Magnaporthe grisea*. Plant Physiol Biochem 73:202–210

Fukazawa J, Sakai T, Ishida S, Yamaguchi I, Kamiya Y, Takahashi Y (2000) Repression of shoot growth, a bZIP transcriptional activator, regulates cell elongation by controlling the level of gibberellins. Plant Cell 12(6):901–915

Fukazawa J, Nakata M, Ito T, Yamaguchi S, Takahashi Y (2010) The transcription factor RSG regulates negative feedback of NtGA20ox1 encoding GA 20-oxidase. Plant J 62(6):1035–1045

Geiger D, Scherzer S, Mumm P, Marten I, Ache P, Matschi S, Liese A, Wellmann C, Al-Rasheid KA, Grill E, Romeis T, Hedrich R (2010) Guard cell anion channel SLAC1 is regulated by CDPK protein kinases with distinct Ca^{2+} affinities. Proc Natl Acad Sci U S A 107(17): 8023–8028

Geng S, Li A, Tang L, Yin L, Wu L, Lei C, Guo X, Zhang X, Jiang G, Zhai W, Wei Y, Zheng Y, Lan X, Mao L (2013) *TaCPK2*-A, a calcium-dependent protein kinase gene that is required for wheat powdery mildew resistance enhances bacterial blight resistance in transgenic rice. J Exp Bot 64(11):3125–3136

Gilroy S, Read ND, Trewavas AJ (1990) Elevation of cytoplasmic calcium by caged calcium or caged inositol triphosphate initiates stomatal closure. Nature 346:769–771

Gleason C, Chaudhuri S, Yang T, Muñoz A, Poovaiah BW, Oldroyd GE (2006) Nodulation independent of rhizobia induced by a calcium-activated kinase lacking autoinhibition. Nature 441:1149–1152

Glinski M, Romeis T, Witte CP, Wienkoop S, Weckwerth W (2003) Stable isotope labeling of phosphopeptides for multiparallel kinase target analysis and identification of phosphorylation sites. Rapid Commun Mass Spectrom 17(14):1579–1584

Goda H, Sasaki E, Akiyama K, Maruyama-Nakashita A, Nakabayashi K, Li W, Ogawa M, Yamauchi Y, Preston J, Aoki K, Kiba T, Takatsuto S, Fujioka S, Asami T, Nakano T, Kato H, Mizuno T, Sakakibara H, Yamaguchi S, Nambara E, Kamiya Y, Takahashi H, Hirai MY, Sakurai T, Shinozaki K, Saito K, Yoshida S, Shimada Y (2008) The AtGenExpress hormone and chemical treatment data set: experimental design, data evaluation, model data analysis and data access. Plant J 55(3):526–42

Godfroy O, Debelle F, Timmers T, Rosenberg C (2006) A rice calcium- and calmodulin-dependent protein kinase restores nodulation to a legume mutant. Mol Plant Microbe Interact 19:495–501

Gong D, Guo Y, Jagendorf AT, Zhu JK (2002) Biochemical characterization of the *Arabidopsis* protein kinase SOS2 that functions in salt tolerance. Plant Physiol 130:256–264

Grandellis C, Giammaria V, Bialer M, Santin F, Lin T, Hannapel DJ, Ulloa RM (2012) The novel Solanum tuberosum calcium dependent protein kinase, StCDPK3, is expressed in actively growing organs. Planta 236(6):1831–1848

Gu Z, Ma B, Jiang Y, Chen Z, Su X, Zhang H (2008) Expression analysis of the calcineurin B-like gene family in rice (*Oryza sativa* L.) under environmental stresses. Gene 415(1–2):1–12

Guo Y, Halfter U, Ishitani M, Zhu JK (2001) Molecular characterization of functional domains in the protein kinase SOS2 that is required for plant salt tolerance. Plant Cell 13:1383–1400

Guo Y, Xiong L, Song CP, Gong D, Halfter U, Zhu JK (2002) A calcium sensor and its interacting protein kinase are global regulators of abscisic acid signaling in *Arabidopsis*. Dev Cell 3(2):233–244

Guo P, Baum M, Grando S, Ceccarelli S, Bai G, Li R, von Korff M, Varshney RK, Graner A, Valkoun J (2009) Differentially expressed genes between drought-tolerant and drought-sensitive barley genotypes in response to drought stress during the reproductive stage. J Exp Bot 60(12):3531–3544

Haley A, Russell AJ, Wood N, Allan AC, Knight M, Campbell AK, Trewavas AJ (1995) Effects of mechanical signaling on plant cell cytosolic calcium. Proc Natl Acad Sci U S A 92:4124–4128

Halfter U, Ishitani M, Zhu JK (2000) The *Arabidopsis* SOS2 protein kinase physically interacts with and is activated by the calcium-binding protein SOS3. Proc Natl Acad Sci U S A 97(7):3735–3740

Hamel LP, Sheen J, Séguin A (2013) Ancient signals: comparative genomics of green plant CDPKs. Trends Plant Sci S1360–1385(13):00250–1

Harmon AC, Yoo BC, McCaffery C (1994) Pseudosubstrate inhibition of CDPK, a protein kinase with a calmodulin-like domain. Biochemistry 33(23):7278–7287

Harmon AC, Gribskov M, Gubrium E, Harper JF (2001) The CDPK superfamily of protein kinases. New Phytol 151:175–183

Harper JF, Sussman MR, Schaller GE, Putnam-Evans C, Charbonneau H, Harmon AC (1991) A calcium-dependent protein kinase with a regulatory domain similar to calmodulin. Science 252(1):951–954

Harper JF, Huang JF, Lloyd SJ (1994) Genetic identification of an autoinhibitor in CDPK, a protein kinase with a calmodulin-like domain. Biochemistry 33(23):7267–7277

Harper JF, Breton G, Harmon A (2004) Decoding Ca^{2+} signals through plant protein kinases. Annu Rev Plant Biol 55:263–288

Hashimoto K, Kudla J (2011) Calcium decoding mechanisms in plants. Biochimie 93(12): 2054–2059

Hashimoto K, Eckert C, Anschütz U, Scholz M, Held K, Waadt R, Reyer A, Hippler M, Becker D, Kudla J (2012) Phosphorylation of calcineurin B-like (CBL) calcium sensor proteins by their CBL-interacting protein kinases (CIPKs) is required for full activity of CBL-CIPK complexes toward their target proteins. J Biol Chem 287(11):7956–7968

Hassan A, Okuta T, Kato M, Hatsugai N, Sano Y, Ishimori T, Okazaki K, Doullah MA, Shah MM (2012) Alternaric acid stimulates phosphorylation of His-tagged RiCDPK2, a calcium-dependent protein kinase in potato plants. Genet Mol Res 11(3):2381–2389

Hassan A, Hatsugai N, Shah MM (2013) A phytotoxin Solanapyrone-A downregulates calcium-dependent protein kinase activity in potato. Genetics and Molecular Research 12 (2): 1540–1545

Hayashi T, Banba M, Shimoda Y, Kouchi H, Hayashi M, Imaizumi-Anraku H (2010) A dominant function of CCaMK in intracellular accommodation of bacterial and fungal endosymbionts. Plant J 63(1):141–54

He L, Yang X, Wang L, Zhu L, Zhou T, Deng J, Zhang X (2013) Molecular cloning and functional characterization of a novel cotton CBL-interacting protein kinase gene (*GhCIPK6*) reveals its involvement in multiple abiotic stress tolerance in transgenic plants. Biochem Biophys Res Commun 435(2):209–215

Hegeman AD, Rodriguez M, Han BW, Uno Y, Phillips GN, Hrabak EM, Uno Y, Phillips GN, Hrabak EM, Cushman JC, Harper JF, Harmon AC, Sussman MR (2006) A phyloproteomic characterization of *in vitro* autophosphorylation in calcium-dependent protein kinases. Proteomics 6:3649–3664

Held K, Pascaud F, Eckert C, Gajdanowicz P, Hashimoto K, Corratgé-Faillie C, Offenborn JN, Lacombe B, Dreyer I, Thibaud JB, Kudla J (2011) Calcium-dependent modulation and plasma membrane targeting of the AKT2 potassium channel by the CBL4/CIPK6 calcium sensor/protein kinase complex. Cell Res 21(7):1116–1130

Ho CH, Lin SH, Hu HC, Tsay YF (2009) CHL1 functions as a nitrate sensor in plants. Cell 138(6):1184–1194

Ho SL, Huang LF, Lu CA, He SL, Wang CC, Yu SP, Chen J, Yu SM (2013) Sugar starvation- and GA-inducible calcium-dependent protein kinase 1 feedback regulates GA biosynthesis and

activates a 14-3-3 protein to confer drought tolerance in rice seedlings. Plant Mol Biol 81(4–5):347–361
Hrabak EM, Chan CW, Gribskov M, Harper JF, Choi JH, Halford N, Kudla J, Luan S, Nimmo HG, Sussman MR, Thomas M, Walker-Simmons K, Zhu JK, Harmon AC (2003) The *Arabidopsis* CDPK-SnRK superfamily of protein kinases. Plant Physiol 132:666–680
Hsu JL, Wang LY, Wang SY, Lin CH, Ho KC, Shi FK, Chang IF (2009) Functional phosphoproteomic profiling of phosphorylation sites in membrane fractions of salt-stressed *Arabidopsis thaliana*. Proteome Sci 10(7):42
Hu HC, Wang YY, Tsay YF (2009) AtCIPK8, a CBL-interacting protein kinase, regulates the low-affinity phase of the primary nitrate response. Plant J 57(2):264–278
Hu DG, Li M, Luo H, Dong QL, Yao YX, You CX, Hao YJ (2012) Molecular cloning and functional characterization of *MdSOS2* reveals its involvement in salt tolerance in apple callus and *Arabidopsis*. Plant Cell Rep 31(4):713–722
Huang JF, Teyton L, Harper JF (1996) Activation of a Ca^{2+}-dependent protein kinase involves intramolecular binding of a calmodulin-like regulatory domain. Biochemistry 35(40): 13222–13230
Huang C, Ding S, Zhang H, Du H, An L (2011) CIPK7 is involved in cold response by interacting with CBL1 in *Arabidopsis thaliana*. Plant Sci 181(1):57–64
Huang TL, Nguyen QT, Fu SF, Lin CY, Chen YC, Huang HJ (2012) Transcriptomic changes and signalling pathways induced by arsenic stress in rice roots. Plant Mol Biol 80(6):587–608
Huang SJ, Chang CL, Wang PH, Tsai MC, Hsu PH, Chang IF (2013a) A type III ACC synthase, ACS7, is involved in root gravitropism in *Arabidopsis thaliana*. J Exp Bot 64(14):4343–4360
Huang Y, Deng T, Zuo K (2013b) Cotton annexin proteins participate in the establishment of fiber cell elongation scaffold. Plant Signal Behav 8(9):e25601
Hubbard KE, Siegel RS, Valerio G, Brandt B, Schroeder JI (2011) Abscisic acid and CO_2 signalling via calcium sensitivity priming in guard cells, new CDPK mutant phenotypes and a method for improved resolution of stomatal stimulus-response analyses. Ann Bot 109(1):5–17
Hwang YS, Bethke PC, Cheong YH, Chang HS, Zhu T, Jones RL (2005) A gibberellin-regulated calcineurin B in rice localizes to the tonoplast and is implicated in vacuole function. Plant Physiol 138(3):1347–1358
Igarashi D, Ishida S, Fukazawa J, Takahashi Y (2001) 14-3-3 proteins regulate intracellular localization of the bZIP transcriptional activator RSG. Plant Cell 13(11):2483–2497
Ishida S, Fukazawa J, Yuasa T, Takahashi Y (2004) Involvement of 14-3-3 signaling protein binding in the functional regulation of the transcriptional activator REPRESSION OF SHOOT GROWTH by gibberellins. Plant Cell 16(10):2641–2651
Ishida S, Yuasa T, Nakata M, Takahashi Y (2008) A tobacco calcium-dependent protein kinase, CDPK1, regulates the transcription factor REPRESSION OF SHOOT GROWTH in response to gibberellins. Plant Cell 20(12):3273–3288
Ito T, Nakata M, Fukazawa J, Ishida S, Takahashi Y (2010) Alteration of substrate specificity: the variable N-terminal domain of tobacco Ca^{2+}-dependent protein kinase is important for substrate recognition. Plant Cell 22(5):1592–1604
Iwanaga T, Tsutsumi R, Noritake J, Fukata Y, Fukata M (2009) Dynamic protein palmitoylation in cellular signaling. Prog Lipid Res 48:117–127
Iwano M, Entani T, Shiba H, Kakita M, Nagai T, Mizuno H, Miyawaki A, Shoji T, Kubo K, Isogai A, Takayama S (2009) Fine-tuning of the cytoplasmic Ca^{2+} concentration is essential for pollen tube growth. Plant Physiol 150:1322–1334
Jain M, Nijhawan A, Arora R, Agarwal P, Ray S, Sharma P, Kapoor S, Tyagi AK, Khurana JP (2007) F-box proteins in rice. Genome-wide analysis, classification, temporal and spatial gene expression during panicle and seed development, and regulation by light and abiotic stress. Plant Physiol 143:1467–1483
Jiang S, Zhang D, Wang L, Pan J, Liu Y, Kong X, Zhou Y, Li D (2013) A maize calcium-dependent protein kinase gene, *ZmCPK4*, positively regulated abscisic acid signaling and enhanced drought stress tolerance in transgenic *Arabidopsis*. Plant Physiol Biochem 71:112–120

Kanchiswamy CN, Mohanta TK, Capuzzo A, Occhipinti A, Verrillo F, Maffei ME, Malnoy M (2013) Differential expression of CPKs and cytosolic Ca^{2+} variation in resistant and susceptible apple cultivars (*Malus* x *domestica*) in response to the pathogen *Erwinia amylovora* and mechanical wounding. BMC Genomics 14:760

Kang CH, Moon BC, Park HC, Koo SC, Chi YH, Cheong YH, Yoon BD, Lee SY, Kim CY (2013) Rice small C2-domain proteins are phosphorylated by calcium-dependent protein kinase. Mol Cells 35(5):381–387

Kiegle E, Moore CA, Haseloff J, Tester MA, Knight MR (2000) Cell-type-specific calcium responses to drought, salt and cold in the *Arabidopsis* root. Plant J 23:267–278

Kilian J, Whitehead D, Horak J, Wanke D, Weinl S, Batistic O, D'Angelo C, Bornberg-Bauer E, Kudla J, Harter K (2007) The AtGenExpress global stress expression data set: protocols, evaluation and model data analysis of UV-B light, drought and cold stress responses. Plant J 50(2):347–363

Kim KN, Lee JS, Han H, Choi SA, Go SJ, Yoon IS (2003) Isolation and characterization of a novel rice Ca2+-regulated protein kinase gene involved in responses to diverse signals including cold, light, cytokinins, sugars and salts. Plant Mol Biol.52(6):1191–1202

Kim BG, Waadt R, Cheong YH, Pandey GK, Dominguez-Solis JR, Schültke S, Lee SC, Kudla J, Luan S (2007) The calcium sensor CBL10 mediates salt tolerance by regulating ion homeostasis in *Arabidopsis*. Plant J 52(3):473–484

Kim TH, Böhmer M, Hu H, Nishimura N, Schroeder JI (2011) Guard cell signal transduction network: Advances in understanding abscisic acid, CO_2, and Ca^{2+} signaling. Annu Rev Plant Biol 61:561–591

Kimura S, Kawarazaki T, Nibori H, Michikawa M, Imai A, Kaya H, Kuchitsu K (2013) The CBL-interacting protein kinase CIPK26 is a novel interactor of *Arabidopsis* NADPH oxidase AtRbohF that negatively modulates its ROS-producing activity in a heterologous expression system. J Biochem 153(2):191–195

Kiselev KV, Grishchenko OV, Zhuravlev YN (2009) CDPK gene expression in salt tolerant *rolB* and *rolC* transformed cell cultures of *Panax ginseng*. Biol Planta 54(4):621–630

Kline KG, Barrett-Wilt GA, Sussman MR (2010) In planta changes in protein phosphorylation induced by the plant hormone abscisic acid. Proc Natl Acad Sci U S A 107(36):15986–15991

Knight H (2000) Calcium signaling during abiotic stress in plants. Int Rev Cytol 195:269–324

Knight H, Knight MR (2001) Abiotic stress signalling pathways: Specificity and cross-talk. Trends Plant Sci 6(6):262–267

Kobayashi M, Ohura I, Kawakita K, Yokota N, Fujiwara M, Shimamoto K, Doke N, Yoshioka H (2007) Calcium-dependent protein kinases regulate the production of reactive oxygen species by potato NADPH oxidase. Plant Cell 19:1065–1080

Kolukisaoglu U, Weinl S, Blazevic D, Batistic O, Kudla J (2004) Calcium sensors and their interacting protein kinases, genomics of the *Arabidopsis* and rice CBL-CIPK signaling networks. Plant Physiol 134(1):43–58

Komatsu S, Yang G, Khan M, Onodera H, Toki S, Yamaguchi M (2007) Over-expression of calcium-dependent protein kinase 13 and calreticulin interacting protein 1 confers cold tolerance on rice plants. Mol Genet Genomics 277:713–723

Kong X, Pan J, Zhang D, Jiang S, Cai G, Wang L, Li D (2013) Identification of mitogen-activated protein kinase kinase gene family and MKK-MAPK interaction network in maize. Biochem Biophys Res Commun 441(4):964–969

Krebs M, Held K, Binder A, Hashimoto K, Den Herder G, Parniske M, Kudla J, Schumacher K (2012) FRET-based genetically encoded sensors allow high-resolution live cell imaging of Ca^{2+} dynamics. Plant J 69(1):181–192

Kudla J, Xu Q, Harter K, Gruissem W, Luan S (1999) Genes for calcineurin B-like proteins in *Arabidopsis* are differentially regulated by stress signals. Proc Natl Acad Sci U S A 96(8):4718–4723

Kumar KG, Ullanat R, Jayabaskaran C (2004) Molecular cloning, characterization, tissue-specific and phytohormone-induced expression of calcium-dependent protein kinase gene in cucumber (*Cucumis sativus* L.). J Plant Physiol 161:1061–1071

Kumar G, Purty RS, Sharma MP, Singla-Pareek SL, Pareek A (2009) Physiological responses among Brassica species under salinity stress show strong correlation with transcript abundance for SOS pathway-related genes. J Plant Physiol 166(5):507–520

Kurusu T, Hamada J, Hamada H, Hanamata S, Kuchitsu K (2010) Roles of calcineurin B-like protein-interacting protein kinases in innate immunity in rice. Plant Signal Behav 5(8):1045–1047

Kushwaha HR, Kumar G, Verma PK, Singla-Pareek SL, Pareek A (2011) Analysis of a salinity induced BjSOS3 protein from Brassica indicate it to be structurally and functionally related to its ortholog from Arabidopsis. Plant Physiol Biochem 49:996–1004

Kwak JM, Moon JH, Murata Y, Kuchitsu K, Leonhardt N, DeLong A, Schroeder JI (2002) Disruption of a guard cell-expressed protein phosphatase 2A regulatory subunit, *RCN1*, confers Abscisic acid insensitivity in *Arabidopsis*. Plant Cell 14:2849–2861

Kawasaki T, Hayashida N, Baba T, Shinozaki K, Shimada H (1993) The gene encoding a calcium-dependent protein kinase located near the sbe1 gene encoding starch branching enzyme I is specifically expressed in developing rice seeds. Gene 129 (2), 183–189

Lan WZ, Lee SC, Che YF, Jiang YQ, Luan S (2011) Mechanistic analysis of AKT1 regulation by the CBL-CIPK-PP2CA interactions. Mol Plant 4(3):527–536

Lee SS, Cho HS, Yoon GM, Ahn JW, Kim HH, Pai HS (2003) Interaction of NtCDPK1 calcium-dependent protein kinase with NtRpn3 regulatory subunit of the 26S proteasome in *Nicotiana tabacum*. Plant J 33(5):825–840

Lee SC, Lan WZ, Kim BG, Li L, Cheong YH, Pandey GK, Lu G, Buchanan BB, Luan S (2007) A protein phosphorylation/dephosphorylation network regulates a plant potassium channel. Proc Natl Acad Sci U S A 104(40):15959–15964

Levy J, Bres C, Geurts R, Chalhoub B, Kulikova O, Duc G, Journet EP, Ane JM, Lauber E, Bisseling T (2004) A putative Ca^{2+} and calmodulin-dependent protein kinase required for bacterial and fungal symbioses. Science 303:1361–1364

Li J, Lee YRJ, Assmann SM (1998) Guard cells possess a calcium-dependent protein kinase that phosphorylates the KAT1 potassium channel. Plant Physiol 116:785–795

Li L., Kim B. G., Cheong Y. H., Pandey G. K., Luan S. (2006). A Ca2+ signaling pathway regulates a K+ channel for low-K response in Arabidopsis. Proc. Natl. Acad. Sci. U.S.A. 103, 12625–12630

Li A, Wang X, Leseberg CH, Jia J, Mao L (2008) Biotic and abiotic stress responses through calcium-dependent protein kinase (CDPK) signaling in wheat (*Triticum aestivum* L.). Plant Signal Behav 3(9):654–656

Li R, Zhang J, Wu G, Wang H, Chen Y, Wei J (2012) HbCIPK2, a novel CBL-interacting protein kinase from halophyte *Hordeum brevisubulatum*, confers salt and osmotic stress tolerance. Plant Cell Environ 35(9):1582–1600

Li ZY, Xu ZS, Chen Y, He GY, Yang GX, Chen M, Li LC, Ma YZ (2013) A novel role for *Arabidopsis* CBL1 in affecting plant responses to glucose and gibberellin during germination and seedling development. PLoS One 8:e56412

Liang Y, Pan J (2013a) Regulation of flagellar biogenesis by a calcium dependent protein kinase in *Chlamydomonas reinhardtii*. PLoS One 8(7):e69902

Liang Y, Pan J (2013b) Regulation of flagellar biogenesis by a calcium dependent protein kinase in Chlamydomonas reinhardtii. PLoS One 8(7):e69902

Liao J, Singh S, Hossain MS, Andersen SU, Ross L, Bonetta D, Zhou Y, Sato S, Tabata S, Stougaard J, Szczyglowski K, Parniske M (2012) Negative regulation of CCaMK is essential for symbiotic infection. Plant J 72(4):572–584

Lima L, Seabra A, Melo P, Cullimore J, Carvalho H (2006a) Post-translational regulation of cytosolic glutamine synthetase of *Medicago truncatula*. J Exp Bot 57(11):2751–2761

Lima L, Seabra A, Melo P, Cullimore J, Carvalho H (2006b) Phosphorylation and subsequent interaction with 14-3-3 proteins regulate plastid glutamine synthetase in *Medicago truncatula*. Planta 223(3):558–567

Liu G, Chen J, Wang X (2006) *VfCPK1*, a gene encoding calcium-dependent protein kinase from *Vicia faba*, is induced by drought and abscisic acid. Plant Cell Environ 29(11):2091–2099

Liu GT, Wang JF, Cramer G, Dai ZW, Duan W, Xu HG, Wu BH, Fan PG, Wang LJ, Li SH (2012) Transcriptomic analysis of grape (*Vitis vinifera* L.) leaves during and after recovery from heat stress. BMC Plant Biol 12:174

Liu LL, Ren HM, Chen LQ, Wang Y, Wu WH (2013) A protein kinase, calcineurin B-like protein-interacting protein Kinase 9, interacts with calcium sensor calcineurin B-like Protein 3 and regulates potassium homeostasis under low-potassium stress in *Arabidopsis*. Plant Physiol 161(1):266–277

Llop-Tous I, Dominquez-Puigjaner E, Vendrel IM (2002) Characterization of a strawberry cDNA clone homologous to calcium-dependent protein kinases that is expressed during ripening and affected by low temperature. J Exp Bot 53:2283–2285

Lu SX, Hrabak EM (2002) An *Arabidopsis* calcium-dependent protein kinase is associated with the endoplasmic reticulum. Plant Physiol 128(3):1008–1021

Lu B, Ding R, Zhang L, Yu X, Huang B, Chen W (2006) Molecular cloning and characterization of a novel calcium-dependent protein kinase gene IiCPK2 Responsive to polyploidy from tetraploid *Isatis indigotica*. J Biochem Mol Biol 39(5):607–617

Luan S, Kudla J, Rodriguez-Concepcion M, Yalovsky S, Gruissem W (2002) Calmodulins and calcineurin B-like proteins: calcium sensors for specific signal response coupling in plants. Plant Cell 14(Suppl):S389–400

Ludwig AA, Romeis T, Jones JDG (2004) CDPK-mediated signalling pathways: Specificity and cross-talk. J Exp Bot 55(395):181–188

Ludwig AA, Saitoh H, Felix G, Freymark G, Miersch O, Wasternack C, Boller T, Jones JD, Romeis T (2005) Ethylene-mediated cross-talk between calcium-dependent protein kinase and MAPK signaling controls stress responses in plants. Proc Natl Acad Sci U S A 102: 10736–10741

Lv F, Zhang H, Xia X, Yin W (2014) Expression profiling and functional characterization of a CBL-interacting protein kinase gene from Populus euphratica. Plant Cell Rep 33(5):807–18

Lyzenga WJ, Liu H, Schofield A, Muise-Hennessey A, Stone SL (2013) *Arabidopsis* CIPK26 interacts with KEG, components of the ABA signalling network and is degraded by the ubiquitin-proteasome system. J Exp Bot 64(10):2779–2791

Ma SY, Wu WH (2007) AtCPK23 functions in *Arabidopsis* responses to drought and salt stresses. Plant Mol Biol 65(4):511–518

Ma F, Lu R, Liu H, Shi B, Zhang J, Tan M, Zhang A, Jiang M (2012a) Nitric oxide-activated calcium/calmodulin-dependent protein kinase regulates the abscisic acid-induced antioxidant defence in maize. J Exp Bot 63(13):4835–4847

Ma TL, Wu WH, Wang Y (2012b) Transcriptome analysis of rice root responses to potassium deficiency. BMC Plant Biol 12:161

Ma P, Liu J, Yang X, Ma R (2013) Genome-wide identification of the maize calcium-dependent protein kinase gene family. Appl Biochem Biotechnol 169(7):2111–2125

Mahajan S, Sopory SK, Tuteja N (2006) Cloning and characterization of CBL-CIPK signalling components from a legume (*Pisum sativum*). FEBS J 273(5):907–925

Mall TK, Dweikat I, Sato SJ, Neresian N, Xu K, Ge Z, Wang D, Elthon T, Clemente T (2011) Expression of the rice *CDPK-7* in sorghum: Molecular and phenotypic analyses. Plant Mol Biol 75(4–5):467–479

Martín ML, Busconi L (2000) Membrane localization of a rice calcium-dependent protein kinase (CDPK) is mediated by myristoylation and palmitoylation. Plant J 24(4):429–435

Martín ML, Busconi L (2001) A rice membrane-bound calcium-dependent protein kinase is activated in response to low temperature. Plant Physiol 125(3):1442–1449

Martínez-Atienza J, Jiang X, Garciadeblas B, Mendoza I, Zhu JK, Pardo JM, Quintero FJ (2007) Conservation of the salt overly sensitive pathway in rice. Plant Physiol 143(2):1001–1012

Matschi S, Werner S, Schulze WX, Legen J, Hilger HH, Romeis T (2013) Function of calcium-dependent protein kinase CPK28 of *Arabidopsis thaliana* in plant stem elongation and vascular development. Plant J 73(6):883–896

McAinsh MR, Webb A, Taylor JE, Hetherington AM (1995) Stimulus-induced oscillations in guard cell cytosolic free calcium. Plant Cell 7:1207–1219

Mehlmer N, Wurzinger B, Stael S, Hofmann-Rodrigues D, Csaszar E, Pfister B, Bayer R, Teige M (2010) The Ca^{2+}-dependent protein kinase CPK3 is required for MAPK-independent salt-stress acclimation in *Arabidopsis*. Plant J 63:484–498

Messinese E, Mun JH, Yeun LH, Jayaraman D, Rouge P, Barre A, Lougnon G, Schornack S, Bono JJ, Cook DR (2007) A novel nuclear protein interacts with the symbiotic DMI3 calcium- and calmodulin-dependent protein kinase of *Medicago truncatula*. Mol Plant Microbe Interact 20:912–921

Meyers BC, Kozik A, Griego A, Kuang H, Michelmore RW (2003) Genome-wide analysis of NBS-LRR-encoding genes in *Arabidopsis*. Plant Cell 15(4):809–34

Milla MAR, Townsend J, Chang IF, Cushman JC (2006a) The *Arabidopsis* AtDi19 gene family encodes a novel type of Cys2/His2 zinc-finger protein implicated in ABA-independent dehydration, high-salinity stress and light signaling pathways. Plant Mol Biol 61:13–30

Milla MAR, Uno Y, Chang IF, Townsend J, Maher EA, Quilici D, Cushman JC (2006b) A novel yeast two-hybrid approach to identify CDPK substrates: Characterization of the interaction between CPK11 and AtDi19, a nuclear zinc finger protein. FEBS Lett 580:904–911

Mitra RM, Gleason CA, Edwards A, Hadfield J, Downie JA, Oldroyd GED, Long SR (2004) A Ca^{2+}/calmodulin-dependent protein kinase required for symbiotic nodule development: gene identification by transcript-based cloning. Proc Natl Acad Sci U S A 101:4701–4705

Monroy AF, Dhindsa RS (1995) Low-temperature signal transduction: Induction of cold acclimation-specific genes of alfalfa by calcium at 25°C. Plant Cell 7(3):321–331

Morello L, Giani S, Coraggio I, Breviario D (1993) Rice membranes contain a calcium-dependent protein kinase activity with biochemical features of animal protein kinase C. Biochem Biophys Res Commun 197:55–61

Mori IC, Murata Y, Yang Y, Munemasa S, Wang YF, Andreoli S, Tiriac H, Alonso JM, Harper JF, Ecker JR, Kwak JM, Schroeder JI (2006) CDPKs CPK6 and CPK3 function in ABA regulation of guard cell S-type anion- and Ca^{2+}-permeable channels and stomatal closure. PLoS Biol 4(10):e327

Murillo I, Jaeck E, Cordero MJ, San Segundo B (2001) Transcriptional activation of a maize calcium-dependent protein kinase gene in response to fungal elicitors and infection.Plant Mol Biol. 45(2):145–58

Munemasa S, Oda K, Watanabe-Sugimoto M, Nakamura Y, Shimoishi Y, Murata Y (2007) The coronatine-insensitive 1 mutation reveals the hormonal signaling interaction between abscisic acid and methyl jasmonate in *Arabidopsis* guard cells: Specific impairment of ion channel activation and second messenger production. Plant Physiol 143:1398–1407

Munemasa S, Mori IC, Murata Y (2011) Methyl jasmonate signaling and signal crosstalk between methyl jasmonate and abscisic acid in guard cells. Plant Signal Behav 6(7):939–941

Muñiz García MN, Giammaria V, Grandellis C, Téllez-Iñón MT, Ulloa RM, Capiati DA (2011) Characterization of StABF1, a stress-responsive bZIP transcription factor from *Solanum tuberosum* L. that is phosphorylated by StCDPK2 *in vitro*. Planta 235(4):761–78

Nagae M, Nozawa A, Koizumi N, Sano H, Hashimoto H, Sato M, Shimizu T (2003) The crystal structure of the novel calcium-binding protein AtCBL2 from Arabidopsis thaliana. J Biol Chem 278:42240–42246

Nakashima K, Yamaguchi-Shinozaki K (2013) ABA signaling in stress-response and seed development. Plant Cell Rep 32(7):959–970

Nozawa A, Koizumi N, Sano H (2001) An Arabidopsis SNF1-related protein kinase, AtSR1, interacts with a calcium-binding protein, AtCBL2, of which transcripts respond to light. Plant Cell Physiol. 42(9):976–81

Oh SI, Park J, Yoon S, Kim Y, Park S, Ryu M, Nam MJ, Ok SH, Kim JK, Shin JS, Kim KN (2008) The *Arabidopsis* calcium sensor calcineurin B-like 3 inhibits the 5′-methylthioadenosine nucleosidase in a calcium-dependent manner. Plant Physiol 148(4):1883–1896

Pandey GK (2008) Emergence of a novel calcium signaling pathway in plants: CBL-CIPK signaling network. Physiol Mol Biol Plants 14:51–68

Pandey S, Tiwari SB, Tyagi W, Reddy MK, Upadhyaya KC, Sopory SK (2002) A Ca^{2+}/CaM-dependent kinase from pea is stress regulated and *in vitro* phosphorylates a protein that binds to AtCaM5 promoter. Eur J Biochem 269(13):3193–3204

Pandey GK, Cheong YH, Kim KN, Grant JJ, Li L, Hung W, D'Angelo C, Weinl S, Kudla J, Luan S (2004) The calcium sensor calcineurin B-like 9 modulates abscisic acid sensitivity and biosynthesis in *Arabidopsis*. Plant Cell 16(7):1912–1924

Pandey GK, Cheong YH, Kim BG, Grant JJ, Li L, Luan S (2007) CIPK9: a calcium sensor-interacting protein kinase required for low-potassium tolerance in *Arabidopsis*. Cell Res 17:411–421

Pandey GK, Grant JJ, Cheong YH, Kim BG, le Li G, Luan S (2008) Calcineurin-B-like protein CBL9 interacts with target kinase CIPK3 in the regulation of ABA response in seed germination. Mol Plant 1(2):238–248

Pang T, Ye CY, Xia X, Yin W (2013) De novo sequencing and transcriptome analysis of the desert shrub, *Ammopiptanthus mongolicus*, during cold acclimation using Illumina/Solexa. BMC Genomics 14(1):488

Patharkar OR, Cushman JC (2000) A stress-induced calcium-dependent protein kinase from *Mesembryanthemum crystallinum* phosphorylates a two-component pseudo-response regulator. Plant J 24(5):679–691

Patharkar OR, Cushman JC (2006) A novel coiled-coil protein co-localizes and interacts with a calcium-dependent protein kinase in the common ice plant during low-humidity stress. Planta 225(1):57–73

Patil S, Takezawa D, Poovaiah BW (1995) Chimeric plant calcium/calmodulin-dependent protein kinase gene with a neural visinin-like calcium-binding domain. Proc Natl Acad Sci U S A 92: 4897–4901

Piao HL, Xuan YH, Park SH, Je BI, Park SJ, Park SH, Kim CM, Huang J, Wang GK, Kim MJ, Kang SM, Lee IJ, Kwon TR, Kim YH, Yeo US, Yi G, Son D, Han CD (2010) OsCIPK31, a CBL-interacting protein kinase is involved in germination and seedling growth under abiotic stress conditions in rice plants. Mol Cells. 2010 30(1):19–27

Pical C, Fredlund KM, Petit PX, Sommarin M, Moller IM (1993) The outer membrane of plant mitochondria contains a calcium-dependent protein kinase and multiple phosphoproteins. FEBS Lett 336:347–351

Plieth C, Hansen UP, Knight H, Knight MR (1999) Temperature sensing by plants: the primary characteristics of signal perception and calcium response. Plant J 18:491–497

Putnam-Evans C, Harmon AC, Palevitz BA, Fechheimer M, Cormier MJ (1989) Calcium-dependent protein kinase is localized with F-actin in plant cells. Cell Motil Cytoskeleton 12:12–22

Qiu QS, Guo Y, Dietrich MA, Schumaker KS, Zhu JK (2002) Regulation of SOS1, a plasma membrane Na+/H+ exchanger in *Arabidopsis thaliana*, by SOS2 and SOS3. Proc Natl Acad Sci U S A 99(12):8436–8441

Quintero FJ, Ohta M, Shi H, Zhu JK, Pardo JM (2002) Reconstitution in yeast of the *Arabidopsis* SOS signaling pathway for Na+ homeostasis. Proc Natl Acad Sci U S A 99(13):9061–9066

Quintero FJ, Martinez-Atienza J, Villalta I, Jiang X, Kim WY, Ali Z, Fujii H, Mendoza I, Yun DJ, Zhu JK, Pardo JM (2011) Activation of the plasma membrane Na/H antiporter Salt-Overly-Sensitive 1 (SOS1) by phosphorylation of an auto-inhibitory C-terminal domain. Proc Natl Acad Sci U S A 108(6):2611–2616

Raices M, Gargantini PR, Chinchilla D, Martin C, Trllez-Inon M, Ulloa RM (2003) Regulation of CDPK isoforms during tuber development. Plant Mol Biol 52:1011–1024

Raichaudhuri A, Bhattacharyya R, Chaudhury S, Chakrabarti P, DasGupta M (2006) Domain analysis of a groundnut cdpk: nuclear localisation sequence in the junction domain is coupled with nonconsensus calcium binding domains. J Biol Chem 281(15):10399–10409

Ray S (2012) Calcium-dependent protein kinase: a tool for plants to crack the calcium code. In: Pandey GK (ed) Plant stress, vol 6. Global Science Books, London, pp 43–59

Ray S, Agarwal P, Arora R, Kapoor S, Tyagi AK (2007) Expression analysis of calcium-dependent protein kinase gene family during reproductive development and abiotic stress conditions in rice (*Oryza sativa* L. ssp. *indica*). Mol Genet Genomics 278:493–505

Ray S, Dansana PK, Giri J, Deveshwar P, Arora R, Agarwal P, Khurana JP, Kapoor S, Tyagi AK (2011) Modulation of transcription factor and metabolic pathway genes in response to water-deficit stress in rice. Funct Intgr Genomics 11:157–178

Ren XL, Qi GN, Feng HQ, Zhao S, Zhao SS, Wang Y, Wu WH (2013) Calcineurin B-like protein CBL10 directly interacts with AKT1 and modulates K+ homeostasis in *Arabidopsis*. Plant J 74(2):258–266

Romeis T, Ludwig AA, Martin R, Jones JD (2001) Calcium-dependent protein kinases play an essential role in a plant defence response. EMBO J 20:5556–5567

Rutschmann F, Stalder U, Piotrowski M, Oecking C, Schaller A (2002) *LeCPK1*, a calcium-dependent protein kinase from tomato: Plasma membrane targeting and biochemical characterization. Plant Physiol 129:156–168

Sachs JN, Engelman DM (2006) Introduction to the membrane protein reviews: the interplay of structure, dynamics, and environment in membrane protein function. Annu Rev Biochem 75:707–712

Sagi M, Fluhr R (2006) Production of reactive oxygen species by plant NADPH oxidases. Plant Physiol 141(2):336–340

Saijo Y, Hata S, Kyozuka J, Shimamoto K, Izui K (2000) Over-expression of a single Ca^{2+}-dependent protein kinase confers both cold and salt/drought tolerance on rice plants. Plant J 23:319–327

Saijo Y, Kinoshita N, Ishiyama K, Hata S, Kyozuka J, Hayakawa T, Nakamura T, Shimamoto K, Yamaya T, Izui K (2001) A Ca^{2+}-dependent protein kinase that endows rice plants with cold- and salt-stress tolerance functions in vascular bundles. Plant Cell Physiol 42:1228–1233

Saito N, Munemasa S, Nakamura Y, Shimoishi Y, Mori IC, Murata Y (2008) Roles of RCN1, regulatory A subunit of protein phosphatase 2A, in methyl jasmonate signaling and signal crosstalk between methyl jasmonate and abscisic acid. Plant Cell Physiol 49(9):1396–1401

Sanchez-Barrena MJ, Martinez-Ripoll M, Zhu JK, Albert A (2005) The structure of the *Arabidopsis thaliana* SOS3: molecular mechanism of sensing calcium for salt stress response. J Mol Biol 345:1253–1264

Sánchez-Barrena MJ, Fujii H, Angulo I, Martínez-Ripoll M, Zhu JK, Albert A (2007) The structure of the C-terminal domain of the protein kinase AtSOS2 bound to the calcium sensor AtSOS3. Mol Cell 26(3):427–35

Sanders D, Brownlee C, Harper JF (1999) Communicating with calcium. Plant Cell 11:691–706

Sanders D, Pelloux J, Brownlee C, Harper JF (2002) Calcium at the crossroads of signalling. Plant Cell 14:S401–S417

Sathyanarayanan PV, Poovaiah BW (2004) Decoding Ca^{2+} signals in plants. CRC Crit Rev Plant Sci 23(1):1–11

Sathyanarayanan PV, Cremo CR, Poovaiah BW (2000) Plant chimeric Ca^{2+}/calmodulin-dependent protein kinase: role of the neural visinin-like domain in regulating autophosphorylation and calmodulin affinity. J Biol Chem 275:30417–30422

Schlücking K, Edel KH, Köster P, Drerup MM, Eckert C, Steinhorst L, Waadt R, Batistic O, Kudla J (2013) A new β-estradiol-inducible vector set that facilitates easy construction and efficient expression of transgenes reveals CBL3-dependent cytoplasm to tonoplast translocation of CIPK5. Mol Plant 6(6):1814–1829

Schnable PS, Ware D, Fulton RS, Stein JC, Wei F, Pasternak S, Liang C, Zhang J, Fulton L, Graves TA, Minx P, Reily AD, Courtney L, Kruchowski SS, Tomlinson C, Strong C, Delehaunty K, Fronick C, Courtney B, Rock SM, Belter E, Du F, Kim K, Abbott RM, Cotton M, Levy A, Marchetto P, Ochoa K, Jackson SM, Gillam B, Chen W, Yan L, Higginbotham J, Cardenas M, Waligorski J, Applebaum E, Phelps L, Falcone J et al (2009) The B73 maize genome: complexity, diversity, and dynamics. Science 326(5956):1112–1115

Sheen J (1996) Ca^{2+}-dependent protein kinases and stress signal transduction in plants. Science 274:1900–1902

Shi B, Ni L, Zhang A, Cao J, Zhang H, Qin T, Tan M, Zhang J, Jiang M (2012) OsDMI3 is a novel component of abscisic acid signaling in the induction of antioxidant defense in leaves of rice. Mol Plant 5(6):1359–1374

Shi B, Ni L, Liu Y, Zhang A, Tan M, Jiang M (2014) OsDMI3-mediated activation of OsMPK1 regulates the activities of antioxidant enzymes in abscisic acid signalling in rice. Plant Cell Environ 37(2):341–352

Shimoda Y, Han L, Yamazaki T, Suzuki R, Hayashi M, Imaizumi-Anraku H (2012) rhizobial and fungal symbioses show different requirements for calmodulin binding to calcium calmodulin-dependent protein kinase in *Lotus japonicas*. Plant Cell 24(1):304–321

Shkryl YN, Veremeichik GN, Bulgakov VP, Zhuravlev YN (2011) Induction of anthraquinone biosynthesis in *Rubia cordifolia* cells by heterologous expression of a calcium-dependent protein kinase gene. Biotechnol Bioeng 108(7):1734–1738

Song WY, Zhang ZB, Shao HB, Guo XL, Cao HX, Zhao HB, Fu ZY, Hu XJ (2008) Relationship between calcium decoding elements and plant abiotic-stress resistance. Int J Biol Sci 4(2):116–125

Stael S, Bayer RG, Mehlmer N, Teige M (2011) Protein N-acylation overrides differing targeting signals. FEBS Lett 585(3):517–522

Staxen II, Pical C, Montgomery LT, Gray JE, Hetherington AM, McAinsh MR (1999) Abscisic acid induces oscillations in guard-cell cytosolic free calcium that involve phosphoinositide-specific phospholipase C. Proc Natl Acad Sci U S A 96:1779–1784

Strahl T, Grafelmann B, Dannenberg J, Thorner J, Pongs O (2003) Conservation of regulatory function in calcium-binding proteins: human frequenin (neuronal calcium sensor-1) associates productively with yeast phosphatidylinositol 4-kinase isoform, Pik1. J Biol Chem 278:49589–49599

Svistoonoff S, Benabdoun FM, Nambiar-Veetil M, Imanishi L, Vaissayre V, Cesari S, Diagne N, Hocher V, de Billy F, Bonneau J, Wall L, Ykhlef N, Rosenberg C, Bogusz D, Franche C, Gherbi H (2013) The independent acquisition of plant root nitrogen-fixing symbiosis in *Fabids* recruited the same genetic pathway for nodule organogenesis. PLoS One 8(5):e64515

Syam PSR, Chelliah J (2006a) Expression and localization of calcium-dependent protein kinase isoforms in chickpea. J Plant Physiol 163(11):1135–1149

Syam PSR, Chelliah J (2006b) Heterologous expression and biochemical characterization of two calcium-dependent protein kinase isoforms CaCPK1 and CaCPK2 from chickpea. J Plant Physiol 63(11):1083–1093

Szczegielniak J, Klimecka M, Liwosz A, Ciesielski A, Kaczanowski S, Dobrowolska G, Harmon AC, Muszynska G (2005) A wound-responsive and phospholipid-regulated maize calcium-dependent protein kinase. Plant Physiol 139:1970–1983

Szczegielniak J, Borkiewicz L, Szurmak B, Lewandowska-Gnatowska E, Statkiewicz M, Klimecka M, Cieśla J, Muszyńska G (2012) Maize calcium-dependent protein kinase (*ZmCPK11*): local and systemic response to wounding, regulation by touch and components of jasmonate signaling. Physiol Plant 146(1):1–14

Takeda N, Maekawa T, Hayashi M (2012) Nuclear-localized and deregulated calcium- and calmodulin-dependent protein kinase activates rhizobial and mycorrhizal responses in *Lotus japonicas*. Plant Cell 24(2):810–822

Tang RJ, Liu H, Bao Y, Lv QD, Yang L, Zhang HX (2010) The woody plant poplar has a functionally conserved salt overly sensitive pathway in response to salinity stress. Plant Mol Biol 74(4–5):367–380

Tang RJ, Liu H, Yang Y, Yang L, Gao XS, Garcia VJ, Luan S, Zhang HX (2012) Tonoplast calcium sensors CBL2 and CBL3 control plant growth and ion homeostasis through regulating V-ATPase activity in *Arabidopsis*. Cell Res 22(12):1650–1665

Taniguchi H (1999) Protein myristoylation in protein-lipid and protein-protein interactions. Biophys Chem 82:129–137

Tillett RL, Ergül A, Albion RL, Schlauch KA, Cramer GR, Cushman JC (2011) Identification of tissue-specific, abiotic stress-responsive gene expression patterns in wine grape (*Vitis vinifera* L.) based on curation and mining of large-scale EST data sets. BMC Plant Biol 11:86

Tirichine L, Imaizumi-Anraku H, Yoshida S, Murakami Y, Madsen LH, Miwa H, Nakagawa T, Sandal N, Albrektsen AS, Kawaguchi M (2006) Deregulation of a Ca^{2+}/calmodulin-dependent kinase leads to spontaneous nodule development. Nature 441:1153–1156

Tominaga M, Harada A, Kinoshita T, Shimazaki K (2010) Biochemical characterization of calcineurin B-like-interacting protein kinase in *Vicia* guard cells. Plant Cell Physiol 51(3):408–421

Toyota M, Furuichi T, Tatsumi H, Sokabe M (2008) Cytoplasmic calcium increases in response to changes in the gravity vector in hypocotyls and petioles of *Arabidopsis* seedlings. Plant Physiol 146:505–514

Tripathi V, Parasuraman B, Laxmi A, Chattopadhyay D (2009) *CIPK6*, a CBL-interacting protein kinase is required for development and salt tolerance in plants. Plant J 58(5):778–790

Tsai TM, Chen YR, Kao TW, Tsay WS, Wu CP, Huang DD, Chen WH, Chang CC, Huang HJ (2007) *PaCDPK1*, a gene encoding calcium-dependent protein kinase from orchid, *Phalaenopsis amabilis*, is induced by cold, wounding, and pathogen challenge. Plant Cell Rep 26(10):1899–1908

Tsou PL, Lee SY, Allen NS, Winter-Sederoff H, Robertson D (2012) An ER-targeted calcium-binding peptide confers salt and drought tolerance mediated by *CIPK6* in *Arabidopsis*. Planta 235(3):539–552

Tuskan GA, Difazio S, Jansson S, Bohlmann J, Grigoriev I, Hellsten U, Putnam N, Ralph S, Rombauts S, Salamov A, Schein J, Sterck L, Aerts A, Bhalerao RR, Bhalerao RP, Blaudez D, Boerjan W, Brun A, Brunner A, Busov V, Campbell M, Carlson J, Chalot M, Chapman J, Chen GL, Cooper D, Coutinho PM, Couturier J et al (2006) The genome of black cottonwood, *Populus trichocarpa* (Torr. & Gray). Science 313(5793):1596–1604

Tuteja N, Mahajan S (2007) Further characterization of calcineurin B-like protein and its interacting partner CBL-interacting protein kinase from *Pisum sativum*. Plant Signal Behav 2(5): 358–361

Urao T, Katagiri T, Mizoguchi T, Yamaguchi-Shinozaki K, Hayashida N, Shinozaki K (1994) Two genes that encode Ca^{2+}-dependent protein kinases are induced by drought and high-salt stresses in *Arabidopsis thaliana*. Mol Gen Genet 244:331–340

Uno Y, Miguel A, Milla R, Maher E, Cushman JC (2009) Identification of proteins that interact with catalytically active calcium-dependent protein kinases from Arabidopsis. Molecular Genetics and Genomics 281, 375–390

Venu RC, Sreerekha MV, Madhav M, Kan SV, Madhan MK, Chen S, Jia Y, Meyers BC, Wang G (2013) Deep transcriptome sequencing reveals the expression of key functional and regulatory genes involved in the abiotic stress signaling pathways in rice. J Plant Biol 56(4):216–231

Vivek PJ, Tuteja N, Soniya EV (2013) CDPK1 from ginger promotes salinity and drought stress tolerance without yield penalty by improving growth and photosynthesis in Nicotiana tabacum. PLoS One 8(10):e76392

Waadt R, Schmidt LK, Lohse M, Hashimoto K, Bock R, Kudla J (2008) Multicolor bimolecular fluorescence complementation reveals simultaneous formation of alternative CBL/CIPK complexes in planta. Plant J 56(3):505–516

Walia H, Wilson C, Condamine P, Liu X, Ismail AM, Zeng L, Wanamaker SI, Mandal J, Xu J, Cui X, Close TJ (2005) Comparative transcriptional profiling of two contrasting rice genotypes under salinity stress during the vegetative growth stage. Plant Physiol 139:822–835

Wan B, Lin Y, Mou T (2007) Expression of rice Ca^{2+}-dependent protein kinases (CDPKs) genes under different environmental stresses. FEBS Lett 581:1179–1189

Wang Y, Zhang M, Ke K, Lu YT (2005) Cellular localization and biochemical characterization of a novel calcium-dependent protein kinase from tobacco. Cell Res 15(8):604–612

Wang M, Gu D, Liu T, Wang Z, Guo X, Hou W, Bai Y, Chen X, Wang G (2007) Overexpression of a putative maize calcineurin B-like protein in *Arabidopsis* confers salt tolerance. Plant Mol Biol 65(6):733–746

Wang RK, Li LL, Cao ZH, Zhao Q, Li M, Zhang LY, Hao YJ (2012a) Molecular cloning and functional characterization of a novel apple MdCIPK6L gene reveals its involvement in multiple abiotic stress tolerance in transgenic plants. Plant Mol Biol 79(1–2):123–135

Wang X, Liu Y, Yang P (2012b) Proteomic studies of the abiotic stresses response in model moss—*Physcomitrella patens*. Front Plant Sci 3:258

Webb A, McAinsh MR, Taylor JE, Hetherington AM (1996) Calcium as a second messenger in plant cells. Adv Bot Res 22:45–96

Weinl S, Kudla J (2009) The CBL-CIPK Ca^{2+}-decoding signaling network: function and perspectives. New Phytol 184:517–528

Williamson RE, Ashley CC (1982) Free Ca^{2+} and cytoplasmic streaming in the alga *Chara*. Nature 296:647–651

Witte CP, Keinath N, Dubiella U, Demoulière R, Seal A, Romeis T (2010) Tobacco calcium-dependent protein kinases are differentially phosphorylated *in vivo* as part of a kinase cascade that regulates stress response. J Biol Chem 285(13):9740–9748

Xiang Y, Huang Y, Xiong L (2007) Characterization of stress-responsive CIPK genes in rice for stress tolerance improvement. Plant Physiol 144(3):1416–1428

Xing T, Wang XJ, Malik K, Miki BL (2001) Ectopic expression of an *Arabidopsis* calmodulin-Like domain protein kinase-enhanced NADPH oxidase activity and oxidative burst in tomato protoplasts. Mol Plant Microbe Interact 14(10):1261–1264

Xu J, Li HD, Chen LQ, Wang Y, Liu LL, He L, Wu WH (2006) A protein kinase, interacting with two calcineurin B-like proteins, regulates K+ transporter AKT1 in *Arabidopsis*. Cell 125(7):1347–1360

Xu J, Tian YS, Peng RH, Xiong AS, Zhu B, Jin XF, Gao F, Fu XY, Hou XL, Yao QH (2010) AtCPK6, a functionally redundant and positive regulator involved in salt/drought stress tolerance in *Arabidopsis*. Planta 231(6):1251–1260

Yang G, Shen S, Yang S, Komatsu S (2003) OsCDPK13, a calcium-dependent protein kinase gene from rice, is induced in response to cold and gibberellin. Plant Physiol Biochem 41:369–374

Yang W, Kong Z, Omo-Ikerodah E, Xu W, Li Q, Xue Y (2008) Calcineurin B-like interacting protein kinase *OsCIPK23* functions in pollination and drought stress responses in rice (Oryza sativa L.). J Genet Genomics 35(9):531–543

Yano K, Yoshida S, Muller J, Singh S, Banba M, Vickers K, Markmann K, White C, Schuller B, Sato S (2008) CYCLOPS, a mediator of symbiotic intracellular accommodation. Proc Natl Acad Sci U S A 105:20540–20545

Ye W, Muroyama D, Munemasa S, Nakamura Y, Mori IC, Murata Y (2013) Calcium-dependent protein kinase CPK6 positively functions in induction by yeast elicitor of stomatal closure and inhibition by yeast elicitor of light-induced stomatal opening in *Arabidopsis*. Plant Physiol 163(2):591–599

Yoo B, Harmon AC (1996) Intramolecular binding contributes to the activation of CDPK, a protein kinase with a calmodulin-like domain. Biochemistry 2960(96):12029–12037

Yoon GM, Cho HS, Ha HJ, Liu JR, Lee HS (1999) Characterization of NtCDPK1, a calcium-dependent protein kinase gene in *Nicotiana tabacum*, and the activity of its encoded protein. Plant Mol Biol 39(5):991–1001

Yoon GM, Dowd PE, Gilroy S, McCubbin AG (2006) Calcium-dependent protein kinase isoforms in *Petunia* have distinct functions in pollen tube growth, including regulating polarity. Plant Cell 18:867–878

Yu XC, Zhu SY, Gao GF, Wang XJ, Zhao R, Zou KQ, Wang XF, Zhang XY, Wu FQ, Peng CC, Zhang DP (2007) Expression of a grape calcium-dependent protein kinase ACPK1 in *Arabidopsis thaliana* promotes plant growth and confers abscisic acid-hypersensitivity in germination, post germination growth, and stomatal movement. Plant Mol Biol 64(5):531–538

Zhang XS, Choi JH (2001) Molecular evolution of calmodulin-like domain protein kinases (CDPKs) in plants and protists. J Mol Evol 53(3):214–224

Zhang T, Wang Q, Chen X, Tian C, Wang X, Xing T, Li Y, Wang Y (2005) Cloning and biochemical properties of CDPK gene *OsCDPK14* from rice. J Plant Physiol 162:1149–1159

Zhang H, Yin W, Xia X (2010) Shaker-like potassium channels in *Populus*, regulated by the CBL-CIPK signal transduction pathway, increase tolerance to low-K+ stress. Plant Cell Rep 29(9):1007–1012

Zhang H, Lv F, Han X, Xia X, Yin W (2013) The calcium sensor PeCBL1, interacting with PeCIPK24/25 and PeCIPK26, regulates Na(+)/K (+) homeostasis in *Populus euphratica*. Plant Cell Rep 32(5):611–621

Zhang H, Yang B, Liu WZ, Li H, Wang L, Wang B, Deng M, Liang W, Deyholos MK, Jiang YQ (2014) Identification and characterization of CBL and CIPK gene families in canola (*Brassica napus* L.). BMC Plant Biol 14(1):8

Zhao J, Sun Z, Zheng J, Guo X, Dong Z, Huai J, Gou M, He J, Jin Y, Wang J, Wang G (2009) Cloning and characterization of a novel CBL-interacting protein kinase from maize. Plant Mol Biol 69(6):661–674

Zhao R, Sun HL, Mei C, Wang XJ, Yan L, Liu R, Zhang XF, Wang XF, Zhang DP (2011) The *Arabidopsis* Ca^{2+}-dependent protein kinase CPK12 negatively regulates abscisic acid signaling in seed germination and post-germination growth. New Phytol 192(1):61–73

Zheng H, von Mollard GF, Kovaleva V, Stevens TH, Raikhel NV (1999) The plant vesicle-associated SNARE AtVTI1a likely mediates vesicle transport from the *trans*-Golgi network to the prevacuolar compartment. Mol Biol Cell 10(7):2251–2264

Zhu JK, Liu J, Xiong L (1998) Genetic analysis of salt tolerance in *Arabidopsis*. Evidence for a critical role of potassium nutrition. Plant Cell 10(7):1181–1191

Zhu SY, Yu XC, Wang XJ, Zhao R, Li Y, Fan RC, Shang Y, Du SY, Wang XF, Wu FQ, Xu YH, Zhang XY, Zhang DP (2007) Two calcium-dependent protein kinases, CPK4 and CPK11, regulate abscisic acid signal transduction in *Arabidopsis*. Plant Cell 19:3019–3036

Zhu JH, Chen X, Chang WJ, Tian WM, Zhang ZL (2010) Molecular characterization of *HbCDPK1*, an ethephon-induced calcium-dependent protein kinase gene of *Hevea brasiliensis*. Biosci Biotechnol Biochem 74(11):2183–2188

Zou JJ, Wei FJ, Wang C, Wu JJ, Ratnasekera D, Liu WX, Wu WH (2010) *Arabidopsis* calcium-dependent protein kinase CPK10 functions in abscisic acid-and Ca^{2+}-mediated stomatal regulation in response to drought stress. Plant Physiol 154(3):1232–1243

Zuo R, Hu R, Chai G, Xu M, Qi G, Kong Y, Zhou G (2013) Genome-wide identification, classification, and expression analysis of CDPK and its closely related gene families in poplar (*Populus trichocarpa*). Mol Biol Rep 40(3):2645–2662

Chapter 6
CBL-Mediated Calcium Signaling Pathways in Higher Plants

Joo Hyuk Cho and Kyung-Nam Kim

Abstract Plants are sessile and unable to move away from unfriendly surrounding conditions. Instead, they have developed during evolution very complicated cellular signaling mechanisms in order to cope with unfavorable or harsh environments. Such signaling involves calcium ions that can serve as a ubiquitous second messenger in mediating a wide range of environmental stresses, including biotic and abiotic stresses. It is clear that plant cells have the ability to generate stimulus-specific calcium signals in the cytosol and possess numerous calcium-binding proteins with different properties, which sense and transduce the specific calcium signals to give rise to appropriate and disparate responses. In this present review, we mainly focus on the calcium signal transduction pathways mediated by calcineurin B-like (CBL) proteins, which were most recently identified and represent a unique family of calcium sensors in higher plants.

Keywords CBL • CIPK • Calcium signaling • Biotic stress • Abiotic stress

6.1 Ca^{2+} Signals and Their Sensors

Calcium ions (Ca^{2+}) in plants are implicated in mediating a wide range of extracellular stimuli such as light, pathogen attacks, and abiotic stresses such as cold, drought, and high salinity (Ehrhardt et al. 1996; Rudd and Franklin-Tong 2001; Sanders et al. 1999). It is an intriguing question, how the simple cation Ca^{2+} acts as a ubiquitous second messenger in a variety of environmental stresses and yet gives rise to appropriate plant responses? The answer to the question can be that the Ca^{2+} signals are very complex and plant cells are equipped with numerous Ca^{2+} sensors with different properties.

Ca^{2+} signals can be complex because they are constituted of not only concentration but also the temporal and spatial parameters, which include frequency, magnitude, duration, and subcellular localization of the transient Ca^{2+} increases in the cytosol (Dolmetsch et al. 1997, 1998; Evans et al. 2001; Rudd and Franklin-Tong

J.H. Cho • K.-N. Kim (✉)
Department of Molecular Biology, PERI, Sejong University, Seoul 143-747, South Korea
e-mail: knkim@sejong.ac.kr

G.K. Pandey (ed.), *Elucidation of Abiotic Stress Signaling in Plants*,
DOI 10.1007/978-1-4939-2211-6_6

2001). In order for the diverse Ca^{2+} signals to elicit stimulus-specific responses, plant cells should harbor various Ca^{2+}-binding proteins that sense and transmit changes in the Ca^{2+} parameters. In fact, plants have numerous Ca^{2+} sensors that possess the EF-hand, a Ca^{2+}-binding motif (Kawasaki et al. 1998). These Ca^{2+}-binding proteins can be classified into three major families: calmodulin (CaM) and CaM-like proteins, Ca^{2+}-dependent protein kinases (CDPKs), and calcineurin B-like (CBL) proteins.

First, the family members of CaM and CaM-like proteins, which do not have any enzymatic activity themselves, act as "sensor relays" (Luan et al. 2002; Zielinski 1998). Upon Ca^{2+} binding, they undergo conformational changes to associate with and regulate various target proteins such as NAD kinase, glutamate decarboxylase, Ca^{2+} ATPase, and protein kinases (Luan et al. 2002; Reddy et al. 2002; Snedden and Fromm 2001; Yang and Poovaiah 2003; Zielinski 1998). The *Arabidopsis* genome contains 50 CaM-like genes in addition to the six genomic loci encoding typical CaMs (Luan et al. 2002; McCormack and Braam 2003).

Second, the CDPK family members are classified as "sensor responders," because they harbor a kinase domain at the N-terminal end along with the C-terminal CaM-like Ca^{2+} sensor (Sanders et al. 2002). It is well known that Ca^{2+} can activate the kinase activity of CDPKs by binding to the CaM-like region (Chandran et al. 2006; Cheng et al. 2002; Harper et al. 2004; Hrabak et al. 2003). The *Arabidopsis* genome was predicted to carry 34 CDPK genes, which appear to possess different properties in terms of expression pattern, subcellular localization, Ca^{2+}-binding affinity, and substrate specificity (Asano et al. 2005; Harper et al. 2004; Hernandez Sebastia et al. 2004; Hrabak et al. 2003; Lee et al. 1998; Rodriguez Milla et al. 2006; White and Broadley 2003).

The third major family of EF-hand Ca^{2+} sensors is CBL proteins, which are most similar to the regulatory B subunit of the protein phosphatase calcineurin (CNB) in animals (Kudla et al. 1999; Liu and Zhu 1998). Following the initial discovery from *Arabidopsis*, many CBLs were identified from various plant species. Therefore, it is believed that the CBL proteins are ubiquitously present in the plant kingdom (Batistic and Kudla 2009). Interestingly, the CBL family is comprised of ten genes in both *Arabidopsis* and rice plants (Kolukisaoglu et al. 2004; Luan et al. 2002). In this present review, we mainly focus on the CBL-mediated Ca^{2+} signaling network in higher plants.

6.1.1 CBLs Interact with a Family of Serine/Threonine Protein Kinases Called CIPKs

The CBL members appeared to be sensor relays like CaM, and therefore they should function by interacting with and modulating their target proteins in a Ca^{2+}-dependent manner. In fact, several research groups originally demonstrated that CBLs have interaction partners which are mainly a group of serine/threonine protein kinases designated as CBL-interacting protein kinases (CIPKs) (Batistic and Kudla 2009; Cheong et al. 2007; Halfter et al. 2000; Huang et al. 2011; Kim et al. 2000;

Kolukisaoglu et al. 2004; Quan et al. 2007; Shi et al. 1999; Waadt et al. 2008). The CIPK proteins are unique to plants in that they contain a distinct regulatory domain in the C-terminus, although their N-terminal kinase domain is most related to those of the yeast SNF1 (sucrose non-fermenting 1) protein kinase and the mammalian AMP-dependent protein kinase (Shi et al. 1999).

Sequence analysis predicted that the CIPK family consists of 25 and 30 genes in the *Arabidopsis* and rice genomes, respectively (Kolukisaoglu et al. 2004). It appears that each CIPK can interact physically and specifically with more than one CBL member at different affinities through the conserved NAF (or FISL) motif present in the C-terminal non-kinase domain (Albrecht et al. 2001; Batistic and Kudla 2004; Halfter et al. 2000; Jeong et al. 2005; Kim et al. 2000; Kolukisaoglu et al. 2004; Luan et al. 2002).

6.1.1.1 CIPKs Phosphorylate Their Interacting CBLs Upon Binding, Enhancing Specificity and Activity of CBL-CIPK Complexes Toward Their Target Proteins

CBL1 formed a complex with CIPK1 only in the presence of Ca^{2+} (Shi et al. 1999). In contrast, other CBL members, such as CBL2 and CBL4 (also known as SOS3), did not display such Ca^{2+} dependency: CBL2 and CBL4 interacted regardless of Ca^{2+} with CIPK14 and CIPK24 (also known as SOS2), respectively (Akaboshi et al. 2008; Halfter et al. 2000). These findings together suggest that the Ca^{2+} requirement for the complex formation can vary among the participating CBL and CIPK members. Nevertheless, it is very important to note that in both cases Ca^{2+} is still necessary for the CBL-CIPK complex to be active and phosphorylate substrates (Halfter et al. 2000) (Fig. 6.1).

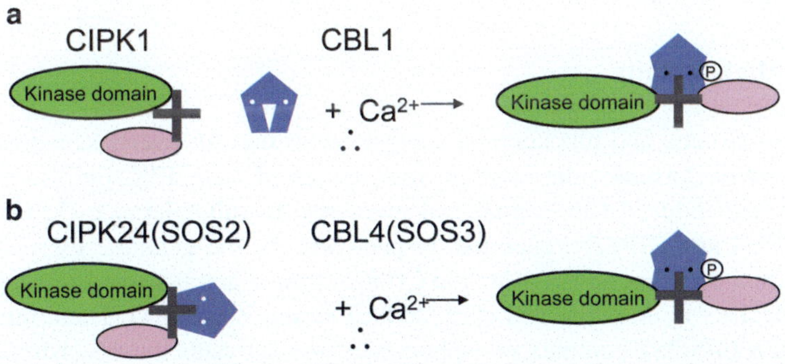

Fig. 6.1 Two hypothetical models for the activation of CIPKs by Ca^{2+}-bound CBLs. (**a**) Ca^{2+}-dependent interaction. Upon Ca^{2+} binding, CBL1 undergoes conformational change and can associate with the NAF motif (*cross*) of CIPK1. The association disrupts the intramolecular interaction of CIPK1, thereby activating the kinase activity. (**b**) Ca^{2+}-independent interaction. CBL4 can interact with the NAF motif of CIPK24 in the absence of Ca^{2+}. However, the interaction itself is not enough to block the auto-inhibitory interaction of CIPK24. CBL4 requires Ca^{2+} to activate the enzyme activity. In both cases, CBLs are phosphorylated by the interacting CIPKs and thereby allow CIPKs to have full kinase activity toward their substrate

Deletion analysis with CIPK24 provided a molecular mechanism by which the CBL-CIPK interaction results in the activation of the CIPK enzyme activity (Guo et al. 2001). The results of this study indicated that the NAF-carrying C-terminal domain serves as an auto-inhibitory module by interacting intramolecularly with the kinase domain and thereby blocks the active site from binding with a substrate. It seems that upon interaction with a Ca^{2+}-bound CBL partner, CIPKs undergo a conformational change, which disrupts the auto-inhibitory effect and allows the kinase domain to gain the phosphorylation activity.

A couple of recent reports provided a new insight into the importance of CBL-CIPK complexes (Du et al. 2011; Hashimoto et al. 2012). According to their finding, CBLs are phosphorylated at a conserved serine residue within the C-terminus by their interacting CIPKs in a CBL-CIPK interaction-dependent manner. Although the phosphorylation did not appear to influence the stability or localization of the calcium sensor proteins, it appeared to be absolutely required for the CBL-CIPK complexes to phosphorylate their substrate proteins (Fig. 6.1).

6.1.2 *Multiple CBL-CIPK Complexes Are Involved in Mediating a Variety of Environmental Stresses*

Extensive studies with *Arabidopsis* mutants have demonstrated that the CBL-CIPK complexes are involved in mediating Ca^{2+} signals elicited by a variety of environmental stresses such as high salinity, low K^+ concentrations, drought, cold, plant hormones, high pH, low nitrate condition, and pathogen attacks. It appears that each pair of the CBL-CIPK complexes plays a specific role in relaying different signals (Fig. 6.2). In addition, it is also notable that multiple pairs of the CBL-CIPK networks mediate the same stress in some cases.

6.1.2.1 High Salinity

A series of genetic and biochemical analyses with the salt overly sensitive (SOS) mutants provided a molecular mechanism by which a CBL-CIPK complex mediates the salt stress-induced Ca^{2+} signal and gives rise to salt tolerance (Liu and Zhu 1998). CBL4 and CIPK24, encoded, respectively, by the genomic loci *SOS3* and *SOS2*, form a complex and activate the CIPK24 kinase activity in a Ca^{2+}-dependent manner (Halfter et al. 2000; Liu et al. 2000; Liu and Zhu 1998). Then, the CBL4/SOS3-CIPK24/SOS2 complex phosphorylates and activates the downstream component SOS1, a Na^+/H^+ antiporter (Shi et al. 2000). The activated SOS1 protein functions to get rid of excess Na^+ in plant cells, thereby conferring salt tolerance (Qiu et al. 2002; Quintero et al. 2002).

Another CBL family member, CBL10, was later included in the salt tolerance pathway. Like CBL4/SOS3, CBL10 also associates with and activates the kinase activity of CIPK24/SOS2 (Kim et al. 2007; Quan et al. 2007). Although both CBL4/

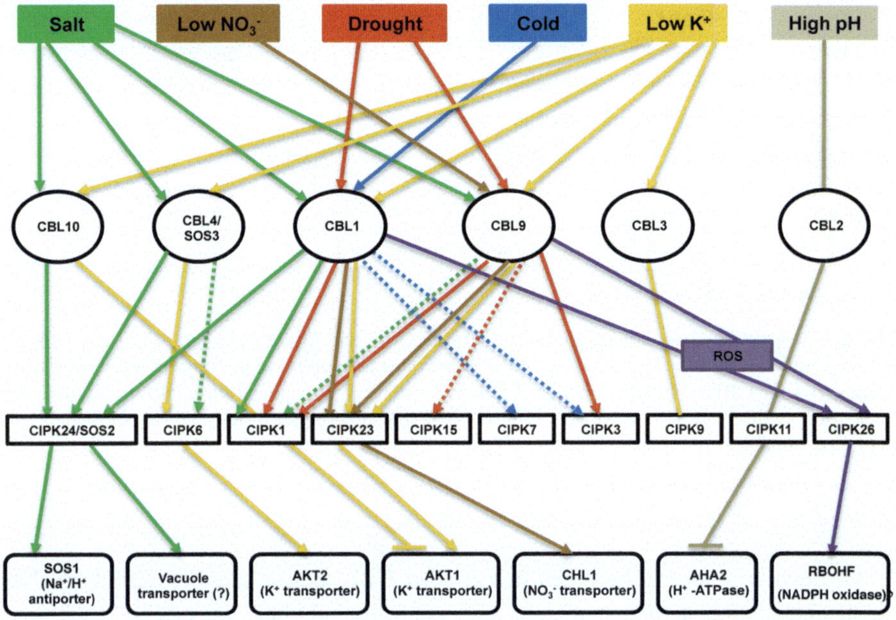

Fig. 6.2 The CBL-CIPK networks mediating various stress responses. The signaling network was generated using the results of genetic and biochemical analyses. The different *line colors* indicate different stress signaling pathways. The *broken lines* represent signaling pathways, which need further solid evidence to be investigated

SOS3 and CBL10 proteins are involved in mediating salt tolerance, they do differ in executing the function due to their distinct expression patterns and subcellular localizations.

CBL4/SOS3 is expressed mainly in the roots and localized at the plasma membrane. Therefore, CIPK24/SOS2 is recruited to the plasma membrane of the root cells where it phosphorylates the Na^+/H^+ antiporter SOS1, thereby enhancing the Na^+ efflux rate (Fig. 6.3a). In contrast, CBL10 is expressed predominantly in the shoots and leaves, and it is localized at the vacuolar membrane (tonoplast). The knockout *Arabidopsis* mutant lacking the CBL10 activity (*cbl10*) showed the salt-sensitive phenotype specifically in the leaves or shoots under the high salt conditions (Kim et al. 2007). Interestingly, the *cbl10* mutant plants accumulated much less Na^+ than the wild type under either normal or high salt conditions. This unique feature of the *cbl10* mutant plants, along with the tonoplast localization of the CBL10 protein, suggests that CBL10 is required for the sequestration of Na^+ into the vacuole. It is likely that CIPK24/SOS2 recruited by CBL10 to the tonoplast may phosphorylate and activate an as-yet-unknown Na^+ channel or transporter, which is tonoplast bound and plays a role in transporting cytosolic Na^+ into the vacuole (Fig. 6.3b).

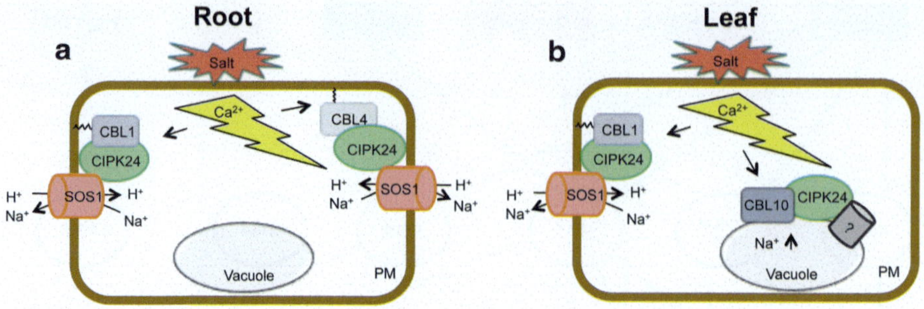

Fig. 6.3 Hypothetical model of CBL-CIPK function in salt stress response. (**a**) In the roots, CIPK24 activates the Na^+/H^+ antiporter SOS1 at the plasma membrane by forming a complex with either CBL1 or CBL4. (**b**) In the leaves, CIPK24 regulates Na^+ sequestration into the vacuole by interacting with the tonoplast-bound CBL10. The *question mark* represents a component of unknown identity

Meanwhile, it should be noted that another CBL family member, CBL1, is also involved in plant response to salt stress. The *cbl1* mutant plants showed less tolerance to the salt stress (Albrecht et al. 2003; Cheong et al. 2003). Like CBL4/SOS3, CBL1 is present in the plasma membrane and interacts with CIPK24/SOS2. Therefore, it is very likely that CBL1 also confers salt tolerance via CIPK24/SOS2-mediated SOS1 regulation (D'Angelo et al. 2006). Unlike CBL4/SOS3, however, CBL1 expression is not restricted in the roots. CBL1 is actually expressed in both root and shoot tissues. Such expression pattern implies that Na^+ extrusion mediated by the CIPK24/SOS2-SOS1 machinery may also occur in the shoots. In summary, CIPK24/SOS2 can regulate two different processes depending on its interacting CBL partners. Through the association with CBL1 or CBL4/SOS3, CIPK24/SOS2 facilitates Na^+ export across the plasma membrane by activating the Na^+/H^+ antiporter SOS1. Furthermore, CIPK24/SOS2 can also play a role in sequestrating Na^+ into the vacuole by forming a complex with CBL10.

Meanwhile, an *Arabidopsis* mutant lacking CIPK6 activity was reported to be more sensitive to salt stress compared to the wild type, suggesting involvement of CIPK6 in salt tolerance (Tripathi et al. 2009). Because CIPK6 was known to associate with CBL4/SOS3 in the yeast two-hybrid system (Kim et al. 2000), it was speculated that CBL4/SOS3 may also target CIPK6 in vivo besides CIPK24/SOS2 to mediate salt tolerance. In fact, Held et al. (2011) provided solid evidence that CBL4 and CIPK6 indeed associate with each other in plant cells. They worked together in a calcium-dependent manner to activate the *Arabidopsis* K^+ channel AKT2 by translocating the AKT2 protein from the endoplasmic reticulum to the plasma membrane. In addition, CIPK16 appeared to be involved in salt tolerance according to the *Arabidopsis* transgenic lines with altered expression levels of the gene (Roy et al. 2013). Therefore, it is clear now that multiple members of the CBL and CIPK families are involved in plant response to high salt stress.

6.1.2.2 Low K^+ Concentrations

Two independent research groups provided solid evidence demonstrating that CBL1 and CBL9 calcium sensors and their target CIPK23 are involved in plant response to low K^+ conditions (Li et al. 2006; Xu et al. 2006). According to their works, the *Arabidopsis* mutant plants with the disrupted CIPK23 gene, *cipk23*, were more sensitive to low K^+ concentrations than the wild-type plants. The similar phenotype was also observed with the *cbl1cbl9* double mutant plants, but not with the *cbl1* or *cbl9* single mutants, which implies that the two CBL members have redundant functions in the low K^+ signaling pathways.

Biochemical assays demonstrated that CIPK23 could be activated by the association with either CBL1 or CBL9, thereby phosphorylating the C-terminal region of K^+ transporter AKT1 in the plasma membrane. Based on these findings, a model was proposed for the K^+ deficiency response in plants: CBL1 and CBL9 detect the Ca^{2+} signatures elicited by low K^+ conditions. Upon binding with Ca^{2+}, they recruit CIPK23 to the plasma membrane and promote the kinase activity. The activated CIPK23 then phosphorylates the C-terminal region of the AKT1 transporter, thereby enhancing its K^+-transporting activity.

It appears that CIPK9 may also play a role in low K^+ stress response. Initially, Pandey et al. (2007) reported that the *Arabidopsis* mutants lacking CIPK9 displayed a hypersensitive phenotype to low K^+ media compared with the wild-type plants. On the contrary, more recent study showed that disruption of CIPK9 rendered the mutant plants tolerant to low K^+ conditions (Liu et al. 2013). Furthermore, transgenic plants overexpressing CIPK9 were more sensitive to low K^+ stress than the wild type. At present, therefore, it is very hard to tell whether CIPK9 serves as a positive or negative regulator in tolerance to low K^+ conditions due to these controversial findings. Anyway, it is clear that CIPK9 is somehow implicated in plant response to low K^+ stress. It is also noteworthy that the CBL3 knockout mutant plants showed the low K^+ -sensitive phenotype like the CIPK9 null mutants. In addition, overexpression of CBL3, an upstream regulator of CIPK9, resulted in the low K^+ -sensitive phenotype similar to the CIPK9 null mutants (Liu et al. 2013).

Meanwhile, the CBL10-overexpressing transgenic plants also exhibited a sensitive phenotype to low K^+ conditions and accumulated significantly reduced K^+ contents compared with the wild type. Moreover, in vivo interaction assays revealed that CBL10 can directly interact with AKT1, and the yeast two-hybrid completion assay further demonstrated that CBL10 could compete with CIPK23 for binding to AKT1 (Ren et al. 2013). These findings indicated that CBL10 might directly regulate AKT1 activity in plant cells without the involvement of CIPK members.

6.1.2.3 Drought

Loss-of-function *Arabidopsis* mutants lacking CBL1, CBL9, or CIPK1 were found to be more sensitive to drought and osmotic stress than the wild-type plants (Pandey et al. 2004; D'Angelo et al. 2006). Moreover, several lines of evidence demonstrated that CIPK1 interacts with either CBL1 or its closest isoform CBL9 at the plasma membrane (Shi et al. 1999; Kim et al. 2000; Albrecht et al. 2001; D'Angelo et al. 2006). These data together suggested that both CBL1 and CBL9 could target CIPK1 to form two distinct complexes, CBL1-CIPK1 and CBL9-CIPK1, thereby mediating the stress responses.

6.1.2.4 Cold

The *cbl1* mutant plants exhibited more tolerance to cold stress and displayed enhanced cold induction of stress genes such as *RD29A* and *Kin1* (Albrecht et al. 2003; Cheong et al. 2003). These results indicate that CBL1 plays a negative role in regulating plant cold response. Recently, a study showed that CBL1 can interact with CIPK7 whose expression is induced by cold stress (Huang et al. 2011). It suggests that CBL1 may target CIPK7 to mediate cold response in plants.

Meanwhile, expression patterns of *RD29A* and *Kin1/Kin2* genes were altered in the *cipk3* mutant plants in response to cold stress (Kim et al. 2003), implying the involvement of CIPK3 in cold response. Because expression of the stress genes is independent of endogenous ABA production (Shinozaki and Yamaguchi-Shinozaki 2000; Thomashow 1999), these findings suggest that CIPK3 can act as a cross-talk node between the ABA-dependent and ABA-independent pathways in stress responses.

6.1.2.5 Phytohormone Abscisic Acid

Abscisic acid (ABA) is a phytohormone involved in plant response to abiotic stresses such as drought and high salt (Yamaguchi-Shinozaki and Shinozaki 2006; Zhu 2002). A specific Ca^{2+} signature is known to be implicated in an early step of the ABA signaling pathways (Allen et al. 2000, 2001; Leung and Giraudat 1998), suggesting involvement of Ca^{2+} sensors in this signaling pathway. In fact, several lines of recent evidence suggested that the CBL-CIPK pathways should play a role in the ABA signaling.

The *Arabidopsis* mutant plants with disrupted CBL9 (*cbl9*) were hypersensitive to ABA in the early developmental stages such as seed germination and post-germination seedling growth (Pandey et al. 2004). The *cbl9* mutant seedlings also accumulated much higher levels of ABA than the wild-type plants under the osmotic stress conditions caused by mannitol and salt. Furthermore, the expression levels of the genes involved in ABA signaling, such as *ABA-INSENSITIVE* 4 and 5, were increased to a greater extent in the *cbl9* mutant plants under the osmotic stress conditions or exogenous ABA (Pandey et al. 2004). These results strongly indicate that CBL9 plays an important role in both the biosynthesis and sensitivity of ABA in *Arabidopsis*. It was demonstrated that CBL9 can form a specific complex with CIPK3 to act together in regulating the ABA responses (Pandey et al. 2008). The CIPK3 knockout *Arabidopsis* mutant (*cipk3*) plants were more sensitive to exogenous ABA during seed germination, and they also expressed significantly lower levels of ABA-induced genes such as *RD22* and *RD29B* (Kim et al. 2003).

CBL1 is the most similar isoform of CBL9 in *Arabidopsis*. The *Arabidopsis* plants lacking the CBL1 activity (*cbl1*) did not show changes in the ABA responsiveness, although they exhibited less tolerance to drought and salt stress (Albrecht et al. 2003; Cheong et al. 2003). These findings indicated that CBL1 is not implicated in the ABA signaling processes unlike CBL9. Meanwhile, it is interesting to note that CIPK1, which can interact with CBL1 and CBL9, mediates ABA responses as well as osmotic stress responses such as drought and salt. Loss of CIPK1 rendered the plants hypersensitive to the osmotic stresses, and it also impaired the ABA responsiveness (D'Angelo et al. 2006). Based on these findings, it is believed that CIPK1 can mediate ABA-independent and ABA-dependent stress responses by forming an alternative complex with either CBL1 or CBL9.

Silencing of the *Arabidopsis* CIPK15 gene (designated PKS3 in the paper) by an RNA interference technique rendered the plants hypersensitive to ABA in seed germination, seedling growth, stomatal closing, and gene expression (Guo et al. 2002). These findings implicated CIPK15 as a global negative regulator of ABA responses. Because yeast two-hybrid assays demonstrated that CIPK15 associates with a group of CBL family members including CBL1, CBL2, CBL3, CBL5, CBL8, and CBL9 (Kim's unpublished data), it is likely that CBL9 may also form a complex with CIPK15 to mediate the ABA responses. However, further genetic analysis is required to demonstrate this speculation. It is also noteworthy that loss of CBL2 function renders plants hypersensitive to ABA during seed germination (Batistic et al. 2012).

6.1.2.6 High pH

The *Arabidopsis* mutant plants lacking the functional protein kinase CIPK11 (designated PKS5 in the paper) displayed the enhanced tolerance to high external pH than the wild-type plants because of the increased rate of H^+ secretion to the extracellular space (Fuglsang et al. 2007). Further studies revealed that CIPK11 phosphorylates the Ser-931 residue in the C-terminal regulatory domain of the plasma membrane proton pump (PM H^+-ATPase 2, AHA2), which results in the inhibition of AHA2 activity by preventing its association with an activating 14-3-3 protein (Fuglsang et al. 2007). Although CBL2 was originally proposed to be responsible for activating CIPK11 in the presence of cytosolic Ca^{2+} signals elicited by external high pH conditions, the involvement of CBL2 was later ruled out due to its exclusive localization at the tonoplast (Batistic et al. 2008). Therefore, it needs to be discovered which member of the CBL family is responsible for Ca^{2+}-dependent modulation of the AHA2 activity *in planta*.

6.1.2.7 Low Nitrate Concentrations

Nitrogen is an essential macronutrient for plant growth and development. For most plants, the major nitrogen source is nitrate (NO_3^-) that also acts as a signaling molecule. It has been well known that nitrate can rapidly induce gene expression of nitrate assimilatory enzymes and nitrate transporter, such as CHL1 and NRT2.1, which is referred to as the primary nitrate response (Wang et al. 2003). However, little was known about the molecular components underlying the nitrate signaling mechanism. Recently, several lines of evidence indicated that CBL-CIPK networks are implicated in the nitrate signaling in plants. Initially, Hu et al. (2009) showed that CIPK8 is responsible for nitrate sensing and acts as a positive regulator in the primary nitrate response. Later, more solid and convincing evidence was provided by Ho et al. (2009). They demonstrated that, in response to low nitrate concentrations, CIPK23 can phosphorylate the amino acid residue T101 of CHL1, resulting in shift from a low- to high-affinity nitrate transporter. Interestingly, co-expression of CBL9, but not CBL1, with CIPK23 increased the CIPK23-mediated phosphorylation of T101 (Ho and Frommer 2014). Together, these findings strongly suggest that the CBL9-CIPK23 complex plays a critical role in regulating the nitrate transporter CHL1 under low nitrate conditions.

6.1.2.8 Biotic Stress

It is well known that pathogen attacks also elicit cytosolic Ca^{2+} signals in plant cells (Rudd and Franklin-Tong 2001). However, it was not clear until recently whether the CBL-CIPK signaling pathway is involved in mediating the Ca^{2+} signals elicited by the biotic stress. A recent study with tomato (*Solanum lycopersicum*) CBL and CIPK genes provided convincing evidence that a member of the CBL-CIPK complexes indeed plays a role in plant response to pathogen attacks (de la Torre et al. 2013). According to the results, tomato CBL10 associated with and activated tomato CIPK6 in a Ca^{2+}-dependent manner, and their interaction was required for programmed cell death elicited not only by *Pseudomonas syringae* effectors, AvrPto or AvrPtoB, but also by other pathogens such as virus, oomycete, and nematode.

6.1.2.9 Production of Reactive Oxygen Species

Reactive oxygen species (ROS) produced by plant NADPH oxidases, known as respiratory burst oxidase homologues (RBOHs), are also involved in biotic and abiotic stress responses (Marino et al. 2012). Recently, a yeast two-hybrid screen revealed CIPK26 as an interaction partner of one of the RBOH family members in *Arabidopsis*, RBOHF (Kimura et al. 2013). Further analyses demonstrated that CIPK26 associates with RBOHF in plant cells and phosphorylates RBOHF in vitro. In addition, expression of CIPK26 together with either CBL1 or CBL9 in vivo drastically increased ROS production by RBOHF (Drerup et al. 2013). Taken together, these findings strongly suggested that Ca^{2+} signaling is directly connected to ROS signaling in plants via CBL1/CBL9-CIPK26 complexes.

6.1.2.10 Other Interactors of CBLs

It is important to note a report describing that CBL3 can interact with and modulate the *Arabidopsis* 5′-methylthioadenosine nucleosidase (AtMTAN) in a

Ca^{2+}-dependent manner (Oh et al. 2008). Since AtMTAN associates only with CBL3 but not with other CBL family members such as CBL1 and CBL4, it is reasonable to speculate that other CBL members may also have as-yet unidentified interaction partners, which do not belong to the CIPK family. In addition, CBL10 was shown to directly interact with AKT1 and negatively modulate its activity (Ren et al. 2013). These findings strongly suggested that CBLs can have diverse interaction partners other than the CIPK family members. Such target diversity confers an additional level of complexity on the CBL-mediated Ca^{2+} signaling pathways, thereby allowing the CBL as Ca^{2+} sensors to regulate a wider range of cellular and physiological processes in higher plants. Anyway, the finding above clearly indicates that the CBL family can relay the Ca^{2+} signals in more diverse ways than currently known. Extensive further studies will be necessary to fully understand the CBL-mediated Ca^{2+} signaling networks in plants.

6.2 Future Perspectives

Identification of the novel Ca^{2+} sensors (CBLs) and their interaction partners (CIPKs) has certainly enhanced our knowledge on the Ca^{2+} signal transduction pathways in higher plants. Furthermore, the findings of the in vivo substrates phosphorylated by the CBL4/SOS3-CIPK24/SOS2 and CBL1 (CBL9)-CIPK23 complexes have provided molecular mechanisms by which plants respond to salt and low K^+ stresses, respectively (Qiu et al. 2002; Quintero et al. 2002; (Li et al. 2006; Xu et al. 2006). Currently, only three CIPK substrates are known, all of which happen to be membrane-bound proteins, SOS1, AKT1, and AHA2 (Fuglsang et al. 2007; Li et al. 2006; Qiu et al. 2002; Quintero et al. 2002; Xu et al. 2006). Therefore, it becomes critical to identify as-yet-unknown in vivo substrates or target molecules for the CIPK family members in order to gain further insight into the CBL-CIPK-mediated Ca^{2+} signaling network.

Analyses of the *CBL* and *CIPK* knockout mutants have indicated that the CBL-CIPK Ca^{2+} signaling pathways are involved in regulating expression of stress genes such as *RD29A/B*, *Kin1/2*, and *DREB1A/B* (Albrecht et al. 2003; Cheong et al. 2003; D'Angelo et al. 2006; Kim et al. 2003; Pandey et al. 2004). However, little is known about how the CBL-CIPK complexes exert their influence on the transcriptional controls. In the cases of the CBL1 and CBL9 proteins, which recruit interacting CIPK partners to the plasma membrane (Albrecht et al. 2003; Cheong et al. 2003; D'Angelo et al. 2006; Pandey et al. 2004), we can speculate the existence of messenger molecules delivering information from the plasma membrane-bound CBL-CIPK complexes into the nucleus, where gene regulation occurs. In this regard, ECT1 can be considered as a strong candidate protein executing such function, because it can associate with CIPK1 in the cytoplasm and can be translocated into the nucleus (Ok et al. 2005). Further analyses will tell whether ECT1 is actually involved in relaying a signal from the CBL1-CIPK1 complex.

Acknowledgements This research was supported by the Basic Science Research Program through the National Research Foundation of Korea (NRF) funded by the Ministry of Education, Science and Technology (2010-0008839). This work was also supported by the National Research Foundation of Korea (NRF) grant funded by the Korean government (MEST) (No. 2011-0029568).

Disclosure This is an updated report of the earlier review article by Kyung-Nam Kim, "Stress responses mediated by the CBL calcium sensors in plants," Plant Biotechnology Journal Reports 2012 (7:1-8).

References

Akaboshi M, Hashimoto H, Ishida H, Saijo S, Koizumi N, Sato M, Shimizu T (2008) The crystal structure of plant-specific calcium-binding protein AtCBL2 in complex with the regulatory domain of AtCIPK14. J Mol Biol 377:246–257

Albrecht V, Ritz O, Linder S, Harter K, Kudla J (2001) The NAF domain defines a novel protein-protein interaction module conserved in Ca^{2+}-regulated kinases. EMBO J 20:1051–1063

Albrecht V, Weinl S, Blazevic D, D'Angelo C, Batistic O, Kolukisaoglu U, Bock R, Schulz B, Harter K, Kudla J (2003) The calcium sensor CBL1 integrates plant responses to abiotic stresses. Plant J 36:457–470

Allen GJ, Chu SP, Schumacher K, Shimazaki CT, Vafeados D, Kemper A, Hawke SD, Tallman G, Tsien RY, Harper JF, Chory J, Schroeder JI (2000) Alteration of stimulus-specific guard cell calcium oscillations and stomatal closing in Arabidopsis det3 mutant. Science 289:2338–2342

Allen GJ, Chu SP, Harrington CL, Schumacher K, Hoffmann T, Tang YY, Grill E, Schroeder JI (2001) A defined range of guard cell calcium oscillation parameters encodes stomatal movements. Nature 411:1053–1057

Asano T, Tanaka N, Yang G, Hayashi N, Komatsu S (2005) Genome-wide identification of the rice calcium-dependent protein kinase and its closely related kinase gene families: comprehensive analysis of the CDPKs gene family in rice. Plant Cell Physiol 46:356–366

Batistic O, Kudla J (2004) Integration and channeling of calcium signaling through the CBL calcium sensor/CIPK protein kinase network. Planta 219:915–924

Batistic O, Kudla J (2009) Plant calcineurin B-like proteins and their interacting protein kinases. Biochim Biophys Acta 1793:985–992

Batistic O, Sorek N, Schultke S, Yalovsky S, Kudla J (2008) Dual fatty acyl modification determines the localization and plasma membrane targeting of CBL/CIPK Ca^{2+} signaling complexes in Arabidopsis. Plant Cell 20:1346–1362

Batistic O, Rehers M, Akerman A, Schlucking K, Steinhorst L, Yalovsky S, Kudla J (2012) S-acylation-dependent association of the calcium sensor CBL2 with the vacuolar membrane is essential for proper abscisic acid responses. Cell Res 22:1155–1168

Chandran V, Stollar EJ, Lindorff-Larsen K, Harper JF, Chazin WJ, Dobson CM, Luisi BF, Christodoulou J (2006) Structure of the regulatory apparatus of a calcium-dependent protein kinase (CDPK): a novel mode of calmodulin-target recognition. J Mol Biol 357:400–410

Cheng SH, Willmann MR, Chen HC, Sheen J (2002) Calcium signaling through protein kinases. The *Arabidopsis* calcium-dependent protein kinase gene family. Plant Physiol 129:469–485

Cheong YH, Kim KN, Pandey GK, Gupta R, Grant JJ, Luan S (2003) CBL1, a calcium sensor that differentially regulates salt, drought, and cold responses in *Arabidopsis*. Plant Cell 15:1833–1845

Cheong YH, Pandey GK, Grant JJ, Batistic O, Li L, Kim BG, Lee SC, Kudla J, Luan S (2007) Two calcineurin B-like calcium sensors, interacting with protein kinase CIPK23, regulate leaf transpiration and root potassium uptake in *Arabidopsis*. Plant J 52:223–239

D'Angelo C, Weinl S, Batistic O, Pandey GK, Cheong YH, Schultke S, Albrecht V, Ehlert B, Schulz B, Harter K, Luan S, Bock R, Kudla J (2006) Alternative complex formation of the Ca-regulated protein kinase CIPK1 controls abscisic acid-dependent and independent stress responses in *Arabidopsis*. Plant J 48:857–872

de la Torre F, Gutierrez-Beltran E, Pareja-Jaime Y, Chakravarthy S, Martin GB, del Pozo O (2013) The tomato calcium sensor Cbl10 and its interacting protein kinase Cipk6 define a signaling pathway in plant immunity. Plant Cell 25:2748–2764

Dolmetsch RE, Lewis RS, Goodnow CC, Healy JI (1997) Differential activation of transcription factors induced by Ca^{2+} response amplitude and duration. Nature 386:855–858

Dolmetsch RE, Xu K, Lewis RS (1998) Calcium oscillations increase the efficiency and specificity of gene expression. Nature 392:933–936

Drerup MM, Schlucking K, Hashimoto K, Manishankar P, Steinhorst L, Kuchitsu K, Kudla J (2013) The calcineurin B-like calcium sensors CBL1 and CBL9 together with their interacting protein kinase CIPK26 regulate the Arabidopsis NADPH oxidase RBOHF. Mol Plant 6:559–569

Du W, Lin H, Chen S, Wu Y, Zhang J, Fuglsang AT, Palmgren MG, Wu W, Guo Y (2011) Phosphorylation of SOS3-like calcium-binding proteins by their interacting SOS2-like protein kinases is a common regulatory mechanism in *Arabidopsis*. Plant Physiol 156:2235–2243

Ehrhardt DW, Wais R, Long SR (1996) Calcium spiking in plant root hairs responding to Rhizobium nodulation signals. Cell 85:673–681

Evans NH, McAinsh MR, Hetherington AM (2001) Calcium oscillations in higher plants. Curr Opin Plant Biol 4:415–420

Fuglsang AT, Guo Y, Cuin TA, Qiu Q, Song C, Kristiansen KA, Bych K, Schulz A, Shabala S, Schumaker KS, Palmgren MG, Zhu JK (2007) Arabidopsis protein kinase PKS5 inhibits the plasma membrane H^+-ATPase by preventing interaction with 14-3-3 protein. Plant Cell 19:1617–1634

Guo Y, Halfter U, Ishitani M, Zhu JK (2001) Molecular characterization of functional domains in the protein kinase SOS2 that is required for plant salt tolerance. Plant Cell 13:1383–1400

Guo Y, Xiong L, Song CP, Gong D, Halfter U, Zhu JK (2002) A calcium sensor and its interacting protein kinase are global regulators of abscisic acid signaling in *Arabidopsis*. Dev Cell 3:233–244

Halfter U, Ishitani M, Zhu JK (2000) The *Arabidopsis* SOS2 protein kinase physically interacts with and is activated by the calcium-binding protein SOS3. Proc Natl Acad Sci U S A 97:3735–3740

Harper JF, Breton G, Harmon A (2004) Decoding Ca^{2+} signals through plant protein kinases. Annu Rev Plant Biol 55:263–288

Hashimoto K, Eckert C, Anschutz U, Scholz M, Held K, Waadt R, Reyer A, Hippler M, Becker D, Kudla J (2012) Phosphorylation of calcineurin B-like (CBL) calcium sensor proteins by their CBL-interacting protein kinases (CIPKs) is required for full activity of CBL-CIPK complexes toward their target proteins. J Biol Chem 287:7956–7968

Held K, Pascaud F, Eckert C, Gajdanowicz P, Hashimoto K, Corratge-Faillie C, Offenborn JN, Lacombe B, Dreyer I, Thibaud JB, Kudla J (2011) Calcium-dependent modulation and plasma membrane targeting of the AKT2 potassium channel by the CBL4/CIPK6 calcium sensor/protein kinase complex. Cell Res 21:1116–1130

Hernandez Sebastia C, Hardin SC, Clouse SD, Kieber JJ, Huber SC (2004) Identification of a new motif for CDPK phosphorylation in vitro that suggests ACC synthase may be a CDPK substrate. Arch Biochem Biophys 428:81–91

Ho CH, Frommer WB (2014) Fluorescent sensors for activity and regulation of the nitrate transceptor CHL1/NRT1.1 and oligopeptide transporters. eLife 3:e01917

Ho CH, Lin SH, Hu HC, Tsay YF (2009) CHL1 functions as a nitrate sensor in plants. Cell 138:1184–1194

Hrabak EM, Chan CW, Gribskov M, Harper JF, Choi JH, Halford N, Kudla J, Luan S, Nimmo HG, Sussman MR, Thomas M, Walker-Simmons K, Zhu JK, Harmon AC (2003) The *Arabidopsis* CDPK-SnRK superfamily of protein kinases. Plant Physiol 132:666–680

Hu HC, Wang YY, Tsay YF (2009) AtCIPK8, a CBL-interacting protein kinase, regulates the low-affinity phase of the primary nitrate response. Plant Journal 57:264–278

Huang CL, Ding S, Zhang H, Du H, An LZ (2011) CIPK7 is involved in cold response by interacting with CBL1 in *Arabidopsis thaliana*. Plant Sci 181:57–64

Jeong HJ, Jwa NS, Kim KN (2005) Identification and characterization of protein kinase that interact with the CBL3 calcium sensor in *Arabidopsis*. Plant Sci 169:1125–1135

Kawasaki H, Nakayama S, Kretsinger RH (1998) Classification and evolution of EF-hand proteins. Biometals 11:277–295

Kim KN, Cheong YH, Gupta R, Luan S (2000) Interaction specificity of *Arabidopsis* calcineurin B-like calcium sensors and their target kinases. Plant Physiol 124:1844–1853

Kim KN, Cheong YH, Grant JJ, Pandey GK, Luan S (2003) CIPK3, a calcium sensor-associated protein kinase that regulates abscisic acid and cold signal transduction in *Arabidopsis*. Plant Cell 15:411–423

Kim BG, Waadt R, Cheong YH, Pandey GK, Dominguez-Solis JR, Schultke S, Lee SC, Kudla J, Luan S (2007) The calcium sensor CBL10 mediates salt tolerance by regulating ion homeostasis in *Arabidopsis*. Plant J 52:473–484

Kimura S, Kawarazaki T, Nibori H, Michikawa M, Imai A, Kaya H, Kuchitsu K (2013) The CBL-interacting protein kinase CIPK26 is a novel interactor of Arabidopsis NADPH oxidase AtRbohF that negatively modulates its ROS-producing activity in a heterologous expression system. J Biochem 153:191–195

Kolukisaoglu U, Weinl S, Blazevic D, Batistic O, Kudla J (2004) Calcium sensors and their interacting protein kinases: genomics of the *Arabidopsis* and rice CBL-CIPK signaling networks. Plant Physiol 134:43–58

Kudla J, Xu Q, Harter K, Gruissem W, Luan S (1999) Genes for calcineurin B-like proteins in *Arabidopsis* are differentially regulated by stress signals. Proc Natl Acad Sci U S A 96:4718–4723

Lee JY, Yoo BC, Harmon AC (1998) Kinetic and calcium-binding properties of three calcium-dependent protein kinase isoenzymes from soybean. Biochemistry 37:6801–6809

Leung J, Giraudat J (1998) Abscisic acid signal transduction. Annu Rev Plant Physiol Plant Mol Biol 49:199–222

Li L, Kim BG, Cheong YH, Pandey GK, Luan S (2006) A Ca^{2+} signaling pathway regulates a K(+) channel for low-K response in *Arabidopsis*. Proc Natl Acad Sci U S A 103:12625–12630

Liu J, Zhu JK (1998) A calcium sensor homolog required for plant salt tolerance. Science 280:1943–1945

Liu J, Ishitani M, Halfter U, Kim CS, Zhu JK (2000) The *Arabidopsis thaliana* SOS2 gene encodes a protein kinase that is required for salt tolerance. Proc Natl Acad Sci U S A 97:3730–3734

Liu LL, Ren HM, Chen LQ, Wang Y, Wu WH (2013) A protein kinase, calcineurin B-like protein-interacting protein Kinase9, interacts with calcium sensor calcineurin B-like Protein3 and regulates potassium homeostasis under low-potassium stress in Arabidopsis. Plant Physiol 161:266–277

Luan S, Kudla J, Rodriguez-Concepcion M, Yalovsky S, Gruissem W (2002) Calmodulins and calcineurin B-like proteins: calcium sensors for specific signal response coupling in plants. Plant Cell 14(Suppl):S389–400

Marino D, Dunand C, Puppo A, Pauly N (2012) A burst of plant NADPH oxidases. Trends Plant Sci 17:9–15

McCormack E, Braam J (2003) Calmodulins and related potential calcium sensors of *Arabidopsis*. New Phytol 159:585–598

Oh SI, Park J, Yoon S, Kim Y, Park S, Ryu M, Nam MJ, Ok SH, Kim JK, Shin JS, Kim KN (2008) The Arabidopsis calcium sensor calcineurin B-like 3 inhibits the 5′-methylthioadenosine nucleosidase in a calcium-dependent manner. Plant Physiol 148:1883–1896

Ok SH, Jeong HJ, Bae JM, Shin JS, Luan S, Kim KN (2005) Novel CIPK1-associated proteins in *Arabidopsis* contain an evolutionarily conserved C-terminal region that mediates nuclear localization. Plant Physiol 139:138–150

Pandey GK, Cheong YH, Kim KN, Grant JJ, Li L, Hung W, D'Angelo C, Weinl S, Kudla J, Luan S (2004) The calcium sensor calcineurin B-like 9 modulates abscisic acid sensitivity and biosynthesis in *Arabidopsis*. Plant Cell 16:1912–1924

Pandey GK, Cheong YH, Kim BG, Grant JJ, Li L, Luan S (2007) CIPK9: a calcium sensor-interacting protein kinase required for low-potassium tolerance in *Arabidopsis*. Cell Res 17:411–421

Pandey GK, Grant JJ, Cheong YH, Kim BG, le Li G, Luan S (2008) Calcineurin-B-like protein CBL9 interacts with target kinase CIPK3 in the regulation of ABA response in seed germination. Mol Plant 1:238–248

Qiu QS, Guo Y, Dietrich MA, Schumaker KS, Zhu JK (2002) Regulation of SOS1, a plasma membrane Na^+/H^+ exchanger in *Arabidopsis thaliana*, by SOS2 and SOS3. Proc Natl Acad Sci U S A 99:8436–8441

Quan R, Lin H, Mendoza I, Zhang Y, Cao W, Yang Y, Shang M, Chen S, Pardo JM, Guo Y (2007) SCABP8/CBL10, a putative calcium sensor, interacts with the protein kinase SOS2 to protect Arabidopsis shoots from salt stress. Plant Cell 19:1415–1431

Quintero FJ, Ohta M, Shi H, Zhu JK, Pardo JM (2002) Reconstitution in yeast of the Arabidopsis SOS signaling pathway for Na^+ homeostasis. Proc Natl Acad Sci U S A 99:9061–9066

Reddy VS, Ali GS, Reddy AS (2002) Genes encoding calmodulin-binding proteins in the *Arabidopsis* genome. J Biol Chem 277:9840–9852

Ren XL, Qi GN, Feng HQ, Zhao S, Zhao SS, Wang Y, Wu WH (2013) Calcineurin B-like protein CBL10 directly interacts with AKT1 and modulates K^+ homeostasis in *Arabidopsis*. The Plant journal : for cell and molecular biology 74:258–266

Rodriguez Milla MA, Uno Y, Chang IF, Townsend J, Maher EA, Quilici D, Cushman JC (2006) A novel yeast two-hybrid approach to identify CDPK substrates: characterization of the interaction between AtCPK11 and AtDi19, a nuclear zinc finger protein. FEBS Lett 580:904–911

Roy SJ, Huang W, Wang XJ, Evrard A, Schmockel SM, Zafar ZU, Tester M (2013) A novel protein kinase involved in $Na^{(+)}$ exclusion revealed from positional cloning. Plant Cell Environ 36:553–568

Rudd JJ, Franklin-Tong VE (2001) Unravelling response-specificity in Ca^{2+} signalling pathways in plant cells. New Phytologist 151:7–33

Sanders D, Brownlee C, Harper JF (1999) Communicating with calcium. Plant Cell 11:691–706

Sanders D, Pelloux J, Brownlee C, Harper JF (2002) Calcium at the crossroads of signaling. Plant Cell 14(Suppl):S401–417

Shi J, Kim KN, Ritz O, Albrecht V, Gupta R, Harter K, Luan S, Kudla J (1999) Novel protein kinases associated with calcineurin B-like calcium sensors in *Arabidopsis*. Plant Cell 11:2393–2405

Shi H, Ishitani M, Kim C, Zhu JK (2000) The *Arabidopsis thaliana* salt tolerance gene SOS1 encodes a putative Na^+/H^+ antiporter. Proc Natl Acad Sci U S A 97:6896–6901

Shinozaki K, Yamaguchi-Shinozaki K (2000) Molecular responses to dehydration and low temperature: differences and cross-talk between two stress signaling pathways. Curr Opin Plant Biol 3:217–223

Snedden WA, Fromm H (2001) Calmodulin as a versatile calcium signal transducers in plants. New Phytologist 151:35–66

Thomashow MF (1999) Plant cold acclimation: freezing tolerance genes and regulatory mechanisms. Annu Rev Plant Physiol Plant Mol Biol 50:571–599

Tripathi V, Syed N, Laxmi A, Chattopadhyay D (2009) Role of CIPK6 in root growth and auxin transport. Plant Signal Behav 4:663–665

Waadt R, Schmidt LK, Lohse M, Hashimoto K, Bock R, Kudla J (2008) Multicolor bimolecular fluorescence complementation reveals simultaneous formation of alternative CBL/CIPK complexes in planta. Plant J 56:505–516

Wang R, Okamoto M, Xing X, Crawford NM (2003) Microarray analysis of the nitrate response in Arabidopsis roots and shoots reveals over 1,000 rapidly responding genes and new linkages to glucose, trehalose-6-phosphate, iron, and sulfate metabolism. Plant Physiol 132:556–567

White PJ, Broadley MR (2003) Calcium in plants. Ann Bot (Lond) 92:487–511

Xu J, Li HD, Chen LQ, Wang Y, Liu LL, He L, Wu WH (2006) A protein kinase, interacting with two calcineurin B-like proteins, regulates K^+ transporter AKT1 in Arabidopsis. Cell 125:1347–1360

Yamaguchi-Shinozaki K, Shinozaki K (2006) Transcriptional regulatory networks in cellular responses and tolerance to dehydration and cold stresses. Annu Rev Plant Biol 57:781–803

Yang T, Poovaiah BW (2003) Calcium/calmodulin-mediated signal network in plants. Trends Plant Sci 8:505–512

Zhu JK (2002) Salt and drought stress signal transduction in plants. Annu Rev Plant Biol 53:247–273

Zielinski RE (1998) Calmodulin and calmodulin-binding proteins in plants. Annu Rev Plant Physiol Plant Mol Biol 49:697–725

Chapter 7
Redox-Regulated Mechanisms: Implications for Enhancing Plant Stress Tolerance and Crop Yield

Ashish Kumar Srivastava and Penna Suprasanna

Abstract In the present scenario of continuously rising world population and climate change, a major thrust in the field of plant biology is to strive for sustainable agriculture. This necessitates a complete understanding of the mechanism underlying plant tolerance against a multitude of stress factors which limit crop productivity. Among other regulators, redox-state homeostasis has recently emerged as the central or core regulator behind stress-induced signaling. At any given point, redox state is defined as the integrated ratio of reduced and oxidized forms of all the redox couples present inside the cell and is governed by the level of reactive oxygen species (ROS) and activities of ROS-producing and ROS-scavenging enzymes. The redox-state homeostasis is a highly dynamic and variable component and is strictly dependent upon the developmental stage as well as the external environment. Considering the significance of redox state in regulating multiple plant processes, a large number of redox-associated genes/transcription factors are currently being characterized using functional genomics-based approach. The present chapter deals with the basic aspect of generating redox signal and then describes selected mechanisms, which are responsible for the viewpoint of plant tolerance and crop yield. The future research directions are also discussed so as to ensure the use of these genetic components in crop improvement program.

Keywords Abiotic stress • Nutrient uptake • Reactive oxygen species • Redox signaling • Root growth • Source-sink relationship • Stomatal movement • Sustainable agriculture

A.K. Srivastava (✉) • P. Suprasanna
Nuclear Agriculture and Biotechnology Division, Bhabha Atomic Research Centre, Mumbai 400 085, India
e-mail: ashish@barc.gov.in; ashishbarc@gmail.com

G.K. Pandey (ed.), *Elucidation of Abiotic Stress Signaling in Plants*,
DOI 10.1007/978-1-4939-2211-6_7

7.1 Introduction

The world population is continuously on the rise, and by 2050 it is expected to reach nine billion people (http://www.fao.org/wsfs/world-summit/en/). In order to meet this ever-increasing demand of feeding the population, plant scientists are striving to facilitate global food production which is estimated to be around approximately 44 million metric tons per year (Tester and Langridge 2010). Although, a steady progress has been seen for major crops in the USA; for example, the production of maize (*Zea mays*), wheat (*Triticum* spp.), and soybean (*Glycine max*) has shown positive linear increase in average yield from 1930 to 2012 (USDA-NASS 2013). However, it has become difficult to maintain this because of decrease in arable land area due to climate change, urbanization, and industrialization. The International Rice Research Institute (IRRI) currently estimates that 1 ha of cultivable land is lost every 7.7 s (http://irri.org/). A decline in crop productivity due to increased episodes of abiotic stresses, such as cold, drought, or high salinity, and biotic stresses, such as bacterial, fungal, and viral infections and insect attacks, places even more pressure on crop productivity and yield. It is thus clear that the challenge is to significantly increase crop production—in both an equitable and sustainable manner. A crucial aspect of achieving this goal is to improve crop performance in terms of tolerance to numerous stresses as well as maintain yield potential. Currently, this is being achieved through conventional crop breeding methods supplemented with modern techniques, such as marker-assisted breeding and transgenic technology. Until 2008, 33 varieties of genetically modified (GM) crops have been commercialized worldwide, and this number is expected to attain 124 in 2015 (Gruskin 2012). In 2011, transgenic crop acreage contributed to 36 % of the entire seed market worldwide (Marshall 2012). Transgenic approaches can rapidly expand gene pools for engineering stress tolerance. Together with high-throughput phenotyping, genetic engineering not only provides a more precise strategy for trait selection but shortens the breeding cycle considerably (Lusser et al. 2012). However, attempts to generate beneficial crop varieties have been hampered due to the lack of knowledge about the molecular mechanisms underlying plant responses toward different abiotic/biotic stresses (Lawlor 2013).

In recent years, various pathways have been identified that can be targeted to enhance the plant stress tolerance and crop yield potential. These processes include osmoregulation, antioxidant defense, water and ion homeostasis, hormone machinery, and metabolite translocation toward sink (Reguera et al. 2012; Ruan et al. 2013). Many studies have been conducted where these mechanisms are manipulated either alone or in combination to improve plant's performance under stress (Chen et al. 2014). However, the success stories are limited especially under field condition. This is because field conditions are entirely different and have combined effect of multiple stresses. Thus, further research is required to identify some common regulatory hubs associated with multi-stress tolerance. Toward this endeavor, redox state has recently emerged as the central or core regulator of stress-induced signaling, and, using functional genomics-based approach, redox-dependent regulation of various processes has been demonstrated. In this context, the present chapter deals with an overview of generating redox signal in plants and the mechanism of redox regulation of selected mechanisms which are important for the viewpoint of plant stress tolerance and crop yield. The future directions are discussed about how to improve our understanding on redox biology so that the identified genes can be used as candidates in crop improvement program to ensure sustainable agriculture.

7.2 Mechanism of ROS Generation and Scavenging Systems in Plants

About 2.7 billion years ago, molecular oxygen (O_2) was introduced in our environment by O_2-evolving photosynthetic organisms, and since then ROS have become important companions of aerobic life (Halliwell 2006). It has been estimated that approximately 2 % of O_2 consumed by plants is sidetracked to produce ROS in various subcellular compartments (Bhattachrjee 2005). The O_2 molecule is a free radical, as it has two unpaired electrons that have the same spin quantum number. This spin restriction makes O_2 a preferred molecule to accept electrons leading to the generation of ROS. Different kinds of ROS are produced as the reduction products of O_2. The single-electron reduction of O_2 results in the generation of superoxide radical ($O_2^{\bullet-}$). Its dismutation results into the formation of H_2O_2. Furthermore, $O_2^{\bullet-}$ can also be protonated to form the perhydroxyl radical ($HO_2^{\bullet}$). In the presence of transition metals, such as copper and iron, further reactions take place, e.g., Haber–Weiss or Fenton-type reactions, giving rise to hydroxyl ion (OH^-) or/and hydroxyl radical ($OH^{\bullet}$), which are the most reactive chemical species in the biological world. Singlet oxygen ($^1O_2$) is another form of ROS in which an electron is elevated to a higher energy orbital, thereby freeing O_2 from its spin-restricted state. Singlet oxygen is generated by the photoexcitation of chlorophyll.

7.2.1 Generation of ROS

In response to different abiotic stresses, many different kinds of ROS are produced. The brief information about the mechanism of their production in different organelles is given below.

7.2.1.1 In Chloroplast

One of the most common and early effects of different abiotic stresses is the closure of stomata, so as to conserve water inside the plants. This causes the limitation in the CO_2 fixation, which is coupled with the over-reduction of electron transport chain (ETC). In over-reduced state, the electron flows from PS-I to O_2 and generates $O^{2\bullet-}$. A membrane-bound copper/zinc superoxide dismutase (Cu/Zn SOD) in the vicinity of PS-I converts $O_2^{\bullet-}$ to H_2O_2. The process of the generation and dismutation of $O_2^{\bullet-}$ at the PS-I site is known as water–water cycle or Mehler reaction (Asada 2006). The singlet oxygen ($^1O_2$) is also generated at PS-II by excited triplet-state chlorophyll at the P_{680} reaction center and in the light-harvesting complex when the ETC is over-reduced. The H_2O_2 produced in the Mehler reaction can positively regulate the formation of $^1O_2$ at PS-II. The H_2O_2 promotes the oxidation of quinone A (primary plastoquinone electron acceptor), which increases the photosynthetic electron transport flow, and hence the level of $^1O_2$ is decreased (Asada 2006; Moller et al. 2007). The sequential events leading to the generation of ROS are depicted in Fig. 7.1.

7.2.1.2 In Mitochondria

Mitochondria are also known as a source of ROS production, although they generate smaller amounts of ROS as compared to chloroplasts and peroxisomes. In mitochondria, complex I and complex III in the respiratory ETC are the major sites of ROS production (Moller et al. 2007). Ubisemiquinone intermediate formed at complexes I and III donates electrons to O_2 and generates $O_2^{\bullet-}$ that is, in turn, reduced to H_2O_2. The ROS production in mitochondria is drastically increased especially under abiotic stress conditions due to very high respiration rate to meet the supply of ATP required for activating plant defense systems.

7.2.1.3 In Peroxisomes

Peroxisomes produce H_2O_2 and $O_2^{\bullet-}$ at high rates through several metabolic processes. Reduced water availability and stomata closure decrease CO_2 to O_2 ratio in mesophyll cells and increase photorespiration and production of glycolate in chloroplasts. Oxidation of glycolate by glycolate oxidase in peroxisomes accounts for the majority of H_2O_2 production during photorespiration (Noctor et al. 2002). Besides, $O_2^{\bullet-}$ is also generated by xanthine oxidase in matrix of leaf peroxisomes.

7.2.1.4 In Apoplast

Apoplast is another important site where H_2O_2 is produced through plasma membrane-located NADPH oxidases and other cell wall-associated oxidases, peroxidases, and polyamine oxidases (Moschou et al. 2008).

7.2.2 Scavenging System of ROS

The ROS detoxification system in plants mainly comprised of low-molecular-weight nonenzymatic [ascorbate (ASC) and glutathione (GSH)] and enzymatic [superoxide dismutase (SOD; EC 1.15.1.1), catalase (CAT; EC 1.11.1.6), ascorbate

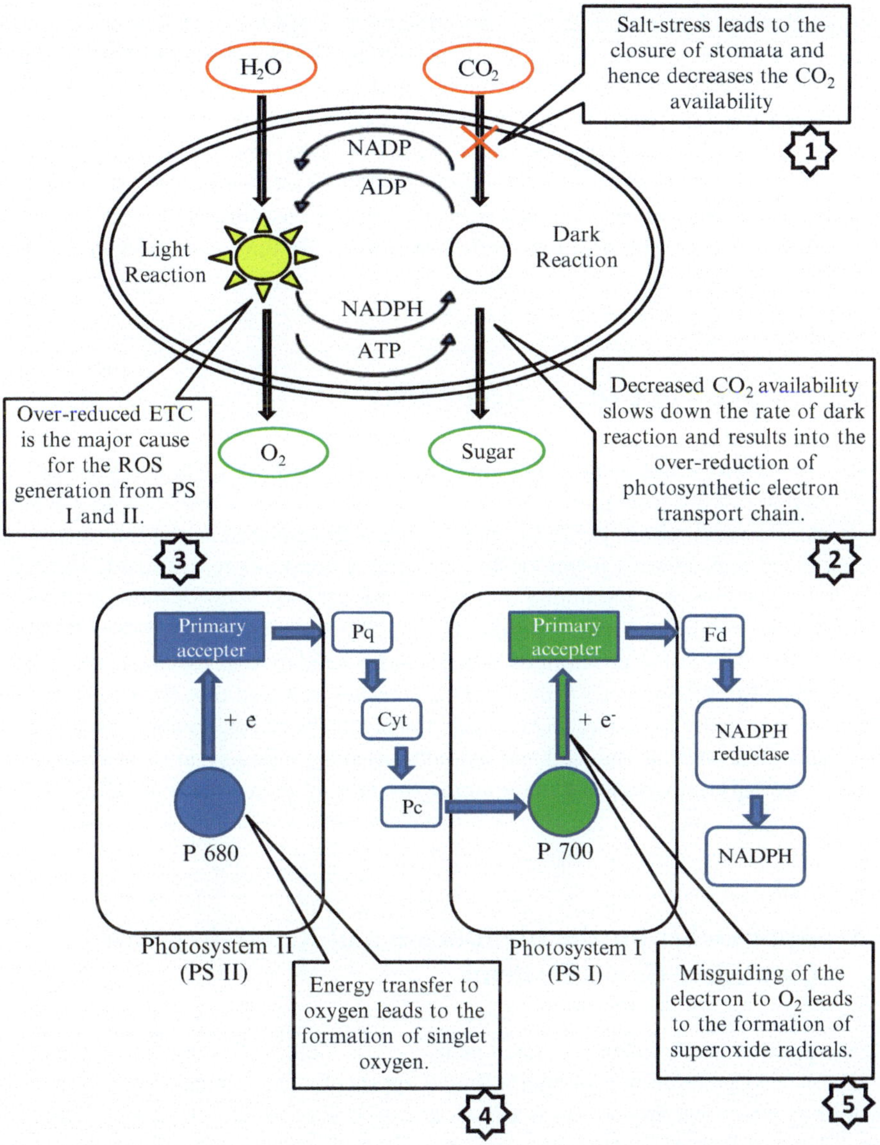

Fig. 7.1 Mechanism of ROS generation in chloroplast. The series of events leading to the generation of ROS under salt stress is depicted. The number represents their sequential occurrence

peroxidase (APX; EC 1.11.1.11)] antioxidants (Miller et al. 2009). The importance of these ROS-regulating mechanisms has been manifested by their presence in various cellular compartments and also by various gain- and loss-of-function mutants and transgenic lines (Tseng et al. 2007; Ashraf 2009). Superoxide dismutase is the major scavenger of superoxide radicals ($O_2^{\bullet-}$), and its enzymatic action results in the formation of hydrogen peroxide (H_2O_2), which is then regulated by CAT and various classes of peroxidases. The detoxification of H_2O_2 by APX in ASC–GSH cycle utilizes ASC as an electron donor, which gets oxidized to dehydroascorbate (DHA) in the process. The regeneration of DHA to ASC is performed by monodehydroascorbate reductase (MDHAR; EC 1.6.5.4) and dehydroascorbate reductase (DHAR; EC 1.8.5.1), utilizing reduced glutathione (GSH) as a reductant, which, in turn, gets converted to oxidized glutathione (GSSG). The final reaction of the cycle is catalyzed by glutathione reductase (GR; EC 1.6.4.2), leading to the conversion of GSSG back to GSH. In addition, both ASC and GSH may also directly quench ROS (Noctor 2006).

7.2.3 Concept of Cellular Redox State

The redox homeostasis is an important process and is now widely accepted as the core regulator of different plant processes both under normal as well as stress conditions. At any given point, the redox state of the cell is defined as the ratio between the oxidized and reduced forms of different redox couples present inside the cell. This is a very complex parameter and is governed by the level of individual ROS and activities of different ROS-producing and ROS-scavenging enzymes (Potters et al. 2010). Any external stimulus which causes oxidative stress shifts the redox status toward oxidizing end. Although it was earlier believed that ROS are toxic to plants as they cause oxidative stress inside the plants, however, now it is established that there exist both the reduced and oxidant-based signaling systems in plants and both are equally important for maintaining the normal plant function. Thus, ROS-mediated effects on plants are strictly dose dependent.

7.3 Approaches for the Identification of Redox-Regulated Candidate Genes/Proteins

At the genome level, redox-regulated genes can be identified either through microarray- or next-generation sequencing-based approach. The comparative expression profiling under changing redox environment can be done to identify redox-regulated candidate genes. Various physiological (cysteine and glutathione) or nonphysiological thiols (thiourea or dithiothreitol) can be used externally to generate different redox regimes. Although physiological thiol can integrate into multiple metabolic pathways to give indirect or redox-independent effects, such a possibility is relatively less for nonphysiological thiols. Supplementation of thiourea (a nonphysiological thiol-based ROS scavenger) has been used successfully for the identification of redox regulatory signaling and effector components of salt (Srivastava et al. 2010a, b) and arsenic (Srivastava et al., 2014) tolerance. Similarly, inhibitor of glutathione biosynthesis like buthionine sulfoximine has also been used for the identification of gene regulatory networks underlying root growth (Koprivova et al. 2010). At protein level, redox-mediated regulation can be studied using various redox proteomics-based approach (Wang et al. 2012; Matallana-Surget et al. 2013; Bykova and Rampitsch 2013; Liu et al. 2014; Rosenwasser et al. 2014). Recently, proteomics-based approach has also been employed for the identification of redox regulatory transcription factors in plants (Dietz 2014). Apart from these individualistic approaches, systems biology-based approach, which integrates the data from transcriptome, proteome, and metabolome, has also been used for the broader understanding of redox-mediated regulation (Srivastava et al. 2013).

7.4 Functional Genomics Study of Redox-Regulated Mechanisms to the Broader Context of Plant Abiotic Stress Tolerance

Considering the flexibility of underlying mechanisms for ROS production and their scavenging, ROS are now established as a highly versatile, effective, and reversible signaling mediator to activate different plant processes. Additionally, due to their diffusible nature, ROS are also important to transfer the signal from the site of stimulus to the distal parts. In recent years, using functional genomics-based approach, significant advancement has come up to delineate the role of different ROS/redox-regulated mechanisms. The fundamental aspects of few such processes, which are considered as an important determinant of plant abiotic stress tolerance, are detailed in the following sections.

7.4.1 Stomata Movement

In order for plants to function efficiently, they must balance the gaseous exchange between inside and outside to maximize CO_2 uptake for photosynthetic carbon assimilation and to minimize water loss through transpiration. These functions are regulated by stomata which have a pair of guard cells that act as "gatekeepers." Guard cells possess a very dynamic signaling system so as to efficiently control the stomatal movement under changing environmental and developmental conditions. The basic step of guard cell signaling is the induction of abscisic acid (ABA), which activates SnRK2 protein kinase OST1 (open stomata 1) that interacts with and phosphorylates NADPH oxidase (AtrbohF), leading to rapid ROS burst. This promotes the release of calcium from internal stores, which is sensed by the calcium-sensing receptors (CAS), resulting in the efflux of K^+ and Cl^- ions. This decreases the internal solute concentration of guard cell, which drives the water out through osmosis, resulting in the loss of turgor pressure and closing of stomata (Xu et al. 2013). ABA-induced ROS burst is an important regulatory step for stomata closing, and hence, the antioxidant status of guard cells is very important, as they are responsible for shaping ROS signature. Flavonols are the major antioxidants, which accumulate specifically in guard cells but not in surrounding pavement cells. The *Arabidopsis* null mutant, *transparent testa4-2*, in CHALCONE SYNTHASE (a flavonol biosynthetic enzyme) synthesizes no flavonol antioxidants and hence showed an increased ROS level in their guard cells. The mutant also showed more rapid ABA-induced closure than the wild type, suggesting that flavonols may shorten either the time of intensity or ABA-dependent ROS burst required for driving the stomata closure. The levels of flavonol antioxidant are also positively regulated by ethylene treatment. In both *ethylene-overproducing1* and ethylene-treated wild-type plants, elevated flavonols lead to decreasing ROS and slower ABA-mediated stomata closure (Watkins et al. 2014). Additionally, an ABA-independent but redox-dependent mode of regulation has also been proposed for stomatal closing. This requires a zinc finger transcription factor (DST; drought and salt tolerance), which negatively regulates H_2O_2-induced stomata closure by directly regulating the expression of genes related to H_2O_2 scavenging. Loss of DST function increases stomata closure and reduces stomatal density, consequently resulting in enhanced drought and salt tolerance in rice (Huang et al. 2009). Besides DST, there are few transcription factors like SNAC1 (stress-responsive NAC 1) and CYCH;1, which modulates the redox status of guard cells. The SNAC1 targets OsSRO1c that actually increases the H_2O_2 accumulation in guard cells. Thus, the loss-of-function mutant of OsSRO1c has increased stomatal aperture and sensitivity to drought and faster water loss as compared to wild type (You et al. 2013). CYCH;1 induces the expression of redox homeostasis genes, such as LIPOXYGENASE3 (LOX3), LOX4, ARABIDOPSIS GLUTATHIONE PEROXIDASE 7 (ATGPX7), EARLY LIGHT-INDUCIBLE PROTEIN1 (ELIP1), and ELIP2, and increased hydrogen peroxide production in guard cells. The down-regulation of CYCH;1 impaired the blue light-induced stomatal opening but did not affect the guard cell development or ABA-mediated stomata closure (Zhou et al. 2013). This clearly suggested that there exist distinct mechanisms for opening stomata under optimal light condition and for closing under stress condition. Along with H_2O_2, nitric oxide (NO) has also emerged as an important regulator of stomata movement (Gayatri et al. 2013). Recently, it is demonstrated that UV-B-induced stomata closure predominantly involves NO and not H_2O_2 (Tossi et al. 2014). Apart from ROS and RNS, Ca^{2+} is yet another regulator of stomata movement. In spite of no change in their stomatal density, the CAS mutants showed higher stomatal conductance due to improper regulation of stomatal aperture. Due to this, CAS mutants demonstrated decreased drought tolerance and water use efficiency (Wang et al. 2014) (Fig. 7.2).

Thus, the proper functioning of stomata requires tight coordination between ABA-, ROS/RNS-, and Ca^{2+}-based signaling systems. Any deficiency in this may disturb the gaseous balance and plant water status, which negatively affects the photosynthetic efficiency and growth of plants.

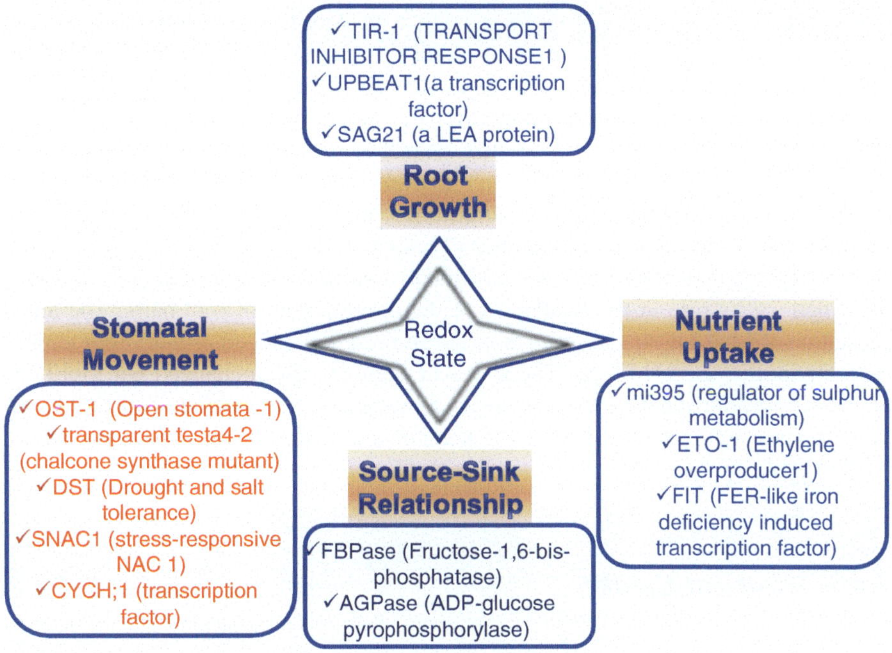

Fig. 7.2 Redox state as central regulator of signaling under stress. The redox state serves as a common regulator behind the induction of different mechanisms, which are important for the viewpoint of stress tolerance as well as crop yield

7.4.2 *Root Growth*

The redox state-mediated regulation of root growth and differentiation was first established when Foreman et al. (2003) demonstrated the requirement of basal ROS level for the expansion of root cells. Later on, auxin was established as the major hormone for regulating root growth. Auxin signaling involves the targeted degradation of transcriptional repressors of Auxin/INDOLE-3-ACETIC ACID (Aux/IAA) protein family, which are generally bound with auxin response factor (ARF) family of transcription factors and, hence, prevent any transcription to occur. Auxin acts as molecular glue to stabilize the interaction between Aux/IAAs and TRANSPORT INHIBITOR RESPONSE1 (TIR1) or closely related proteins of AUXIN SIGNALING F-BOX PROTEIN FAMILY (AFB) (Tan et al. 2007). TIR1 and AFBs are the alternative subunits of SCF (Skp1-Cul1/Cdc53-F-box)-type E3 ubiquitin ligase responsible for targeted degradation of Aux/IAAs through 26S proteasome. Auxin-induced degradation of Aux/IAAs makes the ARFs free to form ARF–ARF dimers, leading to the transcriptional activation of a set of genes responsible for rooting. Auxin response can be enhanced by the transient shift in redox state toward the oxidizing end through the generation of NO. In the presence of NO, TIR1 gets nitrosylated, which enhanced its interaction with Aux/IAAs (Terrile et al. 2012). Besides, auxin itself can also regulate redox state, by inducing the inactivation of APX-1 (ascorbate peroxidase-1) through denitrosylation. Thus, prolonged auxin concentration may lead to the accumulation of H_2O_2, reducing root growth (Correa-Aragunde et al. 2013). Redox-mediated control of root growth is also exerted in a hormone-independent manner through a transcription factor called UPBEAT1 (UPB1) which regulates a set of peroxidases, and its expression gets regulated through redox-state-dependent manner (Tsukagoshi et al. 2010). A redox-related late embryogenesis-associated (LEA) protein family (SAG21/AtLEA5) has been identified for its role in the regulation of the development of root hairs. The antisense line of LEA5 in *Arabidopsis thaliana* showed the reduced primary root length as compared to the wild type. The root hair number was unchanged but root hair length was proportional to SAG21/AtLEA5 expression level, with longer root hairs in overexpressed lines and shorter root hairs in antisense, as compared to wild type (Salleh et al. 2012).

7.4.3 Nutrient Uptake

Sulfur is an essential macronutrient, which is required for the synthesis of amino acids such as cysteine and methionine and is an indispensable component of proteins and peptides, in coenzymes and prosthetic groups, and in a range of secondary metabolites such as glucosinolates. Plants take up sulfur from the soil in the form of inorganic sulfate. This is mediated by a family of sulfate transporters (SULTR; 1–4; Davidian and Koprivova 2010). Inside the plants, sulfate is first activated by ATP sulfurylase (ATPS) to adenosine-5′-phosphosulfate (APS). APS is reduced by APS reductase (APR) to sulfite, which is subsequently reduced to sulfide by sulfite reductase. Sulfide reacts with *O*-acetylserine (OAS) to form cysteine in a reaction catalyzed by OAS-(thiol)lyase. Cysteine is the precursor for methionine, *S*-adenosylmethionine, glutathione (GSH), and glucosinolates. GSH is the major metabolite of *S* assimilation in plants and its concentration ranges between 0.1 and 1.5 mM in leaves and up to 20 mM in chloroplasts. Besides, GSH is also the major redox regulator as it functions as important antioxidants (Takahashi et al. 2011). The uptake of sulfur as well as its assimilation is highly regulated by the internal sulfur status as well the cellular redox state. The redox-dependent regulation of sulfur metabolism is mediated through microRNA395 (miR395) that targets a low-affinity sulfate transporter (SULTR2;1) and three ATP sulfurylases (APS1, APS3, and APS4). Under sulfur-deprived conditions, internal GSH level acts as a sensor to activate necessary strategy so as to maintain the normal sulfur status in plants (Jagadeeswaran et al. 2014). Potassium (K^+) is another essential mineral nutrient in plants. The K^+ uptake is achieved by the orchestrated regulation of a sophisticated network of potassium transport systems which includes the shaker-type and "two-pore" potassium channels; various types of potassium-permeable nonselective cation channels; and KUP/HAK/KT, HKT, and K^+/H^+ transporters (Very and Sentenac 2003). Under K^+-deficient condition, the level of ROS and ethylene is accumulated in roots and activates the genes required for high-affinity K^+ uptake (Shin and Schachtman 2004). Apart from K^+ uptake, recently the significance of redox state and ethylene has been shown to be important for maintaining lower Na^+/K^+ ratio in plants. This was established by analyzing the phenotype of a loss-of-function allele of ETHYLENE OVERPRODUCER1 (ETO1; mutant which causes the increased production of ethylene). The lack of ETO1 function reduces the root Na^+ influx and both stelar and xylem sap Na^+ concentrations, thereby restricting the root-to-shoot delivery of Na. These effects are associated with the increased accumulation of reactive oxygen species in root stele. Additionally, the lack of ETO1 function also led to the significant enhancement of tissue K^+ status through elevated HIGH-AFFINITY K^+ TRANSPORTER5 transcript levels (Jiang et al. 2013). The redox-state-dependent regulation of iron nutrition, metabolism, and homeostasis has also emerged. This is mediated by the level of nitric oxide (NO) in roots. Indeed, NO possesses a high affinity for iron and low-molecular-weight complexes are formed between iron and NO, which are termed as iron–nitrosyl complexes. These compounds consist of both mononitrosyl iron complexes (MNICs) and dinitrosyl iron complexes (DNICs), and their formation and interactions are central for regulating iron mobility and availability especially under low iron concentrations. Besides, NO-mediated regulation of iron homeostasis is also mediated by activating the transcription factor FER-LIKE IRON DEFICIENCY-INDUCED TRANSCRIPTION FACTOR (FIT), which regulates the expression of both iron reductase and iron transporter (Yu et al. 2014).

7.4.4 *Source–Sink Relationship*

In plants, leaves are the prime site for photosynthesis and hence, termed as "source" for the generation of photoassimilatory products, mainly sucrose. Similarly, maturing seeds are responsible for driving sucrose transportation away from leaves and hence termed as "sink." The sucrose translocation from source to sink is considered as the rate-limiting step for regulating the level of photoassimilate in sink. The continuous flux of metabolites toward the sink for an extended time period is considered as the most desirable trait which most of the plant biologists want to achieve so as to enhance plant's harvest index and crop yield (Ainsworth and Bush 2011; Aranjuelo et al. 2013; Patrick et al. 2013). The generation of photoassimilates occurs in chloroplasts through the process of photosynthesis in which light energy is utilized to convert gaseous CO_2 into triose phosphate, which can either be transported into cytosol through triose phosphate transporter for sucrose synthesis or may be retained in chloroplast itself for starch synthesis. The sucrose biosynthesis is a two-step process catalyzed by fructose-1,6-bis-phosphatase (FBPase) and sucrose phosphate synthase (SPS), and, hence, SPS null mutants of *Arabidopsis* have been shown to accumulate a very low level of sucrose in leaves (Sun et al. 2011). Most of the enzymes of Calvin–Benson cycle including FBPase are shown to be regulated in a redox-dependent manner (Michelet et al. 2013). Recently, Pandey et al. (2013) has also demonstrated that the activities of these enzymes can also be regulated by the external application of redox-active molecules like thiourea. The simultaneous overexpression of triose phosphate transporter and cytosolic FBPase has been performed in *Arabidopsis*, and transgenic lines were found to have increased photosynthetic carbon assimilation and growth rate under moderate as well as elevated light conditions as compared to wild-type plants (Cho et al. 2012). Like sucrose, starch biosynthesis process is also redox regulated and is mediated by the rate-limiting action of ADP-glucose pyrophosphorylase (AGPase; Geigenberger 2011). The reducing environment activates the AGPase through NADP-thioredoxin reductase C (NTRC)-mediated monomerization (Sonnewald and Kossmann 2013); however, the *ntrc* mutants have been shown to have comparable level of starch than wild type which also suggests the existence of redox-independent mode of regulation for AGPase (Li et al. 2012).

7.5 Conclusions and Future Perspective

In conclusion, the present chapter highlights redox state as a common regulatory point for various plant processes which are important for the viewpoint of enhancing plant stress tolerance and crop yield. The stomatal opening and closing is regulated by ABA-dependent ROS burst and is important for maintaining gaseous exchange rate and plant water homeostasis. The root growth and uptake of different nutrients are also dependent upon NADPH oxidase-mediated generation of ROS. Apart from these mechanisms, which are more or less associated with growth, the source–sink relationship, which is directly related with crop yield, is also under the strict control of redox state. Since most of the abiotic and biotic stresses cause redox imbalance inside the plants, the functioning of these redox-dependent processes gets disturbed, leading to reduced plant growth and crop yield. Although the underlying molecular mechanisms have been studied by analyzing the phenotype of large number of loss- and gain-of-function lines of associated genes/transcription factor, further research in this direction is required to better understand their regulation under changing redox conditions. Additionally, along with the individual genes, the complete gene regulatory network also needs to be investigated in detail. Since the cellular redox state is a highly dynamic parameter and is dependent upon several factors including plant organ, its developmental stage, nature and magnitude of stress, soil conditions, and other environmental factors, a better understanding of redox-regulated mechanisms using functional genomics-based approach may result in finding novel regulators which can be utilized in crop improvement program for the generation of varieties better suited for changing climate condition.

References

Ainsworth EA, Bush DR (2011) Carbohydrate export from the leaf: a highly regulated process and target to enhance photosynthesis and productivity. Plant Physiol 155:64–69

Aranjuelo I, Sanz-Sáez A, Jauregui I, Irigoyen JJ, Araus JL, Sánchez-Díaz M, Erice G (2013) Harvest index, a parameter conditioning responsiveness of wheat plants to elevated CO_2. J Exp Bot 64:1879–1892

Asada K (2006) Production and scavenging of reactive oxygen species in chloroplasts and their functions. Plant Physiol 141:391–396

Ashraf M (2009) Biotechnological approach of improving plant salt tolerance using antioxidants as markers. Biotechnol Adv 27:84–93

Bhattachrjee S (2005) Reactive oxygen species and oxidative burst: roles in stress, senescence and signal transduction in plant. Curr Sci 89:1113–1121

Bykova NV, Rampitsch C (2013) Modulating protein function through reversible oxidation: redox-mediated processes in plants revealed through proteomics. Proteomics 13:579–596

Chen MX, Lung SC, Du ZY, Chye ML (2014) Engineering plants to tolerate abiotic stresses. Biocatal Agric Biotechnol 3:81–87

Cho MH, Jang A, Bhoo SH, Jeon JS, Hahn TR (2012) Manipulation of triose phosphate/phosphate translocator and cytosolic fructose-1,6-bisphosphatase, the key components in photosynthetic sucrose synthesis, enhances the source capacity of transgenic Arabidopsis plants. Photosynth Res 111:261–268

Correa-Aragunde N, Foresi N, Delledonne M, Lamattina L (2013) Auxin induces redox regulation of ascorbate peroxidase activity by S-nitrosylation/denitrosylation balance resulting in changes of root growth pattern in Arabidopsis. J Exp Bot 64:3339–3349

Davidian JC, Kopriva S (2010) Regulation of sulfate uptake and assimilation–the same or not the same? Mol Plant 3:314–325

Dietz KJ (2014) Redox regulation of transcription factors in plant stress acclimation and development. Antioxid Redox Signal 21(9):1356–1372. doi:10.1089/ars.2013.5672)

Foreman J, Demidchik V, Bothwell JH, Mylona P, Miedema H, Torres MA, Linstead P, Costa S, Brownlee C, Jones JD, Davies JM, Dolan L (2003) Reactive oxygen species produced by NADPH oxidase regulate plant cell growth. Nature 422:442–446

Gayatri G, Agurla S, Raghavendra AS (2013) Nitric oxide in guard cells as an important secondary messenger during stomata closure. Front Plant Sci 4:425

Geigenberger P (2011) Regulation of starch biosynthesis in response to a fluctuating environment. Plant Physiol 155:1566–1577

Gruskin D (2012) Agbiotech. Nat Biotechnol 30:211–214

Halliwell B (2006) Reactive species and antioxidants. Redox biology is a fundamental theme of aerobic life. Plant Physiol 141:312–322

Huang XY, Chao DY, Gao JP, Zhu MZ, Shi M, Lin HX (2009) A previously unknown zinc finger protein, DST, regulates drought and salt tolerance in rice via stomatal aperture control. Genes Dev 23:1805–1817

Jagadeeswaran G, Li YF, Sunkar R (2014) Redox signaling mediates the expression of a sulfate-deprivation-inducible microRNA395 in Arabidopsis. Plant J 77:85–96

Jiang C, Belfield EJ, Cao Y, Smith JA, Harberd NP (2013) An Arabidopsis soil-salinity-tolerance mutation confers ethylene-mediated enhancement of sodium/potassium homeostasis. Plant Cell 25:3535–3552

Koprivova A, Mugford ST, Kopriva S (2010) rabidopsis root growth dependence on glutathione is linked to auxin transport. Plant Cell Rep 29:1157–1167

Lawlor DW (2013) Genetic engineering to improve plant performance under drought: physiological evaluation of achievements, limitations, and possibilities. J Exp Bot 64:83–108

Li J, Almagro G, Munoz FJ, Baroja-Fernandez E, Bahaji A et al (2012) Post-translational redox modification of ADP-glucose pyrophosphorylase in response to light is not a major determinant of fine regulation of transitory starch accumulation in Arabidopsis leaves. Plant Cell Physiol 53:433–444

Liu P, Zhang H, Wang H, Xia Y (2014) Identification of redox-sensitive cysteines in the Arabidopsis proteome using OxiTRAQ, a quantitative redox proteomics method. Proteomics 14:750–762
Lusser M, Parisi C, Plan D, Rodriguez-Cerezo E (2012) Deployment of new biotechnologies in plant breeding. Nat Biotechnol 30:231–239
Marshall A (2012) Existing agbiotech traits continue global march. Nat Biotechnol 30:207
Matallana-Surget S, Cavicchioli R, Fauconnier C, Wattiez R, Leroy B et al (2013) Shotgun redox proteomics: identification and quantitation of carbonylated proteins in the UVB-resistant marine bacterium, Photobacterium angustum S14. PLoS One 8(7):e68112
Michelet L, Zaffagnini M, Morisse S, Sparla F, Pérez-Pérez ME, Francia F, Danon A, Marchand CH, Fermani S, Trost P, Lemaire SD (2013) Redox regulation of the Calvin-Benson cycle: something old, something new. Front Plant Sci 4:470
Miller G, Suzuki N, Ciftci-Yilmaz S, Mittler R (2009) Reactive oxygen species homeostasis and signaling during drought and salinity stresses. Plant Cell Environ 33:453–467
Moller IM, Jensen PE, Hansson A (2007) Oxidative modifications to cellular components in plants. Annu Rev Plant Biol 58:459–481
Moschou PN, Paschalidis KA, Delis ID, Andriopoulou AH, Lagiotis GD, Yakoumakis DI, Roubelakis-Angelakis KA (2008) Spermidine exodus and oxidation in the Apoplast induced by abiotic stress is responsible for H_2O_2 signatures that direct tolerance responses in tobacco. Plant Cell 20:1708–1724
Noctor G (2006) Metabolic signaling in defense and stress: the critical roles of soluble redox couples. Plant Cell Environ 29:409–425
Noctor G, Veljovic-Jovanovic S, Driscoll S, Novitskaya L, Foyer CH (2002) Drought and oxidative load in the leaves of C3 plants: a predominant role for photorespiration? Ann Bot 89:841–850
Pandey M, Srivastava AK, D'Souza SF, Penna S (2013) Thiourea, a ROS scavenger, regulates source-to-sink relationship to enhance crop yield and oil content in Brassica juncea (L.). PLoS One 8(9):e73921
Patrick JW, Botha FC, Birch RG (2013) Metabolic engineering of sugars and simple sugar derivatives in plants. Plant Biotechnol J 11:142–156
Potters G, Horemans N, Jansen MAK (2010) The cellular redox state in plant stress biology: a charging concept. Plant Physiol Biochem 48:292–300
Reguera M, Peleg Z, Blumwald E (2012) Targeting metabolic pathways for genetic engineering abiotic stress-tolerance in crops. Biochim Biophys Acta 1819:186–194
Rosenwasser S, Graff van Creveld S, Schatz D, Malitsky S, Tzfadia O, Aharoni A, Levin Y, Gabashvili A, Feldmesser E, Vardi A (2014) Mapping the diatom redox-sensitive proteome provides insight into response to nitrogen stress in the marine environment. Proc Natl Acad Sci U S A 111:2740–50
Ruan YL, Patrick JW, Shabala S, Slewinski TL (2013) Uptake and regulation of resource allocation for optimal plant performance and adaptation to stress. Front Plant Sci 4:455
Salleh FM, Evans K, Goodall B, Machin H, Mowla SB, Laj M, Runions J, Theodoulou FL, Foyer CH, Rogers HJ (2012) A novel function for a redox-related LEA protein (SAG21/AtLEA5) in root development and biotic stress responses. Plant Cell Environ 35:418–429
Shin R, Schachtman DP (2004) Hydrogen peroxide mediates plant root cell response to nutrient deprivation. Proc Natl Acad Sci U S A 101:8827–8832
Sonnewald U, Kossmann J (2013) Starches-from current models to genetic engineering. Plant Biotechnol J 11:223–232
Srivastava AK, Ramaswamy NK, Suprasanna P, D’Souza SF (2010a) Genome-wide analysis of thiourea modulated salinity-stress responsive transcripts in seeds of Brassica juncea L: Identification of signaling and effector components of stress tolerance. Ann Bot 106:663–674
Srivastava AK, Suprasanna P, Srivastava S, D’Souza SF (2010b) Thiourea mediated regulation in the expression profile of aquaporins and its impact on water homeostasis under salinity stress in Brassica juncea roots. Plant Sci 178:517–522
Srivastava V, Obudulu O, Bygdell J, Löfstedt T, Rydén P, Nilsson R, Ahnlund M, Johansson A, Jonsson P, Freyhult E, Qvarnström J, Karlsson J, Melzer M, Moritz T, Trygg J, Hvidsten TR,

Wingsle G (2013) OnPLS integration of transcriptomic, proteomic and metabolomic data shows multi-level oxidative stress responses in the cambium of transgenic hipI-superoxide dismutase Populus plants. BMC Genomics 14:893

Srivastava AK, Srivastava S, Mishra S, Suprasanna P, D'Souza SF (2014) Identification of redox-regulated components of arsenate (AsV) tolerance through thiourea supplementation in rice. Metallomics 6, 1718–1730

Sun J, Zhang J, Larue CT, Huber SC (2011) Decrease in leaf sucrose synthesis leads to increased leaf starch turnover and decreased RuBP regeneration-limited photosynthesis but not Rubisco-limited photosynthesis in Arabidopsis null mutants of SPSA1. Plant Cell Environ 34:592–604

Takahashi H, Kopriva S, Giordano M, Saito K, Hell R (2011) Sulfur assimilation in photosynthetic organisms: molecular functions and regulations of transporters and assimilatory enzymes. Annu Rev Plant Biol 62:157–184

Tan X, Calderon-Villalobos LI, Sharon M, Zheng C, Robinson CV, Estelle M, Zheng N (2007) Mechanism of auxin perception by the TIR1 ubiquitin ligase. Nature 446:640–645

Terrile MC, Paris R, Calderon-Villalobos LIA, Iglesias MJ, Lamattina L, Estelle M, Casalongue CA (2012) Nitric oxide influences auxin signaling through S-nitrosylation of the Arabidopsis TRANSPORT INHIBITOR RESPONSE 1 auxin receptor. Plant J 70:492–500

Tester M, Langridge P (2010) Breeding technologies to increase crop production in a changing world. Science 327:818–822

Tossi V, Lamattina L, Jenkins GI, Cassia RO (2014) Ultraviolet-B-induced stomata closure in Arabidopsis is regulated by the UV RESISTANCE LOCUS8 photoreceptor in a nitric oxide-dependent mechanism. Plant Physiol 164:2220–2230

Tseng MJ, Liu CW, Yiu JC (2007) Enhanced tolerance to sulfur dioxide and salt stress of transgenic Chinese cabbage plants expressing both superoxide dismutase and catalase in chloroplasts. Plant Physiol Biochem 45:822–833

Tsukagoshi H, Busch W, Benfey PN (2010) Transcriptional regulation of ROS controls transition from proliferation to differentiation in the root. Cell 143:606–616

USDA-NASS (United States Department of Agriculture-National Agricultural Statistics Service) (2013) Field Crops Statistics, USDA-NASS

Very AA, Sentenac H (2003) Molecular mechanisms and regulation of K^+ transport in higher plants. Annu Rev Plant Biol 54:575–603

Wang H, Wang S, Lu Y, Alvarez S, Hicks LM, Ge X, Xia Y (2012) Proteomic analysis of early-responsive redox-sensitive proteins in Arabidopsis. J Proteome Res 11:412–424

Wang WH, Chen J, Liu TW, Chen J, Han AD, Simon M, Dong XJ, He JX, Zheng HL (2014) Regulation of the calcium-sensing receptor in both stomatal movement and photosynthetic electron transport is crucial for water use efficiency and drought tolerance in Arabidopsis. J Exp Bot 65:223–234

Watkins JM, Hechler PJ, Muday GK (2014) Ethylene-induced flavonol accumulation in guard cells suppresses reactive oxygen species and moderates stomatal aperture. Plant Physiol 164:1707–1717

Xu ZY, Kim DH, Hwang I (2013) ABA homeostasis and signaling involving multiple subcellular compartments and multiple receptors. Plant Cell Rep 32:807–813

You J, Zong W, Li X, Ning J, Hu H, Li X, Xiao J, Xiong L (2013) The SNAC1-targeted gene OsSRO1c modulates stomata closure and oxidative stress tolerance by regulating hydrogen peroxide in rice. J Exp Bot 64:569–583

Yu M, Lamattina L, Spoel SH, Loake GJ (2014) Nitric oxide function in plant biology: a redox cue in deconvolution. New Phytol 202:1142–1156

Zhou XF, Jin YH, Yoo CY, Lin XL, Kim WY, Yun DJ, Bressan RA, Hasegawa PM, Jin JB (2013) CYCLIN H;1 regulates drought stress responses and blue light-induced stomatal opening by inhibiting reactive oxygen species accumulation in Arabidopsis. Plant Physiol 162: 1030–1041

Chapter 8
Role of Mitogen-Activated Protein Kinase Cascade in Combating Abiotic Stress in Plants

Hussain Ara and Alok Krishna Sinha

Abstract Plants being sessile have to counter a plethora of stresses, both biotic and abiotic in order to grow and survive. The ability of plants to perceive these stresses at the cell surface and transduce them to the nucleus for appropriate cellular readjustment is one of the most sophisticated mechanisms they have developed during the process of evolution. Among several cascades helping in signal transduction, mitogen-activated protein kinase (MAPK) cascade is one of the most important cascades that is ubiquitously present in all eukaryotes. This unique protein cascade is also involved in several developmental and vital processes in plants. This is essentially a phosphorelay cascade consisting of three components, a MAPK kinase kinase (MAPKKK/MAP3K/MEKK/MKKK), a MAPK kinase (MAPKK/MAP2K/MEK/MKK), and a MAPK (MPK) connected to each other by the event of phosphorylation. All these components of MAPK cascade are multigene family and are involved in efficient transmission of specific stimuli in response to stress signaling. In the present chapter, we will highlight the involvement of different members of this phosphorelay cascade during abiotic stress in plants.

Keywords Mitogen activated protein kinase • MAPK cascade • Abiotic stress • Drought stress • Salt stress

8.1 Introduction

The world population is continuously increasing at an alarming rate, while food production is decreasing simultaneously as threats for crop production are ever increasing. Major threats are various abiotic and biotic stresses. These stresses trigger biochemical, molecular, and physiological changes that influence various cellular processes adversely in plants. Abiotic stress is recognized as the most harmful factor pertaining to the growth, productivity of crops, and worldwide agricultural loss by more than 50 % (Rodriguez et al. 2010; Krasensky and Jonak 2012).

H. Ara • A.K. Sinha (✉)
National Institute of Plant Genome Research, Aruna Asaf Ali Road, New Delhi, India
e-mail: alok@nipgr.ac.in

G.K. Pandey (ed.), *Elucidation of Abiotic Stress Signaling in Plants*,
DOI 10.1007/978-1-4939-2211-6_8

Frequently, plants meet adverse environmental conditions, and being sessile organisms, the options available to them are either to die or to live with restricted growth or with some other disorders. Nevertheless, some plants exhibit tolerance and grow with their defense mechanisms even in the environment consisting of a variety of stresses. The extreme climatic factors, such as high and low temperatures, availability of water, and high salinity in the soils, are major abiotic environmental stresses that limit plant growth and development that eventually affect the agronomical yield of the crop plant. Plants usually have developed sophisticated mechanisms for fast perceiving the signals from a changing environment and for their efficient transduction in the specific component for specific adaptive and/or defensive responses. They possess complicated protection mechanisms to cope with various environmental stresses such as salinity, drought, cold, heat, UV light, ozone, osmotic shock, and mechanical wounding. Accumulating evidences indicate rapid activation of mitogen-activated protein kinases (MAPKs) when exposed to abiotic stress stimuli (Nakagami et al. 2005; Taj et al. 2010; Šamajová et al. 2013). The focal point of current research in the MAPK cascade is to identify the roles of specific components of the cascade under regulation of diverse abiotic stress responses and consequently to uncover their link to defense mechanisms and growth regulation.

A MAPK cascade is a three-tier module consisting of MAP3K, MAP2K, and MAPK subunits. These are linked in various ways to upstream receptors and downstream targets. Receptor-mediated activation of a MAP3K can occur through physical interaction and/or phosphorylation by either the receptor itself or interlinking MAPKs. The peculiarity of MAPK cascade functioning is its specificity within the same cell. This specificity is established through the presence of docking domains found in various components of MAPK modules and scaffold proteins (Nakagami et al. 2005). Members of MAPK cascades have been identified in *Arabidopsis*, rice, and few other plant species, yet the interactions between these components of the cascade have been studied only in *A. thaliana* and rice, and there is very little information of such studies in other plant species. In order to understand MAPK signaling completely, it is essential to understand components of MAPK cascades and their interactions. Surprisingly, there are very few reports on the interaction of MAPKKs–MAPKs in rice either from yeast two-hybrid (Y2H) or phosphorylation assays (Ding et al. 2009; Wankhede et al. 2013a). Recently, Singh et al. (2012) have reported rice MAPK interactome using proteome wide protein–protein interactions. Very recently, Wankhede et al. (2013a) have reported interaction networks of five MAP2Ks and 11 MAPKs by in silico docking and Y2H approaches. Although these studies have provided a good view on interactome map of MAPKs in rice, however, many other components including their substrates are yet to be elucidated. Therefore, it is important to generate more information on the interaction network among the different components of the MAPK cascade.

There are numerous stress perception and signaling pathways in plant cells, some of which are specific while others cross-talk at various steps. These pathways usually work under control of MAP kinases to deal with environmental stress stimuli though each stress stimulus induces kinase activity in a definite manner (Huang et al. 2012). It is well known at this instant that MAP kinase cascades of plants are highly conserved among all eukaryotes and are involved in the regulation of growth, differentiation, proliferation, cell death, and stress responses (Jonak et al. 1994; Hirt

2000; Tena et al. 2001; Sinha et al. 2011; Huang et al. 2012). MAPKs are activated by cold, salt, and drought in both monocot and dicot plants (Jonak et al. 1996; Ichimura et al. 2000; Xiong and Yang 2003). In plants, increased MAPK transcripts and activities have been reported in response to various abiotic stresses, e.g., heat, cold, drought, salinity, touch, ozone, UV irradiation, and oxidative stress (Sinha et al. 2011). Activated MAPKs target genes implicated in stress responses as well as genes that encode regulatory proteins such as protein kinases, phosphatases, and transcription factors (Krasensky and Jonak 2012; Šamajová et al. 2013). Genetic manipulations of MAPK signaling improve plant tolerance to these stresses (Xiong and Yang 2003; Teige et al. 2004; Chinnusamy et al. 2007; Krasensky and Jonak 2012; Šamajová et al. 2013). To generate stress tolerant plants, understanding the mechanisms of stress tolerance along with a plethora of genes involved in stress signaling network is extremely important. The MAPK cascades involved in stress tolerance are very specific in nature, and it is achieved within the same cell by docking domains of scaffold proteins that anchor different MAPK cascades in one complex (Mishra et al. 2006; Rodriguez et al. 2010; Sinha et al. 2011).

Metal toxicity, ozone, and oxidative stresses disturb cellular redox status by the production of reactive oxygen species (ROS) in plant cells. ROS, though react with biomolecules and at times damage them irreversibly, even leading to cell death, are also known to act as signals in diverse biological processes influencing signal transduction pathways and gene expression in plants. The plant cells have thus evolved strategies to utilize ROS as an indicator for controlling various biological processes (Opdenakker et al. 2012). Only hydrogen peroxide (H_2O_2) among ROS member is able to cross plant membranes and directly function in cell-to-cell signaling (Mittler et al. 2004; Pitzschke et al. 2006). In plant cells, H_2O_2 acts as a stimulus in signaling and triggering tolerance to various stresses (Opdenakker et al. 2012). H_2O_2 has also been shown to act as a key regulator in the accumulation of phosphatidic acid and/or activation of MAPK pathways (Miller et al. 2008; Opdenakker et al. 2012).

The cascade is regulated not only by transcriptional and translational regulation but also through posttranscriptional regulation including protein–protein interactions (Taj et al. 2010). The investigations hitherto have confirmed that MAPK cascade plays an important role in growth and development as well as in biotic and abiotic stress signaling. In this chapter, we discuss the latest insights and findings in the framework of MAPK cascade in combating abiotic stresses in plants.

8.2 Mitogen-Activated Protein Cascade

Mitogen-activated protein kinase was first identified in 1986 from animal cells by Sturgill and Ray and named as microtubule-associated protein-2 kinase (Mahajan and Tuteja 2005; Taj et al. 2010). Later, these proteins were found to be related to a set of proteins that phosphorylate at tyrosine residue in response to mitogen/stimuli and then renamed as mitogen-activated protein kinase (MAP kinase). In plants, MAP kinase genes were first reported in 1993 in pea by Stafstrom et al. (1993) and in alfalfa by Duerr et al. (1993). Afterwards, the cDNA of MAPKs have been cloned in many plant species (Pitzschke et al. 2009; Joshi et al. 2011; Sinha et al. 2011). The major milestones in the discovery of MAP kinases and their role in transduction of various stresses are described by Morris (2001), Mishra et al. (2006), Rodriguez et al. (2010), Taj et al. (2010), Sinha et al. (2011), and Opdenakker et al. (2012).

The MAP kinases are located in the cytoplasm and nucleus of the cells and connect multiples of receptors and sensors to cellular and nuclear responses and have emerged as a universal signal transduction apparatus in eukaryotes including yeasts, animals, and plants (Mishra et al. 2006; Taj et al. 2010; Sinha et al. 2011; Joshi et al. 2011). Studies in plants also state that these proteins are involved in fundamental physiological processes such as cell cycle regulation, hormonal responses, cell differentiation, biotic and abiotic stress signaling, and defense mechanisms. These proteins activate habitually in modules (Fig. 8.1). The basic module is a cascade of three functionally interlinked protein kinases in a row. The three kinases of the cascade are (1) MAP kinase kinase kinase (MAPKKK or MAP3K or MEKK or MKKK), (2) MAP kinase kinase (MAPKK or MAP2K or MEK or MKK), and (3) MAP kinase (MAPK or MPK). Colcombet and Hirt 2008 have reported a fourth level of kinases, MAP kinase kinase kinase kinase (MAPKKKK or MAP4K), which serve as an adaptor for linking upstream signaling sensors to the MAPK cascade. A typical cascade turns on when sensors present in the plasma membrane sense any stimulus, which in turn triggers sequential phosphorylation of the three components of the cascade (Fig. 8.1). A MAP3K activates a particular MAP2K through phosphorylation on two serine/threonine residues, which are generally in a conserved S/T-X3-5-S/T motif. Subsequently, the activated MAP2K phosphorylates downstream MAPK on threonine and tyrosine residues, which are located in the invariant sequence TXY (Cobb and Goldsmith 1995; Nakagami et al. 2005; Hamel et al. 2012). This sequential activation of the MAPK cascade eventually transmits signal(s) to the transcription factors, phospholipases or cytoskeletal proteins, microtubule-associated proteins for switching on or off the expression of specific sets of genes in response to the stimuli to adapt organisms to various stress conditions (Popescu et al. 2009; Taj et al. 2010; Danquah et al. 2014). MAPKs are kinases that phosphorylate a variety of substrates, both cytosolic and nuclear that includes transcription factors, splicing factors, and other protein kinases (Mishra et al. 2006; Opdenakker et al. 2012). The substrate specificity of the activated MAPK cascade along with the presence of several synergistic and/or antagonistic MAPK pathways activated upon various external cues regulates effective cell responses. This specificity demonstrates that cells have well-organized rule of gene expression controlling metabolism, cytoskeletal structure, and other regulatory functions.

The activity of kinases is usually calculated by the measurement of efficiency/frequency of the phosphorylation of preferred substrate proteins that contain serine–proline (-SP-) or threonine–proline (-TP-) phosphorylation sites (Rodriguez et al. 2010). Myelin basic protein (MBP) is one of the most preferred substrate used in in-gel kinase assay to study phosphorylation activity of the MAP kinases. To characterize MAPK signaling components, various chemical inhibitors are used (Rodriguez et al. 2010; Sheikh et al. 2012). Site-directed mutagenesis is another technique that can replace specific residues in the ATP-binding pocket to evaluate

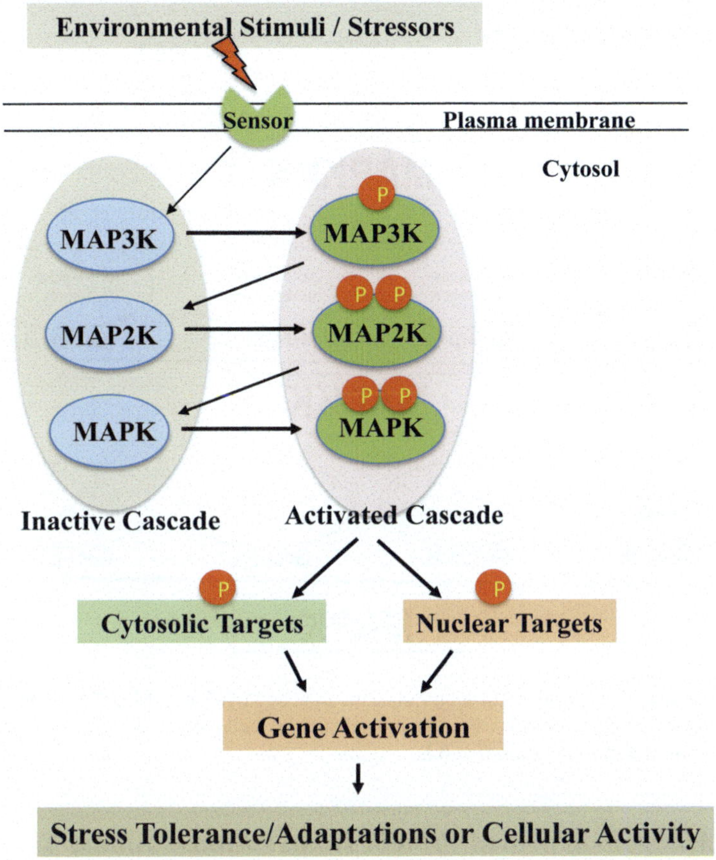

Fig. 8.1 A general MAP kinase pathway: the sensor(s) present in the plasma membrane first receives signal(s) from the environmental stimuli or stressors, which trigger sequential phosphorylation activity of the components of MAPK cascade in cytosol. Through this activity, the inactive cascade becomes active. The activated downstream MAPKs of the cascade then transmit signal(s) either to the cytosolic targets (like cytoskeleton-associated proteins, ribosomal proteins, microtubule binding proteins, etc.) or to the nuclear targets (like transcription factors) for specific gene activation for stress tolerance/adaptations or other cellular activities

enzymatic activity of the MAPKs. In many functional studies of MAPKs, threonine and tyrosine residues were substituted with alanine and phenylalanine in the TXY phosphorylation site or a T-loop generating a non-phosphorylable, inactive kinase (Rodriguez et al. 2010; Sheikh et al. 2012). Many other techniques to detect specific phosphorylation sites are also used. Genome sequencing projects describe sequences of putative MAPK components from several plant species. The genome sequence of *Arabidopsis thaliana* permitted identification of 60–80 MAP3Ks, 10 MAP2Ks, and 20 MAPKs (MAPK Group 2002; Hamel et al. 2006; Rodriguez et al. 2010; Dóczi et al. 2012).

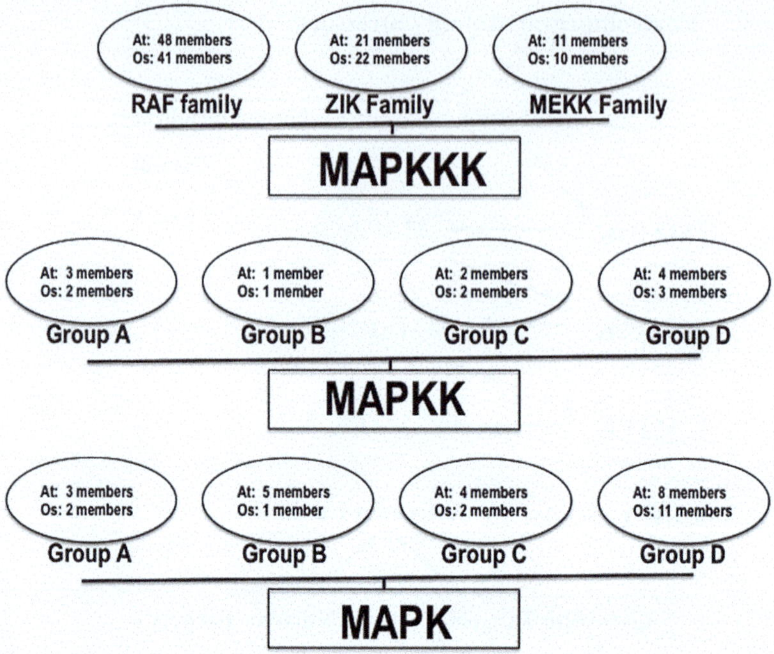

Fig. 8.2 A schematic diagram showing the classification of MAPKKK, MAPKK, and MAPK and the number of members present in each group from eudicot model plant *Arabidopsis thaliana* (At) and monocot model plant *Oryza sativa* (Os)

8.2.1 Mitogen-Activated Protein Kinase Kinase Kinase

Mitogen-activated protein kinase kinase kinases (MAP3Ks) form the upstream component of the MAPK cascade and are the largest group with putative numbers—80 in *Arabidopsis* (MAPK Group 2002; Jonak et al. 2002) and 75 in rice (Rao et al. 2010). They often get activated by G proteins (Sinha et al. 2011). Based on sequence similarities, this largest group is subdivided into three major subgroups: (1) Raf family, (2) MEKK family, and (3) ZIK family (Rao et al. 2010; Sinha et al. 2011) (Fig. 8.2). The number of putative members in these families of *Arabidopsis* and rice has been identified so far (Rao et al. 2010; Sinha et al. 2011). The MEKK-type are the ones for which function is provided, and the Raf-like kinases are those for which MAP3K function is not yet known (Rodriguez et al. 2010; Opdenakker et al. 2012).

The Raf family of MAP3Ks possesses a particular motif GTXX(W/Y)MAPE (Sinha et al. 2011). EDR1 (ENHANCED DISEASE RESISTANCE) and CTR1 (CONSTITUTIVE TRIPLE RESPONSE) are two best-studied MAP3Ks of the Raf family (Rodriguez et al. 2010). In *Arabidopsis*, CTR1 acts as negative regulator of ethylene signaling (Kieber et al. 1993; Huang et al. 2003), while EDR1 participates in defense responses (Frye et al. 2001).

ANP1/ANP2/ANP3 and MEKK1 are MEKK-like kinases of *Arabidopsis*. These are well characterized and are found to be responsive to oxidative stress and are also reported to negatively regulate auxin signal transduction pathway. ANP1 along with ANP2 and ANP3 from *Arabidopsis* is involved in plant cytokinesis (Krysan et al. 2002; Lukowitz et al. 2004). MAPKKKα is involved in defense response (del Pozo et al. 2004), while the role of YODA was identified in stomatal development (Bergmann et al. 2004). G(T/S)PX(F/Y/W)MAPEV forms a conserved signature catalytic domain of MEKK family (Rao et al. 2010; Sinha et al. 2011).

The characteristic conserved signature of ZIK family contains GTPEFMAPE (L/V/M) (Y/F/L) (Rao et al. 2010) and consists of 10 and 11 members in rice and *Arabidopsis*, respectively. WNK1, an *Arabidopsis* ZIK family member, has been shown to be involved in circadian rhythms by phosphorylating a protein involved in the rhythm (Murakami-Kojima et al. 2002). Complete genome sequence from other plants are required for proper annotation of putative MAP3Ks and subsequently for their systematic and complete catheterization.

8.2.2 Mitogen-Activated Protein Kinase Kinase

Mitogen-activated protein kinase kinases (MAP2Ks) are the smallest group and link the upstream component, MAP3K, to the downstream component of the cascade, MAPK. MAP2Ks are, in fact, the controlling switch of the cascade. Plant MAP2Ks have the S/T-X_5-S/T as the phosphorylation site and a putative MAPK-docking domain K/R-K/R-K/R-$X_{1\text{-}6}$-LX-L/V/I (Kumar et al. 2008). They are supposed to activate multiple MAPKs and are subdivided into four groups (A–D) based on sequence alignment (Fig. 8.2). Group A MAP2Ks seem to be involved in multiple abiotic stresses and cell division. They are responsive to pathogens also. MKK1 and MKK2 are group A MAP2Ks of *A. thaliana* that act upstream of the MPK4 (Ichimura et al. 2006). Both MKK1 and MKK2 mediate innate immunity responses (Meszaros et al. 2006) and also cold and salinity stress signals (Teige et al. 2004). In the same group, MKK6 activates downstream MPK13 (Melikant et al. 2004).

Group B MAP2Ks are characterized by a nuclear transfer factor (NTF) domain (Hamel et al. 2006), which enhances the nuclear import of cargo proteins indicating their involvement in cytoplasmic/nuclear trafficking (Sinha et al. 2011). Group B MAP2Ks include *A. thaliana* MKK3, which participates in cascades that are elicited by pathogens and are dependent on jasmonic acid signaling (Doczi et al. 2007; Takahashi et al. 2007).

Group C MAP2Ks include MKK4 and MKK5, whereas group D includes other remaining MKKs (MKK7 to MKK10) from *A. thaliana*. Group C and D MAP2Ks are upstream of group A MAPKs, which play a role in environmental and hormonal responses (Opdenakker et al. 2012). Functional evidences for the role of MAP2Ks are available for MKK1, MKK2, MKK4, MKK5, and MKK6 in *Arabidopsis*, PRKK and SIMKK in *Medicago*, MKK2 and MKK4 in tobacco, MKK5 in parsley, and MEK1 (Nakagami et al. 2005) and MKK6 in rice (Kumar and Sinha 2013; Wankhede et al. 2013a).

8.2.3 Mitogen-Activated Protein Kinase

Mitogen-activated protein kinases (MAPKs) form the downstream and the last component of the MAPK cascade. Based on the phylogenetic analysis of amino acid sequence and phosphorylation motif, plant MAPKs (or MPKs) are classified into two subtypes: (1) presence of TEY motif in T-loop and (2) presence of TDY motif in T-loop. The TEY subtype consists of three groups, A, B, and C, while the TDY subtypes are classified into group D (Fig. 8.2). MAPKs of group A are particularly involved in stress and developmental responses, and MPK3 and MPK6 are the best-studied examples of this group. Group B MAPKs, MPK4 and MPK11, are reported to be involved in environmental stress responses and cell division. These four MAPKs (MPK3, MPK4, MPK6, and MPK11) are considered as immune signaling component (Ahlfors et al. 2004; Droillard et al. 2004; Teige et al. 2004; Sinha et al. 2011; Rasmussen et al. 2012; Danquah et al. 2014). Very little is known about the specific functions of other members of these groups and even less is known about group C MAPKs. Functions of the MAPKs of this group are not well understood, but some are involved in oxidative stress signaling (Danquah et al. 2014). Three members of this group, MPK1 and MPK2 of *Arabidopsis* and PsMPK2 of *Pisum sativum*, are found to be induced by various stresses (Ortiz-Masia et al. 2007, 2008). Recently, Shi et al. (2010) have characterized a group C MAP kinase of *Gossypium* (GhMPK7) and stated its role in biotic stress and also in regulation of plant growth and development.

Interestingly, among group D members having TDY motifs, rice genome contains 11 members, whereas there are 6 members having TEY motifs. *Arabidopsis* on the other hand have more TEY members (12), while there are only eight members with TDY motif (Shi et al. 2011). A distinctive feature of most group D MAPKs is the presence of a C-terminal common docking domain that may act as a docking site for MAP2Ks, phosphatases, and protein substrates (Yoo et al. 2008). Group D MAPKs are shown to be induced by blast fungus and wounding (Opdenakker et al. 2012). *GhMPK16*, encoding a putative group D MAPK gene of *Gossypium*, has been characterized and noticed its importance in regulating pathogen resistance and drought signaling (Shi et al. 2011).

8.3 MAPK Cascade and Abiotic Stress

All plants are usually bound to tolerate variety of environmental changes and to withstand these stresses. Extreme environmental factors such change in temperatures (heat, cold), water stress, and soil salinity are major stressors that limit plant growth and development finally affecting agronomical yield. These unfavorable environmental conditions are generally known as abiotic stresses, which bring many adverse changes in gene expression in plants. Nevertheless, plants survive and grow with their complicated defense strategies involving MAPK cascade. The MAP kinases are acknowledged as one among the most important signaling machineries for regulation of physiological and developmental responses (Jonak et al. 1999; Joshi et al. 2011). Various abiotic stresses such as cold, heat, drought, salt, wounding, osmotic shock, UV rays, ozone, and heavy metal intoxication are known to activate these kinases.

Environmental stresses upset cellular structures and physiological processes (Larcher 2003; Krasensky and Jonak 2012). Salt stress, water stress, salinity, and low temperature impose an osmotic stress leading to turgor loss in plants. Membranes get disorganized, proteins may lose activity or gets denatured, and high levels of ROS are produced leading to oxidative damage. These result into inhibition of photosynthesis, damage of cellular structures, and metabolic dysfunction that contribute to reduced growth or fertility (Krasensky and Jonak 2012). Plant species are often highly variable about their optimum environmental conditions. For example, one condition might be harmful to one plant species but not to others (Munns and Tester 2008). Krasensky and Jonak (2012) described this as a mechanism to avoid and tolerate stress. They reasoned that there are a variety of protective mechanisms in plant that either delay or prevent the adverse impact of a stress factor on a plant. For example, cacti plants have constitutively adjusted to hot and arid climates by suitably adapting their morphology, physiology, and metabolism. This adaptation is stable and heritable. Alternatively, stress tolerance is the potential of a plant to adjust with a stressful condition. Plants increase their resistance with their defense mechanisms to various stresses. And acclimation is plastic and reversible in nature.

To adapt to the responses to environmental stresses, plants adjust their membrane system, modify the cell wall architecture, and change its cell cycle and cell division patterns (Krasensky and Jonak 2012). They also alter their metabolism by the production of compatible solutes (e.g., raffinose, proline, betaine, and glycine) that help in stabilizing proteins and cellular structures and/or sustaining cell turgor by osmotic adjustment and help in redox metabolism by removing excess levels of ROS (Bartels and Sunkar 2005; Munns and Tester 2008; Krasensky and Jonak 2012). On the other hand, at molecular level regulation, gene expression takes place upon stress (Chinnusamy et al. 2007; Shinozaki and Yamaguchi-Shinozaki 2007). Epigenetic regulations are also reported to play an important role in the regulation of gene expression in response to environmental stress (Hauser et al. 2011; Khraiwesh et al. 2012).

Stress-inducible genes comprise of genes that guards against stress, which synthesize osmoprotectants, or detoxifying enzymes, as well as genes that encode regulatory proteins such as protein kinases and phosphatases and transcription factors (Krasensky and Jonak 2012). MAPK cascades regulate abiotic stress responses. A complete MAPK cascade functioning in abiotic stresses is described in *Arabidopsis*, which consists of the MEKK1-MKK2-MPK4/MPK6 module (Teige et al. 2004). A number of MAPKs have been reported to be activated by cold, salt, and drought in both monocots and eudicots (Jonak et al. 1996; Ichimura et al. 2000; Xiong and Yang 2003; Rodriguez et al. 2010; Danquah et al. 2014; Smékalová et al. 2014). Genetic manipulation of MAPK signaling alters plant tolerance to these stresses (Xiong and Yang 2003; Teige et al. 2004; Krasensky and Jonak 2012). The role of MAP kinases in various abiotic stress regulations is discussed separately in the following sections.

8.3.1 Drought Stress

Drought is one of the major environmental stresses that limits growth and yield of plants. Low soil moisture coupled with continuous water extraction by root and subsequent transport of water within the plant results in drought stress. Many drought stress-responsive genes have been identified and characterized (Li et al. 2008; Lee et al. 2009). Alfalfa MAPK kinase, P44MKK4, was shown to be involved in drought stress by in-gel kinase assays and also by immunoprecipitation assay with specific peptide antibodies and was found to act in an ABA-independent manner (Jonak et al. 1996). While in *Arabidopsis* it was reported that the expression of AtMEKK1 and AtMPK3 is induced by drought stress (Mizoguchi et al. 1996), in rice drought resulted in the activation of OsMSRMK2 and OsMAPK5 (Agrawal et al. 2002). Overexpression of DSM1 (a putative rice MAP3K gene) increased tolerance to dehydration (Ning et al. 2010). ZmMPK3 also plays an important role in response to environmental stresses including drought stress (Wang et al. 2010).

8.3.2 Salt Stress

Salinity is a major environmental stress that upset agricultural productivity. High salinity causes both hyperionic and hyperosmotic stresses. Most of the economically important crop species including rice, maize, soybean, and beans are sensitive to soil salinity. Basically, a water shortage condition arises first in plants with salt stress, which later takes the form of a physiological drought (Xiong et al 2002; Mahajan and Tuteja 2005). Salinity greatly affects agriculture in many parts of the world, particularly irrigated lands. The involvement of MAPK cascade in combating salt stress has been observed in some plant species. In *Arabidopsis*, Mizoguchi et al. (1996) demonstrated transient increase in transcript levels for the genes encoding AtMEKK1 (MAP3K) and AtMPK3 (MAPK) within a few minutes of treatment of plants with salinity as well as with touch and low temperature. There are reports indicating of at least two MAPK cascades being activated by the same environmental stress. One such cascade is MPK4 cascade that composes of MEKK1-MEK1/MKK2-MPK4 module, and the other involves MPK6 and p44MAPK module (Ichimura et al. 1998, 2000). MEKK1 was later identified as an upstream activator of MKK2, which subsequently activates downstream MAPKs, MPK4, and MPK6 under salt or cold stress (Teige et al. 2004). Osmotic stress due to salinity has been found to activate expression of AtMPK3, AtMPK4, and AtMPK6 in *Arabidopsis* (Droillard et al. 2004; Kim et al. 2011; Persak and Pitzschke 2013). In particular, the activation of MPK6 has been shown to play a central role in the early stages of sodium detoxification, while in the longer term, it regulates hyperosmotically and salt-induced transcriptional transactivation of responsive genes by directly phosphorylating the ZAT6 zinc finger transcription factor (Liu et al. 2013).

In alfalfa, the activation of a 46-kDa SIMK (salt stress-induced MAPK) was noticed by salt stress (Munnik et al. 1999). A 48-kDa kinase, SIPK (salicylic acid-induced protein kinase) was reported to be activated in tobacco protoplasts exposed to salt and osmotic stress (Mikołajczyk et al. 2000). In rice, however, overexpression of OsMAPK5 transgenic plants showed increased tolerance to various stresses including salt (Xiong and Yang 2003). In *Zea mays*, three salt stress-induced MAPKs, ZmMPK3, ZmMAPK5, and ZmSIMK1, have been identified (Ding et al. 2009; Wang et al. 2010). Evidences for the activation and expression of CbMAPK3 (Zhang et al. 2006) and GhMPK7 (Shi et al. 2010) due to salt stress have been observed, respectively, in *Chorispora* and *Gossypium* species. Overexpression of constitutive active form of OsMKK6 made rice plants more tolerant to salt stress (Kumar and Sinha 2013).

8.3.3 Temperature Stress

Ever-increasing temperature in the environment may seriously affect crop yields particularly in the tropics and subtropics by the end of this century (Battisti 2009; Zou et al. 2011). Plants show evidence of responses to change in temperature. Perception of both high and low temperatures is very important for plants for their survival and regulation of developmental events. Environmental changes are expected to amplify or reduce average temperatures that affect plant growth and productivity adversely.

Studies in *Arabidopsis* indicated the role of MAPKs in temperature stress. Transcriptional induction of AtMEKK1, AtMPK3 (Mizoguchi et al. 1996), AtMPK4, and AtMPK6 (Ichimura et al. 2000) is observed with cold stress. AtMAPKK and AtMKK2 were also upregulated in response to cold stress (Teige et al. 2004). The role of MAPK module consisting of MEKK1-MKK2-MPK4/6 has been confirmed in cold stress in *Arabidopsis* (Ichimura et al. 2000; Sinha et al. 2011). Cold stress also induced the activity of ZmMAPK5 (Ding et al. 2009) and ZmMPK3 (Wang et al. 2010) in *Zea mays*. Moreover, reports are available for the activation of MAPKs with cold stress in other plant species like *Chorispora bungeana* (Zhang et al. 2006), *Gossypium hirsutum* (Shi et al. 2010), and *Salicornia brachiata* (Agarwal et al. 2010).

High temperature activates MAPKs as well. Heat stress in plant is created when there is a sudden increase in ambient maximum temperature by 5–7 °C with corresponding increase in the minimum temperature. Under such condition, the normal physiology of the plant gets affected resulting into accelerated plant maturity or senescence, which causes enormous crop losses. Peng et al. (2004) observed that rice grain yield declined by 10 % for every one degree increase in growing period minimum temperature. Studies on change in proteome profiles of temperature stressed plants have also been reported. Such studies in rice under high temperature conditions have been reported by Jagadish et al. (2010) and Lin et al. (2005, 2010). Earlier, changes in temperature affected the transcript levels of OsMSRMK2 in rice (Agrawal et al. 2002). Sangwan et al. (2002) identified heat shock-activated MAPK (HAMK) from alfalfa cells. It was shown in *Solanum tuberosum* that heat treatment to potato tubers resulted in the elevation of StMPK1 transcript levels (Blanco et al. 2006). Link et al. (2002) have shown phosphorylation of HsfA4 by a heat-activated MAPK in tomato. Such identification of different genes or proteins that are regulated by abiotic stresses is an essential first step toward understanding the molecular mechanisms and also in producing transgenic plants with the genes for enhanced resistance to a particular stress (Zou et al. 2011). Successful attempts for the production of heat-tolerant plant through genetic engineering have been made (Zou et al. 2011). Very recently, it is reported in *Arabidopsis* that during the heat stress MPK6 phosphorylates a major heat stress transcription factor, HsfA2, on its T249 residue (Evrard et al. 2013). A direct physical interaction between MPK6 and HsfA2 was demonstrated by immunoprecipitation, and using *mpk6* mutants, it was revealed that MPK6 is responsible for the nucleocytoplasmic shuttling of HsfA2 (Smékalová et al. 2014).

8.3.4 Heavy Metal Stress

Many heavy metal ions are also essential in many physiological and developmental processes of plants. Metals like iron (Fe), copper (Cu), cobalt (Co), zinc (Zn), and nickel (Ni) are essential micronutrients required for normal plant growth and development; however at higher concentrations they become toxic for plants. Other elements, like lead (Pb), cadmium (Cd), and aluminum (Al), are nonessential, and their presence adversely affects biochemical and physiological processes in plants (Cuypers et al. 2009). Metal contamination in soils is a worldwide problem and is of great concern (Opdenakker et al. 2012). High amount of metals is mainly caused by mining and industrial activities (Ruttens et al. 2011) or due to the agricultural use of different phosphate fertilizers, pesticides and fungicides, and sewage sludge or irrigation with wastewater (Opdenakker et al. 2012).

Metal stress is also known to activate MAPK cascades. Metal stress induces ROS production, which then acts as a secondary messenger and transduces the perception of metals to intracellular signals to the nucleus, where appropriate responses are initiated (Opdenakker et al. 2012). In *Arabidopsis*, Liu et al. (2010) observed the activation of MPK3 and MPK6 in response to $CdCl_2$ (Liu et al. 2010). Involvement of OsMKK4 and OsMKK3 has been reported in rice by arsenic stress (Rao et al. 2011). Transient activation of MAPKs (SIMK, SAMK, MKK2, and MMK3) rapidly induced after exposure to 100 μM $CuCl_2$, whereas treatment with 100 μM $CdCl_2$ delayed this profile was reported in *Medicago sativa* (Jonak et al. 2004). The activation of MAPK and induction of gene of *MSMRK2* by higher amount of copper, cadmium, and mercury in rice is reported by Agrawal et al. (2002). Yeh et al. (2004, 2007) have confirmed the activation of a MAPK gene in rice in response to cadmium stress, and Gupta et al. (2009) reported the activation of MAPK in response to arsenic stress in mustard plant.

However, the function of MAPK cascades during metal stress is poorly understood. The knowledge about metal signaling and more specifically their downstream targets is essential for understanding plant responses to metal stress (Opdenakker et al. 2012). Therefore, it is important to focus future research on the functional analysis of MAP kinases cascade components in plant during metal stress.

8.3.5 *UV Stress*

UV-B rays in tomato have shown the activation of MPK1, MPK2, and MPK3 (orthologues of *Arabidopsis* MPK3 and MPK6) besides other stresses like wounding and various elicitors (Holley et al. 2003). Interestingly, SR160 (a leucine-rich membrane-spanning protein kinase that is identical to the brassinosteroid receptor BRI1) that binds to wounding signaling also responds to UV-B signaling, suggesting that brassinosteroid BRI1 and UV-B might be sensed by the same receptor (Stratmann 2003; Nakagami et al. 2005). However, it still remains unclear whether the same MAPK cascade get activated by all three signaling pathways and that how a cell can differentiate between systemin, brassinosteroids, and UV-B (Nakagami et al. 2005). An *Arabidopsis* mutant (*mkp1*) with increased sensitivity to chemical mutagens and UV-C also encodes a dual specific phosphatase that affects genotoxic stress (Ulm et al. 2002; Nakagami et al. 2005). These mutants were more resistant to salt stress as well. Later, it was shown that MKP1 preferentially interacts and dephosphorylates MPK6 and to a lesser extent also with MPK3 and MPK4 (Ulm et al. 2002). Although the observation indicates that MKP1 controls the genotoxic and salt stress-inducible MPK6 pathway, the role of MPK6 in the adaptation of *Arabidopsis* to these stresses is still unclear. In rice, the role of OsMKK6 has been shown in UV-B-mediated upregulation of phytoalexin biosynthesis genes (Wankhede et al. 2013b).

8.3.6 *Ozone Stress*

Ozone is considered as a strong oxidant atmospheric pollutant. It also activates MAP kinases in plants (Mishra et al. 2006; Rodriguez et al. 2010; Sinha et al. 2011). The activation of MAPK by ozone treatment is reported possibly by the accumulation of H_2O_2, superoxide anion, and hydroxyl radicals in cell cultures and in tobacco plants (Rodriguez et al. 2010). In *Arabidopsis*, MPK3 and MPK6 are activated by ozone treatments, and interestingly, the mutants of these genes are hypersensitive to ozone treatment (Miles et al. 2005). The transcript level of OsMSMRK2 was reported to be induced by ozone treatment in rice (Agrawal et al. 2002). Tobacco plants silenced with NtMPK4 showed enhanced sensitivity to ozone (Gomi et al. 2005). The activation of MAPKs in poplar by ozone treatment depended on the activation of an upstream membrane-localized component, production of ROS, and influx of calcium ions via membrane channels (Rodriguez et al. 2010).

8.3.7 Oxidative Stress

Hydrogen peroxide, superoxide, the hydroxyl radical, and singlet oxygen are the most common ROS, which are generated continuously as a byproduct of aerobic metabolism in plant cells (Huang et al. 2012). ROS once produced inside plant cells are simultaneously degraded by several mechanisms. These generation and degradation of ROS are controlled by sophisticated mechanisms inside the cells (Huang et al. 2012). Almost every abiotic stresses such as drought, salinity, heat and cold, heavy metal ions, UV exposure, and ozone cause overproduction of ROS in plant cells due to metabolic imbalance and oxidation, which create oxidative stress and cause damage to proteins, lipids, carbohydrates, and DNA (Rodriguez et al. 2010; Huang et al. 2012). Scavenger enzymes such as catalases that decompose H_2O_2 help plants to overcome oxidative stress (Rodriguez et al. 2010). MAPK cascades, which get activated by oxidative burst, are also known to regulate ROS accumulation. Such activation of MAPK is mediated by hormones like JA (jasmonic acid), ABA (abscisic acid), and SA (salicylic acid) (Rodriguez et al. 2010).

ROS treatment activates a serine/threonine protein kinase OXI1 (oxidative stress inducible 1) in *Arabidopsis*. This OXI1 in turn activates downstream components, MPK3 and MPK6 (Rentel and Knight 2004). MEKK1-dependent signaling pathway has been studied and implicated in regulation of a variety of processes based on ROS homeostasis in *Arabidopsis* (Pitzschke et al. 2009). The experimental approaches of Nakagami et al. (2006) clearly specify that MEKK1 mediates ROS-induced MPK4 signaling. Earlier in *A. thaliana* protoplasts, it has been reported that MAP3Ks are activated by H_2O_2 that includes ANP1, which activates MPK3 and MPK6 (Kovtun et al. 2000). Later, in alfalfa also a novel MAP3K, OMTK1 (oxidative stress-activated MAP triple-kinase 1), was identified, which activates MAPK, MMK3 (Nakagami et al. 2004). H_2O_2-induced activation of plant MAPK has also been reported in maize (Ding et al. 2009; Zong et al. 2009) and pea (Ortiz-Masia et al. 2008). The findings so give a complex picture of MAPK module activation by oxidative stress responses. The cascade is not only induced and activated by ROS but also plays important role in regulating the production of ROS levels by affecting catalase activity.

8.4 Cross Talk of MAPK Cascade Activation Under Different Abiotic Stress

Detailed studies of MAPK cascades provided evidence for cross talk during abiotic stress signaling in plants. Plant MAP kinase pathways are involved in gene expressions in response to various environmental stimuli, and these signals are not necessarily transduced into separate parallel cascades. Several MAPK cascade components are activated by multiples of stress stimuli suggesting that the cascades act as vital junction in stress signaling. Moreover, they show a lot of overlaps and exhibit cross talks at various levels at the same time. One such example is the activation of

MAP3K (ANP1) by H_2O_2 in *Arabidopsis* that in turn activates a MAPK cascade involving MPK3 and MPK6. This activated MAPK blocks auxin responsive genes, suggesting that there is a cross talk between the oxidative stresses and auxin pathway (Zhang et al. 2006). Similarly in tobacco, NPK1 (a MAP3K), which is mainly involved during cell division, is best characterized and its involvement was observed not only in transduction of oxidative stresses but also in repression of auxin-related signaling (Takahashi et al. 2007). The ANP/NPK-like MAP3Ks target additional MAPKs genes. The specificity for their signal transmission is probably maintained at the lower levels of the MAPK cascade. Yet others may be activated by common extra cellular signals but have a variety of signaling pathways. One such interesting example is the activation of AtMPK4 and AtMPK6 by cold, salinity, and wounding stress where AtMPK4 is activated by AtMEKK1 during wounding and by MKK2 in cold and salt stress (Ichimura et al. 1998, 2000).

Extensive cross talks between different pathways at the level of MAP kinases exist in plants (Zhang et al. 2006). Different MAP2Ks and MAPKs may interact with specific MAP3Ks. Low numbers of MAPKK and high numbers of MAPK genes in plant genomes suggest that a single MAPKK may target multiple MAPKs. Further, it is also possible that an individual MAPK protein may get activated by multiple upstream MAPKK (Dóczi et al. 2012; Wankhede et al. 2013a). There have been ever-increasing reports of cross talk in MAPK cascade from different species. A comprehensive cross talk of MAPK components during abiotic stresses in the model plant *Arabidopsis* has been represented in Fig. 8.3. It is described in earlier

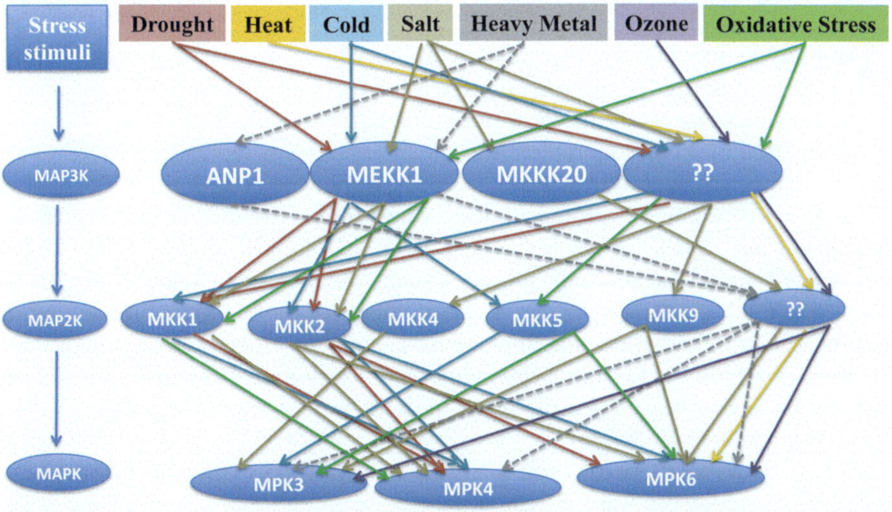

Fig. 8.3 A schematic representation of cross talk in MAPK cascade instigated by abiotic stress in model eudicot plant *Arabidopsis*. The left panel indicates different components of MAPK cascade. *Solid lines* indicate proven component of the cascade, while *dashed lines* indicate postulated interactions

sections that *Arabidopsis* genome encodes approximately 80 MAP3Ks, 10 MAP2Ks, and 20MAPKs. Signals perceived by these 80 MAP3Ks have to be transduced through 10 MAP2Ks to 20 MAPKs, which offer scope for cross talk between different stress signals (Huang et al. 2012).

8.5 Interactome Among Components of MAPK Cascade and Target Proteins

The systematic analysis of the proteome provides a direct link of genome sequence with their possible biological activity (Agrawal and Rakwal 2006). To understand the cellular processes at the protein level, investigation of the proteome at expression levels, posttranslational modifications, and interactions is required (Peck 2005). Two most important approaches for proteome analysis include two-dimensional gel electrophoresis (2DGE) and mass spectrometry (MS). Since 2001, matrix-assisted laser desorption/ionization-time of flight (MALDI-TOF) and electrospray ionization (ESI) have been the most commonly used MS techniques for proteome analysis (Mann and Pandey 2001; Mann et al. 2001). Proteome analysis to study changes in response to drought stress and to study changes in the nuclear proteome in response to cold stress is described first by Salekdeh et al. (2002) and Bae et al. (2003), respectively, in rice and *Arabidopsis* using 2DGE and MALDI-TOF-MS. Since then, protein microarrays for the identification of several MKK/MPK/substrate signaling pathways are being done in many plant species. However, in order to understand MAPK signaling completely, it is important to understand the interaction networks between MAPK components.

All proteins function by interacting with other biomolecules to perform cellular activities. Knowledge of protein–protein interactions is essential for the study of protein functions in cellular processes on a genome-wide level. The collection of information of all protein interactions in an organism is typically referred to as an interactome (Gray 2006; Sinha et al. 2011; Wankhede et al. 2013a). Presently, the analysis of complete protein linkage map (interactome) is a major goal to understand any signaling pathways. The yeast two-hybrid method is the commonly used technique to study protein–protein interactions (Sinha et al. 2011; Wankhede et al. 2013a). At the moment, methods of computational modeling of protein structures and molecular docking have evolved, which holds a great promise in the prediction of protein–protein interactions. Protein–protein docking is an efficient means to study the structure of protein–protein complexes (Kang et al. 2010; Wankhede et al. 2013a).

A large-scale yeast two-hybrid analysis to identify rice proteins involved in stress and development was carried out by Cooper et al. (2003). Later, Ding et al. (2009) generated a rice kinase–protein interaction map of 116 rice kinases and 254 interacting partners. Very recently, Singh et al. (2012) have reported rice MAPK interactome analysis using directed as well as proteome wide protein–protein interactions. These studies reveal that single MAP2K can interact with more than one

MAPK. Similarly, protein interactome network between *Arabidopsis* MAPKs and the upstream MAP2Ks has been studied by Lee et al. earlier in 2008. Many bioinformatics tools are now available for plant proteome analysis. These include PAT (proteins of *A. thaliana* database) (http://www.pat.sdsc.edu/), MAtDB (MIPS *A. thaliana* database) (http://mips.gsf.de/proj/thal/db), and RPD (rice proteome database) (http://gene64.dna.affrc.go.jp/RPD/ main_en.html).

8.6 Conclusions and Future Perspectives

MAPK cascades are recognized as the machinery for signal transduction, which phosphorylate a variety of substrates including transcription factors, other protein kinases, and cytoskeleton-associated proteins after being activated by various environmental stresses in plants. Components of MAPK cascades act as converging points for numerous stress signaling pathways. During the last two decades, significant progress has been made in deciphering signal transduction mechanism using a combination of physiological, biochemical, and genetic approaches. Various signaling components have been identified, but such investigations are restricted to few plant species so far. Moreover, the integral molecular mechanisms of the abiotic stress signal transduction mediated by MAPK cascade are largely unknown yet. It will, therefore, be imminent to identify and functionally characterize MAPK components to define specific functions and mechanisms of signal transduction. It is also compulsory to extend our knowledge about the cross talks of abiotic stress signaling components and their substrates. For this, potentially useful approaches to analyze the stress-stimulated MAP kinase pathways in response to various stimuli are likely to come from integrating publically available transcriptomics datasets and development of targeted phosphoproteomics/proteomics approaches. Enhanced information is required to be generated to understand the mechanisms of environmental signals communication in plant cells that will induce a coordinated metabolic response. This information will play a pivotal role to meet the goal of increased plant stress tolerance and productivity in an ever-changing environment.

Acknowledgments H.A. thanks the Department of Science and Technology, Govt. of India, for Women Scientist Grant. The research AKS lab is supported by grants from the Department of Biotechnology, Govt. of India, and the Department of Science and Technology, Govt. of India.

References

Agarwal PK, Gupta K, Jha B (2010) Molecular characterization of the *Salicornia brachiata* SbMAPKK gene and its expression by abiotic stress. Mol Biol Rep 37:981–986

Agrawal GK, Rakwal R, Iwahashi H (2002) Isolation of novel rice (*Oryza sativa* L.) multiple stress responsive MAP kinase gene, OsMSRMK2, whose mRNA accumulates rapidly in response to environmental cues. Biochem Biophys Res Commun 294:1009–1016

Agrawal GK, Rakwal R (2006) Rice proteomics: a cornerstone for cereal food crop proteomes. Mass Spectrom Rev 25:1–53

Ahlfors R, Macioszek V, Rudd J, Brosché M, Schlichting R, Scheel D, Kangasjärvi J (2004) Stress hormone-independent activation and nuclear translocation of mitogen-activated protein kinases in *Arabidopsis thaliana* during ozone exposure. Plant J 40:512–522

Bae MS, Cho EJ, Choi EY, Park OK (2003) Analysis of the Arabidopsis nuclear proteome and its response to cold stress. Plant J 36:652–663

Bartels D, Sunkar R (2005) Drought and salt tolerance in plants. Crit Rev Plant Sci 24:23–58

Battisti D (2009) Historical warnings of future food insecurity with unprecedented seasonal heat. Science 323:240–244

Bergmann DC, Lukowitz W, Somerville CR (2004) Stomatal development and pattern controlled by a MAPKK kinase. Science 304:1494–1497

Blanco FA, Zanetti ME, Casalongué CA, Daleo GR (2006) Molecular characterization of a potato MAP kinase transcriptionally regulated by multiple environmental stresses. Plant Physiol Biochem 44:315–322

Chinnusamy V, Zhu J, Zhu JK (2007) Cold stress regulation of gene expression in plants. Trends Plant Sci 12:444–451

Cobb MH, Goldsmith EJ (1995) How MAP kinases are regulated. J Biol Chem 270: 14843–14846

Colcombet J, Hirt H (2008) Arabidopsis MAPKs: a complex signalling network involved in multiple biological processes. Biochem J 413:217–226

Cooper B, Clarke JD, Budworth P, Kreps J et al (2003) A network of rice genes associated with stress response and seed development. Proc Natl Acad Sci U S A 100:4945–4950

Cuypers A, Smeets K, Vangronsveld J (2009) Heavy metal stress in plants. In: Hirt H (ed) Plant stress biology: from genomics to systems biology. Wiley-VCH, Weinheim, pp 161–178

Danquah A, de Zelicourt A, Colcombet J, Hirt H (2014) The role of ABA and MAPK signaling pathways in plant abiotic stress responses. Biotechnol Adv 32(1):40–52

del Pozo O, Pedley KF, Martin GB (2004) MAPKKK alpha is a positive regulator of cell death associated with both plant immunity and disease. EMBO J 23:3072–3082

Ding X, Richter T, Chen M, Fujii H, Seo YS, Xie M, Zheng X, Kanrar S, Stevenson RA, Dardick C, Li Y, Jiang H, Zhang Y, Yu F, Bartley LE, Chem M, Bart R, Chen X, Zhu L, Farmerie WG, Gribskov M, Zhu JK, Fromm ME, Ronald PC, Song W-N (2009) A rice kinase-protein interaction map. Plant Physiol 149:1478–1492

Dóczi R, Okrész L, Romero AE, Paccanaro A, Bögre L (2012) Exploring the evolutionary path of plant MAPK networks. Trends Plant Sci 17:1360–1385

Doczi R, Brader G, Pettko-Szandtner A, Rajh I, Djamei A, Pitzschke A, Teige M, Hirt H (2007) The *Arabidopsis* mitogen-activated protein kinase kinase MKK3 is upstream of group C mitogen-activated protein kinases and participates in pathogen signaling. Plant Cell 19: 3266–3279

Droillard MJ, Boudsocq M, Barbier-Brygoo H, Lauriere C (2004) Involvement of MPK4 in osmotic stress response pathways in cell suspensions and plantlets of *Arabidopsis thaliana*: activation by hypoosmolarity and negative role in hyperosmolarity tolerance. FEBS Lett 574:42–48

Duerr B, Gawlenowski M, Ropp T, Jacobs T (1993) MsERK1: A mitogen-activated protein kinase from a flowering plant. Plant Cell 5:87–96

Evrard A, Kumar M, Lecourieux D, Lucks J, von Koskull-Doring P, Hirt H (2013) Regulation of the heat stress response in Arabidopsis by MPK6-targeted phosphorylation of the heat stress factor HsfA2. Peer J 1:e59

Frye CA, Tang D, Innes RW (2001) Negative regulation of defense responses in plants by a conserved MAPKK kinase. Proc Natl Acad Sci U S A 98:373–378

Gomi K, Ogawa D, Katou S, Kamada H, Nakajima N, Saji H (2005) A mitogen-activated protein kinase NtMPK4 activated by SIPKK is required for jasmonic acid signaling and involved in ozone tolerance via stomatal movement in tobacco. Plant Cell Physiol 46:1902–1914

Gray JJ (2006) High resolution protein-protein docking. Curr Opin Struct Biol 16:183–193

Gupta M, Sharma P, Sarin NB, Sinha AK (2009) Differential response of arsenic stress in two varieties of Brassica juncea L. Chemosphere 74:1201–1208

Hamel LP, Nicole MC, Duplessis S, Ellis BE (2012) Mitogen-activated protein kinase signaling in plant- interacting fungi: distinct messages from conserved messengers. Plant Cell 24: 1327–1351

Hamel LP, Nicole MC, Sritubtim S, Morency MJ, Ellis M, Ehlting J, Beaudoin N, Barbazuk B, Klessig D, Lee J et al (2006) Ancient signals: comparative genomics of plant MAPK and MAPKK gene families. Trends Plant Sci 11:192–198

Hauser MT, Aufsatz W, Jonak C, Luschnig C (2011) Transgenerational epigenetic inheritance in plants. Biochim Biophys Acta 1809:459–468

Hirt H (ed) (2000) Results and problems in cell differentiation: map kinases in plant signal transduction. Springer, Heidelberg

Holley SR et al (2003) Convergence of signaling pathways induced by systemin, oligosaccharide elicitors, and ultraviolet-B radiation at the level of mitogen-activated protein kinases in Lycopersicon peruvianum suspension-cultured cells. Plant Physiol 132:1728–1738

Huang GT, Ma SL, Bai LP, Zhang L, Ma H, Jia P, Liu J, Zhong M, Guo ZF (2012) Signal transduction during cold, salt, and drought stresses in plants. Mol Biol Rep 39:969–987

Huang Y, Li H, Hutchison CE, Laskey J, Kieber JJ (2003) Biochemical and functional analysis of CTR1, a protein kinase that negatively regulates ethylene signaling in *Arabidopsis*. Plant J 33:221–233

Ichimura K, Casais C, Peck SC, Shinozaki K, Shirasu K (2006) MEKK1 is required for MPK4 activation and regulates tissue-specific and temperature-dependent cell death in *Arabidopsis*. J Biol Chem 281:36969–36976

Ichimura K, Mizoguchi T, Irie K, Morris P, Giraudat J, Matsumoto K, Shinozaki K (1998) Isolation of ATMEKK1 (a MAP kinase kinase kinase)—interacting proteins and analysis of a MAP kinase cascade in *Arabidopsis*. Biochem Biophys Res Commun 253:532–543

Ichimura K, Mizoguchi T, Yoshida R, Yuasa T, Shinozaki K (2000) Various abiotic stresses rapidly activate Arabidopsis MAP kinases ATMPK4 and ATMPK6. Plant J 24:655–665

Jagadish S, Muthurajan R, Oane R, Wheeler T, Heuer S, Bennett J, Craufurd P (2010) Physiological and proteomic approaches to address heat tolerance during anthesis in rice (Oryza sativa L.). J Exp Bot 61:143–156

Jonak C, Heberle-Bors E, Hirt H (1994) MAP kinases: universal multipurpose signalling tools. Plant Mol Biol 24:407–416

Jonak C, Kiegerl S, Ligterink W, Barker PJ, Huskisson NS, Hirt H (1996) Stress signaling in plants: a mitogen-activated protein kinase pathway is activated by cold and drought. Proc Natl Acad Sci U S A 93:11274–11279

Jonak C, Ligterink W, Hirt H (1999) MAP kinases in plant signal transduction. Cell Mol Life Sci 55:204–213

Jonak C, Nakagami H, Hirt H (2004) Heavy metal stress. Activation of distinct mitogen-activated protein kinase pathways by copper and cadmium. Plant Physiol 136:3276–3283

Jonak C, Okresz L, Bogre L, Hirt H (2002) Complexity, cross talk and integration of plant MAP kinase signalling. Curr Opin Plant Biol 5:415–424

Joshi RK, Kar B, Nayak S (2011) Characterization of mitogen activated protein kinases (MAPKs) in the *Curcuma longa* expressed sequence tag database. Bioinformation 7:180–183

Kang S, Chen S, Dai S (2010) Proteomics characteristics of rice leaves in response to environmental factors. Front Biol 5:246–254

Khraiwesh B, Zhu JK, Zhu J (2012) Role of miRNAs and siRNAs in biotic and abiotic stress responses of plants. Biochim Biophys Acta 1819:137–148

Kieber JJ, Rothenberg M, Roman G, Feldmann KA, Ecker JR (1993) CTR1, a negative regulator of the ethylene response pathway in *Arabidopsis*, encodes a member of the Raf family of protein kinases. Cell 72:427–441

Kim SH, Woo DH, Kim JM, Lee SY, Chung WS, Moon YH (2011) Arabidopsis MKK4 mediates osmotic-stress response via its regulation of MPK3 activity. Biochem Biophys Res Commun 412:150–154

Kovtun Y, Chiu WL, Tena G, Sheen J (2000) Functional analysis of oxidative stress-activated mitogen-activated protein kinase cascade in plants. Proc Natl Acad Sci U S A 97:2940–2945

Krasensky J, Jonak C (2012) Drought, salt, and temperature stress-induced metabolic rearrangements and regulatory networks. J Exp Bot 63:1593–1608

Krysan PJ, Jester PJ, Gottwald JR, Sussman MR (2002) An *Arabidopsis* mitogen-activated protein kinase kinase kinase gene family encodes essential positive regulators of cytokinesis. Plant Cell 14:1109–1120

Kumar K, Sinha AK (2013) Over expression of constitutively active mitogen activated protein kinase kinase 6 enhances tolerance to salt stress in rice. Rice 6(1):25

Kumar K, Rao KP, Sharma P, Sinha AK (2008) Differential regulation of rice mitogen activated protein kinase kinase (MKK) by abiotic stress. Plant Physiol Biochem 46:891–897

Larcher W (2003) Physiological plant ecology, 4th edn. Springer, New York

Lee JS, Wang S, Sritubtim S, Chen JG, Ellis BE (2009) *Arabidopsis* mitogen-activated protein kinase MPK12 interacts with the MAPK phosphatase IBR5 and regulates auxin signaling. Plant J 57:975–985

Li WX, Oono Y, Zhu J, He XJ, Wu JM, Iida K, Lu X-Y, Cui X, Jin H, Zhu J-K (2008) The Arabidopsis NFYA5 transcription factor is regulated transcriptionally and post transcriptionally to promote drought resistance. Plant Cell 20:2238–2251

Lin C-J, Li C-Y, Lin S-K, Yang F-H, Huang J-J, Liu Y-H, Lur H-S (2010) Influence of high temperature during grain filling on the accumulation of storage proteins and grain quality in rice (Oryza sativa L.). J Agric Food Chem 58:10545–10552

Lin SK, Chang MC, Tsai YG, Lur HS (2005) Proteomic analysis of the expression of proteins related to rice quality during caryopsis development and the effect of high temperature on expression. Proteomics 5:2140–2156

Link V, Sinha AK, Vasistha P, Hofmann MG, Proels RK, Ehness R, Roitsch T (2002) Activation of a MAPK by heat in tomato cell suspension cultures. FEBS Lett 531(2):179–183

Liu XM, Kim KE, Kim KC, Nguyen XC, Han HJ, Jung MS, Kim HS, Kim SH, Park HC, Yun DJ, Chung WS (2010) Cadmium activates *Arabidopsis* MPK3 and MPK6 via accumulation of reactive oxygen species. Phytochemistry 71:614–618

Liu XM, Nguyen XC, Kim KE, Han HJ, Yoo J, Lee K, Kim MC, Yun D-J, Chung WS (2013) Phosphorylation of the zinc finger transcriptional regulator ZAT6 by MPK6 regulates Arabidopsis seed germination under salt and osmotic stress. Biochem Biophys Res Commun 430:1054–1059

Lukowitz W, Roeder A, Parmenter D, Somerville C (2004) A MAPKK kinase gene regulates extra-embryonic cell fate in Arabidopsis. Cell 116:109–119

Mahajan S, Tuteja N (2005) Cold, salinity and drought stresses: an overview. Arch Biochem Biophys 444:139–158

Mann M, Pandey A (2001) Use of mass spectrometry-derived data to annotate nucleotide and protein sequence databases. Trends Biochem Sci 26:54–61

Mann M, Hendrickson RC, Pandey A (2001) Analysis of proteins and proteomes by mass spectrometry. Annu Rev Biochem 70:437–473

MAPK Group (2002) Mitogen-activated protein kinase cascades in plants: a new nomenclature. Trends Plant Sci 7:301–308

Melikant B, Giuliani C, Halbmayer-Watzina S, Limmongkon A, Heberle-Bors E, Wilson C (2004) The *Arabidopsis thaliana* MEK AtMKK6 activates the MAP kinase AtMPK13. FEBS Lett 576:5–8

Meszaros T, Helfer A, Hatzimasoura E, Magyar Z, Serazetdinova L, Rios G, Bardóczy V, Teige M, Koncz C, Peck S, Bögre L (2006) The *Arabidopsis* MAP kinase kinaseMKK1participates in defense responses to the bacterial elicitor flagellin. Plant J 48:485–498

Miles GP, Samuel MA, Zhang Y, Ellis BE (2005) RNA interference-based (RNAi) suppression of AtMPK6, an *Arabidopsis* mitogen-activated protein kinase, results in hypersensitivity to ozone and misregulation of AtMPK3. Environ Pollut 138:230–237

Miller G, Shulaev V, Mittler R (2008) Reactive oxygen signaling and abiotic stress. Physiol Plant 133:481–489

Mishra NS, Tuteja R, Tuteja N (2006) Signaling through MAP kinase networks in plants. Arch Biochem Biophys 452:55–68

Mittler R, Vanderauwera S, Gollery M, van Breusegem F (2004) Reactive oxygen gene network of plants. Trends Plant Sci 9:490–498

Mizoguchi T, Irie K, Hirayama T, Hayashida N, Yamaguchi-Shinozaki K, Matsumoto K, Shinozaki K (1996) A gene encoding a mitogen-activated protein kinase kinase kinase is induced simultaneously with genes for a mitogen-activated protein kinase and an S6 ribosomal protein kinase by touch, cold, and water stress in *Arabidopsis thaliana*. Proc Natl Acad Sci U S A 93:765–769

Morris PC (2001) MAP kinase signal transduction pathways in plants. New Phytol 151:67–89

Munnik T, Ligterink W, Meskiene I, Calderini O, Beyerly J, Musgrave A, Hirt H (1999) Distinct osmo-sensing protein kinase pathways are involved in signalling moderate and severe hyperosmotic stress. Plant J 20:381–388

Munns R, Tester M (2008) Mechanisms of salinity tolerance. Annu Rev Plant Biol 59:651–681

Murakami-Kojima M, Nakamichi N, Yamashino T, Mizuno T (2002) Plant Cell Physiol 43: 675–683

Nakagami H, Kiegerl S, Hirt H (2004) OMTK1, a novel MAPKKK, channels oxidative stress signaling through direct MAPK interaction. J Biol Chem 279:26959–26966

Nakagami H, Pitzschke A, Hirt H (2005) Emerging MAP kinase pathways in plant stress signalling. Trends Plant Sci 10:339–346

Nakagami H, Soukupova H, Schikora A, Zarsky V, Hirt H (2006) A mitogen-activated protein kinase kinase kinase mediates reactive oxygen species homeostasis in *Arabidopsis*. J Biol Chem 281:38697–38704

Ning J, Li X, Hicks LM, Xiong L (2010) A Raf-like MAPKKK gene DSM1 mediates drought resistance through reactive oxygen species scavenging in rice. Plant Physiol 152:876–890

Opdenakker K, Remans T, Vangronsveld J, Cuypers A (2012) Mitogen-activated protein (MAP) kinases in plant metal stress: regulation and responses in comparison to other biotic and abiotic stresses. Int J Mol Sci 13:7828–7853

Ortiz-Masia D, Perez-Amador MA, Carbonell J, Marcote MJ (2007) Diverse stress signals activate the C1 subgroup MAP kinases of *Arabidopsis*. FEBS Lett 581:1834–1840

Ortiz-Masia D, Perez-Amador MA, Carbonell P, Carbonell AF, Marcote MJ (2008) Characterization of PsMPK2, the first C1 subgroup MAP kinase from pea (Pisum sativum L.). Planta 227: 1333–1342

Peck SC (2005) Update on proteomics in Arabidopsis. Plant Physiol 138:591–599

Peng S, Huang J, Sheehy JE, Laza RC, Visperas RM, Zhong X, Centeno GS, Khush GS, Cassman KG (2004) Rice yields decline with higher night temperature from global warming. Proc Natl Acad Sci U S A 101:9971–9975

Persak H, Pitzschke A (2013) Tight interconnection and multi-level control of Arabidopsis MYB44 in MAPK cascade signalling. PLoS One 8:e57547

Pitzschke A, Djamei A, Bitton F, Hirt H (2009) A major role of the MEKK1-MKK1/2-MPK4 pathway in ROS signalling. Mol Plant 2:120–137

Pitzschke A, Forzani C, Hirt H (2006) Reactive oxygen species signaling in plants. Antioxid Redox Sign 8:1757–1764

Popescu SC, Popescu GV, Bachan S, Zhang Z, Gerstein M, Snyder M, Dinesh-Kumar SP (2009) MAPK target networks in *Arabidopsis thaliana* revealed using functional protein microarrays. Genes Dev 23:80–92

Rao KP, Richa T, Kumar K, Raghuram B, Sinha AK (2010) In silico analysis reveals 75 members of mitogen-activated protein kinase kinase kinase gene family in rice. DNA Res 17:139–153

Rao KP, Vani G, Kumar K, Wankhede DP, Misra M, Gupta M, Sinha AK (2011) Arsenic stress activates MAP kinase in rice roots and leaves. Arch Biochem Biophys 506:73–82

Rasmussen MW, Roux M, Petersen M, Mundy J (2012) MAP kinase cascades in Arabidopsis innate immunity. Front Plant Sci 3:169

Rentel MC, Knight MR (2004) Oxidative stress-induced calcium signaling in Arabidopsis. Plant Physiol 135:1471–1479

Rodriguez MC, Petersen M, Mundy J (2010) Mitogen-activated protein kinase signaling in plants. Annu Rev Plant Biol 6:621–649

Ruttens A, Boulet J, Weyens N, Smeets K, Adriaensen K, Meers E, van Slycken S, Tack F, Meiresonne L, Thewys T, Witters N, Carleer R, Dupae J, Vangronsveld J (2011) Short rotation coppice culture of willow and poplar as energy crops on metal contaminated agricultural soils. Int J Phytoremediation 13:194–207

Salekdeh GH, Siopongco J, Wade LJ, Ghareyazie B, Bennett J (2002) Proteomic analysis of rice leaves during drought stress and recovery. Proteomics 2:1131–1145

Šamajová O, Plíhal O, Al-Yousif M, Hirt H, Šamaj J (2013) Improvement of stress tolerance in plants by genetic manipulation of mitogen-activated protein kinases. Biotechnol Adv 31:118–128

Sangwan V, Orvar BL, Beyerly J, Hirt H, Dhindsa RS (2002) Opposite changes in membrane fluidity mimic cold and heat stress activation of distinct plant MAP kinase pathways. Plant J 31(5):629–638

Sheikh AH, Ara H, Sinha AK (2012) Mitogen activated protein kinases: A hunt for their physiological substrates in plants. Plant Stress 6:37–42

Shi J, An HL, Zhang L, Gao Z, Guo XQ (2010) GhMPK7, a novel multiple stress-responsive cotton group C MAPK gene, has a role in broad spectrum disease resistance and plant development. Plant Mol Biol 74:1–17

Shi J, Zhang L, An H, Wu C, Guo X (2011) GhMPK16, a novel stress-responsive group D MAPK gene from cotton, is involved in disease resistance and drought sensitivity. BMC Mol Biol 12:22

Shinozaki K, Yamaguchi-Shinozaki K (2007) Gene networks involved in drought stress response and tolerance. J Exp Bot 58:221–227

Singh R, Lee MO, Lee JE, Choi J, Park JH, Kim EH, Yoo RH, Cho JI, Jeon J-S, Rakwal R, Agrawal GK, Moon JS, Jwa N-S (2012) Rice mitogen-activated protein kinase interactome analysis using the yeast two-hybrid system. Plant Physiol 160:477–488

Sinha AK, Jaggi M, Raghuram B, Tuteja N (2011) Mitogen-activated protein kinase signalling in plants under abiotic stress. Plant Signal Behav 6:196–203

Smékalová V, Doskočilová A, Komis G, Šamaj J (2014) Crosstalk between secondary messengers, hormones and MAPK modules during abiotic stress signalling in plants. Biotechnol Adv 32(1):2–11

Stafstrom JP, Altschuler M, Anderson DH (1993) Molecular cloning and expression of a MAP kinase homologue from pea. Plant Mol Biol 22:83–90

Stratmann J (2003) Ultraviolet-B radiation co-opts defense signaling pathways. Trends Plant Sci 8:526–533

Taj G, Agarwal P, Grant M, Kumar A (2010) MAPK machinery in plants: Recognition and response to different stresses through multiple signal transduction pathways. Plant Signal Behav 5:1370–1378

Takahashi F, Yoshida R, Ichimura K, Mizoguchi T, Seo S, Maruyama K, Yamaguchi-Shinozaki K, Shinozakia K (2007) The mitogen-activated protein kinase cascade MKK3-MPK6 is an important part of the jasmonate signal transduction pathway in *Arabidopsis*. Plant Cell 19:805–818

Teige M, Scheikl E, Eulgem T, Doczi R, Ichimura K, Shinozaki K, Dangl JL, Hirt H (2004) The MKK2 pathway mediates cold and salt stress signaling in *Arabidopsis*. Mol Cell 15:141–152

Tena G, Asai T, Chiu WL, Sheen J (2001) Plant mitogen activated protein kinase signaling cascades. Curr Opin Plant Biol 4:392–400

Ulm R, Ichimura K, Mizoguchi T, Peck SC, Zhu T, Wang X, Shinozaki K, Paszkowski J (2002) Distinct regulation of salinity and genotoxic stress responses by *Arabidopsis* MAP kinase phosphatase 1. EMBO J 21:6483–6493

Wang JX, Ding HD, Zhang A, Ma FF, Cao JM, Jiang MY (2010) A novel mitogen-activated protein kinase gene in maize (*Zea mays*), ZmMPK3, is involved in response to diverse environmental cues. J Integr Plant Biol 52:442–452

Wankhede DP, Kundan K, Singh P, Sinha AK (2013a) Involvement of mitogen activated protein kinase kinase 6 in UV induced transcripts accumulation of genes in phytoalexin biosynthesis in rice. Rice 6(1):35

Wankhede DP, Misra M, Singh P, Sinha AK (2013b) Rice mitogen activated protein kinase kinase and mitogen activated protein kinase interaction network revealed by in-silico docking and yeast two-hybrid approaches. PLoS One 8:e65011. doi:10.1371/journal.pone.0065011

Xiong L, Schumaker KS, Zhu JK (2002) Cell Signaling during cold, drought, and salt stress. Plant Cell S:165–183

Xiong L, Yang Y (2003) Disease resistance and abiotic stress tolerance in rice are inversely modulated by an abscisic acid-inducible mitogen-activated protein kinase. Plant Cell 15: 745–759

Yeh CM, Chien PS, Huang HJ (2007) Distinct signalling pathways for induction of MAP kinase activities by cadmium and copper in rice roots. J Exp Bot 58:659–671

Yeh CM, Hsiao LJ, Huang HJ (2004) Cadmium activates a mitogen-activated protein kinase gene and MBP kinases in rice. Plant Cell Physiol 45:1306–1312

Yoo SD, Cho YH, Tena G, Xiong Y, Sheen J (2008) Dual control of nuclear EIN3 by bifurcate MAPK cascades in C2H4 signaling. Nature 451:789–795

Zhang AY, Jiang MY, Zhang JH, Tan MP, Hu XL (2006) Mitogen-activated protein kinase is involved in abscisic acid-induced antioxidant defense and acts downstream of reactive oxygen species production in leaves of maize plants. Plant Physiol 141:475–487

Zong XJ, Li DP, Gu LK, Li DQ, Liu LX, Hu XL (2009) Abscisic acid and hydrogen peroxide induce a novel maize group C MAP kinase gene, ZmMPK7, which is responsible for the removal of reactive oxygen species. Planta 229:485–495

Zou J, Liu C, Chen X (2011) Proteomics of rice in response to heat stress and advances in genetic engineering for heat tolerance in rice. Plant Cell Rep 30:2155–2165

Chapter 9
Small and Large G Proteins in Biotic and Abiotic Stress Responses in Plants

Amita Pandey, Manisha Sharma, and Girdhar K. Pandey

Abstract The sessile nature of plants has enabled them to develop a higher level of plasticity in adaptation against the hostile environmental conditions such as biotic and abiotic stresses. The intricate molecular machineries in plants, consisting of signal sensors, signal transducers, and generation of response, are highly similar with animal's signal transduction components; however, a large number of pathways also show a high level of divergence. G proteins are a part of signal transduction components, which play pivotal role in regulating a large number of physiological and developmental processes in both plants and animals. G proteins or guanine nucleotide-binding proteins serve as molecular switches, existing in two forms: an active conformation, when bound to GTP, and an inactive conformation, when bound to GDP. In active conformation, G proteins interact with downstream effector molecules to initiate various cellular responses. Based on their structure, they have been divided into heterotrimeric and monomeric G proteins, also called large and small G proteins, respectively, where the former is composed of three subunits (α, β, and γ) and latter is composed of single subunit (α). Plant growth and development is partially regulated by the combined activity of large and small G proteins. Besides, identification of profuse number of stress-related genes interacting with monomeric and heterotrimeric G-protein complex in both *Arabidopsis* and rice suggests their involvement in modulation of abiotic stress responses. Further, based on the functional analysis of several G proteins (large and small) regulating cell growth, cell proliferation, seed germination, sugar sensing, and hormonal responses indicate these proteins are equally essential for plant development. Due to increased demand of food crops to sustain growing population in adverse environmental conditions, the role of G proteins and modulation of GTPase signaling has been studied in plants to produce genetically modified crop plants with increased stress tolerance and productivity.

Keywords Abiotic stress • Biotic stress • Plant G proteins • GTPase • Signal transduction

---

A. Pandey (✉) • M. Sharma • G.K. Pandey
Department of Plant Molecular Biology, University of Delhi South Campus, Benito Juarez Road, Dhaula Kuan, New Delhi 110021, India
e-mail: amitap04@gmail.com

G.K. Pandey (ed.), *Elucidation of Abiotic Stress Signaling in Plants*,
DOI 10.1007/978-1-4939-2211-6_9

9.1 Introduction

Plants being sessile have evolved an intricate network of molecular pathways to adapt to changes in their environment, including perception of signals by receptors, signal transduction via transducing molecules and proteins, and gene expression. The natural adaptation is a continuous process occurring by outbreeding over a span of several decades; however, to accelerate this process, plant biologists have adopted various approaches such as breeding and genetic manipulation of plants to generate resistant and tolerant varieties. With the global effort to sequence various plant genomes and availability of full genome sequence of model dicot like *Arabidopsis* (Arabidopsis Genome Initiative 2000) and model monocot crop like rice (Goff et al. 2002), a new field of functional genomics has emerged allowing analysis of the plant transcriptome, proteome, and metabolome employing techniques like epistasis analysis, microarrays, SAGE (serial analysis of gene expression), Y2H (yeast-two-hybrid), AP/MS (affinity purification and mass spectrometry), mutagenesis, and RNAi (RNA interference). Employing high-throughput technology, research in the past decades has identified key molecules acting in abiotic and biotic stress tolerance, including G proteins, also known as guanosine nucleotide-binding proteins. G family proteins are cytosolic proteins involved in transmitting a variety of extracellular signals into the cell through cell surface receptors to modulate cellular activities (Hepler and Gilman 1992). Their activity is regulated by certain cellular factors that control their ability to bind to GTP and hydrolyze it to GDP. G proteins when bound to GTP acquire an on or activated form, whereas GDP-bound form designates an inactivated or off state. Based on their structure, G proteins are mainly divided into two categories: monomeric small GTPases and heterotrimeric large G-protein complexes.

9.1.1 Heterotrimeric G Proteins

Heterotrimeric G protein (hereafter called G protein), also called large G protein, is a cytosolic protein and a complex containing three subunits including G-alpha (Gα), G-beta (Gβ), and G-gamma (Gγ) (Jones and Assmann 2004; Ma 2001). In animals, G proteins are involved in transmitting extracellular signals, such as, hormones, neurotransmitters, chemokines, and light to intracellular signaling components to modulate gene expression, culminating into a response (Gilman 1987; Wettschureck and Offermanns 2005). Animal G protein is a multigene family encoded by 23α, 6β, and 12γ genes (Hamm 1998; McCudden et al. 2005). Additionally, the G-protein signaling module is composed of G-protein-coupled receptors (GPCRs), containing seven-transmembrane α-helical proteins, acting upstream of G proteins and activating G proteins upon binding to a ligand. The human genome encodes for approximately 1,000 GPCRs, falling into six subgroups, comprising the largest receptor superfamily in metazoans (Kawasawa et al. 2003).

In comparison to mammals, plant G-protein family is relatively small; in *Arabidopsis*, Gα is encoded by a single-copy gene, designated as *GPA1* (Ma et al. 1990), and single copy of Gβ, *AGB1* (Weiss et al. 1994), and three genes encoding the Gγ subunits have been isolated: *AGG1*, *AGG2*, and *AGG3* (Mason and Botella 2000, 2001; Chakravorty et al. 2011). The genome of crop plant rice contains one gene each encoding Gα (*RGA1*) and Gβ (*RGB1*) subunits and three genes encoding Gγ subunit (*RGG1*, *RGG2*, and *RGG3*) (Ishikawa et al. 1996; Kato et al. 2004). G-protein signaling components have also been identified in other plants: pea and soybean having two and four Gα-subunits, respectively (Ashikari et al. 1999; Misra et al. 2007; Choudhury et al. 2011). The α-, β-, and γ-subunits of G proteins link extracellular signals to intracellular effector through GPCRs. Interestingly, direct homologs of metazoan GPCR (G-protein-coupled receptors) proteins are altogether absent in plants. However, recently in *Arabidopsis*, *GCR1* has been identified as a likely representative of GPCR-encoding gene because of its significant sequence similarity with them. The *Arabidopsis* GCR1 is predicted heptahelical protein, a hallmark of bona fide GPCRs (Josefsson and Rask 1997; Pandey and Assmann 2004; Chen et al. 2004a, 2004b). Using bioinformatics tools, the *Arabidopsis* genome is predicted to encode for 394 7TMpRs (seven-transmembrane putative receptors) (Moriyama et al. 2006). An additional regulator of G-protein signaling (RGS) reminiscent to GPCRs has been identified in *Arabidopsis; AtRGS1* affects the intrinsic GTPase activity of Gα-subunit, keeping it in inactivated conformation. It appears that the *Arabidopsis* genome contains only one member of the RGS family, *RGS1* (Chen et al. 2003). Additionally, 7 noncanonical, extra-large G (XLG) proteins were identified in plants, including 3 in *Arabidopsis* and 4 in rice (Lee and Assmann 1999). The physiological role of G proteins in plants was determined by the characterization of loss-of-function (lof) mutant and transgenic lines, in stomata opening/closure (Wang et al. 2001; Coursol et al. 2003; Chen et al. 2004a, 2004b), fungal defense (Llorente et al. 2005; Trusov et al. 2006; Trusov et al. 2009), oxidative stress (Joo et al. 2005; Booker et al. 2004), phytohormone signaling (gibberellic acid (GA)), and plant development (Ueguchi-Tanaka et al. 2000; Ullah et al. 2001; Ullah et al. 2002; Lease et al. 2001; Chen et al. 2006; Pandey et al. 2006). In mature leaves, G proteins transmit signals to molecules, including small GTPases, ion channels, and phospholipases, that are effectors in response to various biotic and abiotic stress conditions, including pathogen elicitation, ozone treatment, and water deficit.

9.1.1.1 Mechanism of Regulation of Heterotrimeric G Protein

Mechanism of mammalian G-protein activation involves binding of an activating ligand to its specific GPCR causing a change in its conformation, leading to the conversion of an inactive G protein to its active conformation. The GPCR acts as a guanine nucleotide exchange factor (GEF), causing Gα to exchange GDP for GTP (Gilman 1987). As a result, Gα–GTP separates from the Gβγ dimer, and both Gα–GTP and the Gβγ dimer separate from the receptor and can individually activate downstream effector molecules. Subsequent to signal propagation, the GTP that is bound to Gα is hydrolyzed to GDP, thereby inactivating Gα and allowing its reassociation with the Gβγ dimer to reform the inactive heterotrimeric G-protein complex.

In *Arabidopsis,* the mechanism of G-protein activation is different than in animals involving spontaneous disassociation of heterotrimer into Gαs and Gβγ dimer, which allows Gα to bind to GTP, and both the dissociated subunits separately activate downstream effectors (Urano et al. 2012). A 7TM-RGS protein, AtRGS1, promotes GTP hydrolysis of the Gαs, and this GDP-bound form then interacts with Gβγ dimer and returns to the resting state (Fig. 9.1a) (Johnston et al. 2007; Jones et al. 2011). Here RGS proteins act as GTPase-activating proteins (GAPs) for Gα, attenuating signaling by promoting the return of the G protein to the resting state. Genetic evidence is consistent with D-glucose being the ligand that stops the AtRGS1 GAP activity and allows AtGPA1 to self-activate (Fig. 9.1a) (Chen et al. 2003; Johnston et al. 2007; Chen et al. 2006; Grigston et al. 2008).

9.1.1.2 Domain Structure of Heterotrimeric G Protein

G proteins contain highly conserved G domain/fold, shared by Ras superfamily proteins and translation elongation factors. G domain is composed of 6 β-sheets, 5 α-helices, and 5 regions/loops designated G1–G5, which play a pivotal role in GTP binding and hydrolysis. The organization of G domain from N-terminus is N-β1-G1-α1-G2/SW1-β2-β3/G3-SW2-α2-β4-α3-β5/G4-α4-β6-G5-α5-C, where the G1 region or P-loop has a conserved GxxxxGK(S/T) motif (aka Walker A motif), required to contact α- and β-phosphate of GTP; the G2 region contains a conserved threonine residue for Mg^{2+} coordination important for GTP hydrolysis; G3 region (aka Walker B motif) is identified by sequence DxxG and the conserved aspartate (D) and glycine (G) bind Mg^{2+} and γ-GTP, respectively; G4 region has NKxD motif and the conserved aspartate contacts the guanine ring of GTP; and the G5 region is identified by E(A/C/S/T)(C/S)A(K/L) motif, which helps in guanine base recognition (Wittinghofer and Vetter 2011). Plant Gα-subunit is around 35–46 kDa protein, characterized by functional motifs including myristoylation and S-acetylation sites (for membrane targeting), N- and C-terminal receptor binding sites, and three switch regions (SW) that sense nucleotide binding (Fig. 9.2). The N-terminal helical region, a small interface at switch I (SWI), and a large interface at switch II (SWII) bind Gβγ dimer (Li et al. 2009; Temple and Jones 2007), whereas the helical domain shelters the guanine nucleotide-binding pocket. The P-loop, G_{45}AGESGKS for NTP binding, the DxxGQ motif for GTP hydrolysis, and the NKxD motif for guanine recognition are conserved in plants (Temple and Jones 2007). The Gβs is 35–36 kDa protein, composed of an amino-terminal coiled-coil motif and a carboxyl-terminal WD40 (β-transducin) repeat domain (Fig. 9.2). The amino-terminus of Gβs forms the stable coiled-coil/hydrophobic interaction with the Gγs (Li et al. 2009), and the WD40 repeat domain contains the effector- and Gαs-binding surface. The effector-binding region on Gβs is normally masked by the GDP-bound form of Gαs;

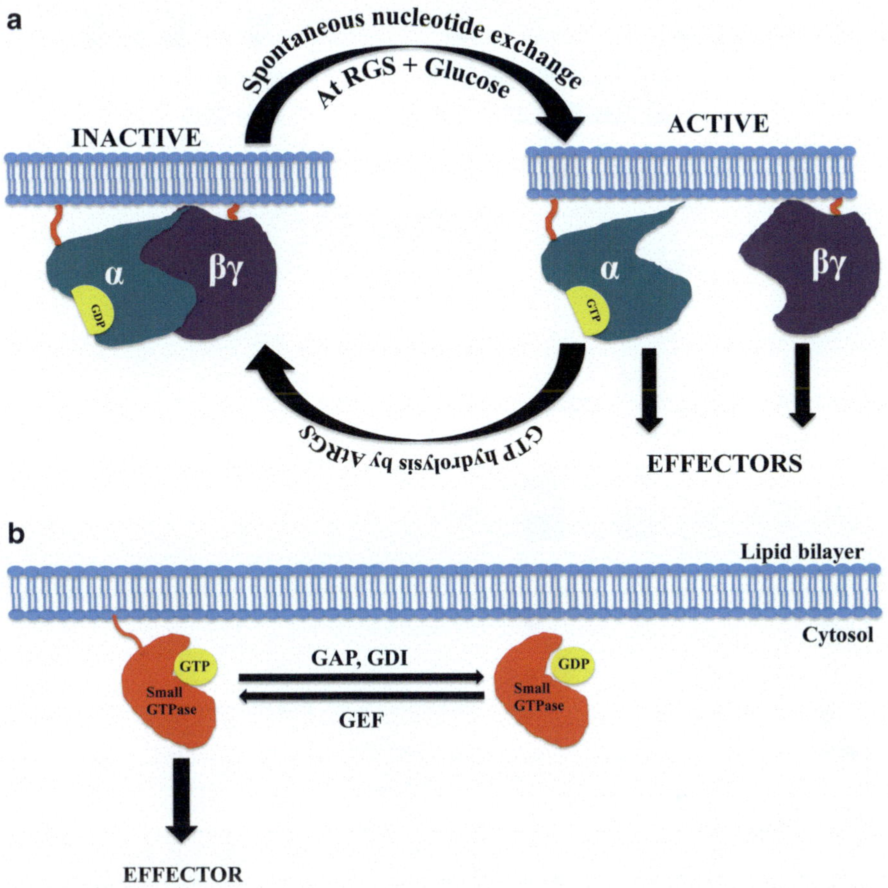

Fig. 9.1 Graphic depicting regulatory mechanism of heterotrimeric and monomeric G protein: (**a**) G proteins or large G proteins act as a molecular switch and regulate many cellular processes. The G protein is made up of three subunits: α, β, and γ. The Gαs binds to the guanosine nucleotide, GDP, in the inactive state and forms a complex with Gβγ dimer. Spontaneous disassociation of heterotrimer into Gαs and Gβγ allows Gαs to bind to GTP, and both the dissociated subunits separately activate downstream effector molecules. A 7TM-RGS (transmembrane-regulator of G-protein signaling) protein, AtRGS1, promotes GTP hydrolysis of the Gαs, and this GDP-bound form then interacts with Gβγ dimer and comes back to the resting state. Here RGS proteins act as GTPase-activating proteins (GAPs) for Gαs, attenuating signaling by hastening the return of the G protein to the resting state. D-glucose acts as the ligand that halts AtRGS1 GAP activity and allows AtGPA1 to self-activate. (**b**) Monomeric GTP-binding proteins or small GTPases are molecular switches that are "activated" by GTP and "inactivated" by the hydrolysis of GTP to GDP. Monomeric GTPase activity is determined by the ratio of GTP/GDP-bound form, regulated by three classes of proteins, including guanine nucleotide exchange factors (GEFs), GTPase-activating proteins (GAPs), and guanine nucleotide dissociation inhibitors (GDIs). Upon stimulation by an upstream signal, a GEF protein converts the GDP-bound inactive form into a GTP-bound active conformation, through GDP/GTP replacement. In the activated state through the effector-binding domain, it interacts with downstream effector proteins to transduce the signal. The small GTPases also have a weak intrinsic GTPase activity for GTP hydrolysis and require a GTPase-activating protein (GAP) for efficient deactivation. Additionally, most small GTPases cycle between membrane-associated and cytosolic states. And only membrane-bound GTPases can be activated by GEF; their removal by cytosolic GDI proteins negatively regulates GTPases

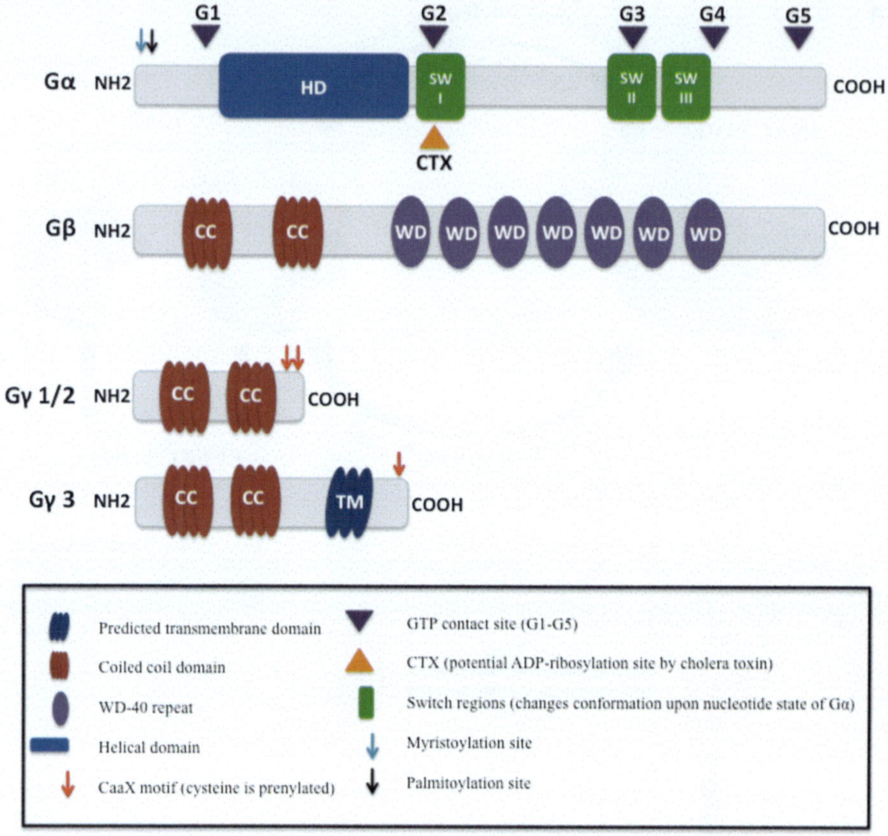

Fig. 9.2 Graphic representation of domain structure α, β, and γ subunits of plant heterotrimeric G protein: plant Gαs is around 35–46 kDa protein, characterized by functional motifs including myristoylation and S-acetylation sites (for membrane targeting), 5 G domains, N- and C-terminal receptor binding sites, and 3 switch regions (SW) that sense nucleotide binding. The N-terminal helical region, a small interface at switch I, and a large interface at switch II bind Gβγ dimer, whereas the helical domain (HD) shelters the guanine nucleotide-binding pocket. The P-loop, G_{45}AGESGKS for NTP binding, the DxxGQ motif for GTP hydrolysis, and the NKxD motif for guanine recognition are conserved in plants. The Gβs is 35–36 kDa protein, composed of an amino-terminal coiled-coil motif and a carboxyl-terminal WD40 repeat domain. The amino-terminus of Gβs forms the stable coiled-coil/hydrophobic interaction with the Gβs, and the WD40 repeat domain contains the effector- and Gαs-binding surface. The effector-binding surface on Gβs is normally masked by the GDP-bound form of Gαs; therefore, Gβγ dimer is active only after dissociating from the Gα-subunit. The Gγs is a small (less than 100 residue, 6–10 kDa) protein containing a coiled-coil region and a prenylation site at its carboxyl-terminus

therefore, Gβγ dimer is active only after dissociating from the α-subunit (Wettschureck and Offermanns 2005). The Gγs is a small (less than 100 residue, 6–10 kDa) protein containing a coiled-coil region and a prenylation site at its carboxyl-terminus (Fig. 9.2) (Gilman 1987), required for its membrane targeting (Wei et al. 2008).

9.1.2 Monomeric G Protein

Monomeric G proteins (21–30 kDa) are the small guanine nucleotide-binding proteins related to α-subunit of heterotrimeric G proteins, belonging to Ras (rat sarcoma oncoprotein) superfamily, named after human Ras genes, discovered as cellular homologs of the viral Ras oncogene (Parada et al. 1982).

The small GTPase superfamily in animals is divided into six distinct families, including Ras, Rho, Rab, Arf, Ran, and Miro (Kahn et al. 1992; Bischoff et al. 1999; Takai et al. 2001; Logan 2010). Members of each family are associated with specific biological function as Ras GTPases regulate cell proliferation; Arf (ADP-ribosylation factor) and Rab (Ras-like protein in brain) primarily regulate various steps in trafficking, where former is required for vesicle budding in the secretory system and latter controls transport and docking of specific vesicles to the plasma membrane. Ran (Ras-like nuclear protein) regulates trafficking of RNA and proteins across the nuclear membrane, and Miro (mitochondria Rho GTPase) GTPases regulate mitochondrial transport and morphology (Ridley et al. 1992; Nobes et al. 1998; Wennerberg et al. 2005; Reis et al. 2009).

In silico analysis identified 93 small GTPase superfamily members in *Arabidopsis*, classified into five small GTPase families: with 57 Rab GTPases, 21 Arf GTPases, 11 Rho GTPases, and 4 Ran GTPases. Interestingly, plant genomes do not encode homologs of Ras GTPases (Regad et al. 1993; Haizel et al. 1997; Vernoud et al. 2003). Furthermore, functional analysis of several members showed that they might regulate similar cellular processes in the plant cells as they do in animal cells. For example, Arf GTPases regulate vesicle trafficking between plasma membrane (PM) and cytoplasm (Geldner et al. 2001); the role of Ran GTPases is not known, but N-terminus of RanGAPs is homologous to nuclear matrix attachment proteins, providing new insight into their function in plant cells (Meier 2000). Rab GTPases comprise the largest family in *Arabidopsis*, functioning in vesicle trafficking (Batoko et al. 2000; Lu et al. 2001). Plants do not have true orthologs of Rho family GTPases, but have evolved a novel family of Rho-like proteins also known as Rho of plants (ROPs) (Zheng and Yang 2000; Shanmugam et al. 2013). Signaling proteins in plants play an important role in plant stress responses and pollen tube outgrowth (Kost et al. 1999; Li et al. 1999). In mammals 14 Rho family GTPases have been identified and divided into 7 subgroups (Nobes and Hall 1994; Aspenstrom 1999). In plants, two groups have been identified of which only one protein is Rho-like homolog while the rest are closely related to Rac (Newman et al. 1994; Yang and Watson 1993; Delmer et al. 1995; Winge et al. 1997).

9.1.2.1 Mechanism of Regulation of Monomeric G Protein

Small GTP-binding proteins are master molecular switches that are "activated" by GTP and "inactivated" by the hydrolysis of GTP to GDP, acting as global regulatory proteins in eukaryotic cells. In the cell, GTPase activity is determined by the ratio of GTP-/GDP-bound form, regulated by three classes of proteins, including guanine nucleotide exchange factors (GEFs), GTPase-activating proteins (GAPs), and guanine nucleotide dissociation inhibitors (GDIs). Upon stimulation by an upstream signal, a GEF protein converts the GDP-bound inactive GTPases to a GTP-bound active conformation, through GDP/GTP exchange. Once in the activated state, small GTPases through their effector-binding domain interacts with downstream effector proteins to transduce the signaling response. The activated conformation possesses weak intrinsic GTPase activity for GTP hydrolysis, requiring a GTPase-activating protein (GAP) for efficient deactivation. Additionally, most of the small GTPases frequently cycle between plasma membrane and cytosolic compartments; only membrane-bound GTPases can be activated by GEF, and removal by cytosolic GDI proteins negatively regulates GTPases. This complex mode of regulation and their ability to interact with variety of upstream and downstream targets create functional diversity in GTPase activity (Fig. 9.1b).

In animals, most RhoGEFs possess a Dbl (diffuse B-cell lymphoma) homology (DH) domain and a pleckstrin homology (PH) domain. However, no proteins with DH/PH domains have been identified in plants so far. Instead, plant-specific Rop nucleotide exchanger (PRONE) domain constitutes a large family in plants (Berken et al. 2005; Gu et al. 2006).

9.1.2.2 Domain Structure of Small GTPases

Members of Ras superfamily share common structural features including guanine nucleotide-binding domains and an effector-binding domain (Zheng and Yang 2000; Takai et al. 2001). Plant small GTPases contain five G domains (G1-G5) of which the G5 domain shows high sequence divergence among the GTPases. The G domains are essential for binding guanosine nucleotide and the associated Mg^{2+} ion and for GTP hydrolysis. The G1 motif associates tightly with the α- and β-phosphate, G2 motif contacts the γ-phosphate, and the G3 motif helps in positioning the water molecule for GTP hydrolysis through highly conserved glutamine residue. Switch 1 and 2 regions bind to the effector or regulatory molecules. Rac/Rop GTPases have a Rho insert region, and all the GTPases have hypervariable region at the C-terminus facilitating localization (Fig 9.3).

9.2 Small GTPases in Biotic and Abiotic Plant Stress

9.2.1 Rac/Rop Small GTPases in Biotic Stress

In plants, Rac/Rop small GTPases participate in diverse signal transduction events including defense responses, root hair development, pollen tube growth, ROS generation, and hormone responses (Nibau et al. 2006; Yang and Fu 2007; Shanmugam et al. 2013). Plants respond to pathogen attack by various ways including the

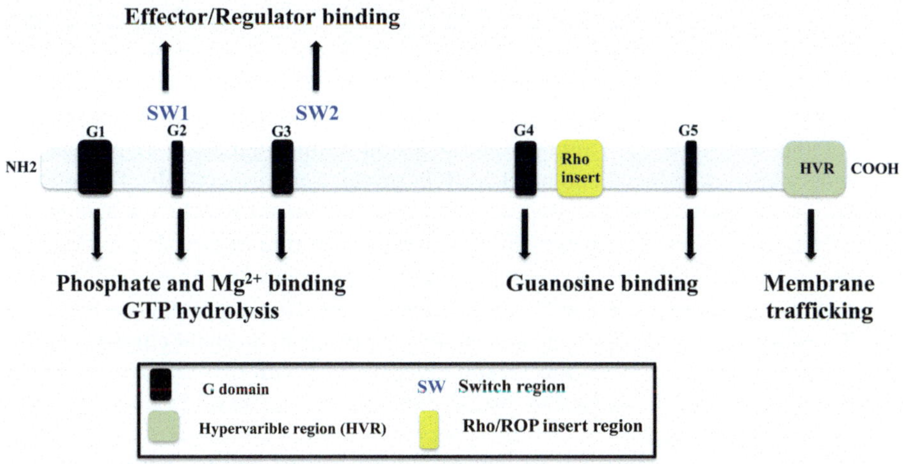

Fig. 9.3 Graphic representation of domain structure of monomeric G protein: plant small GTPases contain five G domains (G1-G5) of which the G5 domain has highly divergent sequence. The G domains are essential for binding guanosine nucleotide and associated Mg^{2+} ion and for GTP hydrolysis. The G1 motif associates tightly with the α- and β-phosphate, the G2 motif contacts the γ-phosphate, and the G3 motif helps in positioning the water molecule for GTP hydrolysis through highly conserved glutamine residue. Switch 1 and 2 (SW1 and SW2) regions bind to effector of regulators. Rac/Rop GTPases have a Rho insert region, and all the GTPases have hypervariable (HVR) region at the C-terminus for localization

activation of defense genes, formation of reactive oxygen species (ROS), synthesis of pathogenesis-related (PR) proteins, localized cell-wall reinforcement, and production of stress-related hormones and antimicrobial compounds. Moreover, plant defense system exists in two layers. The first layer is the PAMP (pathogen-associated molecular patterns)-triggered immunity (PTI) mediated by pattern recognition receptors (PRRs) (Dodds and Rathjen 2010), characterized by ion influx, mitogen-activated protein kinase (MAPK) activation and enhanced gene expression (Zipfel 2008). The second layer of defense is the effector-triggered immunity (ETI); it is in response to effector molecules secreted by pathogens, characterized by hypersensitive response (HR) including reactive oxygen species (ROS) production and programmed cell death (PCD) of infected tissue/cell (Jones and Dangl 2006).

ROS production is associated with PCD during resistance reaction to pathogens, and a plasma membrane NADPH oxidase (Rboh for respiratory burst oxidase homolog) similar to the neutrophil enzyme is suggested to be responsible for ROS production (Morel et al. 2004; Lamb and Dixon 1997; Hammond-Kosack and Jones 1996). In animals the oxidative burst produced by the NADPH oxidase in neutrophils is regulated by Rac2 (Babior 1999). Based on the assumption that the plant NADPH oxidase enzyme complex is similar to its mammalian counterpart, composed of six subunits including Rac (Bokoch 1994), the role of small GTPases was explored in plant defense response in ROS generation.

In rice, among seven Rac/Rop genes, *OsRac1* and *OsRacB* have been associated with defense response (Miki et al. 2005; Ono et al. 2001; Jung et al. 2006). The constitutively active (CA), containing a G19V mutation, and dominant-negative (DN), with T24N mutation, forms of *OsRac1*, when introduced into seed-originated calli of the wild-type cv. Kinmaze and a lesion-mimic mutant of rice plant, *sl* (Sekiguchi lesion), induced and blocked ROS production, respectively. This observation suggests that *OsRac1* mediates ROS production most likely by positive regulation of NADPH oxidase (Kawasaki et al. 1999). The role of NADPH oxidase in Rac-mediated ROS generation was supported by studies in tobacco, where Rac acts as a negative regulator of NADPH oxidase during ETI-triggered immune response. RT-PCR of tobacco cells upon elicitation with cryptogein, a fungal elicitor, exhibited repressed *NtRac5* mRNA levels, and tobacco cells with *CA-NtRac5* transgene showed decreased ROS production, by affecting the regulation of *NtrbohD*, a component of NADPH oxidase enzyme complex (Morel et al. 2004). *OsRac1* regulation of NADPH oxidase is further supported by biochemical studies where OsRac1 interacts with the N-terminus of NADPH oxidase to regulate ROS production (Wong et al. 2007). Recent studies have also implicated PI3K as upstream regulator of *OsRac1* in recruiting OsRac1 to PM to regulate NADPH oxidase activity and accelerate ROS production (Liu et al. 2012b). The role of Rac/Rop proteins in ROS production is conserved within plants and between plants and animals, pointing to existence of conserved signal transduction pathways in both the kingdoms, like maize Rac/Rop induces ROS production in mammalian cells (Hassanain et al. 2000). Similarly, human Rac was found to promote oxidative burst in soybean cell cultures (Park et al. 2000), and *CA-GhRac13* (*Gossypium hirsutum*) transgene increases ROS production in soybean and *Arabidopsis* cell cultures, probably serving as a signal for xylem differentiation (Potikha et al. 1999).

In search for molecules acting in the same signaling pathway as *OsRac1*, Pit, an NLR protein, was identified. *R* gene encodes proteins with a nucleotide-binding site and leucine-rich repeat region (NLR family), which act as intracellular receptors for recognition of pathogen effectors (also called avirulence [Avr] proteins). R-mediated disease resistance results in a host response culminating in a HR and ROS production. Pit, an NLR family protein interaction with OsRac1 at the membrane, is supported by bimolecular fluorescence complementation (BiFC) assays using Vn- (N-terminus region of Venus) and Vc- (C-terminus region of Venus)tagged OsRac1 (*Vn-CA-OsRac1*) and Pit (*PitWT-Vc*) in rice protoplasts. Moreover, Förster resonance energy transfer (FRET) analysis using Raichu (Ras and interacting protein chimeric unit) confirmed activation of OsRac1 by Pit. Their role in HR and ROS production was confirmed by transient expression of autoactivated Pit, *PitD485V*, and *DN-OsRac1* in *N. benthamiana*, where *PitD485V* in the absence of pathogen induce HR and ROS production, whereas, upon co-expression with *DN-OsRac1*, ROS production was visibly reduced. Phenotypic analysis of *japonica* rice cultivar K59, carrying *Pit* resistance gene in which *OsRac1* was silenced by RNAi, showed enhanced fungal growth in RNAi lines, supporting the role of *Pit* acting upstream of *OsRac1* in conferring resistance to rice blast fungus (Kawano et al. 2010). Using FRET probe Raichu-OsRac1 as an intracellular biosensor for monitoring the spatiotemporal activation of *OsRac1* confirmed the induction of *OsRac1* in response to chitin elicitation. The chitin-elicited defense response involves chitin binding and forming a complex with OsCEBiP, a chitin-binding protein, and OsCERK1, a receptor-like kinase at the membrane. A PRONE GEF, *OsRacGEF1*, was shown to regulate *OsRac1* activity in chitin-elicited defense response. Phenotypic analysis of plants confirmed that OsRacGEF1 activation is required for chitin-induced immune response and resistance to rice blast fungus infection. OsRac1, OsRacGEF1, and OsCERK1 were shown to interact, supported by Y2H and in vitro binding assays, and the complex of OsRac1 and OsRacGEF1 was shown to localize to the PM by co-localization studies using OsRacGEF1-Venus and CFP-OsRac1-WT in rice protoplasts. Biochemical studies support the model where OsCERK1 phosphorylates OsRacGEF1 at the S549 residue, where the latter after activation regulates OsRac1. Based on these results, it can be hypothesized that OsCEBiP, OsCERK1, OsRacGEF1, and OsRac1 function as a "defensome" in chitin-triggered immunity (Akamatsu et al. 2013). These study indicate that *OsRac1* function in plant defense response is regulated by GEF proteins acting upstream of Rac/Rops. Another GEF implicated in regulating OsRac1 in chitin-triggered immune response and ROS production is the human SWAP70 homolog in rice, OsSWAP70A, with conventional DH domain, which interacts and exhibits GEF catalytic activity towards OsRac1. OsSWAP70A activity towards OsRac1 was analyzed in vitro, by measuring replacement of unlabeled GDP with fluorescently labeled *N*-methylanthraniloyl (mant)-GTP. Transient expression of *OsSWAP70A* and *OsRac1* together in *N. benthamiana* enhanced ROS production, whereas expressing OsSWAP70A alone had no effect. RNAi-mediated reduction of *OsSWAP70A* mRNA levels results in the suppression of chitin-induced defense response gene and ROS production, thereby indicating that *OsSWAP70A* positively regulates *OsRac1* (Yamaguchi et al. 2012). In another study, the *Arabidopsis SWAP70* was shown to function in both PAMP- and ETI-triggered immune responses, supported by the phenotypic analysis of T-DNA insertion mutant of *AtSWAP70*, which were more susceptible to pathogen growth (Yamaguchi and Kawasaki 2012).

Several independent studies have identified signaling components acting downstream of OsRac1. Expression of *CA-OsRac1* transgene induces phytoalexin synthesis, a secondary metabolite with antimicrobial activity, and defense gene activation leading to resistance towards rice blast fungus (Ono et al. 2001; Fujiwara et al. 2006; Thao et al. 2007). Similarly, introduction of transgene, *CA-gOsRac1*, into rice, enhanced expression of *PAL1* and *PBZ1*, defense-related genes, and induced HR-response like PCD. Moreover, it also activates genes previously known to be induced by *Magnaporthe oryzae* and *Xanthomonas oryzae* pv. *oryzae* (Xoo) in cultured cells (Dang et al. 2013). OsRac1 has also been implicated in defense against the rice pathogen, the brown plant hopper (BPH), along with phytohormones, calcium, and MAPKs (Cheng et al. 2013).

Using proteomics, *OsCCR1* encoding *Oryza sativa* cinnamoyl-CoA reductase 1, an enzyme involved in lignin biosynthesis, was identified as a downstream effector of *OsRac1*. In the cell wall, lignin polymerization requires peroxidase activity using H_2O_2. The role of *OsCCR*1 in plant defense was supported by the fact that

*OsCCR1*expression level is enhanced by sphingolipid elicitor. OsCCR1 and OsRac1 form a complex, evidenced by Y2H and in vitro binding assay, where OsRac1 activates OsCCR1. Transgenic cell cultures expressing *CA-OsRac1* accumulate lignin through enhanced OsCCR1 activity and ROS production. Most likely, OsRac1 regulates lignin synthesis by activating NADPH oxidase and OsCCR1 in rice defense response (Kawasaki et al. 2006). *OsMT2b*, a rice metallothionein gene, was identified as a downstream effector of *OsRac1*. Metallothionein are involved in ROS scavenging and metal homeostasis. *OsMT2b* expression is downregulated synergically by rice blast-derived elicitor and *OsRac1*, supporting its function downstream of *OsRac1. OsMT2b* overexpressing cells showed reduced H_2O_2 production in contrast to homozygous *OsMT2b*:*Tos17* inserted mutant and *OsMT2b* RNAi-silenced transgenic cells. Most likely OsRac1 functions by downregulating *OsMT2b* and activating NADPH oxidase in blast defense response (Wong et al. 2004)

Trancriptomics revealed the signaling components such as transcription factors acting downstream of *OsRac1*. Microarray analysis of *CA-OsRac1* transgenic plants exhibited enhanced levels of RAI1 (Rac immunity 1), a transcription factor. In order to identify genes regulated by RAI1, microarray analysis of cells containing *RAI1-GR* (glucocorticoid receptor), a dexamethasone (DEX)-inducible construct was performed; induction with DEX-induced expression of *OsWRKY19* and *PAL1* (phenylalanine ammonia lyase) defense-related genes also upregulated in response to chitin and sphingolipids. Phenotypic analysis of *RAI1* T-DNA activation-tagged and RNAi transgenic lines, showing increased resistance to rice blast fungus infection, further supported the role of *RAI1* in plant defense response. Moreover, in vitro binding assay supports binding between RAI1 and MAPK3/6, and BiFC assay using *RAI1-Vn* and *OsMAPK3/6-Vc* constructs in rice protoplasts supports in vivo interaction. qPCR of *OsMAPK3/6* overexpression (OE) transgenic lines showed enhanced expression of *PAL1* and *OsWRKY19*, supporting the model that OsRac1 activates RAI1 through OsMAPK3/6 to regulate expression of defense-related genes (Kim et al. 2012). Another transcription factor regulated by OsRac1 in chitin-triggered immunity is OsRap2.6. OsRac1 has been shown to interact with RACK1 (receptor for activated kinase C1). BiFC assay in rice protoplast with Vn- and Vc-tagged OsRap2.6, RACK1, and OsMAPK3/6 showed these proteins interact and localize in nucleus and the cytoplasm. Moreover, *OsRap2.6*, *PAL1*, and *PBZ1* are all induced in response to chitin. Phenotypic analysis of *OsRap2.6* RNAi and OE lines showed that the former are highly susceptible, whereas the latter are more resistant to infection by rice blast fungus, *Magnaporthe oryzae* (Wamaitha et al. 2012), indicating *OsRap2.6* might act downstream of *OsRac1*, *RACK1*, *and MAPK3/6* in plant defense response.

In a study *OsRacB*, a close ortholog of barley *HvRacB* gene, transcript levels were downregulated in response to infection by rice blast fungus. Interestingly, transgenic plants overexpressing *OsRacB* showed increased symptom development. Moreover, OsRacB requires PM localization for initiating response evident from the *GFP*:*OsRacB*-transformed onion cells and *Arabidopsis* protoplasts (Jung et al. 2006). These results indicate that *OsRacB* functions as a negative regulator as opposed to *OsRac1* acting as positive regulator during rice blast fungus defense response.

To find out the role of the remaining Rop genes in defense responses, a tissue-specific expression pattern and subcellular localization of rice Rac/Rop genes were done using semiquantitative RT-PCR and transient expression system using GFP fusion proteins. Moreover, their role in disease resistance by testing single Rac/Rop RNAi plants against the rice blast fungus showed *OsRac2*, *OsRac6*, and *OsRac7* expressions at a very low level in leaf blades. Infection assays showed *OsRac1* as positive regulator of blast resistance, corroborating earlier results (Ono et al. 2001), whereas *OsRac4* and *OsRac5* act as negative regulators of blast resistance (Chen et al. 2010).

Three ROP proteins HvRACB, HvRAC1, and HvRAC3 in barley (*Hordeum vulgare*) are linked to both development and pathogen responses (Schultheiss et al. 2005; Pathuri et al. 2008; Hoefle et al. 2011). *HvRACB* functions as a common signaling component for both biotic and abiotic stresses, supported by phenotypic analysis of *CA-HvRACB* transgenic plants, exhibiting enhanced susceptibility to powdery mildew fungus (*Blumeria graminis* f. sp. *hordei* [*Bgh*]) and increased water loss in detached leaves due to reduced responsiveness to ABA, leading to failure in stomata closing (Schultheiss et al. 2005). In barley defense response, *HvRACB* acts as a negative regulator of disease resistance to *Bgh* (Schultheiss et al. 2002; Schultheiss et al. 2003). Knockdown of *HvRACB* expression by RNAi reduced susceptibility to powdery mildew by increasing resistance to penetration by *Bgh*, whereas the *CA-HvRACB* transgene enhanced susceptibility to *Bgh*. The mechanism of *HvRACB* action depends on the actin cytoskeleton disorganization and the mildew resistance locus (MLO). MLO encodes a seven-transmembrane protein and acts as negative regulator of cell death (Piffanelli et al. 2002; Schultheiss et al. 2002, 2003; Hoefle et al. 2011; Huesmann et al. 2011). HvMAGAP1 (microtubule-associated ROP GTPase-activating protein 1) was identified as a susceptibility factor in barley (Huesmann et al. 2011). Transient-induced gene silencing or OE of *HvMAGAP1* resulted in enhanced or reduced susceptibility to *Bgh*, respectively. Like *HvRACB*, *HvMAGAP1* also influences the polarity of cortical microtubules in interaction with *Bgh*, initiating MT disorganization to accommodate fungus haustoria. Moreover, interaction between HvMAGAP1 and HvRACB is required for the defense response, established in vitro, by Y2H and using FRET, *in planta* (Hoefle et al. 2011). The *Arabidopsis* homologs of *HvMAGAP1*, *AtROPGAP1*, and *AtROPGAP4* also function as plant defense. T-DNA insertion knockout enhanced susceptibility to the virulent powdery mildew fungus *Erysiphe cruciferarum*, indicating functional conservation of ROPGAPs in pathogen response in both monocots and dicots (Huesmann et al. 2011).

Adopting a proteomic approach, *HvRBK1* (*Hordeum vulgare* Rop-binding protein kinase1) was identified as a downstream effector of *HvRACB* and *HvRAC1* signaling pathways involved in MT reorganization in a Y2H assay. In in vitro assay, kinase activity of HvRBK1 was observed to be induced by CA-HvRACB or activated HvRAC1. Transient-induced gene silencing of barley *HvRBK1* supported penetration by *Bgh*, and transient knockdown influenced MT stability in barley epidermal cells; all these results support a model where HvRACB activates HvRBK1 to bring about cytoskeleton reorganization (Huesmann et al. 2012). It is evident from research in both rice and barley that the mechanism of action of Rac/Rop protein function in plant defense is different, where in rice OsRac1 functions by activating enzymes involved in ROS generation and regulating gene expression; HvRACB primarily affects the MT organization in the infected cells by activating HvRBK1.

Phosphatidic acid (PA), a secondary messenger, is implicated in numerous biotic and abiotic stresses including pathogen elicitation, wounding, freezing, hyperosmotic stress, and water deficit (Canonne et al. 2011). Studies have shown that PA mediates PCD and ROS generation in whole leaf and at single-cell level by activating AtROP2, which is shown to enhance ROS production in an in vitro assay. Although, epistasis analysis where in absence of exogenous PA, no spontaneous ROS production was observed in *CA-ROP2* transgenic plants; suggested activation of ROP2 alone is not sufficient to induce ROS generation (Park et al. 2004).

Mutations in plant signaling molecule salicylic acid (SA)-mediated pathways heighten disease susceptibility to pathogens, while induction of SA-dependent defense response reduces pathogen growth (Durrant and Dong 2004). SA-dependent signaling pathway known as systemic acquired resistance (SAR) induces expression of *PR* (pathogenesis-related) genes including phytoalexins and PR proteins. Microarray analysis of *DN-AtROP6* plants revealed a role for AtROP6 in SA-mediated defense response by causing major changes in gene expression associated with the constitutive SA-mediated plant defense responses. The total level of SA in *DN-AtROP6* transgenic plants resembled those of wild-type plants, inoculated with a virulent powdery mildew pathogen. The constitutive SA-associated response in *DN-AtROP-6* was suppressed in mutants defective in SA signaling (norexpressor of PR genes1 (*npr1*)) or biosynthesis (salicylic acid induction deficient 2 (*sid2*)) (Poraty-Gavra et al. 2013), supporting Rac/Rops as negative regulators in SAR.

Similarly, in other plant species, the role of Rac/Rop GTPases in plant defense mechanism is conserved. Rac/Rop GTPases were shown to be involved in both PTI- and ETI-triggered immunity as their expression was found to be induced in pathogen-inoculated chickpea leaf and elicitor-treated cell culture (Ichinose et al. 1999); for example, in tobacco silencing of *NtRop* by expression of *Medicago*, *Ms-Rac1* antisense cDNA causes defects in defense response (Schiene et al. 2000).

9.2.2 Rac/Rop GTPases in Abiotic Stress

Stomata closure in aerial tissue is an important approach adopted by plants to respond to rapid water loss during drought, orchestrated by phytohormone ABA. ABA is also the key mediator of plant responses to other abiotic stresses including salinity and cold, and therefore is known as the stress hormone. The ABA signaling components include ABA receptors (PYL/PYR/RCAR family), a group of type 2C phosphatases (PP2C) (Park et al. 2009), three members of the SNF1-related protein kinase2 family (Fujii et al. 2009), and molecules regulating ABA signaling including NADPH oxidase (Jiang and Zhang 2002; Zhang et al. 2009a, 2009b) and heterotrimeric G protein (Pandey and Assmann 2004).

Genetic and biochemical analysis in *Arabidopsis* points towards the involvement of ROP family GTPases in ABA signaling. Acting as negative regulator, expression of *CA-AtROP6* reduced ABA-mediated stomata closing (Lemichez et al. 2001). Zheng et al. in 2002 demonstrated *AtROP10* as negative regulator of ABA-mediated seed germination, stomata closing, and root elongation. A null *rop10* mutant exhibits enhanced responses to ABA in seed germination, root elongation, and stomata closure, whereas *CA-ROP10* reduces ABA inhibition of seed germination. Li and group in 2012 confirmed that *ROP11* also acts as a negative regulator of ABA-mediated stomata closure, seed germination, and drought response. Expressing *CA-ROP11* allele inhibiting stomata closure and *DN-ROP11* allele exhibited opposite effect. ABA primarily modulates the localization of ROP11, causing it to accumulate in the nucleus for transcriptional regulation. Moreover, phenotypic analysis of *rop10rop11* double mutant showed that they act in parallel pathways in ABA-mediated responses. Li et al. (2012a, 2012b), using luciferase complementation imaging (LCI), Y2H assay, and co-immunoprecipitation assay, showed that activated ROP11 binds ABI1 (ABA insensitive), a PP2C phosphatase, and allows the downstream SnRKs to be activated, which further activate TFs and upregulate stress-related genes like *RAD29*.

Insight into the mechanism of RacRop GTPase action was elucidated by studies providing a link between actin cytoskeleton remodeling in guard cells and ABA-induced stomata closure. Microscopic analysis of WT/*GFP-mTn* (GFP-mouse talin) and WT/*MAP4-GFP* (microtubule-associated protein 4-GFP), expressing GFP fusion proteins targeted to actin (Kost et al. 1998) and tubulin (Mathur and Chua 2000), respectively, supported ABA-mediated actin cytoskeleton reorganization during stomata closure. A role for ABA in actin cytoskeleton reorganization was further supported by studies done in *abi1-1/GFP-mTn* transgenic lines, where ABI1 is an upstream regulator of ABA signaling in stomata closure (Leung et al. 1994). *abi1-1* mutants display a wilty phenotype associated with impaired stomata closure. In *abi1-1/GFP-mTn* transgenic lines, both actin cytoskeleton and stomata closure remain unaffected (Meyer et al. 1994; Gosti et al. 1999). *AtRac1/AtROP3/ARAC3/Rop6AT* was isolated in a reverse genetic approach to identify genes acting in ABA signaling to induce stomata closure in *Arabidopsis* (Winge et al. 1997; Li et al. 1998; Lemichez et al. 2001). A role for *AtRac1/AtROP3* was confirmed by observing actin cytoskeleton organization in *AtRac1-T20N/GFP-mTn*, a DN Rac (Ridley et al. 1992), and *AtRac1-G15V/GFP-mTn*, a CA Rac (Diekmann et al. 1991), transgenic lines, where the former promoted actin cytoskeleton reorganization. Furthermore, WT/*AtRac1-T20N* transgenic lines blocked the ABA-induced stomata closure and restored stomata closure defects of *abi1-1/AtRac1-T20N* transgenic lines. In conclusion, AtRac1 acts as a negative regulator linking the actin cytoskeleton remodeling and ABA signaling during drought response and water homeostasis in *Arabidopsis* (Lemichez et al. 2001).

Hypoxia/oxygen deficiency causes switch from mitochondrial respiration to anaerobic fermentation (Sauter 2000). Ethanolic fermentation relies on activity of pyruvate decarboxylase (PDC) and alcohol dehydrogenase (ADH); both the enzymes are activated upon hypoxia (Fukao et al. 2006). Importance of *ADH* and *PDC* (pyruvate decarboxylase) has been shown by mutant analysis, where *adh* and *pdc* mutants of maize and *Arabidopsis* rapidly succumb to hypoxia, confirming the importance of ethanolic fermentation under oxygen stress (Ismond et al. 2003; Kürsteiner et al. 2003).

In animals, cellular hypoxia triggers H_2O_2 (ROS) production, leading to increased transcription of hypoxia-inducible factor 1α (HIF1α) via a Rho-GTP-dependent mechanism (Turcotte et al. 2003). In plants, hypoxia leads to H_2O_2 production by triggering the activity of ADH (Baxter-Burrell et al. 2002). In *Arabidopsis*, *ROPGAP4* was identified in an attempt to identify genes induced under hypoxia stress, using a *Ds-GUS* transposon gene trap (Baxter-Burrell et al. 2003). Loss-of-function *ropgap4* mutants exhibited increased *ADH* mRNA, indicating ROPGAP4 negatively regulates *ADH*. Moreover, the DN form of ROP2 (*35S:DN-Rop2*) did not show any increase in *ADH* mRNA, whereas the CA-ROP2 (*35S:CA-Rop2*) lines were hypersensitive to oxygen deprivation, supporting ROP2 as positive regulator in signaling during hypoxia. Even though H_2O_2 levels were found to increase significantly in *ropgap4* mutants under hypoxic conditions, no rise in H_2O_2 was detected in *35S:DN-Rop2* seedlings. These experimental results support the hypothesis, where ROPGAP4 acts as a negative feedback regulator of Rop GTPases, which are involved in defense responses and H_2O_2 production required for ADH induction. A feedback loop is formed where H_2O_2 activates RopGAP4, which in turn inhibits Rop2. Rop2 inactivation subsequently leads to reduction in hydrogen peroxide release (Baxter-Burrell et al. 2003).

Further, studies in tobacco revealed the role of Rac/Rops in salinity stress. *NtRop1* of tobacco when expressed in *Arabidopsis* conferred salt sensitivity. It was speculated that *NtRop1*-induced salt sensitivity is caused by increased H_2O_2 production, which acts as a second messenger molecule during salt stress (Cao et al. 2008), inducing PM H^+-ATPase activity resulting in increased K^+ to Na^+ ratio (Zhang et al. 2007).

9.2.3 *Ran GTPases*

Plants respond to various biotic and abiotic stresses by modulating transcriptional activity of various stress-related genes or alternatively by regulating posttranscriptional nucleocytoplasmic trafficking of proteins and noncoding RNA molecules through Ran (Ras-related nuclear protein)-dependent karyopherin (importin β) protein (Mosammaparast and Pemberton 2004; Mazzucoteli et al. 2008; Chinnusamy et al. 2008). Ran GTPase activity is regulated by several proteins including GAPs, GEFs, and RanBP1. Ran in GDP-bound form, generated by action of RanGAP1, exists in the cytoplasm, whereas the GTP-bound form, generated by RCC1 (regulator of chromosome condensation 1), a GEF, exists in the nucleus. This asymmetric distribution of Ran GTPases is pivotal for their functional activities (Bischoff and Ponstingl 1991; Bischoff et al. 1994). Ran-binding protein 1 (RanBP1) acts as an accessory factor to increase RanGAP1-mediated nucleotide hydrolysis by inhibiting the GEF activity of RCC1, thereby promoting inactivated Ran (Bischoff et al. 1995). In plants, the identification of RanBP1 and RanGAP (Ach and Gruissem 1994; Haizel et al. 1997; Rose and Meier 2001; Pay et al. 2002) revealed their high sequence similarity and subcellular localization with animal Ran proteins, pointing to functional conservation in nucleocytoplasmic trafficking and mitotic process (Ach and Gruissem 1994; Haizel et al. 1997).

Ran GTPases are encoded by a family of 4 genes in *Arabidopsis* and 3 genes in rice (Haizel et al. 1997; Vernoud et al. 2003). Several independent studies have implicated Ran GTPases in plant development, for example, overexpression (OE) of wheat Ran, *TaRAN1*, in *Arabidopsis* and rice resulted in an elevated mitotic index and prolonged life cycle (Wang et al. 2006). Besides, Ran GTPases are also affected by oxidative stress, causing reduction in ATP levels, leading to decrease in Ran GTP levels, and their disordered distribution (Yasuda et al. 2006).

Virus-induced gene silencing (VIGS) of *NbRanBP1* in *N. benthamiana* exhibited stress responses, such as reduced mitochondrial membrane potential and excessive production of ROS (Cho et al. 2008). A number of studies in both *Arabidopsis* and rice using transgenic and RT-PCR approaches have shown involvement of Ran in phytohormone signaling and salt stress response (Miche et al. 2006). *OsRAN2* overexpression results in increased sensitivity to salt, osmotic stress, and ABA treatment, and RT-PCR analysis of transgenic lines showed reduced *OsRAN2* trancript levels in salt and osmotic stresses and ABA treatment, supporting their role as negative regulators in abiotic stress response (Zang et al. 2010). Furthermore, three ABA- or stress-responsive genes, *AtNCED3*, *PLC1*, and *AtMYB2*, encoding a key enzyme in ABA synthesis, a phospholipase C homolog, and a putative transcription factor, respectively, showed differentially induced expression in *OsRAN2* OE plants under salt and ABA treatment conditions (Liu and Zhu 1997; Xiong et al. 2002; Verslues et al. 2006). In order to find out the mechanism of OsRAN2 action, *OsRAN2* and *Lc* (leaf color transcription factor) were co-expressed in tobacco epidermal cells, where LC acts as a maize reporter gene for nuclear transport (Shieh et al. 1993), and it was observed that OsRAN2 disturbed the nuclear transport of Lc, suggesting that the *OsRAN2* hypersensitivity phenotype could be due to its effect on nucleocytoplasmic trafficking of regulatory proteins such as transcription factors (Zang et al. 2010).

In *Oryza sativa*, *OsRAN2* was identified to confer resistance towards cold stress. The transcript level of *OsRAN2* increased several fold under cold stress condition. Insight into the mechanism of OsRAN2 action came from the analysis of the mean mitotic index for *OsRAN2* OE and *OsRAN2* RNAi lines, which showed an enhanced and reduced mean mitotic index, suggesting its role in cell cycle to confer cold stress tolerance. Subcellular localization of *OsRAN2:GFP* revealed GFP fluorescence in nucleus at the spindle and tubulin during metaphase and anaphase pointing towards a function in maintaining cell division during cold stress. These results were further corroborated by *OsRAN2* RNAi plants, exhibiting aberrant organization of spindles during mitosis and that OsRAN2 helps in nucleocytoplasmic transport of tubulin and formation of intact nuclear envelop during cold conditions. In conclusion, OsRAN2 regulates cold resistance in rice by maintaining cell division and by promoting accumulation of intranuclear tubulin at the end of mitosis and formation of intact nuclear envelope (Chen et al. 2011).

A recent report has shown that *OsRAN1* is ubiquitously expressed in rice and exhibit enhanced transcript levels upon cold and IAA treatment, where auxin is known for its effects on cell division. *OsRAN1* OE results in improved cold tolerance. It was found that at molecular level, it helps to maintain cell division and cell cycle progression and promote formation of intact nuclear envelope (Xu and Cai 2014). In conclusion, Ran GTPases confer cold stress tolerance by regulating the plant cell cycle; interestingly, cell division has been previously closely related to stress tolerance in plants (Kitsios and Doonan 2011).

9.2.4 RAB GTPases

RABs and SNAREs (soluble N-ethylmaleimide-sensitive (NSF) attachment protein receptors) are conserved regulatory molecules involved in membrane trafficking. RABs promote the tethering of transport vesicles/organelles to target membranes, followed by membrane fusion, which is executed by specific combinations of three glutamine-contributing SNAREs (Qa-, Qb-, and Qc-SNARE) and one arginine-contributing (R-)SNARE in a stable complex (Saito and Ueda 2009; Wickner and Schekman 2008).

Arabidopsis genome encodes 57 Rab GTPases (Pereira-Leal and Seabra 2001), grouped into 8 clades (RabA-RabH) (Segev 2001; Rutherford and Moore 2002; Vernoud et al. 2003). Among them, Rab5 (RabF) forms the largest clade (Stenmark et al. 1994), with three homologs encoded in *Arabidopsis* genome including *AtRabF1* (*Ara6*), *AtRabF2a* (*Rha1*), and *AtRabF2b* (*Ara7*) (Rutherford and Moore 2002; Vernoud et al. 2003). Co-expression of *ARA6-Venus* and *ARA7-GFP* under their native regulatory elements showed overlapping expression pattern in various endosomal population in *Arabidopsis,* confirming their role in endocytosis (Ebine et al. 2012).

Endocytosis, traditionally a constitutive housekeeping process, has recently been implicated in trafficking in stress responses in tobacco (Leyman et al. 1999). It has been shown that in yeast, oxidative stress inhibits vesicle trafficking between cytosol and PM, a phenotype partially restored by OE of *Arabidopsis* synaptobrevin homolog, providing tolerance to lethal concentrations of H_2O_2 (Levine et al. 2001).

Various Rab GTPases have been identified and implicated in biotic and abiotic stresses using SSH, differential display (DD), and northern blot techniques in different plant species. Phenotypic analysis of mutants and various proteomics approach helped to understand the mechanism of Rab GTPase action in conferring stress tolerance.

AtRabG3e (*AtRab7*) was isolated by DD approach for genes induced under abiotic and biotic stresses including salt, superoxide, SA, and necrogenic pathogens (virulent *P. syringae* bacteria or *B. cinerea* fungus) (Mazel et al. 2004). Overexpression of *AtRabG3e* in *Arabidopsis thaliana* leads to less sensitivity to salt and osmotic stress than WT plants. Analysis of general membrane endocytosis in *AtRabG3e* transgenic plants revealed accelerated uptake of membrane dye FM1-43, which could be due to faster vesicle trafficking at the end point, i.e., endosome-vacuole fusion, supporting the hypothesis that in the transgenic plants Na^+ is transported into the vacuoles, reducing the overall cytosolic Na^+ content. Moreover, transgenic plants exhibited reduced ROS content, due to decrease in the cytosolic Na^+ content (Mazel et al. 2004). Another Rab5 member, *ara6*, lof confers hypersensitivity to salt stress, whereas OE transgenic plants showed elevated resistance to salinity stress and high-sorbitol stress (Ebine et al. 2012). Similarly, in rice root cells, ARA6 is secreted in the apoplast in response to salinity stress (Zhang et al. 2009a, 2009b).

PgRab7 of *Pennisetum glaucum*, isolated by subtractive hybridization (SSH) of treated plants, showed enhanced transcript level upon cold, dehydration, and NaCl and IAA treatment. The role of PgRab7 in abiotic stress was confirmed by overexpression of *PgRab7* in heterologous system, enhancing NaCl and mannitol tolerance in transgenic tobacco plants (Agarwal et al. 2008). In a similar study, *Prosopis juliflora*, an abiotic stress-tolerant tree species of *Fabaceae*, was used for isolating genes functioning in abiotic stress tolerance. Northern analysis of leaf tissue revealed upregulation of *PjRAB7* during salt stress conditions. Moreover, expressing *PjRAB7* in heterologous system conferred salt tolerance to tobacco, probably by sequestering sodium in the vacuoles, suggesting functional conservation of Rab GTPase in different plant species (George and Parida 2011). Adopting differential display (DD) approach *Rab2* was identified in the desiccation-tolerant grass *Sporobolus stapfianus* in response to drought and rehydration in leaves. RT-PCR analysis showed *RAB2* transcript accumulation in response to decreased relative water content (RWC), whereas rehydration resulted in decreased levels. Furthermore, ABA was also found to function in this pathway as exogenous ABA slightly increased the *RAB2* transcript accumulation (O'Mahony and Oliver 1999). SmGTP related to *Rab2* gene family was identified in a salt stress SSH expression library of model grass species *Lolium temulentum*, exhibiting strong expression under salt and drought stress, a pattern similar to dehydration-specific gene such as delta 1-pyrroline-5-carboxylate synthetase (P5CS) and cold response gene *COR413* (Dombrowski et al. 2008).

The mechanism of Rab GTPase tolerance to biotic stress was shown in wheat, *TaRab7*, a Rab GTPase gene. It was isolated by reverse transcription technique from wheat cDNA library obtained from leaves infected with *Puccinia striiformis* f. sp. *tritici* (*Pst*), wheat stripe rust pathogen. It showed highest homology to *BdRab7*-like small GTPase from *Brachypodium distachyon*, *OsRab7* (*Oryza sativa*), *AtRabG3b* (Arabidopsis), and *HvRab7* (*Hordeum vulgare*). qRT-PCR analysis of TaRab7 showed enhanced expression levels under both abiotic (salinity, drought, and low temperature) and biotic (*Pst* infection) stresses. Furthermore, knocking down

TaRab7 expression by virus-induced gene silencing (VIGS) enhanced the susceptibility of wheat cv. Suwon 11 towards an avirulent *Pst* race CYR23 (Liu et al. 2012a, 2012b). Previous studies have shown that Rab7 interacts with various downstream effectors (Zerial and McBride 2001) and plays a critical role in vesicular transport from early to late endosomes in the cytoplasm (Feng et al. 1995; Mukhopadhyay et al. 1997). In HeLa cells, Rab7 is critical for lysosomal aggregation and fusion in perinuclear region (Bucci et al. 2000). Transient subcellular localization assays in onion peel cells using *pCaMV35S:TaRab7-GFP* fusion construct revealed the cytoplasmic, perinuclear, and nuclear localization pattern, consistent with its role in regulating lysosome and phagosome to detoxify the toxic material secreted by pathogen into the host cells (Liu et al. 2012a, 2012b).

Studies in *Arabidopsis* have shown the involvement of auxins in stress response and tolerance. RABA (RAB11) group/clade contains 26 members, divided further into six subgroups (RABA1–RABA6). Cell biological and physiological analysis of RABA1 members implicated them in transport between trans-Golgi network and PM and their requirement during salt stress tolerance. Mutant and complementation analysis showed that the four RABs – RABA1a, RABA1b, RABA1c, and RABA1d – act redundantly in salt stress tolerance and are not involved in regulating intracellular distribution of sodium (Asaoka et al. 2012; Asaoka et al. 2013) and probably act by affecting the localization of PIN, auxin efflux transporters (Feraru et al. 2011), supported by the role of auxins in salt stress tolerance (Park et al. 2011). RABA1, subgroup of (RAB11) family members in *Arabidopsis,* shows a different mechanism for salt tolerance; it is involved in salinity stress by regulating the transport of sodium across the PM and accumulation in vacuoles as the quadruple mutants do not affect the sodium content or distribution (Asaoka et al. 2013).

In *Mangifera indica, MiRab5* is upregulated in response to cold, salinity, and PEG stress in addition to its involvement in seed germination (Liu et al. 2014). *Arabidopsis* CPRabA5e (where CP represents chloroplast) is a chloroplast-localized protein found to regulate oxidative stress signaling. The role of CPRabA5e was further validated by Y2H screen where 13 proteins previously predicted to be involved in stress and localized in thylakoids and plastoglobules were identified (Karim et al. 2014). However, the exact mechanism of *CPRab5e* activity is not known yet.

Proteome analysis of rice seedlings during drought stress and recovery upon rewatering revealed that 9 small GTP-binding proteins are highly expressed specifically under severe drought stress, while 3 out of 9 were induced in all stress conditions with highest expression in drought. Among these 9 stress-induced small GTP-binding proteins, 4 belong to Rab GTPases family, indicating Rab GTPases role in plant stress tolerance is conserved throughout the plant kingdom (Mirzaei et al. 2012).

9.2.5 ARF GTPases

High concentrations of salt arrest plant development and lead to plant cell death by disrupting ion and water homeostasis, inhibition of metabolism, and damage to membranes (Huh et al. 2002). Photosynthetic activity is suppressed by salt stress, and the reduction in photosynthetic activity can be accounted for by the decline in chlorophyll content (Al-Aghabary et al. 2004). Plants suffering from salt stress often display symptoms of oxidative damage as indicated by marked accumulation of ROS such as H_2O_2 (Hasegawa et al. 2000). MDA (malondialdehyde) is recognized parameter for lipid peroxidation, leading to increased permeability of membranes to various ions (Mittler 2002). Plants overcome salt stress by either transporting excess Na^+ into the vacuoles or by prohibiting its entry into plant cells, accomplished by various transporters (Apse et al. 1999; Zhang and Blumwald 2001; Rus et al. 2001) or by accumulating osmolytes.

ADP-ribosylation factor (Arf) GTPases are important regulators of membrane-trafficking pathways, initially identified due to their ability to stimulate the ADP-ribosyltransferase activity of cholera toxin A (Moss and Vaughan 1998). They are involved in generation of the three types of vesicle coat proteins (COPI, COPII, and clathrin) (Kirchhausen 2000). Plant genome encodes for Arf-like (Arl) GTPases with protein sequence highly similar to that of Arf GTPases in animals. Knockout of an Arl GTPase, *TITAN*, leads to dramatic alterations in mitosis and cell cycle control during seed development in *Arabidopsis* (McElver et al. 2000). Suppression subtractive hybridization (SSH) in *Medicago falcate*, tolerant to abiotic stresses, identified *MfARL1*, an Arl GTPase, which confers greater salt tolerance. Functional characterization of *MfARL1* in *Arabidopsis* confirmed its role in salt tolerance. The OE transgenic plants showed reduced accumulation of Na^+ and reduced expression levels of *AtHKT1*, a sodium transporter, which probably reduces the influx of Na^+ into the plant thereby conferring resistance to salt (Wang et al. 2013). Function of *AtHKT1* in sodium transport has been previously shown, where *AtHKT1* mutation suppress *sos3* mutant phenotypes. Analysis of ion contents in the *sos3hkt1* mutant demonstrated that *AtHKT1* is involved in mediating Na^+ influx into the plant cells (Rus et al. 2001). Moreover, the photosynthetic activity and chlorophyll content of the *MfARL1* OE transgenic lines were not affected by salt stress when compared to wild-type plants. The MDA and H_2O_2 content of transgenic plants was less, whereas the wild-type plants exhibited increased levels of H_2O_2 and MDA. Increased level of H_2O_2 in wild-type plants was due to decreased CAT (catalase) activity, whereas salt stress has no effect on CAT activity in the transgenic lines. In conclusion, *AtHKT1* provides greater salt tolerance by regulating a sodium transporter and CAT activity.

Similarly, *ARF1* of *Jatropha curcas* is induced in response to abiotic stresses, particularly upon treatment with PEG, which induces osmotic stress, and cold treatment (Qin et al. 2011).

9.2.6 MIRO GTPases

Miro GTPases function in mitochondrial trafficking and morphology in eukaryotes. The role for *AtCBG* (calcium-binding GTPase)/*AtMIRO2*, a Miro GTPase, has been explored in abiotic stress. Expression of *AtMIRO2* was upregulated in salt stress and ABA treatment suggesting its involvement in abiotic stress. Phenotypic analysis of *AtMIRO2* mutant confirmed their sensitivity towards both salt and ABA stress (Jayasekaran et al. 2006).

9.2.7 Seed Plant (Spermatophyte)-Specific Small GTPase

In plants, circadian clock regulates signaling pathways such as light, hormone, and stress (Covington et al. 2008). Plants respond to high salinity stress, by increasing cytosolic calcium, by activating the SOS1-mediated changes in ion homeostasis, and subsequently by generating secondary messengers such as phospholipids and ROS, which activate kinase cascade to regulate transcription of stress-inducible genes. These responses can be ABA dependent and independent and are linked to circadian clock (Sanchez et al. 2011). Kevei et al. in 2007 identified *LIP1* (*LIGHT INSENSITIVE PERIOD 1*), a small GTPase, in a genetic screen for novel circadian clock mutant in Arabidopsis. *lip1* mutant displayed increased sensitivity to salt stress, affecting morphogenesis and ploidy. qRT-PCR analysis confirmed that *LIP1* does not function in ABA-dependent or ABA-independent salt stress signaling pathways leading to expression of osmoprotectant genes like *RAD29A* (*RESPONSIVE TO DESICCATION 29A*), *RAD29B*, or *RAB18* (*RESPONSIVE TO ABA*) and ionic stress genes like *SOS2*, a Na^{+}/H^{+} antiporter and activator of *SOS1*, respectively, though its function at the posttranslational level was not ruled out. Another possible explanation for salt stress sensitivity could be the role of *LIP1* in endoreplication, affecting the ploidy level (Terecskei et al. 2013). Previous studies have implicated endoreplication in seedling development upon exposure to stress including high temperature and drought (Engelen-Eigles et al. 2001). Cookson et al. in 2006 showed that increase in extent of endoreduplication reduced the negative impact of drought on the final leaf size.

9.3 Large G Proteins in Biotic and Abiotic Stress

9.3.1 Biotic Stress

Several independent studies using inhibitors and agonists of G proteins in different plant species supported the role of G proteins in defense signaling (Legendre et al. 1992; Beffa et al. 1995; Gelli et al. 1997). Particularly, changes in cytosolic calcium concentrations, often observed in elicitor-treated plant cells, are regulated by G proteins (Aharon et al. 1998). Calcium then activates the downstream signaling pathway including calcium-activated kinases (Romeis et al. 2000), calmodulin (Heo et al. 1999), and NADPH oxidase (Sagi and Fluhr 2001).

Studies in rice have indicated the specificity of G-protein signaling in plant defense response. Rice *d1* mutants, lacking functional *RGA1* (Gαs), showed reduced defense response to avirulent strains of bacterial blight (*Xanthomonas oryzae* pv. *Oryzae* [*Xoo*]) and rice blast fungus (*Magnaporthe grisea*) (Komatsu et al. 2004). A role supported by elevated *RGA1* mRNA levels upon rice blast infection and sphingolipid elicitor (SE) treatment in leafs and cultured cells. Further analysis supported G-protein function in ETI-triggered defense response, as there was suppression of ROS production and *PBZ1* expression by SE in the *d1* mutant cell cultures. Moreover, Rac signaling was placed genetically downstream of G proteins based on epistasis analysis using *CA-OsRac1*, where *OsRac1* restored the defense signaling, by promoting the *PR* gene expression and ROS production in *d1* mutant plants (Suharsono et al. 2002).

Similarly, *Arabidopsis* mutants of *gpa1* gene, encoding *GPA1*, have slightly increased resistance to several pathogens though the link between G proteins and plant defense is conclusively established through the *Arabidopsis* Gβγ dimer. Mutants deficient in *AGB1* (Gβs) are more susceptible to the fungal pathogens *Alternaria brassicicola*, *Botrytis cinerea*, *Fusarium oxysporum*, and *Plectosphaerella cucumerina* (Llorente et al. 2005; Trusov et al. 2009). Upon infection with *A. brassicicola* or treatment with methyl jasmonate (MeJA), where latter is produced during both biotic and abiotic stress response, *agb1* mutants exhibit delay in the induction of the MeJA-induced *PR* genes, *PDF1.2*, *OPR3*, and *PAD3* (Trusov et al. 2009), whereas expression of the SA-dependent *PR1* was increased after infection with *P. cucumerina* (Llorente et al. 2005). It was concluded that the Gγs confer functional selectivity to Gβγ dimer. Among the three *Arabidopsis* Gγs, *AGG1* has been implicated in plant defense response against *F. oxysporum* and *A. brassicicola* (Trusov et al. 2007), whereas *AGG3* has no defense-related response and the role of *AGG2* is not clear (Delgado-Cerezo et al. 2012). In conclusion, G protein plays an important role in both PTI- and ETI-triggered immune responses. Besides, the canonical G-protein subunits, one of three extra-large α-subunit, *XLG2* (Lee and Assmann 1999), is linked to plant defense. *xlg2* mutant shows enhanced susceptibility to *P. syringae* and reduced induction of the pathogenesis-related gene *PR2* indicating positive regulatory role in plant defense response. Microarray analysis revealed that, aside from *PR2*, other pathogen-inducible genes are downregulated in *xlg2* mutant in response to *P. syringae* infection, whereas overexpression of *XLG2* resulted in accumulation of aberrant transcripts for several defense-related genes. Biochemical analysis revealed that XLG2 physically interacts with AGB1 and that the interaction is restricted to infected tissues (Zhu et al. 2009).

G protein have been shown to be involved in elicitor-induced HR in plants by VIGS-induced silencing of the Gα, Gβ1, and Gβ2 subunits of *Nicotiana benthamiana* G-protein complex. The silenced plants when treated with bacterial and fungal elicitors showed impaired HR including stomata closure, nitric oxide (NO) production, and ROS accumulation in guard cells. Besides, transcription of various plant defense-related genes was also impaired, including *NbrbohA* (NADPH oxidase), NOA1 (associated with NO), and NIA (responsible for NO production) (Zhang et al. 2012).

9.3.2 Abiotic Stress

Sessile nature of plants constantly exposes them to changing environmental conditions giving rise to oxidative stress. A mild oxidative stress induces antioxidant defenses, whereas severe stress leads to rapid necrosis or cell death. The induction and execution of PCD is regulated by various signaling molecules including ROS, jasmonic acid (JA), salicylic acid (SA), and ethylene (Lam et al. 1999). Ozone is a major oxidative stress and a major pollutant that damages crops, forests, and urban vegetation. The main symptoms of ozone damage are foliar lesions and reduction in stomata aperture mediated through ABA signaling (Mansfield 1998), and it shares signaling pathways and gene expression pattern with hypersensitive response (HR) (Tamaoki et al. 2003).

Arabidopsis plants with null mutations in the *gpa1* (Gαs) and *agp1* (Gβγ dimer) are less and more sensitive, respectively, to ozone damage than the wild-type plants. Analysis of *gpa1* mutant plants exhibited ozone-resistant phenotype, indicated by lack of leaf curling in response to ozone (Booker et al. 2004). At molecular level, it was shown that Gαs activates membrane-bound NADPH oxidases, encoded by *AtrbohD* and *AtrbohF* genes, whereas the Gβγ dimer stimulates ROS production in chloroplast. The role of *gpa1* in activating NADPH oxidases was further supported by analysis of *atrbohD* and *atrbohF* double-mutant plants, which are more resistant to ozone than wild-type plants. Moreover, a link between G-protein and Rac signaling has been provided, where Gαs acts through the Rop (Baxter-Burrell et al. 2002; Suharsono et al. 2002) and where Rop is activated by phosphatidic acid (PA), a product of phospholipase Dα (PLDα), which in turn is inhibited by Gαs. A direct interaction was shown between PLDαl and GPA1, where binding inhibited PLDα1 activity, relieved upon GTP addition, suggesting in vivo G-protein activation leads to PLDα1 activation. PA upon activation and dissociation of Gαs inhibits the activity of ABI-1 and ABI-2 (ABA insensitive), negatively regulating ABA signaling through stress-activated MAPK cascade (Zhang et al. 2004). *gpa1* mutant also exhibit reduced ozone sensitivity at single-guard-cell level, indicated by reduced ABA inhibition of guard cell inward K^+ channels and altered ABA promotion of slow anion currents (Wang et al. 2001). ABA is known to mediate its effects by acting through second messenger, S1P (sphingosine-1-phosphate) (Ng et al. 2001; Coursol et al. 2003) and through the effector molecule, PLCs (phosphatidylinositol-phospholipases) (Jacob et al. 1999; Zhang et al. 2004). Guard cells of *gpa1* mutants exhibit insensitivity to inhibition of stomata opening by S1P. Moreover, studies supported existence of two parallel pathways: S1P dependent and independent (Wang et al. 2001; Coursol et al. 2003).

The role for G proteins in ABA-mediated response is further validated by studies in *gcr1* mutants, exhibiting hypersensitivity to ABA and S1P in both stomata opening and promotion of stomata closure, indicating that *GCR1* acts as negative regulator of *GPA1*-mediated ABA responses in guard cells. A hypothesis further supported by higher levels of expression of drought- and ABA-regulated genes after exogenous ABA treatment and inhibited improved recovery following drought stress in *gcr*1 mutant plants (Pandey and Assmann 2004). PLCs, the downstream effectors of ABA, have been shown to both genetically and physically interact with G-protein signaling component, indicated by studies done in BY2 tobacco cells overexpressing *GCR1*, where GCR1 activates DNA synthesis in PLC-dependent manner (Apone et al. 2003).

G proteins have been shown to function in closure of stomata aperture by production of H_2O_2 induced by ExtCaM (extracellular calmodulin) and NO (nitric oxide). G-protein involvement in H_2O_2 production is evident by analysis of *gpa1* mutants; ExtCaM-induced H_2O_2 production is impaired in *gpa1* mutants, indicated by impaired stomata closure. ExtCaM-mediated NO generation is regulated by GPA1, whereas GPA1 activation of NO production depends on H_2O_2 (Chen et al. 2004a, 2004b).

Recently, eATP (extracellular ATP) has been reported as a signaling molecule, existing in the apoplast of plant cells. In *gpa1* mutant plants, eATP-promoted stomata opening, cytoplasmic ROS generation, and Ca^{2+} and H^+ efflux are all suppressed, indicating G-protein involvement during eATP signaling (Hao et al. 2012). Proteomics has identified MAPK (mitogen-activated protein kinase) as a downstream effector molecule for G-protein-mediated signaling in response to ABA and MeJA (Methyl Jasmonate), supported by yeast two-hybrid assays and upregulation of PsGβ and PsMAPK3 (*Pisum sativum* MAPK3), upon treatment with MeJA and ABA (Bhardwaj et al. 2011).

Besides, *Arabidopsis* and rice, the signaling pathways involving G protein during abiotic stress conditions are also conserved in other plants. Overexpression of *AGG3* under the control of CaMV35S promoter in *Camelina sativa* resulted in transgenic seeds with less sensitivity to ABA, osmotic, and salt stress during germination and post germination. In addition, stomata in *CaMV35:AGG3* plants are more sensitive to ABA, imparting better drought recovery (Roy Choudhary et al. 2004).

Some plants respond to hypoxia or oxygen deprivation like underwater loging by enhanced ethylene production (Cohen and Kende 1987; Wang and Arteca 1992; Zarembinski and Theologis 1993). Ethylene signaling regulates adaptation to hypoxic conditions by promoting morphological, anatomical, and metabolic adaptations (He et al. 1996a; He et al. 1996b; Steffens and Sauter 2009). *RGA1*, the rice homolog of Gαs, functions downstream of ethylene and H_2O_2 under oxygen deprivation condition to induce cell death, indicated by reduced cell death in *rga1* (*d1*) mutant plants, with reduced *D1* mRNA (Steffens and Sauter 2010). The cell death phenotype of *rga1* mutant plants could not be restored by treatment with ethylene or H_2O_2, indicating that G proteins act downstream of these signaling molecules. Microarray analysis revealed that the *D1* gene itself was not regulated in epidermal cells by ethylene or H_2O_2, pointing to posttranscriptional regulation of the D1-dependent G-protein activation (Suharsono et al. 2002; Steffens and Sauter 2009).

Studies in maize further support the role of G protein in hypoxia condition. Maize roots incubated with GTPγS, an activator of G proteins, at normoxic conditions, resulted in cell death and aerenchyma formation. This could probably occur due to constitutively activated G proteins, as GTPγS is nonhydrolyzed substrate for GTPases. Moreover, formation of aerenchyma was accompanied by an increase in cellulase activity, where cellulases contribute to the degradation of the cell wall and thus to the removal of the cell corpses. These results demonstrated a central role for G protein in hypoxia by activating cellulases (He et al. 1996a).

ABA, a plant stress hormone, is known to regulate adaptation of plants to environmental stresses including cold, drought, and salt by modulating the expression of stress-responsive genes and controlling the stomatal aperture to regulate water content (Spartz and Gray 2008). *BnGA1*, encoding *Brassica napus* Gαs, is shown to be induced in response to ABA. *BnGA1* is upregulated by salinity and drought stress tolerance while downregulated by temperature stresses (Gao et al. 2010) supporting the role of *BnGA1* in ABA-mediated responses. Similarly, *Arabidopsis GPA1* acts as a regulator of transpiration efficiency (TE). *gpa1* mutant plants, despite having guard cells that are hyposensitive to ABA-induced inhibition of stomata opening, exhibit increased TE under ample water and drought stress conditions and when treated with exogenous ABA, most likely due to increased stomatal density. *GPA1* regulation of stomatal density was further supported by analysis of *gpa1* mutant plants, which showed reduced stomatal density (Nilson and Assmann 2010a). This supports the role of GPA1 downstream of ABA signaling. *AtRGS1* was identified as one of the components of G-protein signaling during salt stress in Arabidopsis. Lof *AtRGS1* (*rgs1-2* mutant) in presence of salt showed better survival and low senescence as opposed to lof Gβs (*agb1-2* mutant), which showed low survival and accelerated senescence under salt stress. Most likely in the presence of Na^+, AtRGS1 undergoes encosytosis due to increased glucose levels, thereby activating the Gαs and releaving the heterotrimer. In conclusion Gαs and Gβs act antagonistically during salt stress affecting the survival and senescence process (Colaneri et al. 2014).

Asakura and Kurosaki (2007) demonstrated that the expression of Gαs in carrot seedlings was transiently decreased by treatment with high concentration of NaCl and decreased by the exposure to high temperature, and no induction was seen at low temperature in carrot. In pea, the expression of Gαs increases upon treatment with high concentration of NaCl and high temperature (Misra et al. 2007).

Functional genomics approach has shown conservation of G-protein signaling across plant species during biotic and abiotic stresses. Transcription profiling of rice *rga1* showed upregulation in the transcript level following NaCl, cold, and drought stresses. Where higher temperatures downregulate *RGA1* transcript levels, heavy metal(loid) stress exhibits upregulation. Similar, transcription profiling of rice *RGG1* and *RGG2* showed that their transcript levels are upregulated following

NaCl, cold, heat, and ABA treatments. However, in drought stress, only *RGG1* is upregulated. Promoter analysis of *RGG1* and *RGG2* revealed presence of stress-related *cis*-regulatory signature motifs, supporting their role in abiotic stress response (Yadav et al. 2012; Yadav et al. 2013). Similar study involving in silico characterization of *cis*-regulatory elements of *RGB1*, encoding the Gβs of Indica rice, revealed the presence of stress-related *cis*-regulatory elements. Transcript profiling revealed enhanced *RGB1* transcript levels under KCl, cold, drought (dehydration), and micronutrient (Mn^{2+} and Zn^{2+}) stresses. Moreover, nuclear localization of RGB1 supports its possible interaction with transcription factors in regulating salt stress-responsive genes (Yadav et al. 2014).

9.4 Conclusions and Future Perspectives

Increased demand of high-yielding food crops and depleting natural resources has resulted in exponential growth in the number of genetically modified crops. To efficiently generate genetically engineered plants, use of high-throughput technologies has led to rapid identification and subsequent manipulation of target genes. Primarily, target genes are involved in signal transduction pathways, which either increase the yield or generate stress-tolerant crop plants. Using high-throughput techniques like proteomics, metabolomics, and phenomics along with functional genomics approach has identified several key signaling molecules in plant stress responses, including large and small G proteins. G proteins are excellent molecular switches in various cellular and biological processes and can serve as excellent targets for generating stress-tolerant crop plants. Monomeric GTPases have been shown to act as molecular switches in various abiotic and biotic stress conditions signaling downstream of heterotrimeric G proteins, elicitor molecules, phytohormones like ABA, ethylene, and others. In GTP-bound form, they undergo conformation change and activate downstream effector molecules like NADPH oxidase, MAPKs, SnRKs, metallothionein, and transcription factors, to generate response either by modulating enzyme activity or gene expression. Similarly, research has shown that heterotrimeric G proteins act in signaling pathways in response to pathogens, ROS, salt, and others. The ease of manipulation of pathways involving large and small G proteins by using constitutively active or dominant-negative forms has made them excellent targets to generate stress-tolerant high-yielding crop plants. Although research done in the past several decades has delineated many signaling pathways involved in both biotic and abiotic stresses as evident from Fig. 9.4, the picture is far from complete, and future efforts have to be focused on establishing the intricate network and cross talk between various networks to find a target pathway for creating genetically modified crop plants.

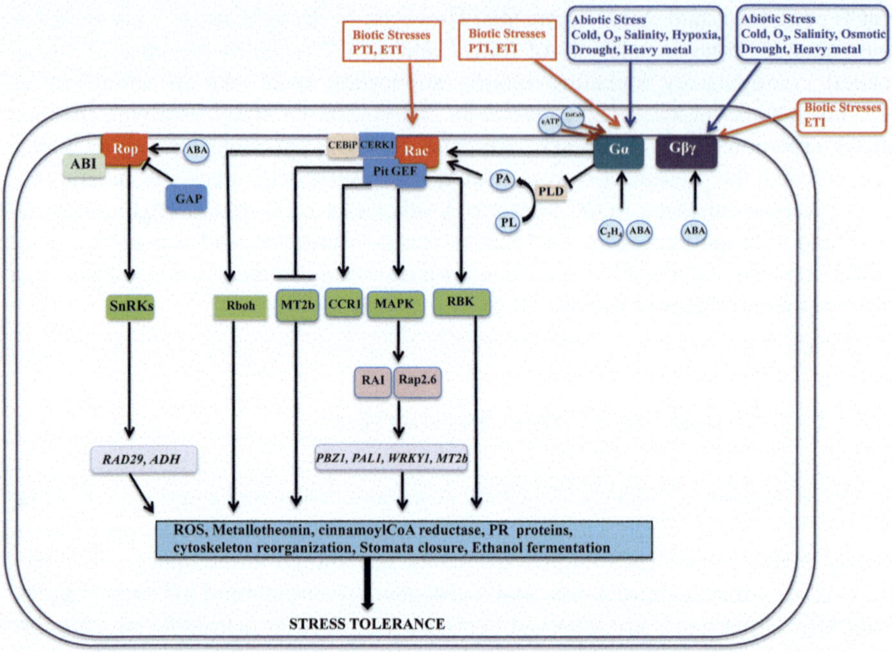

Fig. 9.4 Schematic depiction of various signaling pathways involving small and large G proteins during biotic and abiotic stress responses in plants: the G proteins and small GTPases are activated in response to various abiotic and biotic stimuli and second messenger molecules to activate several signaling pathways. The G-protein subunits activate downstream effectors when activated, and one of the effector molecules is Rac GTPase thus providing a link between G proteins and small GTPases. The Rac GTPase acts in plant defense response by forming a defensome at the membrane consisting CEBiP (chitin-binding protein), CERK (chitin-activated receptor kinase), and Pit (NLR protein). The Rac complex then activates various effector molecules including Rboh (NADPH oxidase), MT2b (metallothionein), CCR1 (cinnamoyl-CoA reductase1), MAPK (mitogen-activated kinase), and RBK (receptor-like kinase). Downstream of effector molecules is transcription factor such as RAI (Rac immunity) and Rap2.6, which regulate transcription of defense-related gene such as *PBZ1*, *PAL1*, *WRKY1*, *and MT2b*, leading to response and stress tolerance

Acknowledgment We are thankful to Delhi University and Department of Biotechnology (DBT), India, for research support in our lab.

References

Ach RA, Gruissem W (1994) A small nuclear GTP-binding protein from tomato suppresses a schizosaccharomyces pombe cell-cycle mutant. Proc Natl Acad Sci U S A 91:5863–5867

Agarwal PK, Agarwal P, Jain P, Reddy HK, Sopory SK (2008) Constitutive overexpression of a stress-inducible small GTP-binding protein PgRab7 from *Pennisetum glaucum* enhances abiotic stress tolerance in transgenic tobacco. Plant Cell Rep 27:105–115

Aharon GS, Gelli A, Snedden WA, Blumwald E (1998) Activation of a plant plasma membrane Ca2+ channel by TGalpha1, a heterotrimeric G protein alpha-subunit homologue. FEBS Lett 424(1–2):17–21

Akamatsu A, Wong HL, Fujiwara M, Okuda J, Nishide K, Uno K, Imai K, Umemura K, Kawasaki T, Kawano Y, Shimamoto K (2013) An OsCEBiP/OsCERK1-OsRacGEF1-OsRac1 module is an essential early component of chitin-induced rice immunity. Cell Host Microbe 13(4):465–476

Al-Aghabary K, Zhu Z, Shi QH (2004) Influence of silicon supply on chlorophyll content, chlorophyll fluorescence, and antioxidative enzyme activities in tomato plants under salt stress. J Plant Nutr 27:2101–2115

Apone F, Alyeshmerni N, Wiens K, Chalmers D, Chrispeels MJ, Colucci G (2003) The G-protein-coupled receptor GCR1 regulates DNA synthesis through activation of phosphatidylinositol-specific phospholipase C. Plant Physiol 133:571–579

Apse MP, Aharon GS, Snedden WA, Blumwald E (1999) Salt tolerance conferred by overexpression of a vacuolar Na+/H+ antiport in Arabidopsis. Science 258:1256–1258

Arabidopsis Genome Initiative (2000) Analysis of the genome sequence of the flowering plant *Arabidopsis thaliana*. Nature 408:796–815

Asakura Y, Kurosaki F (2007) Cloning and expression of *dcga* gene encoding a subunit of GTP-Binding protein in carrot seedlings. Biol Pharm Bull 30:1800–1804

Asaoka R, Uemura T, Ito J, Fujimoto M, Ito E, Ueda T, Nakano A (2012) Arabidopsis RABA1 GTPases are involved in transport between the trans-golgi network and the plasma membrane, and are required for salinity stress tolerance. Plant J 73:240–249

Asaoka R, Uemura T, Nishida S, Fujiwara T, Ueda T, Nakano A (2013) New insights into the role of Arabidopsis RABA1 GTPases in salinity stress tolerance. Plant Signal Behav 8(9):pii, e25377

Ashikari M, Wu J, Yano M, Sasaki T, Yoshimura A (1999) Rice gibberellin-insensitive dwarf mutant gene Dwarf 1 encodes the alpha-subunit of GTP-binding protein. Proc Natl Acad Sci U S A 96:10284–10289

Aspenstrom P (1999) The Rho GTPases have multiple effects on the actin cytoskeleton. Exp Cell Res 246:20–25

Babior BM (1999) NADPH oxidase: an update. Blood 93:1464–1476

Batoko H, Zheng HQ, Hawes C, Moore I (2000) A *rab1* GTPase is required for transport between the endoplasmic reticulum and Golgi apparatus and for normal Golgi movement in plants. Plant Cell 12:2201–2218

Baxter-Burrell A, Yang Z, Springer PS, Bailey-Serres J (2002) RopGAP4-dependent Rop GTPase rheostat control of Arabidopsis oxygen deprivation tolerance. Science 296(5575):2026–2028

Baxter-Burrell A, Chang R, Springer P, Bailey-Serres J (2003) Gene and enhancer trap transposable elements reveal oxygen deprivation-regulated genes and their complex patterns of expression in Arabidopsis. Ann Bot 91 Spec:129–141

Beffa R, Szell M, Meuwly P, Pay A, Vögeli-Lange R, Métraux JP, Neuhaus G, Meins F Jr, Nagy F (1995) Cholera toxin elevates pathogen resistance and induces pathogenesis-related gene expression in tobacco. EMBO J 14(23):5753–5761

Berken A, Thomas C, Wittinghofer A (2005) A new family of RhoGEFs activates the Rop molecular switch in plants. Nature 436:1176–1180

Bhardwaj D, Sheikh AH, Sinha AK, Tuteja N (2011) Stress induced β subunit of heterotrimeric G-proteins from *Pisum sativum* interacts with mitogen activated protein kinase. Plant Signal Behav 6(2):287–292

Bischoff FR, Ponstingl H (1991) Catalysis of guanine nucleotide exchange on Ran by the mitotic regulator RCC1. Nature 54:80–82

Bischoff FR, Klebe C, Kretschmer J, Wittinghofer A, Ponstingl H (1994) RanGAP1 induces GTPase activity of nuclear Ras-related Ran. Proc Natl Acad Sci U S A 91:2587–2591

Bischoff FR, Krebber H, Smirnova E, Dong W, Ponstingl H (1995) Co-activation of RanGTPase and inhibition of GTP dissociation by Ran-GTP binding protein RanBP1. EMBO J 14: 705–715

Bischoff F, Molendijk A, Rajendrakumar CS, Palme K (1999) GTP-binding proteins in plants. Cell Mol Life Sci 55:233–256

Bokoch GM (1994) Regulation of the human neutrophil NADPH oxidase by the Rac GTP-binding proteins. Curr Opin Cell Biol 6(2):212–218

Booker FL, Burkey KO, Overmyer K, Jones AM (2004) Differential response of G-protein *Arabidopsis thaliana* mutants to ozone. New Phytol 162:633–641

Bucci C, Thomsen P, Nicoziani P, McCarthy J, van Deurs B (2000) Rab7: a key to lysosome biogenesis. Mol Biol Cell 11:467–480

Canonne J, Froidure-Nicolas S, Rivas S (2011) Phospholipases in action during plant defense signaling. Plant Signal Behav 6(1):13–88

Cao Y, Li Z, Chen T, Zhang Z, Zhang J, Chen S (2008) Overexpression of a tobacco small G protein gene NtRop1 causes salt sensitivity and hydrogen peroxide production in transgenic plants. Sci China C Life Sci 51(5):383–390

Chakravorty D, Trusov Y, Zhang W, Acharya BR, Sheahan MB, McCurdy DW, Assmann SM, Botella JR (2011) An atypical heterotrimeric G-protein c-subunit is involved in guard cell K^+-channel regulation and morphological development in Arabidopsis thaliana. Plant J 67: 840–851

Chen JG, Willard FS, Huang J, Liang J, Chasse SA, Jones AM, Siderovski DP (2003) A seven-transmembrane RGS protein that modulates plant cell proliferation. Science 301:1728–1731

Chen JG, Pandey S, Huang J, Alonso JM, Ecker JR, Assmann SM, Jones AM (2004a) GCR1 can act independently of heterotrimeric G-protein in response to brassinosteroids and gibberellins in *Arabidopsis* seed germination. Plant Physiol 135:907–915

Chen YL, Huang R, Xiao YM, Lu P, Chen J, Wang XC (2004b) Extracellular calmodulin-induced stomatal closure is mediated by heterotrimeric G protein and H_2O_2. Plant Physiol 136: 4096–4103

Chen Y, Ji F, Xie H, Liang J, Zhang J (2006) The regulator of G-protein signaling proteins involved in sugar and abscisic acid signaling in Arabidopsis seed germination. Plant Physiol 140: 302–310

Chen L, Shiotani K, Togashi T, Miki D, Aoyama M, Wong HL, Kawasaki T, Shimamoto K (2010) Analysis of the Rac/Rop small GTPase family in rice: expression, subcellular localization and role in disease resistance. Plant Cell Physiol 51(4):585–595

Chen N, Xu Y, Wang X, Du C, Du J, Yuan M, Xu Z, Chong K (2011) OsRAN2, essential for mitosis, enhances cold tolerance in rice by promoting export of intranuclear tubulin and maintaining cell division under cold stress. Plant Cell Environ 34:52–64

Cheng X, Zhu L, He G (2013) Towards understanding of molecular interactions between rice and the brown planthopper. Mol Plant 6(3):621–634

Chinnusamy V, Gong Z, Zhu JK (2008) Nuclear RNA export and its importance in abiotic stress response of plants. Curr Top Microbiol Immunol 326:235–255

Cho HY, Park JA, Pai HS (2008) Physiological function of NbRanBP1 in *Nicotiana benthamiana*. Mol Cells 26:270–277

Choudhury SR, Bisht NC, Thompson R, Todorov O, Pandey S (2011) Conventional and novel gamma protein families constitute the heterotrimeric G-protein signaling network in soybean. PLoS One 6:e23361

Cohen E, Kende H (1987) In vivo 1-aminocyclopropane-1- carboxylate synthase activity in internodes of deep-water rice: enhancement by submergence and low oxygen levels. Plant Physiol 84:282–286

Colaneri AC, Tunc-Ozdemir M, Huang JP, Jones AM (2014) Growth attenuation under saline stress is mediated by the heterotrimeric G protein complex. BMC Plant Biol 14:129

Cookson SJ, Radziejwoski A, Granier C (2006) Cell and leaf size plasticity in Arabidopsis: what is the role of endoreduplication? Plant Cell Environ 29:1273–1283

Coursol S, Fan LM, Le Stunff H, Spiegel S, Gilroy S, Assmann SM (2003) Sphingolipid signalling in *Arabidopsis* guard cells involves heterotrimeric G proteins. Nature 423:651–654

Covington MF, Maloof JN, Straume M, Kay SA, Harmer SL (2008) Global transcriptome analysis reveals circadian regulation of key pathways in plant growth and development. Genome Biol 9:R130

Dang TT, Shimatani Z, Kawano Y, Terada R, Shimamoto K (2013) Gene editing a constitutively active OsRac1 by homologous recombination-based gene targeting induces immune responses in rice. Plant Cell Physiol 54(12):2058–2070

Delgado-Cerezo M, Sánchez-Rodríguez C, Escudero V, Miedes E, Fernández PV, Jordá L, Hernández-Blanco C, Sánchez-Vallet A, Bednarek P, Schulze-Lefert P, Somerville S, Estevez JM, Persson S, Molina A (2012) *Arabidopsis* heterotrimeric G-protein regulates cell wall defense and resistance to necrotrophic fungi. Mol Plant 5(1):98–114

Delmer DP, Pear JR, Andrawis A, Stalker DM (1995) Genes encoding small GTP-binding proteins analogous to mammalian rac are preferentially expressed in developing cotton fibers. Mol Gen Genet 248:43–51

Diekmann D, Brill S, Garrett MD, Totty N, Hsuan J, Monfries C, Hall C, Lim L, Hall A (1991) Bcr encodes a GTPase-activating protein for p21rac. Nature 351:400–402

Dodds PN, Rathjen JP (2010) Plant immunity: towards an integrated view of plant-pathogen interactions. Nat Rev Genet 11(8):539–548

Dombrowski JE, Baldwin JC, Martin RC (2008) Cloning and characterization of a salt stress-inducible small GTPase gene from the model grass species *Lolium temulentum*. J Plant Physiol 165(6):651–661

Durrant WE, Dong X (2004) Systemic acquired resistance. Annu Rev Phytopathol 42:185–209

Ebine K, Miyakawa N, Fujimoto M, Uemura T, Nakano A, Ueda T (2012) Endosomal trafficking pathway regulated by ARA6, a RAB5 GTPase unique to plants. Small GTPases 3:23–27

Engelen-Eigles G, Jones RJ, Phillips RL (2001) DNA endoreduplication in maize endosperm cells is reduced by high temperature during the mitotic phase. Crop Sci 41:1114–1121

Feng Y, Press B, Wandinger-Ness A (1995) Rab 7: an important regulator of late endocytic membrane traffic. J Cell Biol 131:1435–1452

Feraru E, Feraru MI, Asaoka R, Paciorek T, De Rycke R, Tanaka H, Nakano A, Friml J (2011) BEX5/RabA1b regulates trans-Golgi network-to-plasma membrane protein trafficking in Arabidopsis. Plant Cell 24:3074–3086

Fujii H, Chinnusamy V, Rodrigues A, Rubio S, Antoni R, Park SY, Cutler SR, Sheen J, Rodriguez PL, Zhu JK (2009) In vitro reconstitution of an abscisic acid signalling pathway. Nature 462:660–664

Fujiwara M, Umemura K, Kawasaki T, Shimamoto K (2006) Proteomics of Rac GTPase signaling reveals its predominant role in elicitor-induced defense response of cultured rice cells. Plant Physiol 140:734–745

Fukao T, Xu K, Ronald PC, Bailey-Serres J (2006) A variable cluster of ethylene response factor-like genes regulates metabolic and developmental acclimation responses to submergence in rice. Plant Cell 18(8):2021–2034

Gao Y, Li T, Liu Y, Ren C, Zhao Y, Wang M (2010) Isolation and characterization of gene encoding G protein α subunit protein responsive to plant hormones and abiotic stresses in *Brassica napus*. Mol Biol Rep 37(8):3957–3965

Geldner N, Friml J, Stierhof YD, Jurgens G, Palme K (2001) Auxin transport inhibitors block PIN1 cycling and vesicle trafficking. Nature 413:425–428

Gelli A, Higgins VJ, Blumwald E (1997) Activation of plant plasma membrane Ca^{2+}-permeable channels by race-specific fungal elicitors. Plant Physiol 113(1):269–279

George S, Parida A (2011) Over-expression of a Rab family GTPase from phreatophyte Prosopis juliflora confers tolerance to salt stress on transgenic tobacco. Mol Biol Rep 38(3):1669–1674

Gilman AG (1987) G proteins: transducers of receptor-generated signals. Annu Rev Biochem 56:615–649

Goff SA, Ricke D, Lan T-H et al (2002) A draft sequence of the rice genome (*Oryza sativa* L. ssp. *Japonica*). Science 296:92–100

Gosti F, Beaudoin N, Serizet C, Webb AAR, Vartanian N, Giraudat J (1999) ABI1 protein phosphatase 2C is a negative regulator of abscisic acid signaling. Plant Cell 11:1897–1910

Grigston JC, Osuna D, Scheible WR, Liu C, Stitt M, Jones AM (2008) D-glucose sensing by a plasma membrane regulator of G signaling protein, AtRGS1. FEBS Lett 582:3577–3584

Gu Y, Li S, Lord EM, Yang Z (2006) Members of a novel class of Arabidopsis Rho guanine nucleotide exchange factors control Rho GTPase- dependent polar growth. Plant Cell 18: 366–381

Haizel T, Merkle T, Pay A, Fejes E, Nagy F (1997) Characterization of proteins that interact with the GTP-bound form of the regulatory GTPase Ran in *Arabidopsis*. Plant J 11:93–103

Hamm HE (1998) The many faces of G protein signaling. J Biol Chem 273(2):669–672

Hammond-Kosack KE, Jones JD (1996) Resistance gene-dependent plant defense responses. Plant Cell 8(10):1773–1791

Hao LH, Wang WX, Chen C, Wang YF, Liu T, Li X, Shang ZL (2012) Extracellular ATP promotes stomatal opening of *Arabidopsis thaliana* through heterotrimeric G protein α subunit and reactive oxygen species. Mol Plant 5(4):852–864

Hasegawa PM, Bressan RA, Zhu JK, Bohnert HJ (2000) Plant cellular and molecular responses to high salinity. Annu Rev Plant Physiol Plant Mol Biol 51:463–499

Hassanain HH, Sharma YK, Moldovan L, Khramtsov V, Berliner LJ, Duvick JP, Goldschmidt-Clermont PJ (2000) Plant Rac proteins induce superoxide production in mammalian cells. Biochem Biophys Res Commun 272:783–788

He CJ, Finlayson SA, Drew MC, Jordan WR, Morgan PW (1996a) Ethylene biosynthesis during aerenchyma formation in roots of *Zea mays* subjected to mechanical impedance and hypoxia. Plant Physiol 112:1679–1685

He CJ, Morgan PW, Drew MC (1996b) Transduction of an ethylene signal is required for cell death and lysis in the root cortex of maize during aerenchyma formation induced by hypoxia. Plant Physiol 112:463–472

Heo WD, Lee SH, Kim MC, Kim JC, Chung WS, Chun HJ, Lee KJ, Park CY, Park HC, Choi JY, Cho MJ (1999) Involvement of specific calmodulin isoforms in salicylic acid-independent activation of plant disease resistance responses. Proc Natl Acad Sci U S A 96(2):766–771

Hepler JR, Gilman AG (1992) G proteins. Trends Biochem Sci 17(10):383–387

Hoefle C, Huesmann C, Schultheiss H, Börnke F, Hensel G, Kumlehn J, Hückelhoven R (2011) A barley ROP GTPase activating protein associates with microtubules and regulates entry of the barley powdery mildew fungus into leaf epidermal cells. Plant Cell 23(6):2422–2439

Huesmann C, Hoefle C, Hückelhoven R (2011) ROPGAPs of arabidopsis limit susceptibility to powdery mildew. Plant Signal Behav 6(11):1691–1694

Huesmann C, Reiner T, Hoefle C, Preuss J, Jurca ME, Domoki M, Fehér A, Hückelhoven R (2012) Barley ROP binding kinase1 is involved in microtubule organization and in basal penetration resistance to the barley powdery mildew fungus. Plant Physiol 159(1):311–320

Huh GH, Damsz B, Matsumoto TK, Reddy MP, Rus AM, Ibeas JI, Narasimhan ML, Bressan RA, Hasegawa PM (2002) Salt causes ion disequilibrium-induced programmed cell death in yeast and plants. Plant J 29:649–659

Ichinose Y, Toyoda K, Barz W (1999) cDNA cloning and gene expression of three small GTP-binding proteins in defense response of chickpea. Biochim Biophys Acta 1489:462–466

Ishikawa A, Iwasaki Y, Asahi T (1996) Molecular cloning and characterization of a cDNA for the b-subunit of a G protein from rice. Plant Cell Physiol 37:223–228

Ismond KP, Dolferus R, de Pauw M, Dennis ES, Good AG (2003) Enhanced low oxygen survival in Arabidopsis through increased metabolic flux in the fermentative pathway. Plant Physiol 132(3):1292–1302

Jacob T, Ritchie S, Assmann SM, Gilroy S (1999) Abscisic acid signal transduction in guard cells is mediated by phospholipase D activity. Proc Natl Acad Sci U S A 96:12192–12197

Jayasekaran K, Kim KN, Vivekanandan M, Shin JS, Ok SH (2006) Novel calcium binding GTPase (AtCBG) involved in ABA-mediated salt stress signaling in Arabidopsis. Plant Cell Rep 25:1255–1262

Jiang M, Zhang J (2002) Involvement of plasma-membrane NADPH oxidase in abscisic acid- and water stress-induced antioxidant defense in leaves of maize seedlings. Planta 215:1022–1030

Johnston CA, Taylor JP, Gao Y, Kimple AJ, Grigston JC, Chen JG, Siderovski DP, Jones AM, Willard FS (2007) GTPase acceleration as the rate-limiting step in Arabidopsis G protein-coupled sugar signaling. Proc Natl Acad Sci U S A 104:17317–17322

Jones AM, Assmann SM (2004) Plants: the latest model system for G-protein research. EMBO Rep 5:572–578

Jones JD, Dangl JL (2006) The plant immune system. Nature 444:323–329

Jones JC, Temple BR, Jones AM, Dohlman HG (2011) Functional reconstitution of an atypical G protein heterotrimer and regulator of G protein signaling protein (RGS1) from *Arabidopsis thaliana*. J Biol Chem 286:13143–13150

Joo JH, Wang S, Chen JG, Jones AM, Fedoroff NV (2005) Different signaling and cell death roles of heterotrimeric G protein alpha and beta subunits in the Arabidopsis oxidative stress response to ozone. Plant Cell 17:957–970

Josefsson LG, Rask L (1997) Cloning of a putative G-protein-coupled receptor from *Arabidopsis thaliana*. Eur J Biochem 249:415–420

Jung YH, Agrawal GK, Rakwal R, Kim JA, Lee MO, Choi PG, Kim YJ, Kim MJ, Shibato J, Kim SH, Iwahashi H, Jwa NS (2006) Functional characterization of OsRacB GTPase—a potentially negative regulator of basal disease resistance in rice. Plant Physiol Biochem 44:68–77

Kahn RA, Der CJ, Bokoch GM (1992) The ras superfamily of GTP-binding proteins: guidelines on nomenclature. FASEB J 6:2512–2513

Karim S, Alezzawi M, Garcia-Petit C, Solymosi K, Khan NZ, Lindquist E, Dahl P, Hohmann S, Aronsson H (2014) A novel chloroplast localized Rab GTPase protein CPRabA5e is involved in stress, development, thylakoid biogenesis and vesicle transport in Arabidopsis. Plant Mol Biol 84(6):675–692

Kato C, Mizutani T, Tamaki H, Kumagai H, Kamiya T, Hirobe A, Fujisawa Y, Kato H, Iwasaki Y (2004) Characterization of heterotrimeric G protein complexes in rice plasma membrane. Plant J 38:320–331

Kawano Y, Akamatsu A, Hayashi K, Housen Y, Okuda J, Yao A, Nakashima A, Takahashi H, Yoshida H, Wong HL, Kawasaki T, Shimamoto K (2010) Activation of a Rac GTPase by the NLR family disease resistance protein pit plays a critical role in rice innate immunity. Cell Host Microbe 7(5):362–375

Kawasaki T, Henmi K, Ono E, Hatakeyama S, Iwano M, Satoh H, Shimoto K (1999) The small GTP binding protein rac is a regulator of cell death in plants. Proc Natl Acad Sci U S A 96:10922–10926

Kawasaki T, Koita H, Nakatsubo T, Hasegawa K, Wakabayashi K, Takahashi H, Umemura K, Umezawa T, Shimamoto K (2006) Cinnamoyl-CoA reductase, a key enzyme in lignin biosynthesis, is an effector of small GTPase Rac in defense signaling in rice. Proc Natl Acad Sci U S A 103(1):230–235

Kawasawa Y, McKenzie LM, Hill DP, Bono H, Yanagisawa M, RIKEN GER Group, GSL Members (2003) G protein-coupled receptor genes in the FANTOM2 database. Genome Res 13(6B):1466–1477

Kevei E, Gyula P, Fehér B, Tóth R, Viczián A, Kircher S, Rea D, Dorjgotov D, Schäfer E, Millar AJ, Kozma-Bognár L, Nagy F (2007) Arabidopsis thaliana circadian clock is regulated by the small GTPase LIP1. Curr Biol 17:1456–1464

Kim SH, Oikawa T, Kyozuka J, Wong HL, Umemura K, Kishi-Kaboshi M, Takahashi A, Kawano Y, Kawasaki T, Shimamoto K (2012) The bHLH Rac immunity1 (RAI1) is activated by OsRac1 via OsMAPK3 and OsMAPK6 in rice immunity. Plant Cell Physiol 53(4):740–754

Kirchhausen T (2000) Three ways to make a vesicle. Nat Rev Mol Cell Biol 1:187–198

Kitsios G, Doonan JH (2011) Cyclin dependent protein kinases and stress responses in plants. Plant Signal. Behav. 6(2):204–209

Komatsu S, Yang G, Hayashi N, Kaku H, Umemura K, Iwasaki Y (2004) Alterations by a defect in a rice G protein a subunit in probenazole and pathogen-induced responses. Plant Cell Environ 27:947–957

Kost B, Spielhofer P, Chua N-H (1998) A GFP-mouse talin fusion protein labels plant actin filaments in vivo and visualizes the actin cytoskeleton in growing pollen tubes. Plant J 16:393–401

Kost B, Lemichez E, Spielhofer P, Hong Y, Tolias K, Carpenter C, Chua NH (1999) Rac homologues and compartmentalized phosphatidylinositol 4, 5-bisphosphate act in a common pathway to regulate polar pollen tube growth. J Cell Biol 145:317–330

Kürsteiner O, Dupuis I, Kuhlemeier C (2003) The pyruvate decarboxylase1 gene of *Arabidopsis* is required during anoxia but not other environmental stresses. Plant Physiol 132(2):968–978

Lam E, Pontier D, del Pozo O (1999) Die and let live: programmed cell death in plants. Curr Opin Plant Biol 2:502–507

Lamb C, Dixon RA (1997) The oxidative burst in plant disease resistance. Annu Rev Plant Physiol Plant Mol Biol 48:251–275

Lease KA, Wen J, Li J, Doke JT, Liscum E, Walker JC (2001) A mutant *Arabidopsis* heterotrimeric G-protein beta subunit affects leaf, flower, and fruit development. Plant Cell 13:2631–2641

Lee YR, Assmann SM (1999) *Arabidopsis thaliana* "extra-large GTP-binding protein" (AtXLG1): a new class of G-protein. Plant Mol Biol 40:55–64

Legendre L, Heinstein PF, Low PS (1992) Evidence for participation of GTP-binding proteins in elicitation of the rapid oxidative burst in cultured soybean cells. J Biol Chem 267(28): 20140–20147

Lemichez E, Wu Y, Sanchez JP, Mettouchi A, Mathur J, Chua NH (2001) Inactivation of AtRac1 by abscisic acid is essential for stomatal closure. Genes Dev 15(14):1808–1816

Leung J, Bouvier-Durand M, Morris P-C, Guerrier D, Chefdor F, Giraudat J (1994) *Arabidopsis* ABA response gene ABI1: Features of a calcium-modulated protein phosphatase. Science 264:1448–1452

Levine A, Belenghi B, Damari-Weisler H, Granot D (2001) Vesicle associated membrane protein of Arabidopsis suppresses bax-induced apoptosis in yeast downstream of oxidative burst. J Biol Chem 276:46284–46289

Leyman B, Geelen D, Blatt MR (1999) Localization and control of expression of Nt-Syr, a tobacco SNARE protein. Plant J 24:369–381

Li H, Wu G, Ware D, Davis KR, Yang Z (1998) *Arabidopsis* Rho-related GTPases: differential gene expression in pollen and polar localization in fission yeast. Plant Physiol 118:407–417

Li H, Lin Y, Heath RM, Zhu MX, Yang Z (1999) Control of pollen tube tip growth by a Rop GTPase-dependent pathway that leads to tip-localized calcium influx. Plant Cell 11: 1731–1742

Li JH, Liu YQ, Lu P, Lin HF, Bai Y, Wang XC, Chen YL (2009) A signaling pathway linking nitric oxide production to heterotrimeric G protein and hydrogen peroxide regulates extracellular calmodulin induction of stomatal closure in *Arabidopsis*. Plant Physiol 150:114–124

Li Z, Kang J, Sui N, Liu D (2012a) ROP11 GTPase is a negative regulator of multiple ABA responses in Arabidopsis. J Integr Plant Biol 54(3):169–179

Li Z, Li Z, Gao X, Chinnusamy V, Bressan R, Wang ZX, Zhu JK, Wu JW, Liu D (2012b) ROP11 GTPase negatively regulates ABA signaling by protecting ABI1 phosphatase activity from inhibition by the ABA receptor RCAR1/PYL9 in Arabidopsis. J Integr Plant Biol 54(3):180–188

Liu J, Zhu JK (1997) Proline accumulation and salt-stress-induced gene expression in a salt-hypersensitive mutant of *Arabidopsis*. Plant Physiol 114:591–596

Liu F, Guo J, Bai P, Duan Y, Wang X, Cheng Y, Feng H, Huang L, Kang Z (2012a) Wheat TaRab7 GTPase is part of the signaling pathway in responses to stripe rust and abiotic stimuli. PLoS One 7(5):e37146

Liu J, Zhou J, Xing D (2012b) Phosphatidylinositol 3-kinase plays a vital role in regulation of rice seed vigor via altering NADPH oxidase activity. PLoS One 7(3):e33817

Liu ZL, Luo C, Dong L, Van Toan C, Wei PX, He XH (2014) Molecular characterization and expression analysis of a GTP-binding protein (MiRab5) in *Mangifera indica*. Gene 540(1):86–91

Llorente F, Alonso-Blanco C, Sanchez-Rodriguez C, Jorda L, Molina A (2005) ERECTA receptor-like kinase and heterotrimeric G protein from *Arabidopsis* are required for resistance to the necrotrophic fungus *Plectosphaerella cucumerina*. Plant J 43:165–180

Logan DC (2010) Mitochondrial fusion, division and positioning in plants. Biochem Soc Trans 38:789–795

Lu C, Zainal Z, Tucker GA, Lycett GW (2001) Developmental abnormalities and reduced fruit softening in tomato plants expressing an antisense Rab11 GTPase gene. Plant Cell 13: 1819–1833

Ma H (2001) Plant G-proteins: the different faces of GPA1. Currr Biol 11:R869–R871

Ma H, Yanofsky MF, Meyerowitz EM (1990) Molecular cloning and characterization of GPA1, a G protein a subunit gene from *Arabidopsis thaliana*. Proc Natl Acad Sci U S A 87:3821–3825

Mansfield TA (1998) Stomata and plant water relations: does air pollution create problems? Environ Pollut 101:1–11

Mason MG, Botella JR (2000) Completing the heterotrimer: isolation and characterization of an *Arabidopsis thaliana* G protein gamma-subunit cDNA. Proc Natl Acad Sci U S A 97(26):14784–14788

Mason MG, Botella JR (2001) Isolation of a novel G-protein gamma-subunit from *Arabidopsis thaliana* and its interaction with Gbeta. Biochim Biophys Acta 1520(2):147–153

Mathur J, Chua N-H (2000) Microtubule stabilization leads to growth reorientation in *Arabidopsis* trichomes. Plant Cell 12:465–477

Mazel A, Leshem Y, Tiwari BS, Levine A (2004) Induction of salt and osmotic stress tolerance by overexpression of an intracellular vesicle trafficking protein AtRab7 (*AtRabG3e*). Plant Physiol 134:118–128

Mazzucoteli E, Mastrangelo AM, Crosatti C, Guerra D, Stanca AM, Cattivelli L (2008) Abiotic stress response in plants: when post-transcriptional and post-translational regulations control transcription. Plant Sci 174:420–431

McCudden CR, Hains MD, Kimple RJ, Siderovski DP, Willard FS (2005) G-protein signaling: back to the future. Cell Mol Life Sci 62:551–577

McElver J, Patton D, Rumbaugh M, Liu C, Yang LJ, Meinke D (2000) The TITAN5 gene of Arabidopsis encodes a protein related to the ADP ribosylation factor family of GTP binding proteins. Plant Cell 12:1379–1392

Meier I (2000) A novel link between ran signal transduction and nuclear envelope proteins in plants. Plant Physiol 124:1507–1510

Meyer K, Leube MP, Grill E (1994) A protein phosphatase 2C involved in ABA signal transduction in *Arabidopsis thaliana*. Science 264:1452–1455

Miche L, Battistoni F, Gernmer S, Belghazi M, Reinhold-Hurek B (2006) Upregulation of jasmonate-inducible defense proteins and differential colonization of roots of *Oryza sativa* cultivars with the endophyte *Azoarcus sp*. Mol Plant Microbe Interact 19:502–511

Miki D, Itoh R, Shimamoto K (2005) RNA silencing of single and multiple members in a gene family of rice. Plant Physiol 138:1903–1913

Mirzaei M, Pascovici D, Atwell BJ, Haynes PA (2012) Differential regulation of aquaporins, small GTPases and V-ATPases proteins in rice leaves subjected to drought stress and recovery. Proteomics 12(6):864–877

Misra S, Wu Y, Venkataraman G, Sopory SK, Tuteja N (2007) Heterotrimeric G-protein complex and 21 G-protein-coupled receptor from a legume (*Pisum sativum*): role in salinity and heat stress and cross-talk with phospholipase C. Plant J 51:656–669

Mittler R (2002) Oxidative stress, antioxidants and stress tolerance. Trends Plant Sci 7:405–410

Morel J, Fromentin J, Blein JP, Simon-Plas F, Elmayan T (2004) Rac regulation of NtrbohD, the oxidase responsible for the oxidative burst in elicited tobacco cell. Plant J 37(2):282–293

Moriyama EN, Strope PK, Opiya SO, Chen Z, Jones AM (2006) Mining the *Arabidopsis thaliana* genome for highly-divergent seven transmembrane receptors. Genome Biol 7:R96

Mosammaparast N, Pemberton LF (2004) Karyopherins: from nuclear-transport mediators to nuclear-function regulators. Trends Cell Biol 14:547–556

Moss J, Vaughan M (1998) Molecules in the ARF orbit. J Biol Chem 273:21431–21434

Mukhopadhyay A, Funato K, Stahl PD (1997) Rab7 regulates transport from early to late endocytic compartments in Xenopus oocytes. J Biol Chem 272:13055–13059

Newman T, de Bruijn FJ, Green P, Keegstra K, Kende H, McIntosh L, Ohlrogge J, Raikhel N, Somerville S, Thomashow M (1994) Genes galore: a summary of methods for accessing results from large-scale partial sequencing of anonymous *Arabidopsis* cDNA clones. Plant Physiol 106:1241–1255

Ng CK, Carr K, McAinsh MR, Powell B, Hetherington AM (2001) Drought-induced guard cell signal transduction involves sphingosine-1-phosphate. Nature 410:596–599

Nibau C, Wu HM, Cheung AY (2006) RAC/ROP GTPases: "hubs" for signal integration and diversification in plants. Trends Plant Sci 11:309–315

Nilson SE, Assmann SM (2010a) The alpha-subunit of the Arabidopsis heterotrimeric G protein, GPA1, is a regulator of transpiration efficiency. Plant Physiol 152(4):2067–2077

Nobes C, Hall A (1994) Regulation and function of the Rho subfamily of small GTPases. Curr Opin Genet Dev 4:77–81

Nobes CD, Lauritzen I, Mattei MG, Paris S, Hall A, Chardin P (1998) A new member of the Rho family, Rnd1, promotes disassembly of actin filament structures and loss of cell adhesion. J Cell Biol 141:187–197

O'Mahony PJ, Oliver MJ (1999) Characterization of a desiccation-responsive small GTP-binding protein (Rab2) from the desiccation-tolerant grass *Sporobolus stapfianus*. Plant Mol Biol 39(4):809–821

Ono E, Wong HL, Kawasaki T, Hasegawa M, Kodama O, Shimamoto K (2001) Essential role of the small GTPase Rac in disease resistance of rice. Proc Natl Acad Sci U S A 98:759–764

Pandey S, Assmann SM (2004) The Arabidopsis putative G protein-coupled receptor GCR1 interacts with the G protein alpha subunit GPA1 and regulates abscisic acid signaling. Plant Cell 16(6):1616–1632

Pandey S, Chen JG, Jones AM, Assmann SM (2006) G-protein complex mutants are hypersensitive to abscisic acid regulation of germination and postgermination development. Plant Physiol 141:243–256

Parada LF, Tabin CJ, Shih C, Weinberg RA (1982) Human EJ bladder carcinoma oncogene is homologue of Harvey sarcoma virus Ras gene. Nature 291:474–478

Park J, Choi HJ, Lee S, Lee T, Yang Z, Lee Y (2000) Rac-related GTP-binding protein in elicitor-induced reactive oxygen generation by suspension-cultured soybean cells. Plant Physiol 124(2):725–732

Park J, Gu Y, Lee Y, Yang Z, Lee Y (2004) Phosphatidic acid induces leaf cell death in Arabidopsis by activating the Rho-related small G protein GTPase-mediated pathway of reactive oxygen species generation. Plant Physiol 134(1):129–136

Park SY, Fung P, Nishimura N, Jensen DR, Fujii H, Zhao Y, Lumba S, Santiago J, Rodrigues A, Chow TF, Alfred SE, Bonetta D, Finkelstein R, Provart NJ, Desveaux D, Rodriguez PL, McCourt P, Zhu JK, Schroeder JI, Volkman BF, Cutler SR (2009) Abscisic acid inhibits type 2C protein phosphatases via the PYR/PYL family of START proteins. Science 324: 1068–1071

Park J, Kim YS, Kim SG, Jung JH, Woo JC, Park CM (2011) Integration of auxin and salt signals by the NAC transcription factor NTM2 during seed germination in Arabidopsis. Plant Physiol 156:537–549

Pathuri IP, Zellerhoff N, Schaffrath U, Hensel G, Kumlehn J, Kogel KH, Eichmann R, Hückelhoven R (2008) Constitutively activated barley ROPs modulate epidermal cell size, defense reactions and interactions with fungal leaf pathogens. Plant Cell Rep 27(12):1877–1887

Pay A, Resch K, Frohnmeyer H, Fejes E, Nagy F, Nick P (2002) Plant Ran GAPs are localized at the nuclear envelope in interphase and associated with microtubules in mitotic cells. Plant J 30:699–709

Pereira-Leal JB, Seabra MC (2001) Evolution of the Rab family of small GTP binding proteins. J Mol Biol 313:889–901

Piffanelli P, Zhou F, Casais C, Orme J, Jarosch B, Schaffrath U, Collins NC, Panstruga R, Schulze-Lefert P (2002) The barley MLP modulator of defense and cell death is responsive to biotic and abiotic stress stimuli. Plant Physiol 129(3):1076–1085

Poraty-Gavra L, Zimmermann P, Haigis S, Bednarek P, Hazak O, Stelmakh OR, Sadot E, Schulze-Lefert P, Gruissem W, Yalovsky S (2013) The Arabidopsis Rho of plants GTPase AtROP6 functions in developmental and pathogen response pathways. Plant Physiol 161(3): 1172–1188

Potikha TS, Collins CC, Johnson DI, Delmer DP, Levine A (1999) The involvement of hydrogen peroxide in the differentiation of secondary walls in cotton fibers. Plant Physiol 119:849–858

Qin X, Lin F, Lii Y, Gou C, Chen F (2011) Molecular analysis of ARF1 expression profiles during development of physic nut (*Jatropha curcas* L.). Mol Biol Rep 38(3):1681–1686

Regad F, Bardet C, Tremousaygue D, Moisan A, Lescure B, Axelos M (1993) cDNA cloning and expression of an *Arabidopsis* GTP-binding protein of the ARF family. FEBS Lett 316:133–136

Reis K, Fransson A, Aspenstrom P (2009) The Miro GTPases: at the heart of the mitochondrial transport machinery. FEBS Lett 583:1391–1398

Ridley AJ, Paterson HF, Johnston CL, Diekmann D, Hall A (1992) The small GTP-binding protein rac regulates growth factor-induced membrane ruffling. Cell 70:401–410

Romeis T, Piedras P, Jones JD (2000) Resistance gene-dependent activation of a calcium-dependent protein kinase in the plant defense response. Plant Cell 12(5):803–816

Rose A, Meier I (2001) A domain unique to plant RanGAP is responsible for its targeting to the plant nuclear rim. Proc Natl Acad Sci U S A 98:15377–15382

Roy Choudhary S, Westfall CS, Laborde JP, Bisht NC, Jez JM, Pandey S (2004) Two chimeric regulators of G-protein signaling (RGS) proteins differentially modulate heterotrimeric G-protein cycle. J Biol Chem 287:17870–17881

Rus A, Yokoi S, Sharkhuu A, Reddy M, Lee BH, Matsumoto TK, Koiwa H, Zhu JK, Bressan RA, Hasegawa PM (2001) AtHKT1 is a salt tolerance determinant that controls Na^{+} entry into plant roots. Proc Natl Acad Sci U S A 98:14150–14155

Rutherford S, Moore I (2002) The Arabidopsis Rab GTPase family-another enigma variation. Curr Opin Plant Biol 5:518–528

Sagi M, Fluhr R (2001) Superoxide production by plant homologues of the gp91(phox) NADPH oxidase. Modulation of activity by calcium and by tobacco mosaic virus infection. Plant Physiol 126(3):1281–1290

Saito C, Ueda T (2009) Chapter 4: functions of RAB and SNARE proteins in plant life. Int Rev Cell Mol Biol 274:183–233

Sanchez A, Shin J, Davis SJ (2011) Abiotic stress and the plant circadian clock. Plant Signal Behav 6:223–231

Sauter M (2000) Rice in deep water: "how to take heed against a sea of troubles". Naturwissenschaften 87:289–303

Schiene K, Puhler A, Niehaus K (2000) Transgenic tobacco plants that express an antisense construct derived from a *Medicago sativa* cDNA encoding a Rac-related small GTP-binding protein fail to develop necrotic lesions upon elicitor infiltration. Mol Gen Genet 263:761–770

Schultheiss H, Dechert C, Kogel KH, Huckelhoven R (2002) A small GTP-binding host protein is required for entry of powdery mildew fungus into epidermal cells of barley. Plant Physiol 128:1447–1454

Schultheiss H, Dechert C, Kogel KH, Huckelhoven R (2003) Functional analysis of barley RAC/ROP G-protein family members in susceptibility to the powdery mildew fungus. Plant J 36:589–601

Schultheiss H, Hensel G, Imani J, Broeders S, Sonnewald U, Kogel KH, Kumlehn J, Hückelhoven R (2005) Ectopic expression of constitutively activated RACB in barley enhances susceptibility to powdery mildew and abiotic stress. Plant Physiol 139(1):353–362

Segev N (2001) Ypt and Rab GTPases: insight into functions through novel interactions. Curr Opin Cell Biol 13:500–511

Shanmugam T, Sharma M, Pandey A, Pandey GK (2013) Small GTPases: Rho of plant (ROP) in development and stress signaling. In: Pandey GK (ed) Stress-mediated signaling in plants II, Plant Stress 7(Special Issue 1)., Global science books pp 10–15

Shieh MW, Wessler SR, Raikhel NV (1993) Nuclear targeting of the maize R protein requires two nuclear localization sequences. Plant Physiol 101:353–361

Spartz AK, Gray WM (2008) Plant hormone receptors: new perceptions. Genes Dev 22:2139–2148

Steffens B, Sauter M (2009) Epidermal cell death in rice is confined to cells with a distinct molecular identity and is mediated by ethylene and H_2O_2 through an auto-amplified signal pathway. Plant Cell 21:184–196

Steffens B, Sauter M (2010) Heterotrimeric G protein signaling is required for epidermal cell death in rice. Plant Physiol 151:732–740

Stenmark H, Valencia A, Martinez O, Ullrich O, Goud B, Zerial M (1994) Distinct structural elements of rab5 define its functional specificity.EMBO J. 13(3):575–583

Suharsono U, Fujisawa Y, Kawasaki T, Iwasaki Y, Satoh H, Shimamoto K (2002) The heterotrimeric G protein subunit acts upstream of the small GTPase Rac in disease resistance of rice. Proc Natl Acad Sci U S A 99(20):13307–13312

Takai Y, Sasaki T, Matozaki T (2001) Small GTP-binding proteins. Physiol Rev 81:153–208

Tamaoki M, Nakajima N, Kubo A, Aono M, Matsuyama T, Saji H (2003) Transcriptome analysis of O3-exposed *Arabidopsis* reveals that multiple signal pathways act mutually antagonistically to induce gene expression. Plant Mol Biol 53:443–456

Temple BRS, Jones AM (2007) The plant heterotrimeric G-protein complex. Annu Rev Plant Biol 58:249–266

Terecskei K, Tóth R, Gyula P, Kevei E, Bindics J, Coupland G, Nagy F, Kozma-Bognár L (2013) The circadian clock-associated small GTPase LIGHT INSENSITIVE PERIOD1 suppresses light-controlled endoreplication and affects tolerance to salt stress in Arabidopsis. Plant Physiol 161(1):278–290

Thao NP, Chen L, Nakashima A, Hara S, Umemura K, Takahashi A, Shirasu K, Kawasaki T, Shimamoto K (2007) RAR1 and HSP90 form a complex with Rac/Rop GTPase and function in innate-immune responses in rice. Plant Cell 19:4035–4045

Trusov Y, Rookes JE, Chakravorty D, Armour D, Schenk PM, Botella JR (2006) Heterotrimeric G proteins facilitate Arabidopsis resistance to necrotrophic pathogens and are involved in jasmonate signaling. Plant Physiol 140:210–220

Trusov Y, Rookes JE, Tilbrook K, Chakravorty D, Mason MG, Anderson D, Chen JG, Jones AM, Botella JR (2007) Heterotrimeric G protein β subunits provide functional selectivity in Gβγ dimer signaling in *Arabidopsis*. Plant Cell 19:1235–1250

Trusov Y, Sewelam N, Rookes JE, Kunkel M, Nowak E, Schenk PM, Botella JR (2009) Heterotrimeric G proteins-mediated resistance to necrotrophic pathogens includes mechanisms independent of salicylic acid-, jasmonic acid/ethylene- and abscisic acid-mediated defense signaling. Plant J 58:69–81

Turcotte S, Desrosiers RR, Béliveau R (2003) HIF-1alpha mRNA and protein upregulation involves Rho GTPase expression during hypoxia in renal cell carcinoma. J Cell Sci 116(Pt 11):2247–2260

Ueguchi-Tanaka M, Fujisawa Y, Kobayashi M, Ashikari M, Iwasaki Y, Kitano H, Matsuoka M (2000) Rice dwarf mutant d1, which is defective in the alpha subunit of the heterotrimeric G protein, affects gibberellin signal transduction. Proc Natl Acad Sci U S A 97(21): 11638–11643

Ullah H, Chen JG, Young JC, Im KH, Sussman MR, Jones AM (2001) Modulation of cell proliferation by heterotrimeric G protein in Arabidopsis. Science 292:2066–2069

Ullah H, Chen JG, Wang S, Jones AM (2002) Role of a heterotrimeric G protein in regulation of Arabidopsis seed germination. Plant Physiol 129:897–907

Urano D, Jones JC, Wang H, Matthews M, Bradford W, Bennetzen JL, Jones AM (2012) G protein activation without a GEF in the plant kingdom. PLoS Genet 8:e1002756

Vernoud V, Horton AC, Yang ZB, Nielsen E (2003) Analysis of the small GTPase gene superfamily of *Arabidopsis*. Plant Physiol 131:1191–1208

Verslues PE, Guo Y, Dong C-H, W-j M, Zhu J-K (2006) Mutation of SAD2, an importin b-domain protein in *Arabidopsis*, alters abscisic acid sensitivity. Plant J 47:776–787

Wamaitha MJ, Yamamoto R, Wong HL, Kawasaki T, Kawano Y, Shimamoto K (2012) OsRap2.6 transcription factor contributes to rice innate immunity through its interaction with receptor for activated kinase-C 1 (RACK1). Rice (N Y) 5(1):35

Wang TW, Arteca RN (1992) Effects of low O(2) root stress on ethylene biosynthesis in tomato plants (*Lycopersicon esculentum* Mill cv Heinz 1350). Plant Physiol 98:97–100

Wang XQ, Ullah H, Jones AM, Assmann SM (2001) G protein regulation of ion channels and abscisic acid signaling in *Arabidopsis* guard cells. Science 292:2070–2072

Wang X, Xu YY, Han Y, Bao SL, Du JZ, Yuan M, Xu ZH, Chong K (2006) Overexpression of RAN1 in rice and *Arabidopsis* alters primordial meristem, mitotic progress, and sensitivity to auxin. Plant Physiol 140:91–101

Wang TZ, Xia XZ, Zhao MG, Tian QY, Zhang WH (2013) Expression of a *Medicago falcata* small GTPase gene, MfARL1 enhanced tolerance to salt stress in *Arabidopsis thaliana*. Plant Physiol Biochem 63:227–235

Wei Q, Zhou WB, Hu GZ, Wei JM, Yang HQ, Huang JR (2008) Heterotrimeric G-protein is involved in phytochrome A-mediated cell death of Arabidopsis hypocotyls. Cell Res 18:949–960

Weiss CA, Garnaat CW, Mukai K, Hu Y, Ma H (1994) Isolation of cDNAs encoding guanine nucleotide-binding protein beta-subunit homologues from maize (ZGB1) and Arabidopsis (AGB1). Proc Natl Acad Sci U S A 91(20):9554–9558

Wennerberg K, Rossman KL, Der CJ (2005) The Ras superfamily at a glance. J Cell Sci 118:843–846

Wettschureck N, Offermanns S (2005) Mammalian G proteins and their cell type specific functions. Physiol Rev 85:1159–1204

Wickner W, Schekman R (2008) Membrane fusion. Nat Struct Mol Biol 15:658–664

Winge P, Brembu T, Bones AM (1997) Cloning and characterization of rac-like cDNAs from *Arabidopsis thaliana*. Plant Mol Biol 35(4):483–495

Wittinghofer A, Vetter IR (2011) Structure-function relationships of the 6 domain, a canonical switch motif. Annu Rev Biochem 80:943–971

Wong HL, Sakamoto T, Kawasaki T, Umemura K, Shimamoto K (2004) Down-regulation of metallothionein, a reactive oxygen scavenger, by the small GTPase OsRac1 in rice. Plant Physiol 135(3):1447–1456

Wong HL, Pinontoan R, Hayashi K, Tabata R, Yaeno T, Hasegawa K, Kojima C, Yoshioka H, Iba K, Kawasaki T, Shimamoto K (2007) Regulation of rice NADPH oxidase by binding of Rac GTPase to its N-terminal extension. Plant Cell 19:4022–4034

Xiong L, Lee H, Ishitani M, Zhu JK (2002) Regulation of osmotic stress responsive gene expression by the LOS6/ABA1 locus in *Arabidopsis*. J Biol Chem 277:8588–8596

Xu P, Cai W (2014) RAN1 is involved in plant cold resistance and development in rice (Oryza sativa). J Exp Bot. 65(12):3277–3287

Yadav DK, Islam SM, Tuteja N (2012) Rice heterotrimeric G-protein gamma subunits (RGG1 and RGG2) are differentially regulated under abiotic stress. Plant Signal Behav 7(7):733–740

Yadav DK, Shukla D, Tuteja N (2013) Rice heterotrimeric G-protein alpha subunit (RGA1): *in silico* analysis of the gene and promoter and its upregulation under abiotic stress. Plant Physiol Biochem 63:262–271

Yadav DK, Shukla D, Tuteja N (2014) Isolation, in silico characterization, localization and expression analysis of abiotic stress-responsive rice G-protein β subunit (RGB1). Plant Signal Behav 9:e28890

Yamaguchi K, Kawasaki T (2012) Function of Arabidopsis SWAP70 GEF in immune response. Plant Signal Behav 7(4):465–468

Yamaguchi K, Imai K, Akamatsu A, Mihashi M, Hayashi N, Shimamoto K, Kawasaki T (2012) SWAP70 functions as a Rac/Rop guanine nucleotide-exchange factor in rice. Plant J 70(3): 389–397

Yang Z, Fu Y (2007) ROP/RAC GTPase signaling. Curr Opin Plant Biol 10:490–494

Yang Z, Watson JC (1993) Molecular cloning and characterization of *rho*, a ras-related small GTP-binding protein from the garden pea. Proc Natl Acad Sci U S A 90:8732–8736

Yasuda Y, Miyamoto Y, Saiwaki T, Yoneda Y (2006) Mechanism of the stress-induced collapse of the Ran distribution. Exp Cell Res 312:512–520

Zang AP, Xu XJ, Neill S, Cai WM (2010) Overexpression of OsRAN2 in rice and *Arabidopsis* renders transgenic plants hypersensitive to salinity and osmotic stress. J Exp Bot 61:777–789
Zarembinski TI, Theologis A (1993) Anaerobiosis and plant growth hormones induce two genes encoding 1-aminocyclopropane-1-carboxylate synthase in rice (*Oryza sativa* L.). Mol Biol Cell 4:363–373
Zerial M, McBride H (2001) Rab proteins as membrane organizers. Nat Rev Mol Cell Biol 2:107–117
Zhang H-X, Blumwald E (2001) Transgenic salt tolerant tomato plants accumulate salt in the foliage but not in the fruits. Nat Biotechnol 19:765–768
Zhang W, Qin C, Zhao J, Wang X (2004) Phospholipase Dα1-derived phosphatidic acid interacts with ABI1 phosphatase 2C and regulates abscisic acid signaling. Proc Natl Acad Sci U S A 101:9508–9513
Zhang F, Wang Y, Wang D (2007) Role of nitric oxide and hydrogen peroxide during the salt resistance response. Plant Signal Behav 2(6):473–474
Zhang L, Tian LH, Zhao JF, Song Y, Zhang CJ, Guo Y (2009a) Identification of an apoplastic protein involved in the initial phase of salt stress response in rice root by two-dimensional electrophoresis. Plant Physiol 149:916–928
Zhang Y, Zhu H, Zhang Q, Li M, Yan M, Wang R, Wang L, Welti R, Zhang W, Wang X (2009b) Phospholipase dalpha1 and phosphatidic acid regulate NADPH oxidase activity and production of reactive oxygen species in ABA-mediated stomatal closure in Arabidopsis. Plant Cell 21:2357–2377
Zhang H, Wang M, Wang W, Li D, Huang Q, Wang Y, Zheng X, Zhang Z (2012) Silencing of G proteins uncovers diversified plant responses when challenged by three elicitors in *Nicotiana benthamiana*. Plant Cell Environ 35(1):72–85
Zheng Z-L, Yang Z (2000) The Rop GTPase: an emerging signaling switch in plants. Plant Mol Biol 44:1–9
Zheng ZL, Nafisi M, Tam A, Li H, Crowell DN, Chary SN, Schroeder JI, Shen J, Yang Z (2002) Plasma membrane-associated ROP10 small GTPase is a specific negative regulator of abscisic acid responses in Arabidopsis. Plant Cell 14:2787–2797
Zhu H, Li GJ, Ding L, Cui X, Berg H, Assmann SM, Xia Y (2009) Arabidopsis extra large G-protein 2 (XLG2) interacts with the Gb subunit of heterotrimeric G protein and functions in disease resistance. Mol Plant 2:513–525
Zipfel C (2008) Pattern-recognition receptors in plant innate immunity. Curr Opin Immunol 20: 10–16

Chapter 10
ABA Receptors: Prospects for Enhancing Biotic and Abiotic Stress Tolerance of Crops

Monika Dalal and Viswanathan Chinnusamy

Abstract The plant stress hormone abscisic acid (ABA) regulates myriad of plant developmental programs such as germination, root development, vegetative growth, seed development, dormancy, and seed desiccation tolerance. ABA, the master controller of transpiration, regulates ion channels and gene expression that are necessary for abiotic stress tolerance of plants and hence popularly called as the plant stress hormone. In the past one decade, the role of ABA in regulation of biotic stress tolerance is emerging. In response to low leaf water status as well as pathogen-associated molecular pattern (PAMP) signaling, ABA induces closure of stomata, which are major gateways of pathogens entry into plant cells. Salicylic acid (SA) promotes systemic acquired resistance (SAR) to biotrophic lifestyle, while jasmonic acid (JA) and ethylene positively regulate induced systemic resistance (ISR) against necrotrophic pathogens and insects. In addition to its role in PAMP-mediated stomatal closure, ABA interacts synergistically or antagonistically with SA, JA, and ethylene to regulate disease resistance pathway. Intense efforts made since 1980s have unraveled the molecular details of ABA signaling, culminating with the breakthrough discovery of the START domain proteins PYR/PYL/RCAR as ABA receptors (ABAR) and identification of core components of ABA signaling in 2009. Recent studies have also revealed the critical role of ABA receptors in plant processes such as fruit ripening and secondary metabolite accumulation. Genetic manipulation of ABA signaling is envisaged as a potential tool for enhancing plant development and biotic and abiotic stress tolerance of crops. This chapter focuses on molecular and structural basis of ABA signaling. Further, it explores the potential of genetic engineering of core components, protein engineering to develop orthogonal receptor, and development of novel synthetic agonists of ABARs for improving crop yield, quality, and stress tolerance.

M. Dalal
National Research Centre on Plant Biotechnology, New Delhi 110 012, Delhi, India

V. Chinnusamy (✉)
Division of Plant Physiology, Indian Agricultural Research Institute, New Delhi 110 012, Delhi, India
e-mail: viswanathan@iari.res.in

G.K. Pandey (ed.), *Elucidation of Abiotic Stress Signaling in Plants*, DOI 10.1007/978-1-4939-2211-6_10

Keywords Abiotic stress • ABA receptors • Agonist • Biotic stress • Crop improvement • Orthogonal receptor • Stress signalling

10.1 Discovery of ABA

The existence of a growth inhibitory substance that moved in paper chromatography with ammoniacal isopropanol between *RF* 0.55 and 0.8 in plant extracts was first reported by Bennet-Clark and Kefford (1953). The substance was named inhibitor-β. Bioassay with seed of *Papaver somniferum* or coleoptile sections of *Triticum vulgare* revealed the presence of this inhibitory substance in many species (Varga and Ferenczy 1957). Later, an auxin antagonist present in young cotton fruit, whose activity was correlated with abscission of the young fruit, was identified. This compound named "abscisin II" was isolated and crystallized, and a tentative molecular formula $C_{15}H_{20}O_4$ was worked out (Ohkuma et al. 1963). Around the same time, a compound "dormin" that causes bud dormancy was isolated from sycamore leaf extracts (Cornforth et al. 1965). Dormin and inhibitor-β were found to be same as abscisin II. Later, in the sixth International Conference on Plant Growth Substances in 1967 at Ottawa, a proper nomenclature was proposed, and the name "abscisic acid" was coined with acronym ABA (Addicott and Lyon 1969). By this time, it was apparent that ABA is involved in many physiological processes including seed germination, dormancy, and abscission.

10.2 Establishment of ABA as the Plant Stress Hormone

The best thing about ABA was that before its formal nomenclature, it was already isolated in crystalline form from various species. The availability of synthetic form further boosted the research on ABA. Wright (1969) reported an increase in acidic ether soluble growth inhibitor in the detached leaves of wheat after a period of wilting. Based on Rf and optical rotatory dispersion, it was identified as ABA (Wright and Hiron 1969). Milborrow and Noddle (1970) showed that severe wilting can induce up to 40-fold increase in levels of endogenous ABA in wheat seedlings within 4 h. The application of exogenous ABA caused stomatal closure and inhibition of transpiration in excised leaves (Mittelheuser and van Steveninck 1969) and the effect was very quick (within 5 min) and ABA specific (Cummins et al. 1971). The high degree of specificity for the naturally occurring form of ABA and requirement in very low concentration for stomatal closure corroborated with the concept of hormonal control of stomatal closure and also appeared a normal physiological function of ABA in leaves (Kriedemann et al. 1972). In water-stressed tissues, there was a concomitant increase in endogenous ABA level, which decreased rapidly on recovery/rewatering (Zeevaart 1980). Moreover, a positive correlation between drought stress tolerance and high free

ABA was also found in maize and sorghum (Larque-Saavedra and Wain 1976). These studies suggested that ABA might be a major regulator of drought tolerance in plants. The molecular genetic evidence for the role of ABA in stress tolerance came from studies with ABA-deficient (*aba*) and ABA-insensitive (*abi*) mutants. The *aba2* mutant of *Nicotiana plumbaginifolia* was characterized by a high tendency to wilt, inability to close stomata, and reduced primary dormancy. The wilty phenotype of *aba2* mutant was overcome by exogenous ABA application. Molecular genetic analysis revealed that *ABA2* gene codes for zeaxanthin epoxidase enzyme that catalyzes the first step of ABA biosynthesis (Marin et al. 1996). The expression of *ABA2* was higher in leaves than in roots. However, expression was upregulated in roots under drought stress, and this increase was correlated with an increased ABA accumulation (Audran et al. 1998). The *viviparous14* (*vp14*) mutant of maize (*Zea mays*) was characterized by precocious germination of embryo. The detached leaves of *vp14* mutant had a higher rate of transpiration and seeds had reduced dormancy due to a lower ABA content. The *Vp14* was expressed in embryos and roots and was strongly induced in leaves by water stress (Tan et al. 1997). The *vp14* gene codes for 9-*cis*-epoxycarotenoid dioxygenase (NCED) enzyme that catalyzes oxidative cleavage of neoxanthin or violaxanthin (Schwartz et al. 1997; Tan et al. 1997). In detached bean leaves, ABA accumulation was preceded by increases in *PvNCED1* expression, while rehydration of stressed leaves caused a rapid decrease in *PvNCED1* expression and ABA levels (Qin and Zeevaart 1999). Transgenic *Arabidopsis* overexpressing *AtNCED3* also showed an increase in endogenous ABA level and improved the drought tolerance of the plants, whereas disruption of *AtNCED3* resulted in drought susceptibility (Iuchi et al. 2001). These results showed that *NCED3* played an important role in ABA biosynthesis, and drought-induced ABA accumulation and enhanced stress tolerance (Qin and Zeevaart 1999; Iuchi et al. 2001). The *aao3* (*Abscisic aldehyde oxidase 3*) mutant deficient in ABA exhibited wilty phenotype at rosette stage (Seo et al. 2000). It was impaired in the *AAO3* gene that catalyzes the last step of ABA biosynthesis specifically in rosette leaves of *Arabidopsis* (Seo et al. 2000). Genetic analysis of the ABA-deficient mutants *los5/aba3* and *los6/aba1* of *Arabidopsis* also showed that ABA plays a pivotal role in osmotic stress-regulated gene expression and stress tolerance (Xiong et al. 2001, 2002). These physiological and molecular genetic evidences proved that ABA is necessary for drought tolerance.

ABA regulates dehydration avoidance and dehydration tolerance mechanisms to confer drought stress tolerance. Genetic analysis of ABA-deficient mutants also established the necessity of ABA signaling in stomatal control of water loss from plants. ABA inhibits potassium influx but promotes anion and potassium efflux from stomatal guard cells and thus closes stomata (Kim et al. 2010). ABA also induces acclimation and tolerance to cold, salt, and heat stresses. Abiotic stress-responsive gene expression is mediated by both ABA-dependent and ABA-independent pathways (Shinozaki and Yamaguchi-Shinozaki 2000; Zhu 2002). Comparative gene expression analysis between WT and ABA-deficient mutants *los5/aba3* and *los6/aba1* revealed that ABA regulates several osmotic stress-regulated genes including genes for osmolyte accumulation, protective proteins (LEAs and LEA-like), ROS detoxification, etc. (Xiong et al. 2001; Shinozaki and Yamaguchi-Shinozaki 2000; Zhu 2002). ABA-dependent pathways of abiotic stress signaling and tolerance have been discussed in several excellent reviews (Shinozaki and Yamaguchi-Shinozaki 2000; Zhu 2002; Yamaguchi-Shinozaki and Shinozaki 2006; Cutler et al. 2010; Finkelstein 2013). ABA-mediated stress tolerance also involves ABA-dependent epigenetic regulation of gene expression (Chinnusamy et al. 2008). Extensive amount of excellent work on genes for ABA synthesis and signaling on imparting abiotic stress tolerance are not covered here due to the paucity of space. Some examples of crop plants engineered with ABA metabolism, signaling pathway, and effector genes and tested under field conditions are given in Table 10.1.

10.3 Role of ABA in Biotic Stress Tolerance of Plants

The field of biotic stress response has been greatly dominated by salicylic acid (SA), jasmonic acid (JA), and ethylene (ET). Salicylic acid and jasmonic acid/ethylene orchestrate response against (hemi)biotrophic pathogens and necrotrophic pathogens, respectively. The first study suggestive of the involvement of ABA in disease response came in 1979 (Mohanty et al. 1979) where bioassays revealed a significantly higher ABA-like activity from the extracts of the rice tango virus (RTV)-infected susceptible rice cultivar than that of the less-susceptible cultivar. There were studies that reported that ABA negatively affects the disease resistance (Rezzonico et al. 1998; McDonald and Cahill 1999; Audenaert et al. 2002). The molecular genetic evidence for the antagonistic interaction of ABA with JA-ethylene and disease response came from the study of Anderson et al. (2004). It was demonstrated that exogenous ABA application in *Arabidopsis* suppressed both basal and JA-ethylene-activated transcription from defense genes. In contrast, ABA mutants defective in ABA biosynthesis or signaling resulted in upregulation of JA-ethylene responsive defense genes (Anderson et al. 2004). Furthermore, mutations in *JIN1* (*Jasmonate-Insensitive1*)/*MYC2*, a positive regulator of ABA signaling, and *ABA2* showed increased resistance to the necrotrophic fungal pathogen *Fusarium oxysporum*, thereby implicating an antagonistic interaction between components of ABA and the JA-ethylene signaling pathways. In ABA-deficient *sitiens*, tomato mutant extracellular hydrogen peroxide accumulation and activation of peroxidases resulted in arrest of pathogen progression (Asselbergh et al. 2007). ROS accumulation and increases in peroxidase activity were associated with induction of SA-inducible defenses in these mutants. This resistance response was reversed by exogenous ABA (Asselbergh et al. 2007; Audenaert et al. 2002). The antagonistic interaction of ABA and SA/JA was observed not only during biotic stress but also under abiotic stress condition. In rice, ABA or low-temperature stress suppressed resistance against *M. oryzae*, while application of ABA biosynthesis inhibitor, fluorine, prevented cold-induced blast susceptibility (Koga et al. 2004). Exogenous ABA also

Table 10.1 Engineering of crops with genes coding for ABA metabolism, signaling pathway, and effector genes for abiotic stress tolerance under field conditions

Gene	Promoter	Transgenic crop	Remarks	Reference
AtNCED2 (9-*cis*-epoxy carotenoid dioxygenase; involved in ABA synthesis)	Actin1 or *OsHVA22*	Rice	Transgenic rice plants showed higher spikelet fertility and yield under drought stress	Xiao et al. (2009)
AtLOS5/ABA3 (molybdenum cofactor sulfurase; involved in ABA synthesis)	Actin1 or *OsHVA22*	Rice	Transgenic rice plants showed higher spikelet fertility and yield under drought stress	Xiao et al. (2009)
AtLOS5/ABA3 (molybdenum cofactor sulfurase; involved in ABA synthesis)	Constitutive super promoter	Soybean	~21 % higher yield than that of wild-type plants under drought stress	Li et al. (2013)
AtERA1 (*Arabidopsis* farnesyltransferase β-subunit; a negative regulator of stomatal closure)	*RD29A* antisense suppression	Canola	~15 % higher seed yield as compared with non-transformed crop under drought stress	Wang et al. (2005)
AtFTA (*Arabidopsis* alpha-subunit of farnesyltransferase; a negative regulator of stomatal closure)	*AtHPR1* RNAi silencing	Canola	~15 % higher seed yield as compared with non-transformed crop under drought stress	Wang et al. (2009)
AtSOS2 (salt overly sensitive 2 kinase; interacts with ABI2 in yeast two-hybrid system)	*Actin1* or *HVA22P*	Rice	Transgenic rice plants showed higher spikelet fertility and yield under drought stress	Xiao et al. (2009)
AtZAT10 (Na^+/H^+ exchanger; upregulated by ABA)	*Actin1* or *HVA22P*	Rice	Transgenic rice plants showed higher spikelet fertility and yield under drought stress	Xiao et al. (2009)
OsSNAC1 (stress-responsive NAC 1; NAM, ATAF, and CUC transcription factor; induced by ABA)	CaMV 35S	Rice	22–34 % higher yield than non-transgenics under severe drought stress	Hu et al. (2006)

(continued)

Table 10.1 (continued)

Gene	Promoter	Transgenic crop	Remarks	Reference
OsNAC5 (NAM, ATAF, and CUC transcription factor; induced by ABA)	*RCc3P* Root-specific promoter	Rice	~22–63 % higher yield than non-transgenics in the under drought stress	Jeong et al. (2013)
OsNAC9 (NAM, ATAF, and CUC transcription factor; induced by ABA)	*RCc3P* Root-specific promoter	Rice	~28–72 % higher yield than non-transgenics under drought stress; *NCED* expression was high	Redillas et al. (2012)
OsNAC10 (NAM, ATAF, and CUC transcription factor; induced by ABA)	*RCc3P* Root-specific promoter	Rice	~25–42 % higher yield than non-transgenics under drought stress	Jeong et al. (2010)
Arabidopsis EDT1/HDG11 (enhanced drought tolerance1/homeodomain glabrous11 TF; *edt1* activation mutant showed enhanced ABA accumulation)	Constitutive	Rice	~16 % higher yield than non-transgenics under drought stress	Yu et al. (2013)
OsLEA3-1 (group 3 late embryogenesis abundant protein; ABA-induced)	*OsLEA3-1*:: or *CaMV35S*	Rice transgenic overexpressing maize	Enhanced drought tolerance over non-transgenic plants under drought stress	Xiao et al. (2007)
Barley *HVA1* (group 3 late embryogenesis abundant protein; ABA-induced)	Maize *UBIQUITIN 1*	Wheat	Enhanced yield and biomass over non-transgenic plants under drought stress	Bahieldin et al. (2005)
AtNHX1 (Na^+/H^+ exchanger; upregulated by ABA and probably negatively regulated by ABI2 PP2C)	*Actin1* or *HVA22P*	Rice	Transgenic rice plants showed higher spikelet fertility and yield under drought stress	Xiao et al. (2009)
AtNHX1 (Na^+/H^+ exchanger; upregulated by ABA and probably negatively regulated by ABI2 PP2C)	Super promoter	Cotton	Enhanced photosynthesis under salt stress and increases fiber yield	He et al. (2005)

compromised the rice resistance to both compatible and incompatible *M. grisea* strains by suppressing SA or blast infection-induced expression of *OsNPR1* and *OsWRKY45*, the two important components of the SA signaling pathway in rice (Jiang et al. 2010). The ABA-responsive gene induction was lesser and expression of *WRKY45* and *OsPR1b* was higher in incompatible rice–*M. grisea* interaction as compared to the compatible interaction. The results indicated that incompatible interactions upregulated SA pathway and downregulated ABA pathway thereby resulting in blast resistance. Interestingly, ABA was also detected in hyphae and conidia of *M. grisea* suggesting that ABA from pathogen can also induce ABA signaling in host and affect the susceptibly of the plant (Jiang et al. 2010). Similarly, susceptibility of rice to leaf blight pathogen *Xanthomonas oryzae* pv. *oryzae* was due to ABA-mediated suppression of SA defense pathway in rice (Xu et al. 2013). In contrast, exogenous ABA was shown to enhance basal resistance of rice against the brown spot-causing ascomycete *Cochliobolus miyabeanus*. The infection was restricted at mesophyll-based growth phase and was independent of salicylic acid, jasmonic acid, or callose-dependent resistance mechanisms. The resistance was attributed to ABA-mediated suppression of pathogen-induced ET action in an ABA-primed mitogen-activated protein kinase gene (*OsMPK*)-dependent manner (De Vleesschauwer et al. 2010).

Though there were more evidences for the negative interaction of ABA with SA and JA/ET. The studies on the mechanism of pathogen defense in plants revealed a synergistic association of these hormones at one or more nodes of the hormone signaling network. Plants close their stomata to avoid water loss during moisture stress conditions as well as to prevent entry of pathogen during pathogen attack. ROS and ion channels play an important role in ABA-mediated guard cell movement (Kwak et al. 2006; Li et al. 2006). ABA mediates ROS production through NADPH oxidases. Interestingly, ROS production during pathogen defense is also catalyzed by NADPH oxidases (Torres and Dangl 2005). The study of defense signaling pathway revealed that ABA, MeJA (methyl jasmonate), and SA adapt a common signaling mechanism for stomatal closure. Both MeJA and ABA induced stomatal closing through ROS and NO production in guard cells. Similar to ABA, MeJA also activated I_{Ca} (Ca^{2+}) channels and S-type anion channels in guard cells (Suhita et al. 2004; Munemasa et al. 2007). MeJA could not induce stomatal closing in *abi2-1* plants but could elicit ROS and NO production in *abi2-1* guard cells. This suggested that MeJA induces stomatal closer via *ABI2*-dependent signal pathway and that the *abi2-1* mutation could disrupt MeJA signaling downstream of production of ROS and NO in *Arabidopsis* guard cells (Munemasa et al. 2007). Similarly, ABA was shown to be involved in innate immunity response in *Arabidopsis*. Innate immunity is the first line of defense in plants, which is based on the recognition of pathogen (or microbe)-associated molecular patterns (PAMPs or MAMPs). In *Arabidopsis*, innate immunity can be activated by salicylic acid (SA)-independent (Hauck et al. 2003; Zipfel et al. 2004) or SA-dependent mechanism (DebRoy et al. 2004). Melotto et al. (2006) demonstrated stomatal closure as a part of innate immunity of plants to restrict entry of pathogen during infection. The SA dependence on stomatal closure was evident by reduced response to bacteria *P. syringae* pv. tomato strain DC3000 (*Pst* DC3000) and lipopolysaccharide (LPS) in SA-deficient *nahG* transgenic plants and SA-biosynthetic mutant (*eds16-2*) plants. Furthermore, neither flg22 (a biologically active peptide derived from flagellin) nor LPS could induce stomatal closure in *ost1-2* kinase mutant or ABA-deficient *aba3-1* mutant plants. In addition, flg22 and LPS both induced NO production in guard cells of wild-type stomata, which was effectively prevented by treatment with Nu-nitro-L-ARGININE (L-NNA), an inhibitor of nitric oxide synthase (NOS) (Melotto et al. 2006). The study revealed that ABA synthesis, NO production, and *OST1* kinase that are required for guard cell/stomatal response under abiotic stress conditions were also required by PAMP/bacterium-induced stomatal closure in plants. Thus ABA-induced signaling in guard cells played a cooperative and common role for SA- and JA-mediated preinvasive penetration resistance in plants.

SA plays an important role in systemic acquired resistance (SAR), while induced systemic resistance (ISR) requires JA and ethylene. However, both are dependent on NPR1 (Pieterse et al. 1998). Besides, SA and JA treatment with nonprotein amino acid β-amino butyric acid (BABA) has been shown to elicit ISR in plants (Cohen 2002). The BABA-induced resistance was independent of JA, SA, and ethylene response. However, ABA-deficient *aba 1-5*, ABA-insensitive mutant *abi4-1*, and callose-deficient (*pmr4-1*) mutants were found incapable of mounting ISR against *P. cucumerina* in *Arabidopsis* (Ton and Mauch-Mani 2004). The analysis revealed that BABA-induced resistance was based on callose formation, and *abi4-1* plants failed to accumulate callose in response to pathogen. Furthermore, exogenous ABA mimicked BABA resulting in tenfold increase in callose formation in infected plants as compared to control plants, which correlated well with increased resistance to pathogen (Ton and Mauch-Mani 2004). This study showed that ABA-dependent callose formation confers resistance against necrotrophic pathogens. Callose deposition was also reported as a defense mechanism against *P. irregulare* (Adie et al. 2007), since susceptibility of callose-deficient mutant to *P. irregulare* was lower than that of ABA-deficient mutants (Adie et al. 2007). It suggested that ABA may have additional roles than just priming of callose formation in plant defense. Moreover, callose was shown to block SA-inducible defense responses (Nishimura et al. 2003). Therefore, ABA can also have an indirect effect on SA-dependent responses through enhancement of callose deposition. A recent study revealed the role of PDLP5 (plasmodesmata-located protein 5) in SA-induced closure of plasmodesmata (PD) and PD callose deposition, and it was proposed as one of the strategy for SA-mediated innate immunity (Wang et al. 2013). However, the role of ABA in PD callose deposition remains to be studied.

The study on pathogenesis of oomycete pathogen *Pythium irregulare* on *Arabidopsis thaliana* defense response opened yet another aspect of hormone cross-talk. In contrast to generally conceived antagonistic role of SA and JA in pathogen response, resistance to *P. irregulare* was found to be predominantly governed by JA, followed by SA though to a lesser extent than JA. Moreover, the meta-analysis of transcriptome profiles showed that more than one-third (39 of 119) of the genes that were upregulated by *P. irregulare* were ABA-responsive genes. Among these upregulated genes, one-half of the genes were also common with that of JA. The analysis of *cis*-regulatory sequences also revealed an overrepresentation of ABA response elements in promoters of *P. irregulare*-responsive genes (Adie et al. 2007). The role of ABA in *P. irregulare* resistance was further supported by the increase in ABA levels in the wild-type and JA-/ET-/SA-/ABA-related mutants (except in *aba2-12*) postinfection and increased susceptibility to *P. irregulare* in ABA mutants, aao3-2/*aba2-12* and *abi4*, that are impaired in ABA biosynthesis or insensitive to ABA, respectively, as compared with the wild type (Adie et al. 2007).

A typical example of bivalent interaction of ABA in pathogen defense was reported in an activation-tagged *Arabidopsis* line, *constitutive disease susceptibility 2-1D* (*cds2-1D*), a dominant mutation that conferred enhanced disease susceptibility to various strains of *Pseudomonas syringae* and compromised SAR (Fan et al. 2009). The phenotype was attributed to the T-DNA insertion adjacent to 9-*cis*-epoxycarotenoid dioxygenase 5 (*NCED5*) gene that caused an increase in *NCED5* transcripts and enhanced ABA level. The susceptibility was reduced in ABA biosynthetic mutant *aba3-1* and increased when ABA was exogenously applied or endogenously accumulated in response to mild water stress. Likewise, ABA accumulation also compromised resistance to biotrophic oomycete *Hyaloperonospora arabidopsidis*. Conversely, *cds2-1D* plants showed enhanced resistance to the fungus *Alternaria brassicicola* and compromised resistance in *aba3-1* plants. Comparison of the accumulation of SA and JA in the wild type, *cds2-1D*, and *aba3-1* plants challenged with *P. syringae* showed that ABA promoted JA accumulation and exhibited an antagonistic relationship with salicylic acid (Fan et al. 2009). Thus ABA increased susceptibility to biotrophic pathogen while conferred resistance against necrotrophic pathogen within the same plant system.

The *nonexpressor of PR* genes 1 (*NPR1*) gene is a key regulator of SA-mediated SAR in *Arabidopsis*. In rice, overexpression of *AtNPR1* conferred resistance against fungal pathogens *Magnaporthe oryzae* and *Fusarium verticillioides*, the causal agents of the blast and bakanae diseases, and *Erwinia chrysanthemi*, a bacterial pathogen that causes the foot rot disease of rice. However, transgenic *AtNPR1* plants showed increased susceptibility to infection by the *Rice yellow mottle virus* (RYMV) (Quilis et al. 2008). In addition, *AtNPR1* negatively regulated the expression of salt stress- and drought stress-responsive genes resulting in a higher sensitivity of transgenic *AtNPR1* rice to these abiotic stresses. Thus *AtNPR1* exhibited both positive and negative regulatory roles for fungal/bacterial and virus/salt/drought stress, respectively.

The transmembrane receptor-like kinases (LysM-containing receptor-like kinase, LYKs) that mediate PAMPs/MAMPs recognition have been characterized in *Arabidopsis*. Among the five putative LYKs encoded by the *Arabidopsis thaliana* genome, LysM RLK1/chitin elicitor receptor kinase 1 is required for response to chitin and peptidoglycan, and *AtLYK4* contributes to chitin perception, while *AtLYK3* has been shown to repress *Arabidopsis* innate immune responses. Paparella et al. (2014) demonstrated that mutants lacking a functional *AtLYK3* show increased expression of defense genes and enhanced pathogen resistance. Furthermore, *AtLYK3* mutant showed increased sensitivity to salt stress, similarly to mutants impaired in ABA synthesis or transduction. These results indicated that *AtLYK3* acts both as a positive regulator of ABA signaling and as a negative regulator of defense responses against the fungal necrotroph *B. cinerea*. Rice *OsERF922*, encoding an APETELA2/ethylene response factor (AP2/ERF)-type transcription factor, was rapidly and significantly induced by ABA and salt treatments, as well as by both virulent and avirulent pathovars of *Magnaporthe oryzae*. *OsERF922-overexpressing* plants showed enhanced susceptibility to *M. oryzae* and decreased tolerance to salt stress (Liu et al. 2012). The ABA levels were high in the overexpressing lines suggesting that OsERF922 is a negative regulator of disease resistance and stress tolerance. The tomato (*Solanum lycopersicum*) *abscisic acid-induced myb1* (*SlAIM1*) gene, a R2R3MYB transcription factor, was induced by pathogens, plant hormones, salinity, and oxidative stress (AbuQamar et al. 2009). Transgenic plants with reduced *SlAIM1* gene expression (RNAi plants) showed an increased susceptibility to the necrotrophic fungus *Botrytis cinerea* and increased sensitivity to salt and oxidative stress. Thus *SlAIM1* is a positive regulator of disease resistance as well as abiotic stress tolerance.

Thus from the studies, it is evident that the role of ABA in biotic stress response is very complex and depending on the pathogen, stage of infection, timing, and mode of invasion of the pathogen. Factors such as the physiological stage of the plant (normal or stress condition) also play an important role in deciding the cooperative or antagonistic role of ABA in plant defense (Fig. 10.1). Moreover, there are pathogens that produce ABA or may produce ABA mimics as virulence factor thereby tripping the hormonal balance of the plant leading to susceptibility of the plant. In spite of the complex interaction, which still needs to be deciphered, few genes have been identified that are working as either negative or positive regulators

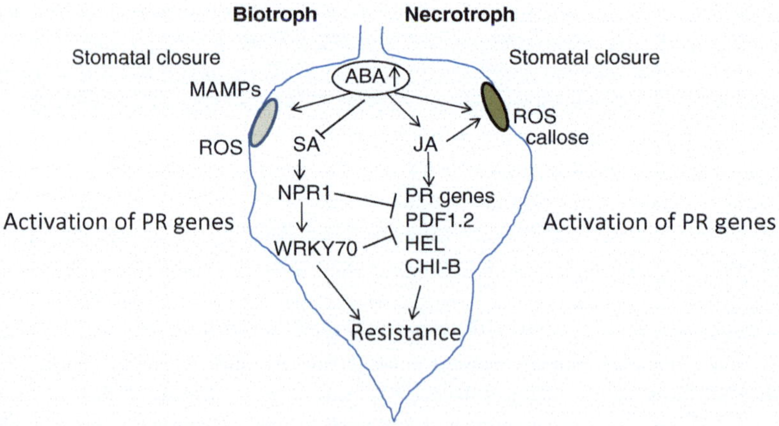

Fig. 10.1 A model for ABA, JA, and SA signaling network of pathogen defense in plants. ABA plays a bivalent role in pathogen defense depending on the pathogen, timing, and mode of invasion of the pathogen. ABA has cooperative role with JA and SA during initial stage, viz., stomatal closure (Munemasa et al. 2007; Suhita et al. 2004; Melotto et al. 2006) and priming of callose deposition (Ton and Mauch-Mani 2004; Adie et al. 2007). ABA suppresses SA-mediated defense gene expression and promotes JA-mediated defense gene expression (Fan et al. 2009). SA in turn has an antagonistic interaction with JA, which is mediated by NPR1 and WRKY

for both biotic and abiotic stress tolerance. These genes can be potentially useful for engineering tolerance to both the stresses.

10.4 ABA Receptors (ABARs) and Core ABA Signaling Pathway

Establishment of ABA as plant stress hormone triggered intensive research in deciphering various biochemical and molecular genetic mechanisms of ABA signaling. This led to the identification of secondary messengers, protein phosphatase 2C (PP2C) and subclass III sucrose non-fermenting-1 (SNF1)-related protein kinase 2 (SnRK2), protein modifiers, RNA processing factors, proteasome components, chromatin remodeling proteins, histone modifying enzymes, ABF/AREB transcription factors, and effector genes (Schroeder et al. 2001; Kuhn and Schroeder 2003; Antoni et al. 2011). PP2Cs belong to Mg^{2+}- or Mn^{2+}-dependent protein phosphatase (PPM) family of protein serine/threonine phosphatases that require Mg^{2+}/Mn^{2+} as cofactor (Schweighofer et al. 2004). *Arabidopsis* genome encodes 80 *PP2C* genes, which are categorized in to 13 subfamilies, namely, A–L (Schweighofer et al. 2004; Xue et al. 2008). Clade A subfamily contains nine proteins, of which six are negative regulators of ABA signal transduction (Gosti et al. 1999; Merlot et al. 2001; Saez et al. 2004; Yoshida et al. 2006; Nishimura et al. 2007). SnRK2 proteins belong to a family of plant-specific serine/threonine kinases that are involved in abiotic stress and ABA responses (Boudsocq et al. 2004; Fujii et al. 2007). The ten members of SnRK2 family are divided into three subclasses, viz., I, II, and III in *Arabidopsis* (Kobayashi et al. 2004). Members of SnRK2 subclass I are activated by osmotic stress, while subclass II and III kinases are activated by both ABA and osmotic stress. Subclass III SnRK2s show higher activation by ABA than subclass II (Boudsocq et al. 2004) and are positive regulators of ABA signal transduction (Mustilli et al. 2002; Yoshida et al. 2002; Fujii and Zhu 2009). ABA-mediated inhibition of the negatively acting PP2Cs leads to activation of a subfamily of SnRK2 kinases (SnRK2.2, SnRK2.3, and SnRK2.6). The active phosphorylated SnRK2s mediate diverse functions including ABA-induced stomatal closure and gene expression. SnRK2-mediated phosphorylation activates *AtRbohF* NADPH oxidase (Sirichandra et al. 2009) and guard cell anion channel slow anion channel -associated 1 (SLAC1) (Geiger et al. 2009) and inhibits potassium channel in *Arabidopsis thaliana* 1 (KAT1) (Sato et al. 2009). SnRK2s activate basic leucine zipper (bZIP) transcription factors such as ABA-responsive element (ABRE) binding (AREB)/ABRE-binding factor (AREB1, AREB2, ABF3, and ABI5) by phosphorylating these TFs (Fujita et al. 2012). The ABFs bind to ABA-responsive promoter elements (ABRE) to induce the expression of ABA-responsive genes involved in seed development and stress responses (Johnson et al. 2002; Furihata et al. 2006; Fujii and Zhu 2009).

The quest for ABA receptor started about three decades ago, and many candidate proteins were proposed as ABA receptors. Yet none of these proteins could be proved as bona fide ABA receptor (McCourt and Creelman 2008). Chloroplast-localized

Mg-chelatase H subunit (Shen et al. 2006) and plasma membrane-localized G protein-coupled receptor (GPCR)-type G proteins (GTGs) (Pandey et al. 2009) were also reported as receptors for ABA in *Arabidopsis*. However, the targets, detailed characterization of signaling pathway, and their conservation across plant species remain to be analyzed. The enigmatic search for the ABA receptor culminated in 2009, when two research groups independently reported the discovery of soluble receptors of ABA. Ma et al. (2009) employed yeast two-hybrid screens and found an ABI2 (a member of clade A PP2C)-interacting protein that was named as regulatory component of ABA receptor 1 (RCAR1). The team led by Sean R. Cutler applied a chemical genetic strategy and used pyrabactin, a synthetic seed germination inhibitor, to identify the ABA receptors (Park et al. 2009). Hence, the receptor was named as pyrabactin resistance 1 (PYR1), and its 13 other members in *Arabidopsis* genome were named as PYL1–PYL13 (for PYR1-like). These scientists demonstrated ABA-dependent inhibition of clade A PP2Cs (ABI1 or ABI2) by PYR/PYL/RCARs (Ma et al. 2009; Park et al. 2009). Based on these results, it was proposed that ABA-bound receptors (PYR/PYLs/RCARs) inhibit the activity of clade A PP2Cs, which relieves repression of SnRK2s. This proposed pathway for signal perception and transduction was successfully reconstituted in vitro demonstrating that ABA receptors (hereafter will be collectively referred as PYLs or ABARs), PP2Cs, and SnRK2s constitute the core components of ABA signaling for stress-responsive gene expression (Fujii et al. 2009).

The atomic details of mechanisms of ABA–ABAR binding, ABA–ABAR–PP2C binding, and PP2C–SnRK2 binding were elucidated by several crystallographic studies combined with cell biological experiments to validate the biological significance of the observed interactions (Melcher et al. 2009; Miyazono et al. 2009; Nishimura et al. 2009; Santiago et al. 2009a; Yin et al. 2009; Soon et al. 2012). PYL family of ABARs belongs to the START/Bet v I superfamily of proteins with a common fold enclosing large hydrophobic ligand-binding pocket. ABARs have a helix-grip structure with a large C-terminal α-helix (a4) surrounded by a seven-strand antiparallel β-sheet and two small α-helices. A large open ligand-binding pocket is present between the C-terminal helix and the β-sheet resides. The entry of the ligand-binding pocket is flanked by two β-loops, namely, gate (SGLPA) and latch (HRL). PYLs bind to ABA in 1:1 stoichiometry. ABA binding induces conformational changes in the PYLs that lead to closure of the gate and latch over the ABA-bound hydrophobic ligand-binding pocket and forms a new surface for subsequent interaction with PP2Cs. A conserved tryptophan residue from the PP2C inserts between the cap and lock of the PYL–ABA complex and forms a novel hydrogen bond between the indole ring of the tryptophan and the ketone group of ABA. The interaction between ABA–PYL–PP2C blocks the catalytic site of the PP2Cs and thus inhibits PP2C activity (Melcher et al. 2009; Miyazono et al. 2009; Nishimura et al. 2009; Santiago et al. 2009a; Yin et al. 2009). PP2Cs dephosphorylate S175 in the activation loop of SnRK2.6 that is necessary for SnRK2.6 activity. The crystal structure of the SnRK2.6–HAB1 complex revealed that the insertion of HAB1 W385 into the SnRK2.6 catalytic cleft acts as a physical barrier for the substrates (TFs and ion channels) to access the kinase active site. Thus, PP2Cs inhibit

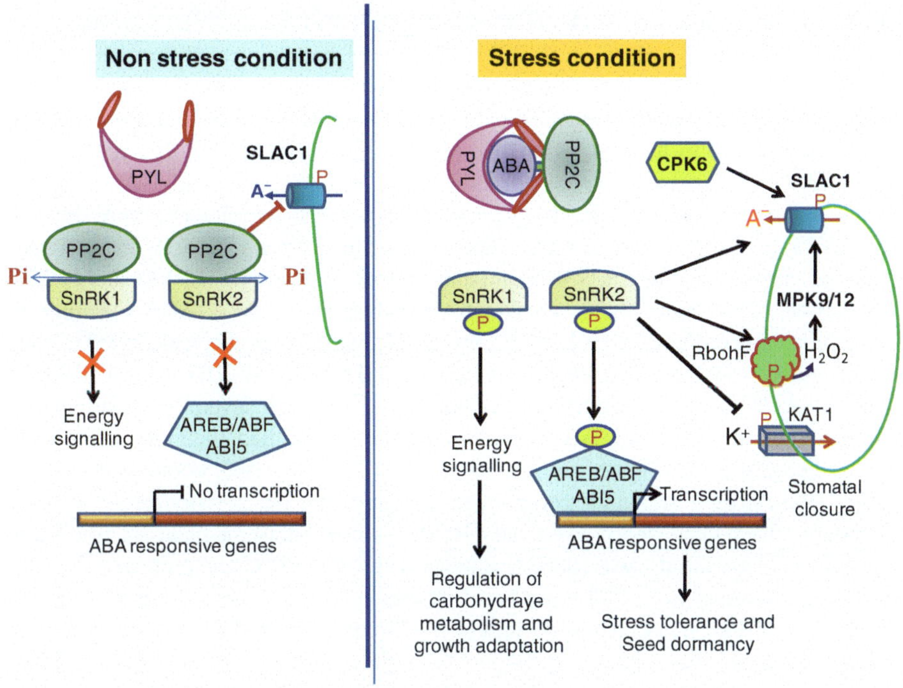

Fig. 10.2 ABA signaling pathway of abiotic stress tolerance. Under non-stress condition when the ABA concentration is below physiological threshold, PP2Cs are active and dephosphorylate SnRK2s and block substrate access to the kinase active site by binding to the catalytic cleft of SnRK2. PP2Cs dephosphorylate and inhibit SnRK1 activity. ABI1 PP2C also directly dephosphorylates and inhibits SLAC1. In stress conditions, ABA accumulates to a higher concentration. ABA binds to the PYL receptor, and the ABA–PYL receptor complex in turn bind to and inhibit PP2Cs. In the absence of PP2C-mediated inhibition, SnRK2s are activated by autophosphorylation. Active SnRK2s phosphorylate and activate SLAC1 and RobhF but inhibit KAT1 activity. RbohF produces H_2O_2, which activates MPK9/MPK12. These MAP kinases and calcium-dependent protein kinases (CPKS) have also been shown to activate SLAC1. SnRK2s also phosphorylate ABFs/ABI5 transcription factors, which activate transcription of ABA-responsive genes. The active SnRK1-mediated energy signaling leads to altered carbohydrate metabolism, energy conservation, and stress tolerance

SnRK2s activity by dephosphorylation of S175 in the activation loop and blocking the catalytic cleft of SnRK2s (Soon et al. 2012). PP2C can also directly dephosphorylate and inhibit SLAC1 activity. Further, in the absence of ABA, PP2Cs (ABI1 and PP2CA) dephosphorylate and inhibit SnRK1-mediated energy signaling (Rodrigues et al. 2013). As the PP2Cs are inhibited by PYLs in the presence of ABA, SnRK1 is activated, which leads to energy signaling and growth adaptation to abiotic stresses (Fig. 10.2).

The receptor family can be classified into different subtypes based on the sequence similarity, ABA sensitivity, and oligomeric state (Dupeux et al. 2011; Miyakava et al. 2013). *Arabidopsis* ABA receptor family consists of four dimeric (PYR1, PYL1 to

PYL3) and ten monomeric PYLs (PYL4 to PYL13; PYL7 and PYL13 are not confirmed experimentally) (Hao et al. 2011). There are biochemical differences in the receptor–PP2C interaction, indicating possible functional specialization among the PYR/RCAR family members. Initially, of the 14 PYL family members in *Arabidopsis*, 13 were demonstrated to function as ABA receptors (Fujii et al. 2009). The function of AtPYL13, which lacks some of the conserved residues in these domains, was not deciphered, and it was not considered to play a role in ABA signaling. Recent study showed that AtPYL13 is not a receptor but interacts specifically with PP2CA in an ABA-independent manner. It interacts with and antagonizes the function of PYLs and PP2Cs (Zhao et al. 2013). Since the establishment of core components of ABA signaling in *Arabidopsis*, the orthologs of PYL, PP2C, and SnRK2 family have been identified in rice, tomato, grape, strawberry, and sorghum (Chai et al. 2011; Sun et al. 2011; Kim et al. 2012; Boneh et al. 2012a, b; Dalal and Inupakutika 2014). Although several signaling molecules and proteins involved in ABA signaling have been identified, their interaction with core component or how core components regulate these signaling components needs further studies. Few signaling components that regulate the activity/levels of core components have been identified recently. ROP11, a member of the plant-specific Rho-like small GTPase family, was found to interact with ABI1 PP2C and protects the ABI1 phosphatase activity from PYL9-mediated inhibition (Li et al. 2012). The FERONIA (FER) receptor-like kinase, a positive regulator of auxin-promoted growth, was found to activate ABI2 and thus suppress the signal throughput of the core complex. The FER kinase interacts with guanine exchange factors (GEF1, GEF4, and GEF10), which activate ROP11 GTPase, which in turn interacts with ABI2 and enhances its activity (Yu et al. 2012). In *Arabidopsis*, DET1-, DDB1-ASSOCIATED1 (DDA1), as part of the COP10–DET1–DDB1 (CDD) complex was identified as CULLIN4-RING E3 ubiquitin ligases (CRL4s) substrate. DDA1 binds to PYL4, PYL8, and PYL9 and facilitates the proteasomal degradation of these PYLs. ABA bound to the PYL8 limits DDA1-mediated polyubiquitination and thus regulates ABA receptor stability (Irigoyen et al. 2014). Further characterization of ABA core components and regulation of core components by other signaling proteins are required for detailed understanding of ABA signaling and their effective utilization in crop improvement program.

10.5 Genetic Engineering of Core Components of ABA Signaling

10.5.1 Overexpression of ABA Receptors for Enhancement of Stress Tolerance

The transgenic *Arabidopsis* plants overexpressing PYL9/RCAR1 or PYL8/RCAR3 were hypersensitive to ABA with respect to both seed germination and root elongation (Ma et al. 2009; Saavedra et al. 2010). The transgenic *Arabidopsis* overexpressing *PYL5* (*PYL5-OE*) showed higher sensitivity to ABA-mediated inhibition of seed germination than wild type. *PYL5-OE* lines also showed reduced water loss and enhanced drought resistance than the wild type (Santiago et al. 2009b). The overexpression of *OsPYL/RCAR5* driven by maize ubiquitin promoter in rice also revealed the role of *OsPYL5* in modulating seed germination and early seedling growth in rice (Kim et al. 2012). Interestingly, PYL13, which due to lack of certain conserved residues, does not confirm to function as ABA receptor, when overexpressed in *Arabidopsis* resulted in enhanced drought tolerance, photosynthetic rate, and water-use efficiency in transgenic plants (Zhao et al. 2013). It was found that PYL13 interacts with clade A PP2Cs specifically PP2CA in ABA-independent manner and induces several stress-responsive gene expression. Zhang et al. (2013) characterized *AaPYL9*, from *Artemisia annua* in both *Arabidopsis* and *A. Annua*. Transgenic *Arabidopsis* plants overexpressing *AaPYL9* were more sensitive to ABA and exhibited reduced seedling growth under ABA, osmotic, and salt stress treatments, while overexpression of *AaPYL9* in *Artemisia* improved drought tolerance by inducing stomatal closure and reducing rates of transpiration.

Among PP2Cs, constitutive expression of *ZmPP2C* in *Arabidopsis* resulted in decreased tolerance to salt and drought stress at seed germination and vegetative growth stage (Liu et al. 2009). The overexpression of members of *SnRK2* family, viz., *TaSnRK2.8* and *TaSnRK2.4* in *Arabidopsis*, showed enhanced tolerance to drought, salt, and cold stresses. *TaSnRK2.8* belongs to class III SnRK2, while *TaSnRK2.4* is a class I SnRK2. However, *TaSnRK2.8* and *TaSnRK2.4* transgenics exhibited similar physiological characteristics such as high relative water content, strengthened cell membrane stability, more chlorophyll content, and enhanced PSII activity as compared to wild-type plants (Mao et al. 2010; Zhang et al. 2010). In case of *TaSnRK2.8* overexpressing plants, the primary roots were significantly longer than that of wild type (Zhang et al. 2010). The heterologous overexpression of *ZmSAPK8*, an orthologue of rice *SAPK8* and *Arabidopsis SnRK2.6*, a class III SnRK2 in *Arabidopsis* significantly increased the salt tolerance of the plants. Under salt treatment, transgenic plants overexpressing *ZmSAPK8* exhibited higher germination rate and proline content, low electrolyte leakage, and higher survival rate than wild type (Ying et al. 2011).

10.5.2 Vaccination of Plants Against Biotic and Abiotic Stresses

Crop production is one of the largest consumers of freshwater and uses about 70 % of freshwater worldwide. Since the freshwater availability to agriculture is decreasing drastically, we need to improve the water-use efficiency of agriculture to sustain food, fiber, feed, fodder, and fuel security of the world. Further, about 80 % of the global cultivated area is under rainfed cultivation and is the major cause of low agricultural productivity and poverty in these areas. Therefore, enhancing WUE of crops in irrigated agriculture and drought tolerance in rainfed agriculture are major targets for sustaining and enhancing agricultural productivity in the future. More than 98 % of water taken up by the plant is lost through transpiration. Hence, minimization of transpiration can enhance WUE of agriculture significantly. As ABA is the master regulator of transpiration and drought tolerance, ABA signaling pathway is a potential target for agronomic management of WUE and drought tolerance. Since 1970s, once ABA was established as the major regulator of transpiration and drought tolerance, significant efforts have been made to use ABA, ABA mimics, and chemicals that affect ABA signaling for agronomic management of WUE and drought tolerance. As early as 1971, Jones and Mansfield showed that methyl and phenyl esters of ABA can be used as antitranspirants. Later, few other chemicals were tried. These chemicals were not used in large scale due to chemical instability, catabolism by plant enzymes, less effectiveness, and off-target effects (adverse effect on processes other than stomatal regulation).

The discovery of ABA receptors and the atomic level structural details of ABA receptor–ABA–PP2C complex now enabled the scientists to identify chemical agonists or antagonists of these ABA receptors that can be used for agronomic management of WUE and tolerance to various biotic and abiotic stresses. Timely application of ABA can induce acclimation and tolerance of crops to these stresses. Since ABA is costly, significant efforts are being made to generate synthetic small molecule agonists of ABA receptor for agricultural use.

As we have discussed earlier, pyrabactin played a key role in the discovery of PYR/PYL ABA receptors in *Arabidopsis*. Protoplast and in vitro ABA signaling pathway reconstitution assays revealed that pyrabactin is an agonist of PYR1 and PYL1, but it antagonizes ABA activation of PYL2. Crystal structures of PYL1–pyrabactin–ABI1 and PYL2–pyrabactin complexes revealed that a single isoleucine (PYR1 and PYL1) versus valine (PYL2) position in the ligand-binding pockets is responsible for pyrabactin agonistic and antagonistic properties. Nonproductive binding of pyrabactin occurs without gate closure and prevents receptor activation. This knowledge was used to engineer I137V that converts PYL1 to a pyrabactin-inhibited receptor, A93F in PYL2 that converts PYL2 to a pyrabactin-activated receptor, and identification of pyrabactin-based ABA receptor agonists (Melcher et al. 2010; Peterson et al. 2010; Yuan et al. 2010, Table 10.2). To identify effect of ABA receptor agonists, Cao et al. (2013) screened ~12,000 chemicals for their ability to promote PYR1–HAB1 interactions by an AlphaScreen assay. This led to the identification of three related sulfonamide compounds named as ABA mimic 1 (AM1), AM2, and AM3 (Table 10.2). Among these compounds, AM1 had highest activity than ABA and other agonists in promoting PYR1–HAB1 interaction, and AM1 also activated most ABA receptors. Similar to ABA, treatment with AM1 also inhibited seed germination and minimized excised leaf water loss. *Arabidopsis* plants sprayed with 0.1 % DMSO, AM1, AM2, or (+)-ABA solutions were subjected to drought stress. The survival rate of DMSO-sprayed plants were only <5 %, while ABA- and AM1-sprayed plants showed 100 and >80 % survival, respectively. However, AM2 sprayed plants showed only about 15 % survival. Further, UV-stability was significantly higher for AM1 than that of ABA (Cao et al. 2013). Okamoto et al. (2013) screened ~57,000 compounds for their ability to activate five different ABA receptors (PYR1, PYL1, PYL2, PYL3, and PYL4) in yeast two-hybrid assay and identified quinabactin as agonist that preferentially activates

Table 10.2 Structure and activity of ABA receptor agonists/antagonists

Name	Structure	ABA receptor activated	Remarks	Reference
ABA	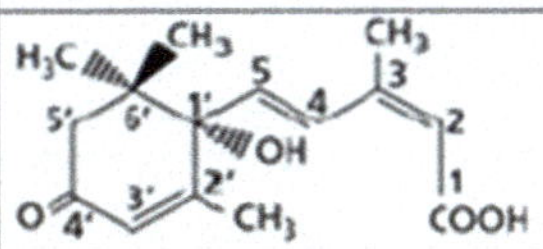	All ABA receptors	EC_{50} for different combinations of receptor–PP2C combinations varies. EC_{50} for inhibition of HAB1 by PYR1, PYL1, and PYL2 are 2.00, 1.01, and 0.32 μM, respectively	Cao et al. (2013)
Pyrabactin	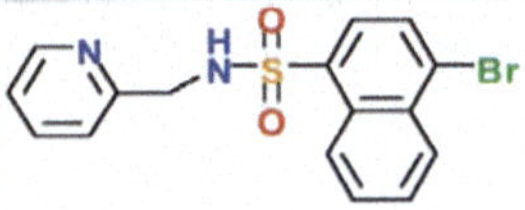	PYR1, PYL1	EC_{50} for inhibition of ABI1, ABI2, and HAB1 by PYR1 are 5.5, 3.5, and 7.5 μM, respectively	Melcher et al. (2010)
			Acts as antagonist of PYL2	
#32	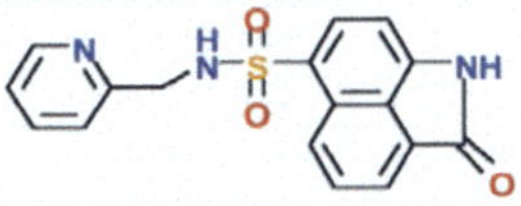	PYR1	EC_{50} for inhibition of ABI1, ABI2, and HAB1 by PYR1 are 14.1, 56.1, and 42.2 μM, respectively	Melcher et al. (2010)
#68	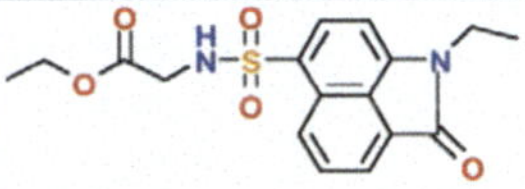	PYR1	EC_{50} for inhibition of ABI1, ABI2, and HAB1 by PYR1 are 11.3, 29.4, and 7.0 μM, respectively	Melcher et al. (2010)
#71	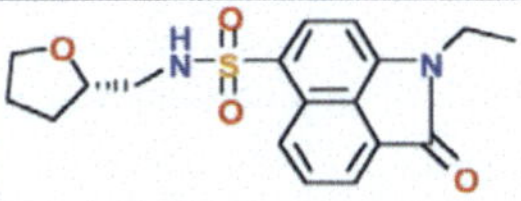	PYR1	EC_{50} for inhibition of ABI1, ABI2, and HAB1 by PYR1 are 16.1, 56.8, and 6.8 μM, respectively	Melcher et al. (2010)
#98	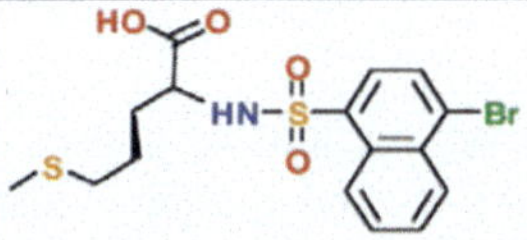	PYR1	EC_{50} for inhibition of ABI1, ABI2, and HAB1 by PYR1 are 1.4, 4.2, and 1.8 μM, respectively	Melcher et al. (2010)

(continued)

Table 10.2 (continued)

Name	Structure	ABA receptor activated	Remarks	Reference
ABA mimic 1 (AM1)	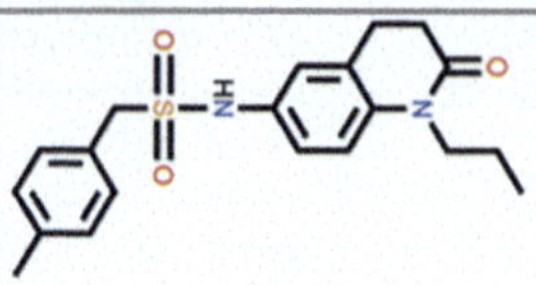	PYR1, PYL1 to PYL10	EC_{50} for inhibition of HAB1 by PYR1, PYL1, and PYL2 are 0.23, 0.42, and 0.48 μM, respectively Foliar spray enhances drought tolerance of *Arabidopsis*	Cao et al. (2013)
AM2	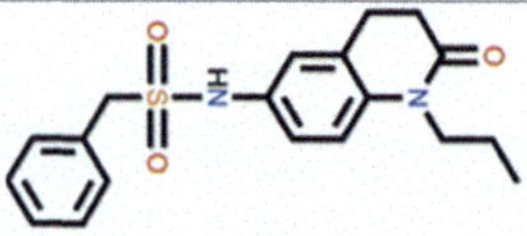	PYR1, PYL1, PYL2	EC_{50} for inhibition of HAB1 by PYR1, PYL1, and PYL2 are 0.28, 1.66, and 2.38 μM, respectively	Cao et al. (2013)
AM3	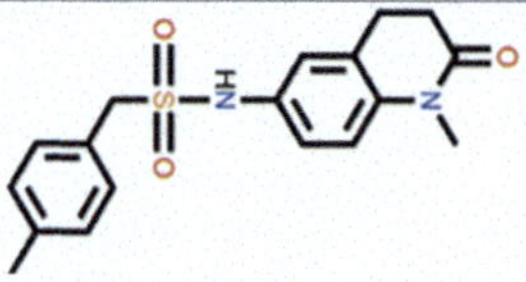	PYR1, PYL1	EC_{50} for inhibition of HAB1 by PYR1 and PYL1 are 3.02 and 1.49 μM, respectively	Cao et al. (2013)
Quinabactin	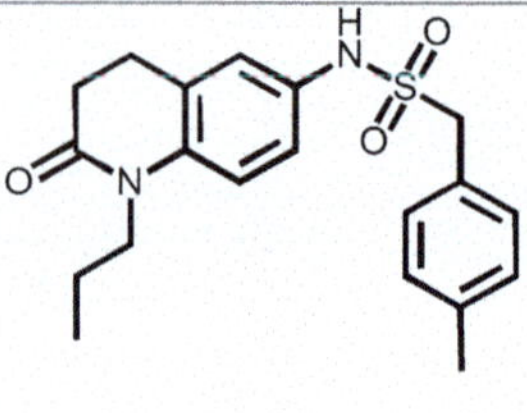	PYR1, PYL1, PYL2, PYL3, PYL5	EC_{50} for inhibition of HAB1 by PYR1, PYL1, PYL2, PYL3, and PYL5 are 0.10, 0.25, 0.27, 0.72, 0.65 μM, respectively Foliar spray enhances drought tolerance of *Arabidopsis*, soybean, and barley	Okamoto et al. (2013)

dimeric ABA receptors. In *Arabidopsis*, quinabactin elicits responses similar to that of ABA. Microarray analysis revealed that ABA- and quinabactin-induced gene expression in 10-day old *Arabidopsis* seedlings have about 0.90 correlation, suggesting that quinabactin induces transcriptional responses similar to that of ABA. Quinabactin treatments also minimized excised leaf water loss and induced stress-responsive gene expression similar to that of ABA in soybean, barley, and maize. Further, quinabactin foliar spray (50 μM) prior to subjecting the plants to drought stress conferred drought tolerance to *Arabidopsis*, soybean, and barley, but the level of tolerance provided by quinabactin varies in different plant species (Okamoto et al. 2013). Crystal structures of these chemical agonists and antagonists with ABA receptors (Melcher et al. 2010; Peterson et al. 2010; Yuan et al. 2010; Cao et al. 2013; Okamoto et al. 2013) are a repository of knowledge that will be useful for synthesis of novel chemical agonists as well as engineer orthogonal receptors.

Evidences from these studies showed the possibilities of:

1. Designing agonists or antagonists to a specific ABA receptor or group of ABA receptors
2. Differences in metabolism, uptake, or activity of chemical agonists
3. Designing agonists which have higher physiological activity than ABA
4. Designing agonists which have higher stability in environment

10.5.3 Engineering Orthogonal Receptors for Crop Management

Orthogonal receptor is an engineered receptor that can specifically bind to synthetic ligand (which cannot bind to natural receptor) and then regulate cellular processes, while functionally, orthogonal receptor is an engineered receptor, which is not activated by natural level concentrations of the endogenous ligand. The functionally orthogonal receptor is also very useful, as they will be activated by synthetic ligands at very low concentrations, while the natural receptors may require a higher concentration of synthetic ligand for activation. Orthogonal receptors can be used to engineer crop plants, where synthetic agonists or antagonists can be used to induce a specific physiological process such as metabolite production, fruit ripening, or tolerance to biotic and abiotic stresses. An orthogonal receptor and ligand pair is highly useful to avoid undesired pleiotropic effect of treatment of crops with antagonists or agonists that activate several receptors and thus general response. Further, these orthogonal receptors can be expressed in transgenic crops by using tissue- or stress-specific or chemical-inducible promoters. In the later case, a combination of chemicals for inducing the expression and activities of orthogonal receptors can yield effective and specific desirable results. Thus orthogonal receptor engineering offers a mean for genetic reprogramming of a specific cellular process or stress response.

As discussed in the earlier sections, ABA regulates several developmental programs and tolerance to biotic and abiotic stresses. ABA-mediated enhancement in abiotic stress tolerance may lead to susceptibility to some diseases. Constitutive overexpression of *OsPYL5* resulted in drought and other abiotic stress tolerance of transgenic rice plants, but these transgenic rice plants produced very low yield (Kim et al. 2014). These pleiotropic effects can be avoided by using orthogonal or functionally orthogonal ABA receptors, considering the diverse role of ABA in plant development and stress responses. The diverse role of ABA in plants and large size of ABA receptor family in plants suggest that ABA receptors or subfamilies of ABA receptors may have specific function in cellular process and their response to biotic and abiotic stresses.

Melcher et al. (2010) demonstrated that it is possible to design and engineer ABA receptors, which can be activated or inhibited by a synthetic ligand. Pyrabactin activates PYL1 but inhibits PYL2. Site-directed mutagenesis of I137V in PYL1 resulted in a pyrabactin-inhibited receptor, while A93F and V114I in PYL2 resulted in a pyrabactin-activated receptor (Melcher et al. 2010; Yuan et al. 2010). Further, Mosquna et al. (2011) showed that site-directed mutagenesis can be used to engineer constitutively active receptors, which show high-affinity binding and inhibition of PP2Cs in the absence of ABA. PYL13 shows very low activity with ABA. It could be converted in to an ABA-activated receptor by engineering three amino acids namely Q38K-F71L-T135N (Zhao et al. 2013). These studies show the possibilities of engineering ABA receptors, which can be activated by synthetic chemical (A93F and V114I in PYL2), inhibited by synthetic chemical (I137V in PYL1), and activated by natural ligand (Q38K-F71L-T135N in PYL13) and constitutively active ABA receptors. These studies give impetus to engineer and to use orthogonal or functionally orthogonal ABA receptors for enhancing crop production in the near future.

10.6 Conclusion and Perspectives

Genetic and agronomic manipulation of plant growth, development, and stress responses are necessary to sustain and improve agricultural production. One of the targets for genetic and agronomic manipulation of crop yield is ABA, the master controller of transpiration, plant development, and tolerance to biotic and abiotic stresses. Hence, significant research efforts have been made to understand and utilize ABA pathways. These efforts culminated in the identification of PYL family of ABARs and the core components of ABA signaling in *Arabidopsis* and their conservation in diverse plant species. Reconstitution studies and crystal structure analyses have proved that PYL family of proteins is a bona fide ABA receptor that regulates several aspects of ABA signaling involved in plant development, transpiration, and abiotic stress response. The core ABA signaling components comprising the ABAR–ABA–PP2C complex and SnRK2s control ABA-mediated stomatal closure via ion transports in guard cells and transcriptional regulations required for development and stress responses. PP2Cs, the negative regulator of ABA signaling pathway, have been shown to be regulated by ROP11 GTPase-FER kinase pathway. PP2Cs also regulate SnRK1 and SnRK3 kinases and effector genes such as *SLAC1*. Several CPKs and MAPKs have been found to regulate ABA-target ion channels and transcription factors. Thus, further studies are required to illustrate how the core

ABA signaling pathway components and effector genes are regulated by developmental and environmental cues.

PYL ABAR family members vary in their tissue-specific and stress-responsive expression, dimerization, binding efficiency to ABA, and ability to inhibit different clade A PP2Cs. Hence, the physiological relevances of different combinations of ABAR–ABA–PP2C need to be understood. ABA regulates biotic stress responses, in general positively regulates JA pathways, and negatively regulates SA pathways. Hence, genetic or agronomic manipulation of ABA pathway for improving abiotic stress tolerance may result in enhanced tolerance to certain pathogens and enhanced susceptibility to other pathogens. Hence, it is necessary to understand how PYL–PP2C–SnRK1 pathways regulate biotic stress responses. Several other questions to be answered include (1) whether natural antagonists of PYL receptors present and regulate homeostasis, (2) whether ABA antagonism of gibberellic acid responses is mediated by PYL–PP2C–SnRK2 pathways, (3) whether PYL–PP2C–SnRK2 pathways interact with other ABA pathway mediated by Mg-chelatase H subunit (CHLH) and GPCR-type G proteins (GTG1 and GTG2), (4) whether PYLs have substrates other than PP2Cs, and (5) whether ABA-induced epigenetic modifications are mediated directly by PYLs.

Genetic engineering of crops with genes coding for ABA synthesis, catabolism, signaling pathway, and effector genes has demonstrated that genetic modification of ABA pathways can significantly enhance crop yield under abiotic stress conditions. Efforts for identification ABAR agonists and antagonists have also led to enhanced knowledge on ABAR-ligand-binding and identification of novel synthetic molecules that regulate ABARs and drought tolerance. Further, the potential for engineering orthogonal receptors have been demonstrated. These studies on PYL–PP2C–SnRK pathways have laid a strong foundation for genetic engineering and agronomic management of crops for enhanced productivity, quality, and tolerance to biotic and abiotic stresses in the near future.

Acknowledgements The authors thank NRCPB (ICAR) and IARI (ICAR), New Delhi, for the facilities. Research work in VC's Lab is supported by NFBSFARA (ICAR), New Delhi.

References

AbuQamar S, Luo H, Laluk K, Mickelbar MV, Mengiste T (2009) Crosstalk between biotic and abiotic stress responses in tomato is mediated by the AIM1 transcription factor. Plant J 58:347–360

Addicott FT, Lyon JL (1969) Physiology of abscisic acid and related substances. Annu Rev Plant Physiol 20:139–164

Adie BAT, Perez-Perez J, Perez-Perez MM, Godoy M, Sanchez- Serrano JJ, Schmelz EA, Solano R (2007) ABA is an essential signal for plant resistance to pathogens affecting JA biosynthesis and the activation of defenses in *Arabidopsis*. Plant Cell 19:1665–1681

Anderson JP, Badruzsaufari E, Schenk PM, Manners JM, Desmond OJ, Ehlert C, Maclean DJ, Ebert PR, Kazan K (2004) Antagonistic interaction between abscisic acid and jasmonate-

ethylene signaling pathways modulates defense gene expression and disease resistance in *Arabidopsis*. Plant Cell 16:3460–3479
Antoni R, Rodriguez L, Gonzalez-Guzman M, Pizzio GA, Rodriguez PL (2011) News on aba transport, protein degradation, and ABFs/WRKYs in ABA signaling. Curr Opin Plant Biol 14:547–553
Asselbergh B, Curvers C, França SC, Audenaert K, Vuylsteke M, Van Breusegem F, Hofte M (2007) Resistance to *Botrytis cinerea* in *sitiens*, an abscisic acid-deficient tomato mutant, involves timely production of hydrogen peroxide and cell wall modifications in the epidermis. Plant Physiol 144:1863–1877
Audenaert K, De Meyer GB, Hofte M (2002) Abscisic acid determines basal susceptibility of tomato to *Botrytis cinerea* and suppresses salicylic acid-dependent signaling mechanisms. Plant Physiol 128:491–501
Audran C, Borel C, Frey A, Sotta B, Meyer C, Simonneau T, Marion-Poll A (1998) Expression studies of the zeaxanthin epoxidase gene in *Nicotiana plumbaginifolia*. Plant Physiol 118:1021–1028
Bahieldin A, Mahfouz HT, Eissa HF, Saleh OM, Ramadan AM, Ahmed IA, Dyer WE, El-Itriby HA, Madkour MA (2005) Field evaluation of transgenic wheat plants stably expressing the HVA1 gene for drought tolerance. Physiol Plant 123:421–427
Bennet-Clark TA, Kefford NP (1953) Chromatography of the growth substances in plant extracts. Nature 171:645–647
Boneh U, Biton I, Schwartz A, BenAri G (2012a) Characterization of the ABA signal transduction pathway in *Vitis vinifera*. Plant Sci 187:89–96
Boneh U, Biton I, Zheng C, Schwartz A, BenAri G (2012b) Characterization of potential ABA receptors in *Vitis vinifera*. Plant Cell Rep 31:311–321
Boudsocq M, Barbier-Brygoo H, Lauriere C (2004) Identification of nine sucrose nonfermenting 1-related protein kinases 2 activated by hyperosmotic and saline stresses in *Arabidopsis thaliana*. J Biol Chem 279:41758–41766
Cao M, Liu X, Zhang Y, Xue X, Zhou XE, Melcher K, Gao P, Wang F, Zeng L, Zhao Y, Zhao Y, Deng P, Zhong D, Zhu J-K, Xu HE, Xu Y (2013) An ABA-mimicking ligand that reduces water loss and promotes drought resistance in plants. Cell Res 23:1043–1054
Chai YM, Jia HF, Li CL, Dong QH, Shen YY (2011) FaPYR1 is involved in strawberry fruit ripening. J Exp Bot 62:5079–89
Chinnusamy V, Gong Z, Zhu JK (2008) Abscisic acid-mediated epigenetic processes in plant development and stress responses. J Integr Plant Biol 50:1187–1195
Cohen YR (2002) β amino butyric acid induced resistance against plant pathogens. Plant Dis 86:448–457
Cornforth JW, Milborrow BV, Ryback G, Wareing PF (1965) Chemistry and physiology of dormins in sycamore: identity of sycamore dormin with abscisin II. Nature 205:1269–1270
Cummins WR, Kende H, Raschke K (1971) Specificity and reversibility of the rapid stomatal response to abscisic acid. Planta 99:347–351
Cutler SR, Rodriguez PL, Finkelstein RR, Abrams SR (2010) Abscisic acid: emergence of a core signaling network. Annu Rev Plant Biol 61:651–679
Dalal M, Inupakutika M (2014) Transcriptional regulation of ABA core signaling component genes in sorghum (*Sorghum bicolor* L. Moench). Mol Breed 34:1517–1525
De Vleesschauwer D, Yang Y, Cruz CV, Hofte M (2010) Abscisic acid-induced resistance against the brown spot pathogen *Cochliobolus miyabeanus* in rice involves MAP kinase mediated repression of ethylene signaling. Plant Physiol 152:2036–2052
DebRoy S, Thilmony R, Kwack YB, Nomura K, He SY (2004) A family of conserved bacterial effectors inhibits salicylic acid-mediated basal immunity and promotes disease necrosis in *Arabidopsis*. Proc Natl Acad Sci U S A 101:9927–9932
Dupeux F, Santiago J, Betz K, Twycross J, Park S-Y, Rodriguez L, Gonzalez-Guzman M, Jensen RM, Krasnogor N, Blackledge M, Holdsworth M, Cutler SR, Rodriguez PL, Marquez JA (2011) A thermodynamic switch modulates abscisic acid receptor sensitivity. EMBO J 30:4171–4184

Fan J, Hill L, Crooks C, Doerner P, Lamb C (2009) Abscisic acid has a key role in modulating diverse plant-pathogen interactions. Plant Physiol 150:1750–1761

Finkelstein R (2013) Abscisic acid synthesis and response. Arabidopsis Book 11:e0166. doi:10.1199/tab.0166

Fujii H, Zhu JK (2009) *Arabidopsis* mutant deficient in 3 abscisic acid-activated protein kinases reveals critical roles in growth, reproduction, and stress. Proc Natl Acad Sci U S A 106:8380–8385

Fujii H, Verslues PE, Zhu JK (2007) Identification of two protein kinases required for abscisic acid regulation of seed germination, root growth, and gene expression in *Arabidopsis*. Plant Cell 19:485–494

Fujii H, Chinnusamy V, Rodrigues A, Rubio S, Antoni R, Park SY, Cutler SR, Sheen J, Rodriguez PL, Zhu JK (2009) *In vitro* reconstitution of an abscisic acid signalling pathway. Nature 462:660–664

Fujita Y, Yoshida T, Yamaguchi-Shinozaki K (2012) Pivotal role of the AREB/ABF-SnRK2 pathway in ABRE-mediated transcription in response to osmotic stress in plants. Physiol Plant 147:15–27

Furihata T, Maruyama K, Fujita Y, Umezawa T, Yoshida R, Shinozaki K, Yamaguchi-Shinozaki K (2006) Abscisic acid dependent multisite phosphorylation regulates the activity of a transcription activator AREB1. Proc Natl Acad Sci U S A 103:1988–1993

Geiger D, Scherzer S, Mumm P, Stange A, Marten I, Bauer H, Ache P, Matschi S, Liese A, Al-Rasheid KAS, Romeis T, Hedrich R (2009) Activity of guard cell anion channel SLAC1 is controlled by drought-stress signaling kinase-phosphatase pair. Proc Natl Acad Sci U S A 106:21425–21430

Gosti F, Beaudoin N, Serizet C, Webb AA, Vartanian N, Giraudat J (1999) ABI1 protein phosphatase 2C is a negative regulator of abscisic acid signaling. Plant Cell 11:1897–1910

Hao Q, Yin P, Li W, Wang L, Yan C, Lin Z, Wu JZ, Wang J, Yan SF, Yan N (2011) The molecular basis of ABA-independent inhibition of PP2Cs by a subclass of PYL proteins. Mol Cell 42:662–672

Hauck P, Thilmon R, He SY (2003) A *Pseudomonas syringae* type III effector suppresses cell wall-based extracellular defense in susceptible *Arabidopsis* plants. Proc Natl Acad Sci U S A 100:8577–8582

He C, Yan J, Shen G, Fu L, Holaday AS, Auld D, Blumwald E, Zhang H (2005) Expression of an Arabidopsis vacuolar sodium/proton antiporter gene in cotton improves photosynthetic performance under salt conditions and increases fiber yield in the field. Plant Cell Physiol 46:1848–1854

Hu H, Dai M, Yao J, Xiao B, Li X, Zhang Q, Xiong L (2006) Overexpressing a NAM, ATAF, and CUC (NAC) transcription factor enhances drought resistance and salt tolerance in rice. Proc Natl Acad Sci U S A 103:12987–12992

Irigoyen ML, Iniesto E, Rodriguez L, Puga MI, Yanagawa Y, Pick E, Strickland E, Paz-Ares J, Wei N, De Jaeger G, Rodriguez PL, Deng XW, Rubio V (2014) Targeted degradation of abscisic acid receptors is mediated by the ubiquitin ligase substrate adaptor DDA1 in *Arabidopsis*. Plant Cell 26:712–728

Iuchi S, Kobayashi M, Taji T, Naramoto M, Seki M, Kato T, Tabata S, Kakubari Y, Yamaguchi-Shinozaki K, Shinozaki K (2001) Regulation of drought tolerance by gene manipulation of 9-cis-epoxycarotenoid dioxygenase, a key enzyme in abscisic acid biosynthesis in *Arabidopsis*. Plant J 27:325–333

Jeong JS, Kim YS, Baek KH, Jung H, Ha SH, Do Choi Y, Kim M, Reuzeau C, Kim JK (2010) Root-specific expression of OsNAC10 improves drought tolerance and grain yield in rice under field drought conditions. Plant Physiol 153:185–197

Jeong JS, Kim YS, Redillas MC, Jang G, Jung H, Bang SW, Choi YD, Ha SH, Reuzeau C, Kim JK (2013) OsNAC5 overexpression enlarges root diameter in rice plants leading to enhanced drought tolerance and increased grain yield in the field. Plant Biotechnol J 11:101–114

Jiang CJ, Shimono M, Sugano S, Kojima M, Yazawa K, Yoshida R, Inoue H, Hayashi N, Sakakibara H, Takatsuji H (2010) Abscisic acid interacts antagonistically with salicylic acid signaling pathway in rice–*Magnaporthe grisea* interaction. Mol Plant Microbe Interact 23:791–798

Johnson RR, Wagner RL, Verhey SD, Walker-Simmons MK (2002) The abscisic acid-responsive kinase PKABA1 interacts with a seed-specific abscisic acid response element-binding factor, TaABF, and phosphorylates TaABF peptide sequences. Plant Physiol 130:837–846

Jones RJ, Mansfield TA (1971) Antitranspirant activity of the methyl and phenyl esters of abscisic acid. Nature 231:331–332

Kim TH, Böhmer M, Hu H, Nishimura N, Schroeder JI (2010) Guard cell signal transduction network: advances in understanding abscisic acid, CO_2, and Ca^{2+} signaling. Annu Rev Plant Biol 61:561–591

Kim H, Hwang H, Hong J-W, Lee Y-N, Ahn IP, Yoon IS, Yoo S-D, Lee S, Lee SC, Kim B-G (2012) A rice orthologue of the ABA receptor, OsPYL/RCAR5, is a positive regulator of the ABA signal transduction pathway in seed germination and early seedling growth. J Exp Bot 63: 1013–1024

Kim H, Lee K, Hwang H, Bhatnagar N, Kim DY, Yoon IS, Byun MO, Kim ST, Jung KH, Kim BG (2014) Overexpression of PYL5 in rice enhances drought tolerance, inhibits growth, and modulates gene expression. J Exp Bot 65:453–464

Kobayashi Y, Yamamoto S, Minami H, Kagaya Y, Hattori T (2004) Differential activation of the rice sucrose nonfermenting1-related protein kinase 2 family by hyperosmotic stress and abscisic acid. Plant Cell 16:1163–1177

Koga H, Dohi K, Mori M (2004) Abscisic acid and low temperature suppress the whole plant-specific resistance reaction of rice plants to the infection of *Magnaporthe grisea*. Physiol Mol Plant Pathol 65:3–9

Kriedemann PE, Loveys BR, Fuller GL, Leopold AC (1972) Abscisic acid and stomatal regulation. Plant Physiol 49:842–847

Kuhn JM, Schroeder JI (2003) Impacts of altered RNA metabolism on abscisic acid signaling. Curr Opin Plant Biol 6:463–469

Kwak JM, Nguyen V, Schroeder JI (2006) The role of reactive oxygen species in hormonal responses. Plant Physiol 141:323–329

Larque-Saavedra A, Wain RL (1976) Studies on plant growth-regulating substances. XLII. Abscisic acid as a genetic character related to drought tolerance. Ann Appl Biol 83:291–297

Li S, Assmann SM, Albert R (2006) Predicting essential components of signal transduction networks: A dynamic model of guard cell abscisic acid signaling. PLOS Biol 4:1732–1748

Li Z, Li Z, Gao X, Chinnusamy V, Bressan R, Wang ZX, Zhu JK, Wu JW, Liu D (2012) ROP11 GTPase negatively regulates ABA signaling by protecting ABI1 phosphatase activity from inhibition by the ABA receptor RCAR1/PYL9 in *Arabidopsis*. J Integr Plant Biol 54:180–188

Li Y, Zhang J, Zhang J, Hao L, Hua J, Duan L, Zhang M, Li Z (2013) Expression of an Arabidopsis molybdenum cofactor sulphurase gene in soybean enhances drought tolerance and increases yield under field conditions. Plant Biotechnol J 11:747–758

Liu L, Hu X, Song J, Zong X, Li D, Li D (2009) Over-expression of a *Zea mays* L. protein phosphatase 2C gene (*ZmPP2C*) in *Arabidopsis thaliana* decreases tolerance to salt and drought. J Plant Physiol 166:531–542

Liu D, Chen X, Liu J, Ye J, Guo Z (2012) The rice ERF transcription factor *OsERF922* negatively regulates resistance to *Magnaporthe oryzae* and salt tolerance. J Exp Bot 63:3899–3912

Ma Y, Szostkiewicz I, Korte A, Moes D, Yang Y, Christmann A, Grill E (2009) Regulators of PP2C phosphatase activity function as abscisic acid sensors. Science 324:1064–1068

Mao X, Zhang H, Tian S, Chang X, Jing R (2010) TaSnRK2.4, an SNF1-type serine/threonine protein kinase of wheat (*Triticum aestivum* L.), confers enhanced multistress tolerance in *Arabidopsis*. J Exp Bot 61:683–696

Marin E, Nussaume L, Quesada A, Gonneau M, Sotta B, Hugueney P, Frey A, Marion-Poll A (1996) Molecular identification of zeaxanthin epoxidase of *Nicotiana plumbaginifolia*, a gene involved in abscisic acid biosynthesis and corresponding to the ABA locus of *Arabidopsis thaliana*. EMBO J 15:2331–2342

McCourt P, Creelman R (2008) The ABA receptors—we report you decide. Curr Opin Plant Biol 11:474–478

McDonald KL, Cahill DM (1999) Influence of abscisic acid and the abscisic acid biosynthesis inhibitor, norflurazon, on interactions between *Phytophthora sojae* and soybean (*Glycine max*). Eur J Plant Pathol 105:651–658

Melcher K, Ng LM, Zhou XE, Soon FF, Xu Y, Suino-Powell KM, Park SY, Weiner JJ, Fujii H, Chinnusamy V, Kovach A, Li J, Wang Y, Li J, Peterson FC, Jensen DR, Yong EL, Volkman BF, Cutler SR, Zhu JK, Xu HE (2009) A gate-latch-lock mechanism for hormone signalling by abscisic acid receptors. Nature 462:602–608

Melcher K, Xu Y, Ng LM, Zhou XE, Soon FF, Chinnusamy V, Suino-Powell KM, Kovach A, Tham FS, Cutler SR, Li J, Yong EL, Zhu JK, Xu HE (2010) Identification and mechanism of ABA receptor antagonism. Nat Struct Mol Biol 17:1102–1108

Melotto M, Underwood W, Koczan J, Nomura K, He SY (2006) Plant stomata function in innate immunity against bacterial invasion. Cell 126:969–980

Merlot S, Gosti F, Guerrier D, Vavasseur A, Giraudat J (2001) The ABI1 and ABI2 protein phosphatases 2C act in a negative feedback regulatory loop of the abscisic acid signaling pathway. Plant J 25:295–303

Milborrow BV, Noddle RC (1970) Conversion of 5-(1,2-epoxy-2,6,6- tri methylcyclohexyl)-3-methylpenta-cis-2-trans-4-dienoic acid into abscisic acid in plants. Biochem J 119:727–734

Mittelheuser CJ, van Steveninck RFM (1969) Stomatal closure and inhibition of transpiration induced by (RS)-abscisic acid. Nature 221:281–282

Miyakava T, Fujita Y, Yamaguchi-Shinozaki K, Tanokura M (2013) Structure and function of abscisic acid receptors. Trends Plant Sci 18:259–266

Miyazono K, Miyakawa T, Sawano Y, Kubota K, Kang HJ, Asano A, Miyauchi Y, Takahashi M, Zhi Y, Fujita Y, Yoshida T, Kodaira KS, Yamaguchi-Shinozaki K, Tanokura M (2009) Structural basis of abscisic acid signalling. Nature 462:609–614

Mohanty SK, Mohanty SK, Anjaneyulu A, Sridhar R (1979) Physiology of rice tungro virus disease: involvement of abscisic acid-like substance in susceptible host-virus interactions. Physiol Plant 45:132–136

Mosquna A, Peterson FC, Park SY, Lozano-Juste J, Volkman BF, Cutler SR (2011) Potent and selective activation of abscisic acid receptors in vivo by mutational stabilization of their agonist-bound conformation. Proc Natl Acad Sci U S A 108:20838–20843

Munemasa S, Oda K, Watanabe-Sugimoto M, Nakamura Y, Shimoishi Y, Murata Y (2007) The *coronatine-insensitive 1* mutation reveals the hormonal signaling interaction between abscisic acid and methyl jasmonate in *Arabidopsis* guard cells. Specific impairment of ion channel activation and second messenger production. Plant Physiol 143:1398–1407

Mustilli AC, Merlot S, Vavasseur A, Fenzi F, Giraudat J (2002) *Arabidopsis* OST1 protein kinase mediates the regulation of stomatal aperture by abscisic acid and acts upstream of reactive oxygen species production. Plant Cell 14:3089–3099

Nishimura MT, Stein M, Hou BH, Vogel JP, Edwards H, Somerville SC (2003) Loss of a callose synthase results in salicylic acid-dependent disease resistance. Science 301:969–972

Nishimura N, Yoshida T, Kitahata N, Asami T, Shinozaki K, Hirayama T (2007) ABA-Hypersensitive Germination1 encodes a protein phosphatase 2C, an essential component of abscisic acid signaling in *Arabidopsis* seed. Plant J 50:935–949

Nishimura N, Hitomi K, Arvai AS, Rambo RP, Hitomi C, Cutler SR, Schroeder JI, Getzoff ED (2009) Structural mechanism of abscisic acid binding and signaling by dimeric PYR1. Science 326:1373–1379

Ohkuma K, Lyon JL, Addicott FT, Smith OE (1963) Abscisin II, an abscission-accelerating substance from young cotton fruit. Science 142:1592–1593

Okamoto M, Peterson FC, Defries A, Park SY, Endo A, Nambara E, Volkman BF, Cutler SR (2013) Activation of dimeric ABA receptors elicits guard cell closure, ABA-regulated gene expression, and drought tolerance. Proc Natl Acad Sci U S A 110:12132–12137

Pandey S, Nelson DC, Assmann SM (2009) Two novel GPCR-Type G Proteins are abscisic acid receptors in Arabidopsis. Cell 136:136–148

Paparella C, Savatin DV, Marti L, De Lorenzo G, Ferrari S (2014) The *Arabidopsis thaliana* LYSM-CONTAINING RECEPTOR-LIKE KINASE 3 regulates the cross talk between immunity and abscisic acid responses. Plant Physiol. doi:10.1104/pp. 113.233759

Park SY, Fung P, Nishimura N, Jensen DR, Fujii H, Zhao Y, Lumba S, Santiago J, Rodrigues A, Chow TFF, Alfred SE, Bonetta D, Finkelstein R, Provart NJ, Desveaux D, Rodriguez PL, McCourt P, Zhu JK, Schroeder JI, Volkman BF, Cutler SR (2009) Abscisic acid inhibits type 2C protein phosphatases via the PYR/PYL family of start proteins. Science 324:1068–1071

Peterson FC, Burgie ES, Park SY, Jensen DR, Weiner JJ, Bingman CA, Chang CE, Cutler SR, Phillips GN Jr, Volkman BF (2010) Structural basis for selective activation of ABA receptors. Nat Struct Mol Biol 17:1109–1113

Pieterse CMJ, van Wees SCM, van Pelt JA, Knoester M, Laan R, Gerrits H, Weisbeek PJ, van Loona LC (1998) A novel signaling pathway controlling induced systemic resistance in *Arabidopsis*. Plant Cell 10:1571–1580

Qin X, Zeevaart JAD (1999) The 9-cis-epoxycarotenoid cleavage reaction is the key regulatory step of abscisic acid biosynthesis in water-stressed bean. Proc Natl Acad Sci U S A 96:2615354–2615361

Quilis J, Penas G, Messeguer J, Brugidou C, Segundo BS (2008) The *Arabidopsis AtNPR1* inversely modulates defense responses against fungal, bacterial, or viral pathogens while conferring hypersensitivity to abiotic stresses in transgenic rice. Mol Plant Microbe Interact 21:1215–1231

Redillas MC, Jeong JS, Kim YS, Jung H, Bang SW, Choi YD, Ha SH, Reuzeau C, Kim JK (2012) The overexpression of OsNAC9 alters the root architecture of rice plants enhancing drought resistance and grain yield under field conditions. Plant Biotechnol J 10:792–805

Rezzonico E, Flury N, Meins F, Beffa R (1998) Transcriptional down-regulation by abscisic acid of pathogenesis-related beta-1,3-glucanase genes in tobacco cell cultures. Plant Physiol 117:585–592

Rodrigues A, Adamo M, Crozet P, Margalha L, Confraria A, Martinho C, Elias A, Rabissi A, Lumbreras V, González-Guzmán M, Antoni R, Rodriguez PL, Baena-González E (2013) ABI1 and PP2CA phosphatases are negative regulators of Snf1-Related Protein Kinase1 signaling in Arabidopsis. Plant Cell 25:3871–3884

Saavedra X, Modrego A, Rodriguez D, Gonzalez-Garcia MP, Sanz L, Nicolas G, Oscar L (2010) The nuclear interactor PYL8/RCAR3 of Fagus sylvatica FsPP2C1 is a positive regulator of abscisic acid signaling in seeds and stress. Plant Physiol 152:133–150

Saez A, Apostolova N, Gonzalez-Guzman M, Gonzalez-Garcia MP, Nicolas C, Lorenzo O, Rodriguez PL (2004) Gain-of-function and loss-of-function phenotypes of the protein phosphatase 2CHAB1 reveal its role as a negative regulator of abscisic acid signaling. Plant J 37:354–369

Santiago J, Dupeux F, Round A, Antoni R, Park SY, Jamin M, Cutler SR, Rodriguez PL, Márquez JA (2009a) The abscisic acid receptor PYR1 in complex with abscisic acid. Nature 462:665–668

Santiago J, Rodrigues A, Saez A, Rubio S, Antoni R, Dupeux F, Park S-Y, Marquez JA, Cutler SR, Rodriguez Pedro L (2009b) Modulation of drought resistance by the abscisic acid receptor PYL5 through inhibition of clade A PP2Cs. Plant J 60:575–588

Sato A, Sato Y, Fukao Y, Fujiwara M, Umezawa T, Shinozaki K, Hibi T, Taniguchi M, Miyake H, Goto DB, Uozumi N (2009) Threonine at position 306 of the KAT1 potassium channel is essential for channel activity and is a target site for ABA-activated SnRK2/OST1/SnRK2.6 protein kinase. Biochem J 424:439–448

Schroeder JI, Allen GJ, Hugouvieux V, Kwak JM, Waner D (2001) Guard cell signal transduction. Annu Rev Plant Physiol Plant Mol Biol 52:627–658

Schwartz SH, Tan BC, Gage DA, Zeevaart JAD, McCarty DR (1997) Specific oxidative cleavage of carotenoids by *VP14* of maize. Science 276:1872–1874

Schweighofer A, Hirt H, Meskiene I (2004) Plant PP2C phosphatases: emerging functions in stress signalling. Trends Plant Sci 9:236–243

Seo M, Peeters AJM, Koiwai H, Oritani T, Marion-Poll A, Zeevaart JAD, Koornneef M, Kamiyai Y, Koshiba T (2000) The *Arabidopsis* aldehyde oxidase 3 (*AAO3*) gene product catalyzes the final step in abscisic acid biosynthesis in leaves. Proc Natl Acad Sci U S A 97:12908–12913

Shen YY, Wang XF, Wu FQ, Du SY, Cao Z, Shang Y, Wang XL, Peng CC, Yu XC, Zhu SY, Fan RC, Xu YH, Zhang DP (2006) The Mg-chelatase H subunit is an abscisic acid receptor. Nature 443:823–826

Shinozaki K, Yamaguchi-Shinozaki K (2000) Molecular response to dehydration and low temperature: differences and cross-talk between two stress signalling pathways. Curr Opin Plant Biol 3:217–223

Sirichandra C, Gu D, Hu HC, Davanture M, Lee S, Djaoui M, Valot B, Zivy M, Leung J, Merlot S, Kwak JM (2009) Phosphorylation of the *Arabidopsis* AtrbohF NADPH oxidase by OST1 protein kinase. FEBS Lett 583:2982–2986

Soon FF, Ng LM, Zhou XE, West GM, Kovach A, Tan MH, Suino-Powell KM, He Y, Xu Y, Chalmers MJ, Brunzelle JS, Zhang H, Yang H, Jiang H, Li J, Yong EL, Cutler S, Zhu JK, Griffin PR, Melcher K, Xu HE (2012) Molecular mimicry regulates ABA signaling by SnRK2 kinases and PP2C phosphatases. Science 335:85–88

Suhita D, Raghavendra AS, Kwak JM, Vavasseur A (2004) Cytoplasmic alkalization precedes reactive oxygen species production during methyl jasmonate- and abscisic acid-induced stomatal closure. Plant Physiol 134:1536–1545

Sun L, Wang YP, Chen P, Ren J, Ji K, Li Q, Li P, Dai SJ, Leng P (2011) Transcriptional regulation of SlPYL, SlPP2C, and SlSnRK2 gene families encoding ABA signal core components during tomato fruit development and drought stress. J Exp Bot 62:5659–5669

Tan BC, Schwartz SH, Zeevaart JAD, McCarty DR (1997) Genetic control of abscisic acid biosynthesis in maize. Proc Natl Acad Sci U S A 94:12235–12240

Ton J, Mauch-Mani B (2004) β-Amino-butyric acid-induced resistance against necrotrophic pathogens is based on ABA dependent priming for callose. Plant J 38:119–130

Torres MA, Dangl JL (2005) Functions of the respiratory burst oxidase in biotic interactions, abiotic stress and development. Curr Opin Plant Biol 8:397–403

Varga M, Ferenczy L (1957) Paper-chromatographic examination of inhibiting substances in fleshy fruits. Naturwissenschaften 44:398–399

Wang Y, Ying J, Kuzma M, Chalifoux M, Sample A, McArthur C, Uchacz T, Sarvas C, Wan J, Dennis DT, McCourt P, Huang Y (2005) Molecular tailoring of farnesylation for plant drought tolerance and yield protection. Plant J 43:413–424

Wang Y, Beaith M, Chalifoux M, Ying J, Uchacz T, Sarvas C, Griffiths R, Kuzma M, Wan J, Huang Y (2009) Shoot-specific down-regulation of protein farnesyltransferase (alpha-subunit) for yield protection against drought in canola. Mol Plant 2:191–200

Wang X, Sager R, Cui W, Zhang C, Lu H, Lee JY (2013) Salicylic acid regulates Plasmodesmata closure during innate immune responses in *Arabidopsis*. Plant Cell 25:2315–2329

Wright STC (1969) An increase in the "inhibitor-β" content of detached wheat leaves following a period of wilting. Planta 86:10–20

Wright STC, Hiron RWP (1969) Abscisic acid, the growth inhibitor induced in detached wheat leaves by a period of wilting. Nature 224:719–720

Xiao B, Huang Y, Tang N, Xiong L (2007) Over-expression of a LEA gene in rice improves drought resistance under the field conditions. Theor Appl Genet 115:35–46

Xiao BZ, Chen X, Xiang CB, Tang N, Zhang QF, Xiong LZ (2009) Evaluation of seven function-known candidate genes for their effects on improving drought resistance of transgenic rice under field conditions. Mol Plant 2:73–83

Xiong L, Ishitani M, Lee H, Zhu JK (2001) The *Arabidopsis LOS5/ABA3* locus encodes a molybdenum cofactor sulfurase and modulates cold and osmotic stress responsive gene expression. Plant Cell 13:2063–2083

Xiong L, Lee H, Ishitani M, Zhu JK (2002) Regulation of osmotic stress-responsive gene expression by the *LOS6/ABA1* locus in *Arabidopsis*. J Biol Chem 277:8588–8596

Xu J, Audenaert K, Hofte M, De Vleesschauwer D (2013) Blight pathogen *Xanthomonas oryzae pv oryzae* by suppressing salicylic acid-mediated defenses. PLoS One 8:e67413

Xue T, Wang D, Zhang S, Ehlting J, Ni F, Jakab S, Zheng C, Zhong Y (2008) Genome-wide and expression analysis of protein phosphatase 2C in rice and *Arabidopsis*. BMC Genomics 9:550

Yamaguchi-Shinozaki K, Shinozaki K (2006) Transcriptional regulatory networks in cellular responses and tolerance to dehydration and cold stresses. Annu Rev Plant Biol 57:781–803

Yin P, Fan H, Hao Q, Yuan X, Wu D, Pang Y, Yan C, Li W, Wang J, Yan N (2009) Structural insights into the mechanism of abscisic acid signaling by PYL proteins. Nat Struct Mol Biol 16:1230–1236

Ying S, Zhang DF, Li HY, Liu YH, Shi YS, Song YC, Wang TY, Li Y (2011) Cloning and characterization of a maize SnRK2 protein kinase gene confers enhanced salt tolerance in transgenic *Arabidopsis*. Plant Cell Rep 30:1683–1699

Yoshida R, Hobo T, Ichimura K, Mizoguchi T, Takahashi F, Aronso J, Ecker JR, Shinozaki K (2002) ABA-activated SnRK2 protein kinase is required for dehydration stress signaling in *Arabidopsis*. Plant Cell Physiol 43:1473–1483

Yoshida T, Nishimura N, Kitahata N, Kuromori T, Ito T, Asami T, Shinozaki K, Hirayama T (2006) ABA-hypersensitive germination3 encodes a protein phosphatase 2C (AtPP2CA) that strongly regulates abscisic acid signaling during germination among *Arabidopsis* protein phosphatase 2Cs. Plant Physiol 140:115–126

Yu F, Qian L, Nibau C, Duan Q, Kita D, Levasseur K, Li X, Lu C, Li H, Hou C, Li L, Buchanan BB, Chen L, Cheung AY, Li D, Luan S (2012) FERONIA receptor kinase pathway suppresses abscisic acid signaling in *Arabidopsis* by activating ABI2 phosphatase. Proc Natl Acad Sci U S A 109:14693–14698

Yu L, Chen X, Wang Z, Wang S, Wang Y, Zhu Q, Li S, Xiang C (2013) Arabidopsis enhanced drought tolerance1/HOMEODOMAIN GLABROUS11 confers drought tolerance in transgenic rice without yield penalty. Plant Physiol 162:1378–1391

Yuan X, Yin P, Hao Q, Yan C, Wang J, Yan N (2010) Single amino acid alteration between Valine and Isoleucine determines the distinct pyrabactin selectivity by PYL1 and PYL2. J Biol Chem 285:28953–28958

Zeevaart JAD (1980) Changes in the levels of abscisic acid and its metabolites in excised leaf blades of *Xanthium strumarium* during and after water stress. Plant Physiol 66:672–678

Zhang H, Mao X, Wang C, Jing R (2010) Overexpression of a common wheat gene *TaSNRK2.8* enhances tolerance to drought, salt and low temperature in *Arabidopsis*. PLoS One 5:e1604

Zhang F, Lu X, Lv Z, Zhang L, Zhu M, Jiang W, Wang G, Sun X, Kexuan T (2013) Overexpression of the Artemisia orthologue of ABA receptor, AaPYL9, enhances ABA sensitivity and improves Artemisinin Content in *Artemisia annua* L. PLoS One 8:e56697

Zhao Y, Chan Z, Xing L, Liu X, Hou YJ, Chinnusamy V, Wang P, Duan C, Zhu JK (2013) The unique mode of action of a divergent member of the ABA-receptor protein family in ABA and stress signaling. Cell Res 23:1380–1395

Zhu JK (2002) Salt and drought stress signal transduction in plants. Annu Rev Plant Biol 53: 247–273

Zipfel C, Robatzek S, Navarro L, Oakeley EJ, Jones JD, Felix G, Boller T (2004) Bacterial disease resistance in *Arabidopsis* through flagellin perception. Nature 428:764–767

Chapter 11
Emerging Roles of Auxin in Abiotic Stress Responses

Eshan Sharma, Raghvendra Sharma, Pratikshya Borah, Mukesh Jain, and Jitendra P. Khurana

Abstract Auxin is among the key growth regulators that play diverse roles in virtually all aspects of plant growth and development. Intensive investigations during the past two decades have helped in elucidation of auxin perception and signal transduction mechanisms operative in plants. In addition to its primary role in regulating plant development, several studies in recent years have provided unflinching evidence for the involvement of auxin in abiotic stress responses. Functional genomics studies and genome-wide expression analysis have revealed altered expression of auxin-responsive genes, such as *Aux/IAA*, *GH3*, *SAURs*, and *ARFs*, under abiotic stress conditions. Variations in endogenous levels of auxin at global and local levels under various abiotic stress conditions have been associated with phenotypic changes and provided intriguing evidences regarding its role in response to environmental changes. Modulation of reactive oxygen species (ROS) levels in response to exogenous auxin as well as to drought, salinity, and ABA have indicated towards a complex relationship network between auxin, ROS, and abiotic stresses in plants. The advent of recent functional genomics technologies has led to identification of several candidate genes that may modulate crosstalk between auxin and abiotic stresses. This chapter discusses auxin homeostasis, signal transduction mechanisms, and how these processes are modulated under abiotic stresses, thus emphasizing on the emerging roles of auxin as a key integrator of abiotic stress pathways and plant development.

Keywords Abiotic stress • Auxin • Transcription factor • Regulatory network • Signal transduction • ROS

E. Sharma • P. Borah • J.P. Khurana (✉)
Interdisciplinary Centre for Plant Genomics and Department of Plant Molecular Biology, University of Delhi South Campus, New Delhi 110 021, Delhi, India
e-mail: khuranaj@genomeindia.org

R. Sharma • M. Jain
Functional and Applied Genomics Laboratory, National Institute of Plant Genome Research, Aruna Asaf Ali Marg, New Delhi 110 067, Delhi, India

G.K. Pandey (ed.), *Elucidation of Abiotic Stress Signaling in Plants*,
DOI 10.1007/978-1-4939-2211-6_11

11.1 Introduction

Auxin is an essential regulator of plant growth and development and plays a critical role in regulating cell elongation, cell division, differentiation, root initiation, apical dominance, and tropic responses, besides a multitude of other developmental events. The role of auxin in generation and maintenance of primary meristems and formation of axillary meristems is indeed quite critical. During embryogenesis too, auxin regulates organ polarity and specification of root meristem initials. Although auxin was the first plant hormone discovered, it is only in the past nearly 20 years that complexity associated with auxin-regulated plant development is beginning to be realized. It has largely been accomplished through the analysis of mutants of *Arabidopsis* altered in auxin biosynthesis, transport, or sensitivity. Significant progress has thus been made in elucidation of the role of auxin metabolism, tissue distribution, and signal transduction cascades in regulating plant growth and development (Chapman and Estelle 2009; Zhao 2010). Perhaps the most dramatic development has been the discovery of the F-box protein TIR1 as auxin receptor and how its interaction with auxin releases the repression of auxin-induced gene expression by degradation of the Aux/IAA proteins through 26S proteasome (Chapman and Estelle 2009). In fact, a few other F-box proteins have been identified (named AFBs), which also act as auxin receptors and may account for diversity of functions auxin performs during different stages of plant life cycle.

In recent years, it is becoming evident that auxin also plays a role as an integral part of plant responses towards unfavorable abiotic and biotic stress conditions. Genes involved in various auxin-related pathways invariably express differentially under different environmental stresses (Cheong et al. 2002; Jain and Khurana 2009; Song et al. 2009; Ha et al. 2013). In fact, elucidation of the functional roles of auxin-related genes is providing useful clues in establishing the important role auxin plays under osmotic stress imposed by salinity, drought, and low temperature conditions. Loss of membrane stability, protein denaturation accompanied with oxidative damage due to excessive reactive oxygen species (ROS) can disrupt cellular structures and impair physiological functions (Hirayama and Shinozaki 2010). Here, auxin appears to integrate stress signals from different plant hormones and modulate redox pathways to regulate root development as a stress-induced morphological response. Although emphasis in this article is on the role of auxin in abiotic stress responses, but it will be essential to first understand how auxin levels are regulated and what precisely is current understanding about its mechanism of action at the molecular level.

11.2 Auxin Homeostasis

The pool of active auxin in plant is coordinately regulated by its biosynthesis, degradation, conjugation, and directional cell-to-cell transport. Owing to its contribution in almost all aspects of plant development, auxin appears to be a master regulator that coordinates crosstalk and regulates plant development at multiple levels. Auxin is also unique among other plant growth regulators that influence development through pattern of its production and spatio-temporal distribution. Thus, the mechanism of auxin action is determined by cell-type-dependent response and concentration of the hormone. Indole-3-acetic acid (IAA) is the major naturally occurring auxin in plants along with indole-3-butyric acid (IBA), 4-chloroindole-3-acetic acid (4-Cl-IAA) and phenylacetic acid (PAA), which have been detected both as free acids and in conjugated forms (Ludwig-Müller 2011). Although the details regarding signalling components of auxin are now better understood, its biosynthesis has remained obscure for years and centered on tryptophan-dependent IAA biosynthesis. Traditionally, auxin is known to synthesize in young aerial leaves and meristems and then transported to other parts of the plant. However, local auxin biosynthesis has been demonstrated in other tissues, such as the meristematic region of the primary root or the tips of emerging lateral roots (Ljung et al. 2005). The maintenance of IAA gradient thus occurs through a combination of IAA transport and localized IAA synthesis.

IAA biosynthesis is much more complex and involves multiple pathways. Broadly, IAA biosynthesis occurs through tryptophan-dependent and tryptophan-independent pathways. Although tryptophan-dependent pathway is the major route for auxin biosynthesis in plants, independent pathway involving indole-3-glycerol phosphate has also been shown operational in *Arabidopsis* and maize (Wright et al. 1991; Normanly et al. 1993). In fact, the clinching evidence for the existence of tryptophan-independent pathway came from the observation that the orange pericarp mutant of maize defective in tryptophan biosynthesis accumulated 50-fold higher IAA than the normal seedlings (Wright et al. 1991). However, the corresponding genes involved in tryptophan-independent pathway have not been identified in monocots or dicots. In addition, indole-3-butyric acid (IBA) has also been suggested to function as an endogenous IAA precursor, being converted to IAA in peroxisomes by β-oxidation (Strader and Bartel 2011).

Tryptophan-dependent pathways can be categorized into indole-3-acetaldoxime (IAOx), indole-3-acetamide (IAM), tryptamine (TAM), and indol-3-pyruvic acid (IPA) pathways based on their major intermediates. Plant pathogens like *Agrobacterium* use IAM as an intermediate to synthesize IAA. The two-step pathway consists of tryptophan-2-monoxygenase (iaaM) that converts tryptophan to IAM which is subsequently hydrolyzed by iaaH into IAA (Camilleri and Jouanin 1991). IAM is present in many plant species, including *Arabidopsis*, maize, rice, and tobacco (Sugawara et al. 2009; Novák et al. 2012). IAM hydrolases (AtAMI1, NtAMI1), which can convert IAM to IAA, have been isolated in *Arabidopsis* and tobacco. Evidences have shown that IAM may also act as an intermediate in IAA biosynthesis from indole-3-acetaldoxime (IAOx) (Sugawara et al. 2009). Cytochrome P450 monoxygenases, CYP79B2 and CYP79B3, catalyze the conversion of tryptophan into IAOx which has been proposed to hydrolyze into IAA (Hull et al. 2000). The conversion of IAOx to IAA is poorly understood. Evidences have shown that IAOx is the precursor of indole-3-acetonitrile (IAN) and indole-3-acetaldehyde, which can then be used to generate IAA by nitrilases and aldehyde oxidases (Zhao 2010). Genes involved in IAOx conversion into IAA are yet to be identified. Analysis of auxin overproduction mutants in *Arabidopsis* led to the identification of key genes involved in Indole-3-pyruvate (IPA) pathway for IAA biosynthesis. Indole-3-pyruvate (IPA) pathway is the main pathway for auxin biosynthesis in both monocots and dicots (McSteen 2010). Characterization of mutants that are defective in shade avoidance and ethylene responses led to identification of *WEAK ETHYLENE INSENSITIVE8 (WEI8)/TRYPTOPHAN AMINOTRANSFERASE OF ARABIDOPSIS1 (TAA1)* gene important for the production of indole-3-pyruvate (IPA) (Stepanova et al. 2008; Tao et al. 2008). The *taa1* mutants including *sav3/taa1*, *wei8, and tir2* show a decreased IAA synthesis and reduced expression of the auxin-responsive genes, phenotypes that could be rescued by synthetic auxin picloram or IAA (Tao et al. 2008; Stepanova et al. 2008). *YUCCA (YUC)* gene family encodes flavin monooxygenase-like proteins, and genetic studies suggested its function downstream of TAA1, which catalyzes conversion of IPA into IAA (Dai et al. 2013). *YUCCA* genes were initially considered to be associated with tryptamine (TAM) pathway of IAA synthesis where YUCCA catalyzes the rate-limiting step in conversion of TAM to *N*-hydroxyl tryptamine (HTAM) (Zhao 2010). HTAM can further proceed to IAA using IAOx pathway or other intermediates. These multiple pathways appear to be differentially regulated in response to changing environmental conditions and their integration seems to be a key for regulating developmental pathways. In this regard, compartmentalization of different biosynthetic enzymes and subcellular localization of IAA pools can be speculated to be important for auxin metabolism.

Much of the auxin found in the plant is stored as conjugates to amino acids, sugars, peptides, or proteins (Normanly 2010). Conjugated auxin can be stored easily or can be transported over long distances also. The auxin-inducible *GRETCHEN HAGEN3 (GH3)* gene family coding for IAA-amido synthases facilitates ATP-dependent conjugation of IAA to amino acids (Hagen and Guilfoyle 1985; Park et al. 2007). IAA–alanine (IAA–Ala), IAA–leucine (IAA–Leu), IAA–aspartate (IAA–Asp), and IAA–glutamate (IAA–Glu) are the most abundant amino acid auxin conjugates in *Arabidopsis*. The orthologues of *AtGH3* genes, which code for enzymes involved in conjugating IAA to amino acids, have also been identified in rice (Jain et al. 2006b). Auxin conjugation with sugar molecules is less understood. IAA can also be conjugated and stored in seeds and used for early seedling growth. Free IAA produced at the top of maize coleoptile is hydrolyzed from IAA–inositol conjugates in seeds (Ludwig-Müller 2011). Apart from temporary reversible conjugation of IAA, oxidative degradation of auxin appears to be another pathway to attenuate excess auxin levels. The metabolites, 2-oxoindole-3-acetic acid (oxIAA) and oxIAA-glucose (oxIAA-Glc), are the major degradation products of IAA (Novák et al. 2012). The genes involved in the IAA catabolism are yet to be identified, and the molecular mechanisms underlying IAA degradation are poorly understood.

The spatio-temporal distribution of auxin is critical for plant development. The polar, cell-to-cell transport of auxin provides further complexity to auxin homeostasis culminating in different auxin gradients across the plant. A number of auxin transporters have been identified in *Arabidopsis*, maize and rice (McSteen 2010).

These include auxin influx carriers AUXIN-RESISTANT 1/LIKE AUX1 (AUX1/LAX), PIN-FORMED (PIN) auxin efflux carriers, and ABC transporters of MULTIDRUG RESISTANCE (MDR) subfamily. ABCB transporters facilitate non-directional auxin efflux, whereas AUX1/LAX catalyze its influx at the expense of H^+ influx. In *Arabidopsis*, PIN family consists of eight members grouped into canonical long "PIN1-type" proteins (PIN1, PIN2, PIN3, PIN4, PIN7) and short PIN proteins (PIN5, PIN6, PIN8). Long PIN proteins are all localized to the plasma membrane, short PIN proteins localize to the endomembranes, e.g., PIN5 is localized to the ER. PIN auxin efflux carriers also cycle continuously between endosomal compartments and the plasma membrane. Further, three homologs of the PIN1 auxin efflux carrier have been reported in maize and rice (McSteen 2010). The transcription of *PIN* and *ABCB1* genes encoding auxin carriers has also been shown to be under direct control of auxin (Rosquete et al. 2012). Thus, the feedback mechanisms to control its polar transport are also essential for developmentally controlled growth regulation by auxin. Finally, it still remains obscure, how multiple auxin metabolic pathways, carrier dependent auxin transport, and signalling pathways integrate to maintain a steady-state of auxin levels that can finetune the growth and developmental responses.

11.3 Auxin Signal Transduction

During last three decades, significant progress has been made in understanding how auxin regulates transcriptional responses. The application of exogenous auxin upregulates the transcript levels of several genes in plants within 10–30 min of exposure (Hagen and Guilfoyle 1985). Some of these genes are referred as early auxin-responsive genes and are categorized in three classes, namely auxin/indole-3-acetic acid (*Aux/IAA*), *GH3*, and small auxin-up RNAs (*SAURs*) (Guilfoyle 1999; Hagen and Guilfoyle 2002; Jain et al. 2006a, b, c). A rather large complement of auxin-responsive genes has been identified from different plant species, including pea, soybean, *Arabidopsis*, cucumber, mung bean, and rice (Theologis et al. 1985; Hagen and Guilfoyle 2002; Jain et al. 2006a, b, c; Jain and Khurana 2009). In the promoter region of auxin-responsive genes, auxin-responsive elements (AuxREs) have been identified (Guilfoyle 1999; Hagen and Guilfoyle 2002; Jain et al. 2006a). The DNA-binding domain of auxin response factors (ARFs) bind to the auxin response elements (AuxREs) present in the promoter of auxin-responsive genes and regulate their expression (Kim et al. 1997; Ulmasov et al. 1997; Tiwari et al. 2003).

At low auxin levels, Aux/IAA class of transcriptional repressors along with co-repressor proteins, such as TOPLESS, inhibit the genes targeted by ARFs (Tiwari et al. 2001; Long et al. 2006). ARFs directly bind DNA and may activate or repress transcription of genes depending on the type of ARF. When auxin levels rise, the Aux/IAA proteins are degraded by the 26S proteasome, resulting in de-repression of ARFs and activation of transcriptional responses (Weijers et al. 2005; Chapman and Estelle 2009). Aux/IAA proteins are short-lived nuclear proteins with half-lives ranging from ~10 min for *Arabidopsis* AXR2/IAA7 and AXR3/IAA17 to more than 60 min for IAA28 which is reduced twofold or more on application of auxin (Zenser et al. 2001; Thakur et al. 2005; Dreher et al. 2006). *Aux/IAA* gene family comprises of 29 members in *Arabidopsis* and 31 members in rice (Jain et al. 2006a). Most *Aux/IAA* genes are themselves highly auxin-inducible and encode proteins with four characteristic domains. Domain I binds TOPLESS and is required for transcriptional repression. Domain II contains the degron motif, a sequence of 13 amino acids that is required for the characteristic instability of Aux/IAA proteins. Domains III and IV of Aux/IAA proteins mediate homo- and heterodimerization, including interactions with ARF proteins (Overvoorde et al. 2005).

For a biological response, auxin must be perceived by the plant and converted into a signal. To date, three auxin-binding/receptor proteins, ABP1, TIR1/AFB, and SKP2A, have been identified. TRANSPORT INHIBITOR RESISTANT 1 (TIR1) was the first protein to be recognized as an auxin receptor. Auxin-dependent degradation of the Aux/IAAs occurs through the action of a SKP1–Cullin–F-box (SCF)-type E3 ligase called $SCF^{TIR1/AFB1-5}$. The F-box protein TIR1 and the related proteins AFB1, AFB2, AFB3, AFB4, and AFB5 provide substrate specificity to the SCF. In *Arabidopsis*, structural analysis has shown that auxin interacts with TIR1–IAA7 complex at the pocket formed by interaction of TIR1 with Aux/IAA protein. The pocket occupied by auxin is internal within structure of TIR1–IAA7 and is formed by faces of both proteins (Chapman and Estelle 2009). Since Aux/IAA form part of the receptor–ligand complex, they can be considered to function as co-receptors. Evidences indicate that all six members of the TIR1/AFB family in *Arabidopsis* can function as auxin receptors; however, individual TIR1/AFB proteins have distinct biochemical properties and biological functions. TIR1 and AFB2 have a much stronger interaction with Aux/IAA proteins than AFB1 and AFB3 (Parry et al. 2009). AFB4 has been reported as a negative regulator of auxin responses in seedlings (Greenham et al. 2011). In comparison to 23 ARFs in *Arabidopsis*, there are 25 OsARF genes in rice (McSteen 2010). Among the characterized ones in *Arabidopsis*, some are transcriptional activators (ARF5–ARF8 and ARF19), whereas others are transcriptional repressors (ARF2, ARF4, and ARF9). Also, no *TIR1*-like genes have been functionally characterized in maize and rice. Finally, as multi-gene families encode for these proteins, a number of combinatorial interactions are possible especially between Aux/IAA and ARFs. Also, there are strong evidences of occurrence of independent auxin perception pathways than TIR/AFB-mediated responses.

Another F-box protein, SKP2a (S-PHASE KINASE-ASSOCIATED PROTEIN 2a) has been shown to bind auxin the same way as TIR1/AFB and degrade transcription factors that repress cell division (Jurado et al. 2010). *Arabidopsis* SKP2a is a positive regulator of the cell cycle, which bears homology to mammalian SCF^{SKP2} E3 ubiquitin ligase, an important player for the degradation of cell-cycle factors. Similar to TIR1, high-auxin levels promote the interaction between SKP2a and cell-cycle factors, DPB or E2FC, but SKP2a is itself degraded in ubiquitin-dependent manner both in vivo and in vitro. Thus, the hormone appears to act as a signal to trigger SKP2a proteolysis. Mutated versions of the putative auxin-binding site in SKP2a are no longer able to bind auxin. Although SKP2a appears to fulfill basic requirements as an auxin receptor, its biological relevance as a link between auxin and cell-cycle control still remains equivocal.

Independent of TIR1 family, AUXIN-BINDING PROTEIN1 (ABP1) has been speculated to be an auxin receptor since long time. ABP1 was first discovered in maize as an auxin-binding protein in the 1970s; however, its acceptance as an auxin receptor is a much recent event. Biochemical analysis proved that ABP1 can specifically bind auxin (Jones and Venis 1989). ABP1, despite carrying a KDEL-endoplasmic reticulum (ER) retention motif, is secreted to some extent to the extracellular space, where it is active (Jones and Herman 1993). It is thus likely that ABP1 is transferred to the plasma membrane/apoplast interface and regulates number of auxin responses at plasma membrane or cytoplasm not involving gene transcription (see Sauer and Kleine-Vehn 2011 for detailed review). Further, ABP1 has been shown to negatively regulate $SCF^{TIR1/AFB1-5}$, indicating that it may regulate gene transcription by coordinating with TIR/AFB pathway. Recent evidences have indicated that the ABP1 transmits the auxin signal through ROP-GTPase (guanidine triphosphate hydrolases of plants (Rho)-related GTPases of plants) and their associating RICs (ROP Interactive CRIB motif-containing proteins) (Xu et al. 2010; Sauer and Kleine-Vehn 2011). When exposed to auxin, ABP1 can rapidly activate ROPs to inhibit ROP–RIC-mediated regulation of PIN endocytosis. ABP1 can also mediate auxin inhibition of clathrin-dependent endocytosis in *Arabidopsis*. ABP1 has also been shown to physically interact with a transmembrane kinase, TMK1, member of receptor-like kinases (Xu et al. 2014). Auxin binding to ABP1 thus appears to regulate membrane hyperpolarization, cell expansion, inhibition of endocytosis, and activation of ROPs. After remaining enigma for more than 40 years, ABP1 needs further investigations as a key player in auxin signalling. However, several questions regarding its mechanism of action, both independent of TIR1 and together with TIR1, still remain to be answered.

Lastly, auxin signal attenuation is key to regulate such highly complex signalling network requiring precise cellular responses. Conjugation of auxin with amino acids or its compartmentalization in subcellular compartments or apoplast appears to be important for bringing down concentration according to environmental signals. Auxin accumulation also coordinates with ROS and in this regard oxidation of auxin also appears important towards elimination of differential auxin levels at local tissue levels (Peer et al. 2013). Thus, exceedingly high auxin concentrations detected by signalling mechanisms could lead to its oxidation under different environmental conditions.

11.4 Changes in Endogenous Level of Auxin Under Various Abiotic Stresses

The endogenous level of various phytohormones, including auxin, changes in response to abiotic stresses. The maintenance of endogenous pool of auxin at an appropriate level is important for plants to coordinate various cellular processes. Auxin exists in free and conjugated (to a variety of amino acids, peptides, and carbohydrates) forms in plants (Seidel et al. 2006). Biosynthesis, degradation, and conjugation are the processes, which regulate auxin homeostasis in plants (Normanly 2010). For IAA biosynthesis, tryptophan-dependent and tryptophan-independent routes have been proposed based on genetic and biochemical studies (Wright et al. 1991; Bartel et al. 2001). The biosynthesis and distribution of auxin in conjugated form to different plant organs are the most important mechanisms for maintenance of endogenous auxin pool (Hagen and Guilfoyle 2002; Woodward and Bartel 2005). IAA conjugates help in the transportation of IAA and serves as the temporary stock of inactive IAA that can be hydrolyzed to provide the plant with active hormone. These conjugates also protect IAA from enzymatic destruction and control homeostatic concentration of IAA in plants.

Changes in the level of IAA and/or its redistribution in response to environmental signals regulate plant growth and development (Figs. 11.1 and 11.2). Local auxin concentration and auxin distribution may be regulated by changes in auxin transport under abiotic stresses (Shibasaki et al. 2009). In rice, investigations have shown that after 3 days of drought stress, IAA level was reduced to 72 %. On the contrary, after 3 days of cold stress, IAA level was estimated to be 1.6-fold higher and after 6 h of heat stress, IAA level was 1.3-fold higher as compared to control. Overall, these observations suggest that abiotic stresses modulate endogenous levels of IAA (Du et al. 2013a). The expression of a *GH3* family gene, *OsGH3-2* (catalyzing IAA conjugation

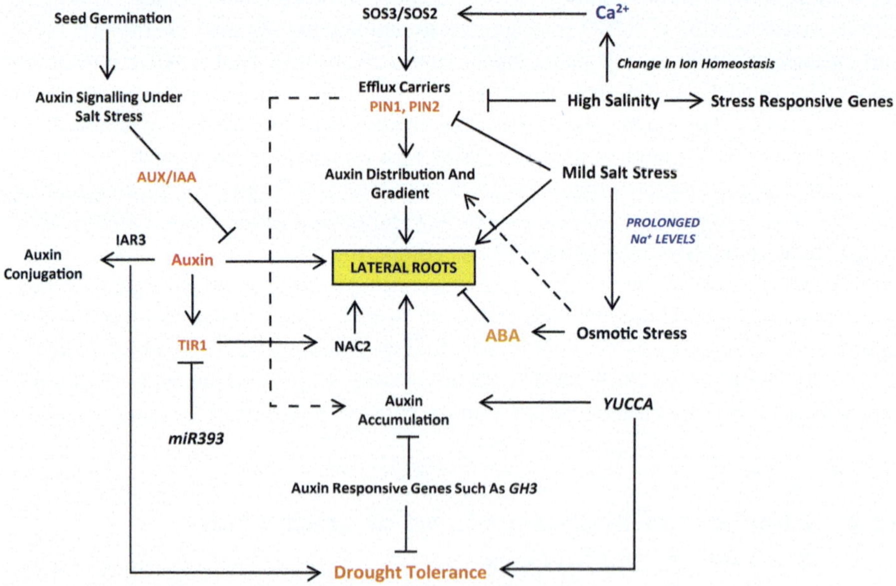

Fig. 11.1 Schematic diagram depicting crosstalk between auxin, salt and drought stress signalling pathways. Salt stress negatively regulates auxin accumulation, distribution and thereby adversely affects lateral roots production. However, mild salt stress has positive effect on lateral roots. Salt-induced Ca^{2+} fluxes can regulate auxin carriers through SOS pathway and also induce expression of stress-responsive genes. Drought also adversely affects lateral roots through ABA production. Auxin conjugation, on other hand, has been found to have positive effect on drought tolerance

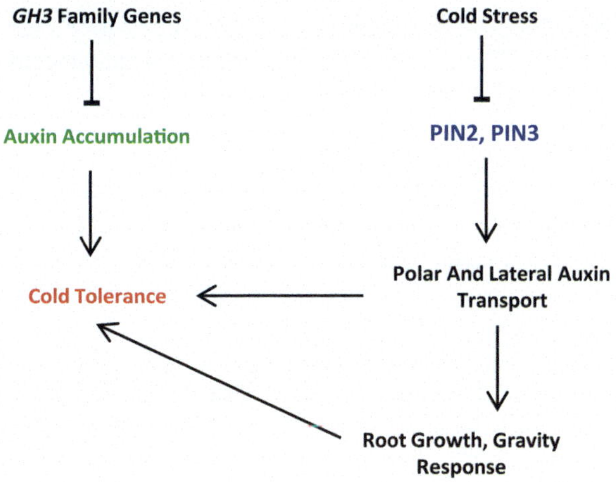

Fig. 11.2 Schematic diagram depicting crosstalk between auxin and cold stress. Cold stress inhibits action of auxin carriers such as PINs and thereby it negatively regulates auxin transport. However, in contrast, decreased levels of free auxin due to conjugation have been shown to have a positive effect on cold tolerance. Auxin conjugation and transport may regulate cold tolerance differentially through different stress pathways

to amino acids), was induced under drought stress, but was suppressed during cold stress (Du et al. 2012). The overexpression of *OsGH3-2* in rice showed morphological aberrations related to IAA deficiency and reduced carotene, ABA, and free IAA levels, which led to cold tolerance. This effect was due to modulation of endogenous free IAA levels and ABA homeostasis (Du et al. 2012). Carotenoid-deficient mutant lines with lower auxin level exhibited tolerance to cold stress (Du et al. 2013b). Yuan et al. (2013) also documented the decline in endogenous auxin level under cold and salinity stresses in leaves and roots of *Malus sieversii.* Overexpression of *Arabidopsis YUCCA6* gene, encoding flavin monooxygenase of the tryptophan-dependent auxin biosynthetic pathway, showed high auxin phenotypes, such as narrow downward curled leaves, increased height, erect stature, and longevity (Kim et al. 2013). Transgenic potato overexpressing *YUCCA6* displayed elevated endogenous auxin level and enhanced drought stress tolerance (Kim et al. 2013). In contrast, the activation of *TLD1* gene, *GH3.13*, in rice, reduced the IAA maxima at the lamina joint, shoot base, and nodes, resulting in altered plant architecture and tissue patterning, and improved drought tolerance (Zhang et al. 2009). Similarly, activation of another gene encoding flavin monooxygenase, *YUCCA7*, resulted in elevated level of total auxin and displayed plant phenotypes similar to those observed in auxin overproducing mutants (Lee et al. 2012). During Fe stress, increased level of IAA was observed in isolated protoplast of *Alfa alfa* (Pasternak et al. 2002). These observations suggest a correlation between endogenous auxin levels and abiotic stress responses in plants.

11.5 Differential Expression of Auxin-Responsive Genes Under Abiotic Stress Conditions

Several studies based on genome-wide transcriptome analyses in plants revealed that the expression of several auxin-related genes is altered during abiotic stress response (Cheong et al. 2002; Jain and Khurana 2009; Song et al. 2009; Yuan et al. 2013). The members of *Aux/IAA* and *ARF* class showed altered expression during cold acclimation in *Arabidopsis* (Hannah et al. 2005). Expression analysis of 31 rice *OsIAA* genes revealed that at least 15 of them are induced under drought stress and eight *OsIAA* genes are induced during salinity stress (Song et al. 2009). In a previous study, Cheong et al. (2002) showed differential expression of auxin-related genes, including *IAA2*, *IAA3*, *GH3*-like protein, and SAUR-AC1, after wounding. This study suggested crosstalk between wounding and auxin signalling pathway. Interestingly, all these genes were positively regulated by auxin and negatively regulated by wounding. Two IAA glucosyl transferases encoding genes were highly induced after wounding, which suggested that due to conjugation of IAA with sugars, the endogenous free auxin level was reduced. It was also reported that *NPK1-like* gene, having a negative role in auxin signalling was upregulated during wounding (Kovtun et al. 1998). During oxidative stress, the expression of components of polar auxin transport, PIN1 and PIN3, was found to be downregulated (Pasternak et al. 2005). In another study, Huang et al. (2008) documented the downregulation of auxin-responsive genes, including *IAA1*, *IAA4*, At4g34760, At4g38860, and At1g29430, by drought stress as well as ABA and methyl jasmonate treatment.

A genome-wide analysis of glutathione-*S*-transferase (GST) gene family in rice revealed that at least 31 *GST* genes respond to plant hormones, auxin, and cytokinin, and many of them exhibited differential expression under abiotic stress conditions too (Jain et al. 2010). The overexpression of an auxin-responsive GST gene, *OsGSTU4*, enhanced tolerance to salinity and oxidative stresses in transgenic *Arabidopsis* plants (Sharma et al. 2014). In another genome-wide analysis of glutaredoxin (GRX) gene family, among the 48 members, at least 11 *GRX* genes exhibited differential expression pattern under different hormone treatments. Among them, *OsGRX8*, a member of plant-specific CC-type group of GRX proteins, was differentially expressed during abiotic stress conditions and phytohormone auxin (Garg et al. 2010). The overexpression of *OsGRX8* in *Arabidopsis* enhanced tolerance to multiple abiotic stresses and several auxin-responsive genes were found to be upregulated in the transgenics (Sharma et al. 2013). Based on the microarray and quantitative PCR analysis, differential expression of IAA biosynthesis and metabolism-related genes in different rice varieties have also been reported (Du et al. 2013a). IAA biosynthesis-related genes were found to be upregulated by cold and heat stress, but suppressed by drought stress. The expression of IAA signalling and polar transport-related genes was also affected by abiotic stresses. The changes observed in the transcript level of genes involved in IAA biosynthesis and signalling under these stress conditions were in good agreement with the endogenous levels of IAA in plants (Du et al. 2013a). The overexpression of Cys2/His2 zinc-finger proteins, AZF1 and AZF2, showed severe growth retardation and reduced viability of plants. The transcriptome analysis of transgenic plants revealed that various genes including many auxin-responsive genes were downregulated during osmotic stress and ABA treatments (Kodaira et al. 2011).

A genome level microarray analysis in rice identified 315 probe sets representing 298 (225 upregulated and 73 downregulated) unique genes as auxin-responsive

(Jain and Khurana 2009). Among these, at least 154 (auxin-induced) and 50 (auxin-suppressed) probe sets exhibited altered expression under one or more of the stress conditions (desiccation, salt and cold) also. Similarly, 41 members of auxin-related gene families (*GH3*, *Aux/IAA*, *SAUR*, and *ARF*) were identified as differentially expressed under at least one abiotic stress condition. Based on these results, it was suggested that auxin could also act like a stress hormone, which may directly or indirectly modulate the expression of several stress-responsive genes (Jain and Khurana 2009). Further, several auxin-responsive genes were commonly regulated during plant developmental processes and abiotic stress conditions, suggesting their roles as mediator of plant growth response to stress conditions during development (Jain and Khurana 2009).

The differential expression of *Aux/IAA* genes was observed in soybean, pea, *Arabidopsis* and rice in various tissues and in response to exogenous auxin (Theologis et al. 1985; Ainley et al. 1988; Thakur et al. 2001; Jain et al. 2006a). In expression analysis of auxin-responsive genes, including *Aux/IAA*, *ARF*, *GH3*, *SAURs*, and lateral organ boundaries (*LBD*) genes, differential expression was observed in leaf/root of sorghum when treated with IAA and brassinosteroid (BR), and exposed to salt and drought stress (Wang et al. 2010). Altogether, these reports suggest that auxin-responsive genes are differentially expressed under various abiotic stress conditions, further indicating/corroborating crosstalk between auxin and abiotic stress signalling.

11.6 Auxin and ROS Homeostasis

Environmental constraints affect both auxin and ROS. Elevated level of ROS is a common indicator of stress conditions in plants, which cause oxidative damage at both cellular and molecular levels. Because the overproduction of ROS leads to oxidative damage, it needs to be maintained at basal level. Antioxidant scavenging systems detoxify the toxicity mediated by ROS and help to maintain the ROS homeostasis (Vranova et al. 2002; Gill and Tuteja 2010). Plant growth and development is affected by the interaction between ROS and auxin, under various environmental stresses (Fig. 11.3).

Auxin modulates ROS homeostasis indirectly by affecting the stability of DELLA proteins or directly by inducing ROS detoxification enzymes (Fu and Harberd 2003; Paponov et al. 2008). The role of ROS in auxin biosynthesis, transport, and metabolism and signalling has been reported earlier (Tognetti et al. 2012). Both ROS and auxin networks are important for abiotic stress adaptation (De Tullio et al. 2010; Tognetti et al. 2010). The integration of ROS and auxin signalling pathways can unravel mechanisms that allow plants to adapt and survive in adverse environmental stress conditions.

Auxin is involved in the regulation of various processes like cell viability, cell-cycle progression, and programmed cell death, which in fact depend on ROS signalling (Xia et al. 2005). Earlier reports showed the production of ROS molecules, like

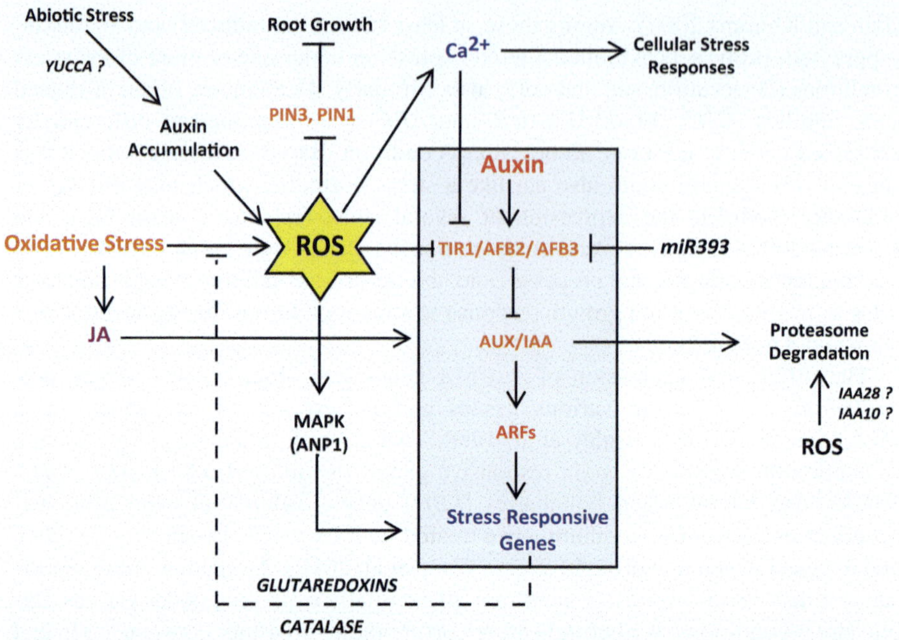

Fig. 11.3 Schematic diagram showing crosstalk between auxin and oxidative stress. ROS produced as a result of different abiotic stresses leads to a common oxidative stress. ROS production has been shown to negatively regulate auxin transport and signalling. However, due to cellular antioxidant mechanisms, stress-responsive genes are induced which can also be regulated by ARFs. The ARFs bind to auxin-responsive elements in upstream promoter regions of auxin-responsive genes. Few glutaredoxins are induced by exogenous auxin. ROS has also been proposed to regulate degradation of Aux/IAA proteins, thereby promoting ARF action. Reciprocally, Auxin accumulation can have a positive effect on ROS accumulation

superoxide anion, hydroxyl radical, singlet oxygen, and hydrogen peroxide, in various cellular compartments, mainly in the chloroplast, endoplasmic reticulum, plasma membrane, cell wall, and apoplast (Gill and Tuteja 2010). Among them, H_2O_2 is one of the main ROS compounds generated outside and within the cell in response to abiotic and biotic stresses and is also induced on application of exogenous auxin to plants (Peer et al. 2013). H_2O_2 is thought to be an essential component of stomatal closure in response to multiple stress-derived input signals, including ABA, light, CO_2, pathogen infection, and SA (Neill et al. 2002; Gill and Tuteja 2010). In a recent study, the application of exogenous auxin reduced the level of H_2O_2 in tomato roots by the altered expression and activity of H_2O_2 scavenging enzymes, such as catalase, Cu-Zn-superoxide dismutase (SOD), and peroxidase (Jaroslaw et al. 2009). In another study, the auxin signalling mutant, *tir1afb2*, was found to accumulate reduced levels of H_2O_2 and superoxide anion, and increased activity of catalase and ascorbic peroxidase (Iglesias et al. 2010). Both these studies suggested that the level of H_2O_2 in specific tissues is regulated by the altered expression of antioxidant enzymes and auxin homeostasis. The application of exogenous auxin was responsible for the production of H_2O_2 and irreversible accumulation of an inactive auxin, 2-oxindole-3-acetic acid (oxIAA) (Peer et al. 2013). Krishnamurthy and Rathinasabapathi (2013) demonstrated that the auxin mutant, *aux1*, is more susceptible towards arsenate stress in comparison to wild-type. The level of H_2O_2 was found to be more in wild-type plants during arsenate stress, but not in *aux1* mutant, indicating a positive role of *AUX1*, an auxin transporter in ROS generation. Likewise, Cheng et al. (2011) showed the involvement of *AtGRXS17* in auxin signalling and suggested that it maintains ROS homeostasis to mediate temperature-dependent postembryonic growth.

Application of 2, 4-D on pea (*Pisum sativum* L.) plants induced severe disturbance in mesophyll cell structure and proliferation of vascular tissue in young leaves and increased acyl-CoA oxidase (ACX), xanthine oxidase (XOD), and lipoxygenase (LOX) activities in young and/or adult leaves. The modulation in activities of these enzymes resulted in overaccumulation of ROS, increasing oxidative damage to proteins and membrane lipids (Pazmiño et al. 2011). Auxin-induced H_2O_2 production also acts as signal in the root gravitropism and stomatal opening response, suggesting that auxin can trigger ROS production (Joo et al. 2001; Song et al. 2006).

By oxidizing IAA via induction of peroxidase activities, oxidative stress can affect auxin stability (Kawano 2003) as observed in tobacco and duckweed species exposed to ultraviolet radiation (Jansen et al. 2001) or in *Medicago truncatula* under cadmium stress (Chaoui and El Ferjani 2005). Auxin has been found to be responsible for the induction of stress-related genes (Huang et al. 2008) in interaction with abiotic stress-induced anthocyanin production (Pasternak et al. 2002; Winkel-Shirley 2002) and regulation of ROS homeostasis (Pfeiffer and Höftberger 2001; Cheong et al. 2002; Guan and Scandalios 2002; Pasternak et al. 2002; Winkel-Shirley 2002; Joo et al. 2005). Data obtained from these studies suggest that there is signalling reciprocity between auxin and ROS pathways.

11.7 Auxin and Abiotic Stress Crosstalk

11.7.1 Auxin and Salinity Stress

Soil Salinity induced by accumulation of sodium chloride (NaCl) is a growing agricultural problem worldwide leading to loss of land and yield of crops (Munns and Tester 2008). Increased accumulation of NaCl in soil leads to reduced water uptake initially followed by impaired cellular functions due to Na^+ and Cl^- ions taken up by the roots. Thus, salinity stress can be divided into osmotic and ionic components, both of which are needed to be sensed by roots at the same time to maximize water uptake and reduce Na^+ concentration by its export or sequestration in vacuoles. The sensory processes thus need to distinguish salinity from stress imposed mainly by hyperosmotic conditions.

The root system architecture (RSA) is a key component for adaptive plant responses towards changing environmental conditions. Root branching facilitated through lateral root development enables plant to withstand unfavorable soil conditions such as salinity, moisture deficit, presence of heavy metals, etc. Branching not only affects the water and nutrient uptake from soil but also provides strong anchorage to the plant. Recent advances in our understanding of lateral root (LR) development, mainly in model plant *Arabidopsis*, have established auxin as a central component that integrates a number of developmental processes in roots (Fig. 11.1) (Lavenus et al. 2013). Young seedlings contain only primary roots formed from radicle present in the embryo. Auxin stimulates lateral root emergence by promoting lateral root primordia (LRP) initiation by local activation of root pericycle cells located opposite to xylem (Lavenus et al. 2013). The auxin control of lateral root development involves multiple IAAs and ARFs (Okushima et al. 2007). Blocking of polar auxin transport inhibits lateral root initiation and overproduction or exogenous application of auxin increases number of lateral roots (Malamy 2005). Changes in the expression levels of auxin efflux transporters, PIN proteins, also lead to changes in the number of lateral root initiation events (Benkova et al. 2003). An analogous response by roots is believed to occur in response to high salinity or osmotic stress leading to repression of LR formation in *Arabidopsis* and *Medicago* (Galvan-Ampudia and Testerink 2011). On the contrary, mild salt stress leads to reduction in LR elongation and an increase in the number of LRs (Zolla et al. 2010). This stress-induced morphogenic response seems to occur only by salt stress but not by osmotic stress. In addition, the increased LR numbers due to salt stress are reduced in auxin signalling mutant *axr1*, *axr4*, *and tir1*, and completely blocked in the auxin influx mutant, *aux1*. Salt stress promotes the formation of more lateral roots by progression of LRPs from preemergence to emergence stage (Zolla et al. 2010). Increasing salt concentration can affect auxin distribution in WT roots also (Wang et al. 2009). Low concentration of NaCl can repress expression of auxin efflux carrier, *PIN2*, and also reduce the abundance of PIN2:GFP in elongation zone of root tips, suggesting that salt stress modulates this key regulator of auxin transport both transcriptionally and posttranscriptionally (Zhao et al. 2011). Reduced PIN2 levels not only affect root growth but also alter root growth direction (Sun et al. 2008). Salt exposure can cause rapid degradation of amyloplasts in root columella cells thereby reducing gravity response, an auxin-mediated process; amyloplasts are considered to be involved in sensing gravity stimulus. Mutants for SOS (Salt overly sensitive) pathway show reduced degradation of amyloplasts (Sun et al. 2008). The SOS signalling pathway has been reported to export Na^+ out of the cell and is a key regulator of ionic homeostasis in salinity (Deinlein et al. 2014). The development of lateral roots by SOS3 overexpression under low salinity has been reported to act through auxin redistribution (Zhao et al. 2011). SOS3 mutant plants do not accumulate auxin in aerial tissues as elucidated by examining DR5:GUS activity in the mutant background. Decreased PIN1 activity in stressed *sos3* mutants also indicates reduced acropetal auxin transport from shoot to root. Therefore, in stressed plants, defects in LRP initiation lead to growth arrest before LR emergence. It appears possible that a spatio-temporal distribution and accumulation of auxin can modulate lateral root development by crosstalk between auxin and salt stress.

Recent studies in several plant species have indicated towards role of auxin in seed germination (Liu et al. 2013) and this has been supported by global gene expression studies. Crosstalk between auxin and salt stress also occurs at the level of seed germination and in young seedlings also. Although under normal conditions, auxin does not greatly influence germination process per se, it negatively regulates germination under high salinity. Germination of YUCCA3 overexpressing transgenic plants containing higher levels of auxin was found to be much more repressed than wild-type plants. Park and co-workers (2011) found a membrane-bound NAC domain containing transcription factor NTM2 that links the salt signalling cascade with auxin via *Aux/IAA30* gene. NTM2 and *IAA30* genes are induced by high salinity and the salt induction of *IAA30* was repressed in *ntm2* mutants. NTM2 can act as a transcriptional activator and bind directly to *IAA30* gene promoter (Park et al. 2011).

The NAC transcription factors constitute a large family of transcription factors with role in plant growth, floral development, and stress responses. Few NAC proteins, such as NAC1 and NAC2, promote LR formation in *Arabidopsis*. Expression of *AtNAC2* gene was induced by ethylene, auxin, and ABA, and suppressed in the auxin-receptor mutant *tir1* under salt stress. Synthetic auxin NAA can also induce *AtNAC2* expression (He et al. 2005). NAC1 is also known to act downstream of TIR1 (Xie et al. 2000). TIR1 along with AFB1–5 are known receptors for auxin perception and downstream signalling in response to auxin. TIR1 may play critical roles in the expression of NAC proteins acting downstream under salt stress conditions. However, evidence regarding functional roles of these auxin signalling components under abiotic stresses is lacking.

11.7.2 Auxin and Oxidative Stress

Under various biotic and abiotic stress conditions, such as drought, salinity, heat, cold, heavy metals, and pathogen attack, the levels of ROS can rise disproportionately leading to oxidative stress and damage of cellular structures. The ROS homeostasis in cell is thus maintained by an array of antioxidant metabolites and enzymes so that oxidative damage can be minimized under stress conditions. However, ROS molecules such as H_2O_2 can also function as signalling components that enable plants to respond appropriately and adjust their growth and development, such as seed germination, root elongation, auxin-mediated gravitropism, stomatal closure, and abiotic/biotic responses (Gao et al. 2014). The crosstalk of ROS and hormones has been exemplified mainly by the role of H_2O_2 and ABA in stomatal closure. However, evidence has also emerged for the role of auxin in stress-related hormonal networks where it can modulate ROS levels (Fig. 11.3) (Tognetti et al. 2012).

In genome-wide expression analysis under different abiotic stresses, invariably, genes involved in both biotic stress and auxin signal transduction pathways are differentially regulated (Jain and Khurana 2009; author's unpublished results). Oxidative stress pathways are commonly induced under both abiotic and biotic stresses. Thus, under both these environmental stresses, auxin responses and oxidative stress pathways may share common elements. Oxidative stress could induce broad spectrum of auxin-like effects in seedlings (including the inhibition of root elongation and inhibition of cotyledon and leaf expansion), and these effects were consistent with alterations in auxin levels and/or distribution (Pasternak et al. 2005). Many plant pathogens can produce auxin as a part of invasion strategy. Evidences indicate that auxin signalling is therefore downregulated under biotic stress (Kazan and Manners 2009). A conserved flagellin flg22 from *Pseudomonas syringae* is recognized by *Arabidopsis* leading to inhibition of auxin receptors TIR1, AFB2, and AFB3 through microRNA 393 activation. This inhibition of auxin receptors leads to stabilization of Aux/IAA repressors thereby repressing expression of auxin-responsive genes and consequently inhibition of auxin-induced responses, including lateral root development (Navarro et al. 2006). Functional analysis of genes encoding TIR1/AFB auxin receptors also showed that they are involved in tolerance to oxidative and salinity stress (Iglesias et al. 2010). *Arabidopsis tir1/afb* mutants displayed reduced H_2O_2 accumulation, enhanced antioxidant enzyme activity, enhanced ascorbic acid and chlorophyll content in plants (Iglesias et al. 2010). In tomato, auxin can regulate the level of H_2O_2 in the root tips, so increasing the auxin level triggers accumulation of H_2O_2 leading to inhibition of root cell elongation and root growth (Ivanchenko et al. 2013). Conversely, oxidative stress may also lead to increased levels of auxin either through tryptophan-dependent auxin biosynthesis or through β-oxidation of precursors resulting from lipid peroxidation of membranes (Woodward and Bartel 2005). The latter process is linked to jasmonic acid production also, which can be induced by wounding or pathogen attack leading to oxidative damage of cellular structures (Sun et al. 2009). Plants deficient in antioxidant enzyme catalase showing high H_2O_2 were believed to modulate auxin levels by two different mechanisms: either by auxin conjugation or by increasing auxin transport (Park et al. 2007; Tognetti et al. 2012). Thus, oxidative stress can also attenuate high auxin levels by its oxidation. Pasternak et al. (2005) showed that expression of genes encoding auxin efflux carriers, PIN3 and PIN1, decreases under oxidative stress-related growth conditions induced by alloxan. Recent reports have shown that changes in glutathione redox status can also mediate crosstalk between auxin and H_2O_2 in catalase mutants (Gao et al. 2014).

Heavy metals such as cadmium interfere with the antioxidant machinery and lead to generation of oxidative damage leading to membrane damage by lipid peroxidation and disruption of cellular redox homeostasis. Cd treatment was shown to reduce the IAA content significantly accompanied by increase in the activity of the IAA oxidase and alter the expression of several putative auxin biosynthetic and catabolic genes (Hu et al. 2013).

MAPK pathways in plants are induced in response to a number of abiotic and biotic stresses. MAPKs also link auxin-dependent cell-cycle regulation and oxidative stress signalling in plants and mediate stress tolerance responses (Teale et al. 2006). Cold and H_2O_2 can activate AtMPK3 and AtMPK6 pathways through either an unknown MAPKKK or ANP1, respectively. Constitutively active ANP1 mimics the H_2O_2 effect and initiates the MAPK cascade that induces specific stress-responsive genes, but it blocks the action of auxin (Kovtun et al. 2000). Additionally, calcium signalling component CAMTA1 (calcium/calmodulin-binding transcription activator) provides another link in the ROS–auxin interface (Galon et al. 2010).

Using pollutant ozone as an inducer of apoplastic ROS, Blomster et al. (2011) pointed towards a novel mechanism of auxin crosstalk with ROS. Studies with *DR5-uidA* auxin reporter construct showed that auxin signalling is transiently suppressed by ROS treatment. Transcripts of several auxin receptors and auxin/indole-3-acetic acid (Aux/IAA) transcriptional repressors were reduced in response to apoplastic ROS (Blomster et al. 2011). The ROS-derived changes in the expression of auxin signalling genes partially overlapped with abiotic stress, pathogen responses, and salicylic acid signalling. Exclusion of known suppressors of auxin signalling, such as SA, ethylene, miR393, IAA28, or MAPKs, during biotic stress from microarray data analysis indicated ROS regulation of auxin responses via a novel mechanism. Altered leaf morphology as a result of either chronic ROS treatment or loss of function mutation in *tir1/afb2* suggested that auxin is a negative regulator of stress-induced morphogenic response. Future studies for the identification and characterization of ROS network components involved in this crosstalk will shed light on the molecular convergence points between ROS and the auxin signalling pathways.

11.7.3 Auxin and Drought Stress

Plants have evolved diverse mechanisms to counteract water deficit and act in response to drought stress. These include stomata closure, root growth adjustment, increased epicuticular waxes, and accumulation of osmolytes. At hormonal level, ABA and ethylene are known markers of drought stress responses. Under drought, ABA levels can go up more than 40-folds and regulate a number of stress-responsive genes (Verslues et al. 2006). However, the role of auxin in abiotic stress responses, including drought, is being increasingly realized. Genome-wide expression analysis also indicates regulation of auxin-responsive genes under abiotic stresses such as drought (Jain and Khurana 2009; Ha et al. 2013). Although several hormones have been shown to regulate response of roots under unfavorable conditions, the cross-talk between them is still not completely understood. Despite the fact that evidences are mounting for the role of auxin in controlling root growth under stress, its interaction with ABA under water deficit still remains a question. The positive control of lateral root growth by auxin is contradicted by LR growth inhibition by ABA. Although, few reports have shown that ABA may control auxin distribution to regulate root growth under water stress (Yamaguchi and Sharp 2010; Xu et al. 2013). ABA can modulate auxin transport to modulate proton secretion at root tips and control root growth under moderate water deficit (Fig. 11.1). Also, ABA-mediated inhibition of LR growth is controlled by reducing polar auxin transport (Shkolnik-Inbar and Bar-Zvi 2010). At molecular level, ABA *insensitive3* (ABI3) interacts with ARF or Aux/IAA proteins and LR growth is inhibited in ABI3 mutants by attenuation of auxin responses. In contrast ABI4 overexpression impairs LR development.

WES1, a member of auxin-responsive *GH3* gene family encoding auxin conjugating enzyme, is induced by cold, drought, and heat, and also by ABA and SA (Park et al. 2007). *GH3* family genes are early auxin-responsive genes and, unlike many other auxin-responsive gene families, their number is rather small in rice in comparison to *Arabidopsis*; in fact, one of the subclasses is totally absent from rice as also other monocots (Jain et al. 2006b). The *GH3* genes encode IAA-amido synthetases, the conjugating enzymes, and thus negatively regulate free auxin concentration. Strikingly, auxin accumulation can stimulate *GH3* gene expression. The gain-of-function mutant plants of WES1 were found to be auxin-deficient but resistant to drought. Stress-responsive genes such as *RD29A*, *CBFs*, *ERD1*, and *P5CS1* were upregulated in the mutant. T-DNA mutants of *WES1* showed reduced resistance to biotic and abiotic stresses. Similarly, *TLD1*/*OsGH3.13* activation mutant showed increased tolerance to drought in rice (Zhang et al. 2009). Another *GH3* family member in rice, *OsGH3-2*, was induced by drought. Overexpression of *OsGH3-2* resulted in IAA-deficient phenotype with reduced free IAA, reduced carotene, and ABA (Du et al. 2012). Thus, stomatal apertures were found larger with increased water loss and a drought hypersensitive phenotype. However, the overexpression lines showed greater tolerance to cold. Carotenoid-deficient rice *phs1*, *phs2*, *phs3-1*, *phs4*, and PDS-RNAi transgenic rice showed ABA and IAA biosynthesis deficiency and thus showed similar phenotypes to *OsGH3-2* overexpression plants against drought and cold (Du et al. 2013b). Thus, a possible crosstalk modulated by ABA and IAA levels can affect drought responses of plants. Involvement of other downstream signalling partners that mediate stress responses here still remains unclear. IAA biosynthetic pathway has been well characterized in *Arabidopsis*. Members of YUCCA family of flavin monoxygenases have been shown to catalyze rate-limiting steps in tryptophan-dependent auxin biosynthesis. Recent reports have shown overexpression of YUCCA genes lead to drought resistance in *Arabidopsis* and rice. Rice mutants for CONSTITUTIVELY WILTED 1 (OsCOW1), encoding a member of YUCCA protein family, showed water-deficient phenotype indicating decreased water uptake (Woo et al. 2007). As expected, overexpression of YUCCA6 and YUCCA7 resulted in drought resistance in potato and *Arabidopsis*, respectively (Kim et al. 2013; Lee et al. 2012). Activation tagged *yuc*7-1D mutants not only showed enhanced lateral root production under drought stress but also an increase in transcript abundance of drought-responsive genes, such as *RD29A* and *COR15A* (Lee et al. 2012).

The spatio-temporal distribution and accumulation of auxin under different environmental conditions is important prerequisite for a number of plant growth responses. Expression profiles for a number of auxin transport genes showed differential regulation under ABA, drought and salt treatments (Shen et al. 2010). Moreover, auxin transport inhibitors and water deficit have synergistic effects on leaf abscission. Although, molecular understanding of the crosstalk between polar auxin transport and drought stress is poorly understood, recent studies have indicated towards involvement of auxin transporters in drought stress. Zhang et al.

(2012) showed involvement of auxin efflux carrier, OsPIN3t, in drought stress response and drought tolerance. Overexpression of *OsPIN3t* in rice led to upregulation of drought responsive genes, *OsDREB2A* and *OsAP37*, and increased tolerance to drought. Functional analysis of a *ZINC-INDUCED FACILITATOR LIKE-1 (ZIFL1)* revealed two isoforms of the gene produced through alternative splicing that have distinct roles in polar auxin transport and drought tolerance (Remy et al. 2013). *ZIFL1* belongs to major facilitator superfamily (MFS) transporters, which are believed to be secondary transporters that use proton gradients as an energy source. In *Arabidopsis*, ZIF1.1 appears to function in the tonoplast of root cells, whereas ZIF1.3 is found in the plasma membrane of stomatal guard cells. *zifl1* loss of function mutant plants were hypersensitive to drought stress and showed defects in lateral root emergence and root gravitropism. ZIF-like 1 (ZIFL1), in *Saccharomyces cerevisiae*, confers resistance to the synthetic auxin 2, 4-dichlorophenoxyacetic acid (2, 4-D) by reducing its concentration inside the yeast cell (Cabrito et al. 2009). ZIFL1.1 could also modulate auxin transport at root tips by regulating plasma membrane PIN2 abundance. In contrast, overexpression of *ZIFL1.3* had no effect on auxin-related processes and instead these lines appeared to close stomata more efficiently, thereby regulating drought tolerance (Remy et al. 2013).

Involvement of miRNA has also been shown under drought, salinity, and ABA. Using deep sequencing approach and RNA gel blot analysis, Kinoshita et al. (2012) found that accumulation of miR167a/b and its precursor pri-miR167a was reduced under high osmotic stress. In contrast, levels of its target gene IAR3, which encodes an indole-3-acetic acid (IAA)-Ala hydrolase, was found to be upregulated. IAR3 hydrolyses an inactive form of auxin (indole-3-acetic acid [IAA]-alanine) and releases bioactive auxin (IAA). *iar3* mutants showed hypersensitivity towards drought; and transgenic plants expressing a miR167-resistant form of IAR3 without changing its encoded amino acid sequence exhibited a constitutive stress response. These plants showed high levels of IAR3 mRNA accumulation and increased lateral root development compared with plants expressing the wild-type form of IAR3. Also, miR167a downregulation by target mimicry technique resulted in the upregulation of IAR3 transcripts as well as lateral root development (Kinoshita et al. 2012). In another report, stress-regulated miR393-guided cleavage of the transcripts encoding two auxin receptors, TIR1 and AFB2, was required for inhibition of lateral root growth by ABA and drought stress (Chen et al. 2012).

11.7.4 Auxin and Cold Stress

Cold stress can be classified as chilling (<20 °C) and freezing (<0 °C) stress, both restricting plant growth and development directly by inhibition of metabolic reactions and indirectly through cold-induced osmotic and oxidative stress (Chinnusamy et al. 2007). Thus, signalling under cold stress is much more dynamic and involves crosstalk with other stress signalling pathways to cope up with adverse conditions. Plasma membrane has been primarily suggested as the site for perception of cold stress due to changes in membrane fluidity caused by unsaturation of fatty acids and changes in lipid–protein composition in membranes (Wang et al. 2006). Rigidification of membrane activates CBFs/DREBs through calcium signalling and histidine kinases, eventually inducing expression of a number of downstream target genes.

Among plant hormones, ABA is known to play adaptive roles against abiotic stresses including cold stress. However, links between auxin and cold stress have also started emerging now. Global gene expression analysis has shown that the expression of *Aux/IAA* and *ARF* gene family members is altered during cold acclimation in *Arabidopsis* and rice (Hannah et al. 2005; Jain and Khurana 2009). However, reports have shown that cold stress modulates auxin response by altering its polar and lateral transport in a reversible manner rather than auxin biosynthesis or signalling (Fig. 11.2) (Morris 1979; Shibasaki et al. 2009). Shibasaki and co-workers showed that cold stress inhibited root growth and delayed gravity response indicating cold stress affects auxin responses. Analysis of auxin-responsive IAA2-GUS marker suggested that cold stress inhibits the gravity response by enhancing the accumulation of endogenous auxin in root meristem by inhibiting the intracellular auxin transport. Further, inhibition of auxin transport was linked to blocking of intracellular trafficking of the auxin efflux carrier, PIN2. Cold stress also blocks the asymmetric redistribution and intracellular cycling of PIN3 (Shibasaki et al. 2009). PIN3 is suggested to work in asymmetric auxin distribution and early gravity response. It was also shown that immobilization of PINs during cold stress is not because of a global slowdown of trafficking or a change in the membrane structure but instead represents a selective process to regulate the activity of specific proteins, which provides a mechanistic basis to explain the role of auxin in regulating plant growth and development under cold stress.

Earlier studies have shown a reversible inhibition of inflorescence gravitropism in *Arabidopsis*, an auxin-mediated response. The gravity response and rootward auxin transport were restored when plants returned to room temperature. Analysis of gravity persistent signal (*gps*) mutants by Wyatt et al. (2002) showed an altered gravity response of mutants at room temperature after cold stress. Further analysis of *gps* mutants revealed that these mutants fail to establish a proper auxin gradient in the inflorescence after gravistimulation and also show altered polar and lateral auxin transport (Nadella et al. 2006). Altogether, these studies show that cold stress affects the polar transport of auxin by selectively inhibiting the intracellular trafficking of a pool of proteins that include the auxin efflux carriers. Moreover, the effect of changes in membrane fluidity and actin cell cytoskeleton is limited in cold-induced inhibition of protein trafficking.

As stated earlier, members of auxin-responsive *GH3* gene family encoding for auxin conjugating enzymes (IAA-amido synthetases) have been shown to be also involved in cold-stress responses. *OsGH3-2* was found to modulate both endogenous free IAA and ABA homeostasis and differentially affect drought and cold tolerance in rice. Overexpression of *OsGH3-2* resulted in IAA-deficient phenotype but resistant to cold stress (Du et al. 2012). Similarly, carotenoid-deficient rice *phs1*, *phs2*, *phs3-1*, *phs4*, and PDS-RNAi transgenic rice showed ABA and IAA biosynthesis deficiency and showed similar phenotypes to *OsGH3-2* overexpression plants against drought and cold (Du et al. 2013b). To conclude, it can be suggested that local auxin gradients link auxin responses and developmental changes under cold stress. However, questions still remain regarding auxin crosstalk with other stress signal transduction and hormonal pathways under cold stress.

11.8 Auxin and Abiotic Stress: A Functional Genomics Perspective

With the advent of recent technologies in the area of functional genomics, many of the stress-related genes at genome-wide level have been identified. Genome-wide transcriptome analysis showed the altered expression of several auxin-related genes under abiotic stress conditions (Cheong et al. 2002; Jain and Khurana 2009; Song et al. 2009; Yuan et al. 2013). The analysis of auxin-responsive genes (identified at the whole genome level) has also been carried out during various reproductive developmental stages and abiotic stress conditions (Jain and Khurana 2009). In *Arabidopsis*, comprehensive microarray data have been used to establish relationship among the expression profiles for hormone and stress responses (Goda et al. 2008). This study revealed several candidate genes having a potential role in crosstalk between auxin and abiotic stress signalling pathways. Similarly, in another study, the members of *Aux/IAA* family were found to express differentially during cold acclimation in *Arabidopsis* (Hannah et al. 2005). Likewise, GST gene family was analyzed through genome-wide analysis and 31 GST genes were found to be responsive to auxin and cytokinin and many of them showed differential expression under abiotic stress conditions (Jain et al. 2010). The role of GST genes in mediating crosstalk between auxin and abiotic stress responses has also been reported (Smith et al. 2003; Sharma et al. 2014). In another study, the transcriptome analysis of AZF1 and AZF2 overexpression transgenic lines showed downregulation of many auxin-responsive genes during osmotic stress and ABA treatments (Kodaira et al. 2011). Genome-wide analysis of rice glutaredoxin genes showed their differential expression under abiotic stress and hormone (auxin) treatment conditions (Garg et al. 2010). The overexpression of one of the auxin-responsive CC-type GRX genes (*OsGRX8*) imparted tolerance to multiple abiotic stresses (Sharma et al. 2013). Microarray analysis of transgenic lines expressing *OsGRX8* showed that genes involved in detoxification processes, metabolite production, and other auxin-responsive genes are upregulated (Sharma et al. 2013).

The role of many of the auxin-responsive genes in abiotic stress responses has been elucidated by mutant or overexpression approach in model systems and crop plants. OsGH3-2 overexpression in rice imparted tolerance to cold and showed morphological aberrations, IAA deficiency and reduced carotene, ABA and free IAA levels and ABA homeostasis (Du et al. 2012). Carotenoid-deficient mutant lines exhibited tolerance to cold stress with lower auxin level (Du et al. 2013b). Overexpression of *YUCCA6* in transgenic potato showed increased endogenous auxin level and enhanced drought stress tolerance (Kim et al. 2013). The involvement of auxin in salinity stress has been shown by genetic analysis of auxin mutants in *Arabidopsis* (Tiryaki 2007). The involvement of auxin receptor, *TIR1/AFB*, was also found in oxidative and salinity stress tolerance (Iglesias et al. 2010). Based on molecular and cellular analyses of auxin signalling mutants, it has been shown that cold stress affects the polar transport of auxin by blocking the intracellular trafficking of auxin efflux carrier (Shibasaki et al. 2009). Transgenic plants overexpressing *AZF1* and *AZF2* genes caused severe damage to plant growth and viability. The transcriptome analysis of these transgenic plants showed that *AZF1* and *AZF2* repressed the genes that were otherwise downregulated by osmotic stress and abscisic acid. Additionally, these transgenic plants showed downregulation of auxin-responsive genes (Kodaira et al. 2011). Overall, the functional genomics approaches seem to be quite promising in deciphering the role of auxin in abiotic stress responses. It is anticipated that the use of emerging technologies in near future may provide novel insights into the exact mechanism underlying this crosstalk between two important signalling pathways.

11.9 Conclusion and Future Perspective

Auxin is one of the essential hormones that play important roles in regulating plant growth and development. Its levels are modulated by developmental cues as well as by environmental signals. The tissue distribution of auxin is also a critical role for the diversity of functions it performs. Thus, regulations of auxin biosynthesis, metabolism, transport mechanisms, tissue sensitivity, signal perception, and transduction have been the subject of intensive investigations.

Although biosynthesis of auxin through a tryptophan-independent pathway has been established beyond doubt, its synthesis in most plants essentially occurs through tryptophan. However, the regulation of the steps involved in auxin biosynthesis at the molecular level still remains to be understood in its entirety. The understanding of these regulatory steps involved in auxin biosynthesis (and also homeostasis) would aid in providing greater understanding of the role of auxin in abiotic stress responses as well. Auxin levels at global and local levels are controlled differently, where the roles of auxin transporters and carrier proteins predominate in controlling local auxin gradients. How auxin biosynthesis is integrated with local auxin maxima generation under abiotic stress conditions remains to be elucidated. Endogenous auxin concentration appears to play much wider role in abiotic stress responses as it has been found to have variable effects in response to different abiotic stresses, such as drought and cold.

The changes in auxin levels are sensed by a diversity of receptors, such as TIR1/AFBs, ABP1, and SKP2A. ABP1, although identified 40 years ago in maize, has gained wider acceptance as a receptor in *Arabidopsis* only recently, while studies on SKP2A are still in its infancy. Among the downstream signalling components, modulation of interaction of Aux/IAA and ARFs by auxin bound TIR1 (an F-box protein) has been at the center-stage of auxin action. The diversity of Aux/IAA and

ARFs family members and their mutual interaction possibilities add further complexity to auxin signalling. Their role specifically under abiotic stress conditions needs to be examined in greater detail.

Genome-wide expression studies in both dicot and monocot systems have shown differential regulation of auxin-responsive genes, such as *Aux/IAA*, *GH3* and *SAURs*, and ARFs and PIN proteins under abiotic stresses. Quite a number of such genes are also differentially regulated on exposure to oxidative stress. The differential expression of genes involved in cellular detoxification and redox homeostasis, such as GSTs and glutaredoxins, on auxin treatment, also provides further impetus to a crosstalk between auxin and oxidative stress pathways that may be involved in regulating cellular detoxification under abiotic stresses. Recent studies have also shown alteration in ROS levels on application of exogenous auxin. Thus, it appears that auxin and ROS share certain reciprocity, which needs to be better understood. Although most of these studies have been confined to TIR1/AFB mutants of *Arabidopsis*, role of other signalling components, especially in monocot systems like rice, under abiotic stresses need to be addressed.

Lastly, since auxin is a known central regulator of plant growth and development, its role in providing developmental plasticity to plants under abiotic stresses needs greater exploration. The understanding of crosstalk between auxin and stress signalling would be key in application of biotechnological approaches towards developing abiotic stress tolerance in crops, especially when climate change is emerging as a major concern globally.

Acknowledgements We acknowledge the financial support provided by the Department of Biotechnology, Government of India (J.P.K. and M.J.) and Indian Council of Agricultural Research (J.P.K.) for conducting research in respective laboratories. M.J. also acknowledges the in-house financial support given by the National Institute of Plant Genome Research. E.S. thanks the Council of Scientific and Industrial Research (CSIR), New Delhi, for the award of a Research Fellowship. J.P.K. also thanks the University Grants Commission, New Delhi, for providing infrastructural support.

References

Ainley WM, Walker JC, Nagao RT, Key JL (1988) Sequence and characterization of two auxin-regulated genes from soybean. J Biol Chem 263:10658–10666

Bartel B, LeClere S, Magidin M, Zolman BK (2001) Inputs to the active indole-3-acetic acid pool: de novo synthesis, conjugate hydrolysis, and indole-3-butyric acid β-oxidation. J Plant Growth Regul 20:198–216

Benkova E, Michniewicz M, Sauer M, Teichmann T, Seifertova D, Jurgens G, Friml J (2003) Local, efflux-dependent auxin gradients as a common module for plant organ formation. Cell 115:591–602

Blomster T, Salojärvi J, Sipari N, Brosché M, Ahlfors R, Keinänen M, Overmyer K, Kangasjärvi J (2011) Apoplastic reactive oxygen species transiently decrease auxin signaling and cause stress-induced morphogenic response in Arabidopsis. Plant Physiol 157:1866–1883

Cabrito TR, Teixeira MC, Duarte AA, Duque P, Sa-Correia I (2009) Heterologous expression of a Tpo1 homolog from Arabidopsis thaliana confers resistance to the herbicide 2,4-D and other chemical stresses in yeast. Appl Microbiol Biotechnol 84:927–936

Camilleri C, Jouanin L (1991) The TR-DNA region carrying the auxin synthesis genes of the Agrobacterium rhizogenes agropine-type plasmid pRiA4: nucleotide sequence analysis and introduction into tobacco plants. Mol Plant Microbe Interact 4:155–162
Chaoui A, El Ferjani E (2005) Effects of cadmium and copper on antioxidant capacities, lignification and auxin degradation in leaves of pea (Pisum sativum L.) seedlings. C R Biol 328:23–31
Chapman EJ, Estelle M (2009) Mechanism of auxin-regulated gene expression in plants. Annu Rev Genet 43:265–285
Chen H, Li Z, Xiong L (2012) A plant microRNA regulates the adaptation of roots to drought stress. FEBS Lett 586:1742–1747
Cheng N-H, Liu J-Z, Liu X, Wu Q, Thompson SM, Lin J, Cheng J, Whitham SA, Park S, Cohen JD, Hirschi KD (2011) Arabidopsis monothiol glutaredoxin, AtGRXS17, is critical for temperature-dependent postembryonic growth and development via modulating auxin response. J Biol Chem 286:20398–20406
Cheong YH, Chang HS, Gupta R, Wang X, Zhu T, Luan S (2002) Transcriptional profiling reveals novel interactions between wounding, pathogen, abiotic stress, and hormonal responses in Arabidopsis. Plant Physiol 129:661–677
Chinnusamy V, Zhu J, Zhu JK (2007) Cold stress regulation of gene expression in plants. Trends Plant Sci 12:444–451
Dai X, Mashiguchi K, Chen Q, Kasahara H, Kamiya Y, Ojha S, DuBois J, Ballou D, Zhao Y (2013) The biochemical mechanism of auxin biosynthesis by an arabidopsis YUCCA flavin-containing monooxygenase. J Biol Chem 288:1448–1457
De Tullio MC, Jiang K, Feldman LJ (2010) Redox regulation of root apical meristem organization: connecting root development to its environment. Plant Physiol Biochem 48:328–336
Deinlein U, Stephan AB, Horie T, Luo W, Xu G, Schroeder JI (2014) Plant salt-tolerance mechanisms. Trends Plant Sci 19(6):371–379. doi:10.1016/j.tplants.2014.02.001
Dreher KA, Brown J, Saw RE, Callis J (2006) The Arabidopsis Aux/IAA protein family has diversified in degradation and auxin responsiveness. Plant Cell 18:699–714
Du H, Wu N, Fu J, Wang S, Li X, Xiao J, Xiong L (2012) A GH3 family member, OsGH3-2, modulates auxin and abscisic acid levels and differentially affects drought and cold tolerance in rice. J Exp Bot 63:6467–6480
Du H, Liu H, Xiong L (2013a) Endogenous auxin and jasmonic acid levels are differentially modulated by abiotic stresses in rice. Front Plant Sci 4:397
Du H, Wu N, Chang Y, Li X, Xiao J, Xiong L (2013b) Carotenoid deficiency impairs ABA and IAA biosynthesis and differentially affects drought and cold tolerance in rice. Plant Mol Biol 83:475–488
Fu X, Harberd NP (2003) Auxin promoters Arabidopsis root growth by modulating gibberellins response. Nature 421:740–743
Galon Y, Aloni R, Nachmias D, Snir O, Feldmesser E, Scrase-Field S, Boyce JM, Bouché N, Knight MR, Fromm H (2010) Calmodulin-binding transcription activator 1 mediates auxin signaling and responds to stresses in Arabidopsis. Planta 232:165–178
Galvan-Ampudia CS, Testerink C (2011) Salt stress signals shape the plant root. Curr Opin Plant Biol 14:296–302
Gao X, Yuan H-M, Hu Y-Q, Li J, Lu Y-T (2014) Mutation of Arabidopsis CATALASE2 results in hyponastic leaves by changes of auxin levels. Plant Cell Environ 37:175–188
Garg R, Jhanwar S, Tyagi AK, Jain M (2010) Genome-wide survey and expression analysis suggests diverse roles of glutaredoxin gene family members during development and response to various stimuli in rice. DNA Res 17:353–367
Gill SS, Tuteja N (2010) Reactive oxygen species and antioxidant machinery in abiotic stress tolerance in crop plants. Plant Physiol Biochem 48:909–930
Goda H, Sasaki E, Akiyama K, Maruyama-Nakashita A, Nakabayashi K, Li W et al (2008) The AtGenExpress hormone and chemical treatment data set: experimental design, data evaluation, model data analysis and data access. Plant J 55:526–542
Greenham K, Santner A, Castillejo C, Mooney S, Sairanen I, Ljung K, Estelle M (2011) The AFB4 auxin receptor is a negative regulator of auxin signaling in seedlings. Curr Biol 21:520–525

Guan LM, Scandalios JG (2002) Catalase gene expression in response to auxin-mediated developmental signals. Physiol Plantarum 114:288–295

Guilfoyle TJ (1999) Auxin-regulated genes and promoters. In: Hooykaas PJJ, Hall MA, Libbenga KR (eds) Biochemistry and molecular biology of plant hormones. Elsevier, Amsterdam, pp 423–459

Ha CV, Le DT, Nishiyama R, Watanabe Y, Sulieman S, Tran UT, Mochida K, Dong N Van, Yamaguchi-Shinozaki K, Shinozaki K, Tran L-SP (2013) The auxin response factor transcription factor family in soybean: genome-wide identification and expression analyses during development and water stress. DNA Res 20:511–524

Hagen G, Guilfoyle TJ (1985) Rapid induction of selective transcription by auxins. Mol Cell Biol 5:1197–1203

Hagen G, Guilfoyle T (2002) Auxin-responsive gene expression: genes, promoters, and regulatory factors. Plant Mol Biol 49:373–385

Hannah MA, Heyer AG, Hincha DK (2005) A global survey of gene regulation during cold acclimation in Arabidopsis thaliana. PLoS Genet 1:179–196

He X-J, Mu R-L, Cao W-H, Zhang Z-G, Zhang J-S, Chen S-Y (2005) AtNAC2, a transcription factor downstream of ethylene and auxin signaling pathways, is involved in salt stress response and lateral root development. Plant J 44:903–916

Hirayama T, Shinozaki K (2010) Research on plant abiotic stress responses in the post-genome era: past, present and future. Plant J 61:1041–1052

Hu YF, Zhou G, Na XF, Yang L, Nan WB, Liu X, Zhang YQ, Li JL, Bi YR (2013) Cadmium interferes with maintenance of auxin homeostasis in Arabidopsis seedlings. J Plant Physiol 170:965–975

Huang Y-C, Chang Y-L, Hsu J-J, Chuang H-W (2008) Transcriptome analysis of auxin-regulated genes of Arabidopsis thaliana. Gene 420:118–124

Hull AK, Vij R, Celenza JL (2000) Arabidopsis cytochrome P450s that catalyze the first step of tryptophan-dependent indole-3-acetic acid biosynthesis. Proc Natl Acad Sci U S A 97:2379–2384

Iglesias MJ, Terrile MC, Bartoli CG, D'Ippólito S, Casalongué CA (2010) Auxin signaling participates in the adaptative response against oxidative stress and salinity by interacting with redox metabolism in Arabidopsis. Plant Mol Biol 74:215–222

Ivanchenko MG, den Os D, Monshausen GB, Dubrovsky JG, Bednarova A, Krishnan N (2013) Auxin increases the hydrogen peroxide (H_2O_2) concentration in tomato (Solanum lycopersicum) root tips while inhibiting root growth. Ann Bot 112:1107–1116

Jain M, Khurana J (2009) Transcript profiling reveals diverse roles of auxin-responsive genes during reproductive development and abiotic stress in rice. FEBS J 276:3148–3162

Jain M, Kaur N, Garg R, Thakur JK, Tyagi AK, Khurana JP (2006a) Structure and expression analysis of early auxin-responsive Aux/IAA gene family in rice (Oryza sativa). Funct Integr Genomics 6:47–59

Jain M, Kaur N, Tyagi AK, Khurana JP (2006b) The auxin-responsive GH3 gene family in rice (Oryza sativa). Funct Integr Genomics 6:36–46

Jain M, Tyagi AK, Khurana JP (2006c) Genome-wide analysis, evolutionary expansion, and expression of early auxin-responsive SAUR gene family in rice (Oryza sativa). Genomics 88:360–371

Jain M, Ghanshyam C, Bhattacharjee A (2010) Comprehensive expression analysis suggests overlapping and specific roles of rice glutathione S-transferase genes during development and stress responses. BMC Genomics 11:73

Jansen MAK, van den Noort RE, Tan MYA, Prinsen E, Lagrimini LM, Thorneley RNF (2001) Phenol-oxidizing peroxidases contribute to the protection of plants from ultraviolet radiation stress. Plant Physiol 126:1012–1023

Jaroslaw T, Dunajska K, Mazurek P, Piotrowska B, Tretyn A (2009) Exogenous auxin regulates H_2O_2 metabolism in roots of tomato (Lycopersicon esculentum Mill.) seedlings affecting the expression and activity of CuZn-superoxide dismutase, catalase, and peroxidase. Acta Physiol Plant 31:249–260

Jones AM, Venis MA (1989) Photoaffinity labeling of indole-3-acetic acid-binding proteins in maize. Proc Natl Acad Sci USA 86: 6153–6156

Jones AM, Herman EM (1993) KDEL-containing auxin-binding protein is secreted to the plasma membrane and cell wall. Plant Physiol 101:595–606

Joo JH, Bae YS, Lee JS (2001) Role of auxin-induced reactive oxygen species in root gravitropism. Plant Physiol 126:1055–1060

Joo JH, Yoo HJ, Hwang I, Lee JS, Nam KH, Bae YS (2005) Auxin-induced reactive oxygen species production requires the activation of phosphatidylinositol 3-kinase. FEBS Lett 579:1243–1248

Jurado S, Abraham Z, Manzano C, López-Torrejón G, Pacios LF, Del Pozo JC (2010) The Arabidopsis cell cycle F-box protein SKP2A binds to auxin. Plant Cell 22:3891–3904

Kawano T (2003) Roles of the reactive oxygen species-generating peroxidase reactions in plant defense and growth induction. Plant Cell Rep 21:829–837

Kazan K, Manners JM (2009) Linking development to defense: auxin in plant-pathogen interactions. Trends Plant Sci 14:373–382

Kim J, Harter K, Theologis A (1997) Protein–protein interactions among the Aux/IAA proteins. Proc Natl Acad Sci U S A 94:11786–11791

Kim JI, Baek D, Park HC, Chun HJ, Oh DH, Lee MK, Cha JY, Kim WY, Kim MC, Chung WS, Bohnert HJ, Lee SY, Bressan RA, Lee SW, Yun DJ (2013) Overexpression of Arabidopsis YUCCA6 in potato results in high-auxin developmental phenotypes and enhanced resistance to water deficit. Mol Plant 6:337–349

Kinoshita N, Wang H, Kasahara H, Liu J, Macpherson C, Machida Y, Kamiya Y, Hannah MA, Chua NH (2012) IAA-Ala Resistant3, an evolutionarily conserved target of miR167, mediates Arabidopsis root architecture changes during high osmotic stress. Plant Cell 24: 3590–3602

Kodaira K-S, Qin F, Phan Tran L-S, Maruyama K, Kidokoro S, Fujita Y, Shinozaki K, Shinozaki KY (2011) Arabidopsis Cys2/His2 zinc-finger proteins AZF1 and AZF2 negatively regulate abscisic acid-responsive and auxin-inducible genes under abiotic stress conditions. Plant Physiol 157:742–756

Kovtun Y, Chiu WL, Zeng W, Sheen J (1998) Suppression of auxin signal transduction by a MAPK cascade in higher plants. Nature 395:716–720

Kovtun Y, Chiu WL, Tena G, Sheen J (2000) Functional analysis of oxidative stress-activated mitogen-activated protein kinase cascade in plants. Proc Natl Acad Sci U S A 97:2940–2945

Krishnamurthy A, Rathinasabapathi B (2013) Auxin and its transport play a role in plant tolerance to arsenite induced oxidative stress in Arabidopsis thaliana. Plant Cell Environ 36:1838–1849

Lavenus J, Goh T, Roberts I, Guyomarc'h S, Lucas M, De Smet I, Fukaki H, Beeckman T, Bennett M, Laplaze L (2013) Lateral root development in Arabidopsis: fifty shades of auxin. Trends Plant Sci 18:450–458

Lee M, Jung JH, Han DY, Seo PJ, Park WJ, Park CM (2012) Activation of a flavin monooxygenase gene YUCCA7 enhances drought resistance in Arabidopsis. Planta 235:923–938

Liu X, Zhang H, Zhao Y, Feng Z, Li Q, Yang HQ, Luan S, Li J, He ZH (2013) Auxin controls seed dormancy through stimulation of abscisic acid signaling by inducing ARF-mediated ABI3 activation in Arabidopsis. Proc Natl Acad Sci U S A 110:15485–15490

Ljung K, Hull AK, Celenza J, Yamada M, Estelle M, Normanly J, Sandberg G (2005) Sites and regulation of auxin biosynthesis in Arabidopsis roots. Plant Cell 17:1090–1104

Long JA, Ohno C, Smith ZR, Meyerowitz EM (2006) TOPLESS regulates apical embryonic fate in Arabidopsis. Science 312:1520–1523

Ludwig-Müller J (2011) Auxin conjugates: their role for plant development and in the evolution of land plants. J Exp Bot 62:1757–1773

Malamy JE (2005) Intrinsic and environmental response pathways that regulate root system architecture. Plant Cell Environ 28:67–77

McSteen P (2010) Auxin and monocot development. Cold Spring Harb Perspect Biol 2:a001479

Morris DA (1979) The effect of temperature on the velocity of exogenous auxin transport in intact chilling-sensitive and chilling-resistant plants. Planta 146:603–605

Munns R, Tester M (2008) Mechanisms of salinity tolerance. Annu Rev Plant Biol 59:651–681

Nadella V, Shipp MJ, Muday GK, Wyatt SE (2006) Evidence for altered polar and lateral auxin transport in the gravity persistent signal (gps) mutants of Arabidopsis. Plant Cell Environ 29:682–690

Navarro L, Dunoyer P, Jay F, Arnold B, Dharmasiri N, Estelle M, Voinnet O, Jones JDG (2006) A plant miRNA contributes to antibacterial resistance by repressing auxin signaling. Science 312:436–439

Neill S, Desikan R, Hancock J (2002) Hydrogen peroxide signalling. Curr Opin Plant Biol 5:388–395

Normanly J (2010) Approaching cellular and molecular resolution of auxin biosynthesis and metabolism. Cold Spring Harb Perspect Biol 2:a001594

Normanly J, Cohen JD, Fink GR (1993) Arabidopsis thaliana auxotrophs reveal a tryptophan-independent biosynthetic pathway for indole-3-acetic acid. Proc Natl Acad Sci U S A 90:10355–10359

Novák O, Hényková E, Sairanen I, Kowalczyk M, Pospíšil T, Ljung K (2012) Tissue-specific profiling of the Arabidopsis thaliana auxin metabolome. Plant J 72:523–536

Okushima Y, Fukaki H, Onoda M, Theologis A, Tasaka M (2007) ARF7 and ARF19 regulate lateral root formation via direct activation of LBD/ASL genes in Arabidopsis. Plant Cell 19:118–130

Overvoorde PJ, Okushima Y, Alonso JM, Chan A, Chang C, Ecker JR, Hughes B, Liu A, Onodera C, Quach H, Smith A, Yu G, Theologis A (2005) Functional genomic analysis of the AUXIN/INDOLE-3-ACETIC ACID gene family members in Arabidopsis thaliana. Plant Cell 17:3282–3300

Paponov IA, Paponov M, Teale W, Menges M, Chakrabortee S, Murray JAH, Palme K (2008) Comprehensive transcriptome analysis of auxin responses in Arabidopsis. Mol Plant 1:321–337

Park J-E, Park J-Y, Kim Y-S, Staswick PE, Jeon J, Yun J, Kim S-Y, Kim J, Lee Y-H, Park C-M (2007) GH3-mediated auxin homeostasis links growth regulation with stress adaptation response in Arabidopsis. J Biol Chem 282:10036–10046

Park J, Kim YS, Kim SG, Jung JH, Woo JC, Park CM (2011) Integration of auxin and salt signals by the NAC transcription factor NTM2 during seed germination in Arabidopsis. Plant Physiol 156:537–549

Parry G, Calderon-Villalobos LI, Prigge M, Peret B, Dharmasiri S, Itoh H, Lechner E, Gray WM, Bennett M, Estelle M (2009) Complex regulation of the TIR1/AFB family of auxin receptors. Proc Natl Acad Sci U S A 106:22540–22545

Pasternak TP, Prinsen E, Ayaydin F, Miskolczi P, Potters G, Asard H, Van Onckelen HA, Dudits D, Fehér A (2002) The role of auxin, pH, and stress in the activation of embryogenic cell division in leaf protoplast-derived cells of alfalfa. Plant Physiol 129:1807–1819

Pasternak TP, Potters G, Caubergs R, Jansen MAK (2005) Complementary interactions between oxidative stress and auxins control plant growth responses at plant, organ and cellular level. J Exp Bot 56:1991–2001

Pazmiño DM, Rodríguez-Serrano M, Romero-Puertas MC, Archilla-Ruiz A, Del Río LA, Sandalio LM (2011) Differential response of young and adult leaves to herbicide 2,4-dichlorophenoxyacetic acid in pea plants: role of reactive oxygen species. Plant Cell Environ 34:1874–1889

Peer WA, Cheng Y, Murphy AS (2013) Evidence of oxidative attenuation of auxin signaling. J Exp Bot 64:2629–2639

Pfeiffer W, Höftberger M (2001) Oxidative burst in Chenopodium rubrum suspension cells: induction by auxin and osmotic changes. Physiol Plant 111:144–150

Remy E, Cabrito TR, Baster P, Batista RA, Teixeira MC, Friml J, Sá-Correia I, Duque P (2013) A major facilitator superfamily transporter plays a dual role in polar auxin transport and drought stress tolerance in Arabidopsis. Plant Cell 25:901–926

Rosquete MR, Barbez E, Kleine-Vehn J (2012) Cellular auxin homeostasis: gatekeeping is housekeeping. Mol Plant 5:772–786

Sauer M, Kleine-Vehn J (2011) AUXIN BINDING PROTEIN1: the outsider. Plant Cell 23:2033–2043

Seidel C, Walz A, Park S, Cohen JD, Ludwig-Muller J (2006) Indole-3-acetic acid protein conjugates: novel players in auxin homeostasis. Plant Biol (Stuttg) 8:340–345

Sharma R, Priya P, Jain M (2013) Modified expression of an auxin-responsive rice CC-type glutaredoxin gene affects multiple abiotic stress responses. Planta 238:871–884

Sharma R, Sahoo A, Devendran R, Jain M (2014) Over-expression of a rice tau class glutathione S-transferase gene improves tolerance to salinity and oxidative stresses in Arabidopsis. PLoS One. doi:10.1371/journal.pone.0092900

Shen C, Bai Y, Wang S, Zhang S, Wu Y, Chen M, Jiang D, Qi Y (2010) Expression profile of PIN, AUX/LAX and PGP auxin transporter gene families in Sorghum bicolor under phytohormone and abiotic stress. FEBS J 277:2954–2969

Shibasaki K, Uemura M, Tsurumi S, Rahman A (2009) Auxin response in Arabidopsis under cold stress: underlying molecular mechanisms. Plant Cell 21:3823–3838

Shkolnik-Inbar D, Bar-Zvi D (2010) ABI4 mediates abscisic acid and cytokinin inhibition of lateral root formation by reducing polar auxin transport in Arabidopsis. Plant Cell 22:3560–3573

Smith AP, Nourizadeh SD, Peer WA, Xu J, Bandyopadhyay A, Murphy AS, Goldsbrough PB (2003) Arabidopsis AtGSTF2 is regulated by ethylene and auxin, and encodes a glutathione S-transferase that interacts with flavonoids. Plant J 36:433–442

Song X-G, She X-P, He J-M, Huang C, Song T (2006) Cytokinin- and auxin-induced stomatal opening involves a decrease in levels of hydrogen peroxide in guard cells of Vicia faba. Funct Plant Biol 33:573–583

Song Y, Wang L, Xiong L (2009) Comprehensive expression profiling analysis of OsIAA gene family I developmental processes and I responses to phytohormone and stress treatments. Planta 229:577–591

Stepanova AN, Robertson-Hoyt J, Yun J, Benavente LM, Xie D-Y, Dolezal K, Schlereth A, Jürgens G, Alonso JM (2008) TAA1-mediated auxin biosynthesis is essential for hormone crosstalk and plant development. Cell 133:177–191

Strader LC, Bartel B (2011) Transport and metabolism of the endogenous auxin precursor indole-3-butyric acid. Mol Plant 4:477–486

Sugawara S, Hishiyama S, Jikumaru Y, Hanada A, Nishimura T, Koshiba T, Zhao Y, Kamiya Y, Kasahara H (2009) Biochemical analyses of indole-3-acetaldoxime-dependent auxin biosynthesis in Arabidopsis. Proc Natl Acad Sci U S A 106:5430–5435

Sun F, Zhang W, Hu H, Li B, Wang Y, Zhao Y, Li K, Liu M, Li X (2008) Salt modulates gravity signaling pathway to regulate growth direction of primary roots in Arabidopsis. Plant Physiol 146:178–188

Sun J, Xu Y, Ye S, Jiang H, Chen Q, Liu F, Zhou W, Chen R, Li X, Tietz O, Wu X, Cohen JD, Palme K, Li C (2009) Arabidopsis ASA1 is important for jasmonate-mediated regulation of auxin biosynthesis and transport during lateral root formation. Plant Cell 21:1495–1511

Tao Y, Ferrer J-L, Ljung K, Pojer F, Hong F, Long JA, Li L, Moreno JE, Bowman ME, Ivans LJ, Cheng Y, Lim J, Zhao Y, Ballaré CL, Sandberg G, Noel JP, Chory J (2008) Rapid synthesis of auxin via a new tryptophan-dependent pathway is required for shade avoidance in plants. Cell 133:164–176

Teale WD, Paponov IA, Palme K (2006) Auxin in action: signalling, transport and the control of plant growth and development. Nat Rev Mol Cell Biol 7:847–859

Thakur JK, Tyagi AK, Khurana JP (2001) *OsIAA1*, an Aux/IAA cDNA from rice, and changes in its expression as influenced by auxin and light. DNA Res 8:193–203

Thakur JK, Jain M, Tyagi AK, Khurana JP (2005) Exogenous auxin enhances the degradation of a light down-regulated and nuclear-localized OsiIAA1, an Aux/IAA protein from rice, via proteasome. Biochim Biophys Acta 1730:196–205

Theologis A, Huynh TV, Davis RW (1985) Rapid induction of specific mRNAs by auxin in pea epicotyl tissue. J Mol Biol 183:53–68

Tiryaki I (2007) The role of auxin-signaling gene *axr1* in salt stress and jasmonic acid inducible gene expression in *Arabidopsis thaliana*. J Cell Mol Biol 6:189–195

Tiwari SB, Wang XJ, Hagen G, Guilfoyle TJ (2001) AUX/IAA proteins are active repressors, and their stability and activity are modulated by auxin. Plant Cell 13:2809–2822

Tiwari SB, Hagen G, Guilfoyle T (2003) The roles of auxin response factor domains in auxin-responsive transcription. Plant Cell 15:533–543

Tognetti VB, Van Aken O, Morreel K, Vandenbroucke K, van de Cotte B, De Clercq I, Chiwocha S, Fenske R, Prinsen E, Boerjan W, Genty B, Stubbs KA, Inzé D, Van Breusegem F (2010) Perturbation of indole-3-butyric acid homeostasis by the UDP-glucosyltransferase UGT74E2 modulates Arabidopsis architecture and water stress tolerance. Plant Cell 22: 2660–2679

Tognetti VB, Mühlenbock P, Van Breusegem F (2012) Stress homeostasis - the redox and auxin perspective. Plant Cell Environ 35:321–333

Ulmasov T, Hagen G, Guilfoyle TJ (1997) ARF1, a transcription factor that binds to auxin response elements. Science 276:1865–1868

Verslues PE, Agarwal M, Katiyar-Agarwal S, Zhu J, Zhu JK (2006) Methods and concepts in quantifying resistance to drought, salt and freezing, abiotic stresses that affect plant water status. Plant J 45:523–539

Vranova E, Inze D, Breusegem FV (2002) Signal transduction during oxidative stress. J Exp Bot 53:1227–1236

Wang X, Li W, Li M, Welti R (2006) Profiling lipid changes in plant response to low temperatures. Physiol Plant 126:90–96

Wang Y, Li K, Li X (2009) Auxin redistribution modulates plastic development of root system architecture under salt stress in Arabidopsis thaliana. J Plant Physiol 166:1637–1645

Wang S, Bai Y, Shen C, Wu Y, Zhang S, Jiang D, Guilfoyle TJ, Chen M, Qi Y (2010) Auxin-related gene families in abiotic stress response in Sorghum bicolor. Funct Integr Genomics 10:533–546

Weijers D, Benkova E, Jäger KE, Schlereth A, Hamann T, Kientz M, Wilmoth JC, Reed JW, Jürgens G (2005) Developmental specificity of auxin response by pairs of ARF and Aux/IAA transcriptional regulators. EMBO J 24:1874–1885

Winkel-Shirley B (2002) Biosynthesis of flavonoids and effects of stress. Curr Opin Plant Biol 5:218–223

Woo YM, Park HJ, Su'udi M, Yang JI, Park JJ, Back K, Park YM, An G (2007) Constitutively wilted 1, a member of the rice YUCCA gene family, is required for maintaining water homeostasis and an appropriate root to shoot ratio. Plant Mol Biol 65:125–136

Woodward AW, Bartel B (2005) Auxin: regulation, action, and interaction. Ann Bot 95:707–735

Wright AD, Sampson MB, Neuffer MG, Michalczuk L, Slovin JP, Cohen JD (1991) Indole-3-Acetic Acid Biosynthesis in the Mutant Maize orange pericarp, a Tryptophan Auxotroph. Science 254:998–1000

Wyatt SE, Rashotte AM, Shipp MJ, Robertson D, Muday GK (2002) Mutations in the gravity persistence signal loci in Arabidopsis disrupt the perception and/or signal transduction of gravitropic stimuli. Plant Physiol 130:1426–1435

Xia Q-Z, Zhang X-L, Nie Y-C, Guo X-P (2005) Withdrawal of exogenous auxin induces programmed cell death of cotton embryonic suspension cultures. J Plant Physiol Mol Biol 31:78–84

Xie Q, Frugis G, Colgan D, Chua NH (2000) Arabidopsis NAC1 transduces auxin signal downstream of TIR1 to promote lateral root development. Genes Dev 14:3024–3036

Xu T, Wen M, Nagawa S, Fu Y, Chen J-G, Wu M-J, Perrot-Rechenmann C, Friml J, Jones AM, Yang Z (2010) Cell surface- and rho GTPase-based auxin signaling controls cellular interdigitation in Arabidopsis. Cell 143:99–110

Xu W, Jia L, Shi W, Liang J, Zhou F, Li Q, Zhang J (2013) Abscisic acid accumulation modulates auxin transport in the root tip to enhance proton secretion for maintaining root growth under moderate water stress. New Phytol 197:139–150

Xu T, Dai N, Chen J, Nagawa S, Cao M, Li H, Zhou Z, Chen X, De Rycke R, Rakusová H, Wang W, Jones AM, Friml J, Patterson SE, Bleecker AB, Yang Z (2014) Cell surface ABP1-TMK auxin-sensing complex activates ROP GTPase signaling. Science 343:1025–1028

Yamaguchi M, Sharp RE (2010) Complexity and coordination of root growth at low water potentials: recent advances from transcriptomic and proteomic analyses. Plant Cell Environ 33:590–603

Yuan H, Zhao K, Lei H, Shen X, Liu Y, Liao X, Li T (2013) Genome-wide analysis of the GH3 family in apple (Malus × domestica). BMC Genomics 14:297
Zenser N, Ellsmore A, Leasure C, Callis J (2001) Auxin modulates the degradation rate of Aux/IAA proteins. Proc Natl Acad Sci U S A 98:11795–11800
Zhang SW, Li CH, Cao J, Zhang YC, Zhang SQ, Xia YF, Sun DY, Sun Y (2009) Altered architecture and enhanced drought tolerance in rice via the down-regulation of indole-3-acetic acid by TLD1/OsGH3.13 activation. Plant Physiol 151:1889–1901
Zhang Q, Li J, Zhang W, Yan S, Wang R, Zhao J, Li Y, Qi Z, Sun Z, Zhu Z (2012) The putative auxin efflux carrier OsPIN3t is involved in the drought stress response and drought tolerance. Plant J 72:805–816
Zhao Y (2010) Auxin biosynthesis and its role in plant development. Annu Rev Plant Biol 61:49–64
Zhao Y, Wang T, Zhang W, Li X (2011) SOS3 mediates lateral root development under low salt stress through regulation of auxin redistribution and maxima in Arabidopsis. New Phytol 189:1122–1134
Zolla G, Heimer YM, Barak S (2010) Mild salinity stimulates a stress-induced morphogenic response in Arabidopsis thaliana roots. J Exp Bot 61:211–224

Chapter 12
Biotic and Abiotic Stress Signaling Mediated by Salicylic Acid

Dhirendra Kumar, Danda Chapagai, Phillip Dean, and Mackenzie Davenport

Abstract Biotic and abiotic stresses are significant factors limiting the production of food and other supporting materials required to sustain increasing world population. Plant health is directly related to human health and is increasingly becoming significant and demands more attention towards limiting the damages caused by biotic and abiotic stresses. Significant progress has been made towards our understanding of the processes, which mediate both biotic and abiotic stress signaling in plants. Significant role is played by various plant hormones, e.g., salicylic acid (SA) and jasmonic acid (JA) in biotic stress and abscisic acid (ABA) in abiotic stress (Annu Rev Cell Dev Biol 28:489–521, 2012). Other hormones with minor role include the cytokinins (CK), auxins (indole 3 acetic acid. IAA), and the brassinosteroids (BR) (Annu Rev Cell Dev Biol 28:489–521, 2012). Cross talk between these plant hormones is significant and may result in either synergistic or antagonistic effect on stress responses (Annu Rev Cell Dev Biol 28:489–521, 2012). In recent years, extensive research carried out in various laboratories has implicated cross talk between the ABA and the SA in abiotic stress response. This is significant in light of SA being key player in biotic stress responses in plants. This review will discuss the role of SA in biotic and abiotic stress signaling and its cross talk with other hormones in mediating abiotic stress signaling in plants.

Keywords SA • SAR • NPR1 • SABP2 • Biotic stress • Abiotic stress • ABA

D. Kumar (✉) • D. Chapagai • P. Dean • M. Davenport
Department of Biological Sciences, East Tennessee State University, Johnson City, TN 37614, USA
e-mail: kumard@etsu.edu

G.K. Pandey (ed.), *Elucidation of Abiotic Stress Signaling in Plants*,
DOI 10.1007/978-1-4939-2211-6_12

12.1 Introduction

The SA, a small phenolic compound is widely studied for its role in plant disease resistance signaling pathways (Vlot et al. 2009; Shah 2003; Boatwright and Pajerowska-Mukhtar 2013; An and Mou 2011; Delaney et al. 1994). Plants resisting pathogen infection, synthesize and accumulate high levels of the SA and express a battery of SA-responsive defense genes (Delaney et al. 1994). Accumulation of the SA is required for the activation of resistance response. Plants, which either do not synthesize the SA or fail to accumulate the SA, are defective in mounting an effective resistance response (Delaney et al. 1994; Gaffney et al. 1993; Wildermuth et al. 2001). The SA-mediated defense pathway is mostly active against biotrophic and hemibiotrophic pathogens (Vlot et al. 2009). The SA accumulations in infected and surrounding cells result in hypersensitive response, which is characterized by the localized cell death and expression of wide array of *pathogenesis-related* genes (Ward et al. 1991). Signals originating from infected tissue travel to the uninfected parts of plant and initiate resistance response to defend plants against any future attack. This heightened state of plant resistance in systemic tissues is commonly known as systemic acquired resistance (SAR) (Kuc 1982; Ryals et al. 1994; Ross 1961). SAR is dependent upon biosynthesis and accumulation of SA in systemic uninfected tissue (Gaffney et al. 1993).

Plants are in constant combat with adverse environmental conditions. Their state of sessility sets them apart from animals and defines their defenses against biotic and abiotic stressors. Immobility in plants led to the evolution of highly complex defense mechanisms that ranges from physiological changes to signaling pathways to changes on a transcriptomic level. A wide range of defense strategies allows plants to quickly recognize a unique stressor and deploy a resistance tailored to a specific attack. While biotic stressors include living organisms such as viruses, bacteria, insects, and herbivores; abiotic stressors include nonliving stressors such as temperature, drought, salinity, and wind. The sensing of these stressors induces signaling cascades that are largely defined by phytohormones such as the SA, jasmonic acid (JA), abscisic acid (ABA), and the ethylene (ET). The selection and direction of these hormones maintain the ability to induce a distinct subset of genes that together comprise an overall defense response. Although each hormone serves as a predominant signal for either an abiotic or biotic stressor, there is cross talk between them to serve purpose in both instances. It is important to view these hormones not as single entities declaring certain paths but as overlapping signals forming specific defense gradients.

12.2 SA in Biotic Stress

Exogenous treatments with SA and its active analogs make plants more resistant to infection by microbial pathogen (Shah 2003; White 1979). Levels of SA increase in response to attack by microbial pathogens (Enyedi et al. 1992a, b). Mutant/transgenic plants, which fail to accumulate SA, exhibit enhanced susceptibility (Gaffney et al. 1993; Nawrath and Metraux 1999; Wildermuth et al. 2001). Plants express a battery of defense-related genes in response to attack by pathogens and exogenous treatments with SA and its active analogs (Ward et al. 1991). Role of many of these expressed genes in imparting resistance phenotype is still unclear. Plants, which do not accumulate SA are defective in expression of these defense-related genes. Most studied defense-related genes are *pathogenesis-related* genes (Delaney et al. 1994). Synthetic analogs of SA also induce expression of similar set of defense genes as microbial pathogens (Shah 2003). Infection by microbial pathogens also leads to activation of SAR in distal tissues, which also accumulate SA (Ross 1961). Mutant or transgenic plants which fail to accumulate SA in distal tissues are defective in mounting effective SAR response.

12.2.1 SA Biosynthetic Pathway

Plants are known to synthesize and accumulate SA. Given the important role played by SA in various physiological processes including defense response against microbial pathogens, it became imperative to understand the biochemical pathways responsible for the synthesis of SA and regulation of this pathway during plants responses. Currently, two routes for the SA synthesis in plants are known to exist, a phenylalanine ammonia lyase (PAL) catalyzed pathway and an isochorismate synthase (ICS) catalyzed pathway. Interestingly, both PAL- and ICS-mediated pathways use primary metabolite chorismate, a product of shikimate pathway to synthesize SA (Fig. 12.1). The synthesis of the SA through the PAL-mediated pathway follows two possible routes. One pathway is through the conversion of the *trans*-cinnamic acid to *ortho*-coumaric acid to SA (Fig. 12.1) while the other route from the *trans*-cinnamic acid is through the production of benzoic acid (BA), which is then catalytically

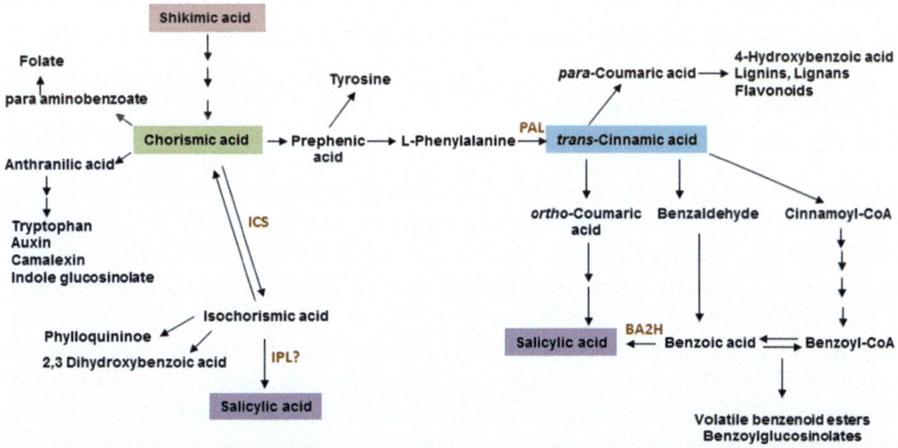

Fig. 12.1 SA biosynthetic pathways in plants. The phenylalanine ammonia lyase (PAL)-mediated salicylic acid (SA) pathway is well-studied pathway in tobacco and other plants and the isochorismate synthase (ICS) catalyzed pathway is major pathway for SA biosynthesis in *Arabidopsis* and other plants (This figure is adapted from Dempsey et al., 2011, Salicylic acid Biosynthesis and Metabolism, The Arabidopsis Book)

converted into SA (Fig. 12.1). Both routes appear to be active in plants (El-Basyouni et al. 1964; Chadha and Brown 1974). Conversion of BA to SA is catalyzed by the benzoic acid 2-hydroxylase (BA2H). Activity of BA2H is induced upon TMV infection and is likely to play a role in defense mounted by the tobacco plant (Leon et al. 1995). Expression of *PAL* gene expression is induced upon pathogen infection while treatment with 2-aminoindan-2-phosphonic acid, a PAL inhibitor made potato, cucumber, and *Arabidopsis* plants completely susceptible (Mauch-Mani and Slusarenko 1996; Meuwly et al. 1995; Coquoz et al. 1998). The transgenic tobacco plants suppressed in *PAL* gene expression accumulate lower levels of SA both in infected and systemic leaves (Pallas et al. 1996). The pepper plants silenced in *PAL* (*CaPAL1*) expression showed enhanced susceptibility to *Xanthomonas campestris* pv. *vesicatoria*. While overexpression of *CaPAL1* in *Arabidopsis* plants made them more resistant to *Pseudomonas syringae* pv. *tomato* (Kim and Hwang 2014). The *PAL* gene expression is also induced upon various abiotic stresses including wounding, nutrient depletion, UV irradiations, and extreme temperatures (Jin et al. 2013; Liang et al. 1989a, b; Huang et al. 2010; Payyavula et al. 2012).

The PAL-mediated pathway was the only known pathway for SA synthesis in plants till 2001 when another pathway mediated by ICS was discovered in *Arabidopsis* plants (Wildermuth et al. 2001). The *Arabidopsis sid2* (*eds16*) mutants defective in *ICS1* gene exhibited reduction in SA accumulation and more importantly failed to mount a robust SAR response (Wildermuth et al. 2001). The s*id2* mutants only accumulated 5–10 % of SA compared to the wild type Col-0 *Arabidopsis* plants. In wild type *Arabidopsis* plants, *ICS1* gene expression was systemically induced in response to infection by avirulent *P. s.* pv. *maculicola* (Wildermuth et al. 2001). Existence of *ICS* gene homologs are now reported in many other plants including *Nicotiana benthamiana, Solanum lycopersicon, Catharanthus roseus,* and *Populus trichocarpa* indicates that this pathway is not specific to *Arabidopsis* (van Tegelen et al. 1999; Catinot et al. 2008; Uppalapati et al. 2007; Yuan et al. 2009).

12.2.2 SA-Binding Proteins

Several soluble proteins, which interact with SA or are directly affected by increase in the SA levels have been identified and characterized (Kumar 2014; Kumar et al. 2013). The SA-binding protein, a catalase was first such SA-interacting protein that was identified from the tobacco leaves (Chen et al. 1993a, b; Conrath et al. 1995). It was proposed that the SA binding with catalase likely inhibits its catalytic activity resulting in accumulation of cellular ROS leading to induction of disease resistance (Chen et al. 1993b). In vitro studies using mammalian catalases supported the SA-mediated inhibition of its H_2O_2 degrading activity (Durner and Klessig 1996). One serious flaw in this proposed hypothesis was the assumption that SA levels increase proceeding to increase in ROS levels. This hypothesis was subsequently questioned with the observation that increase in cellular ROS levels leads to increase in SA levels and not vice versa (Tenhaken and Rubel 1997). Later studies also questioned inhibition of catalase by the SA with the observation that levels of the SA required to inhibit catalase activity are never reached in the cells. Role of SA in the inhibition of H_2O_2 degrading activity of catalases is still debatable and so is its role in SA-mediated biotic and abiotic responses.

Search for other high affinity SA-binding protein resulted in identification of two additional tobacco proteins. One was a ~25 kDa soluble leaf protein, which bound SA with significantly higher affinity compared to SABP (catalase) (Du and Klessig 1997). The SABP2 bound SA with $K_D = 90$ nM compared to SABP, which bound SA with $K_D = 14$ μM. The SABP2 was purified, its gene cloned and characterized from tobacco leaves (Kumar and Klessig 2003). SABP2 is a methyl salicylate esterase, which catalyzes the conversion of methyl salicylate (MeSA) into SA (Forouhar et al. 2005). Interestingly, SA binds to SABP2 with high affinity. The crystal structure analysis of SABP2 bound with SA revealed that SA binds in its active site (Forouhar et al. 2005). This observation suggested that SA might be a potential inhibitor of its enzymatic activity. Indeed, SA inhibited the esterase activity of SABP2 (Forouhar et al. 2005). Transgenic tobacco plants silenced in SABP2 expression were shown to be more susceptible and it also failed to induce SAR (Kumar and Klessig 2003). This observation suggested that MeSA esterase activity of SABP2 is critical for SAR induction in tobacco plants (Kumar et al. 2006). The MeSA was earlier shown to be a volatile signal molecule in plant to plant communication to induce resistance (Shulaev et al. 1997). Later, it was shown that MeSA also moves through phloem tissues (Seskar et al. 1998). Studies involving SABP2 silenced plants, transgenic chimera plants with combination of wild type root stock and silenced scion and vice versa, combined with biochemical analysis showed that the MeSA is a phloem mobile signal for SAR in plants (Park et al. 2007). Another SA-binding protein, SABP3 is a chloroplastic carbonic anhydrase, which binds SA with $K_D = 3.7$ μM (Du and Klessig 1997; Slaymaker et al. 2002). Silencing of SABP3 suppressed hypersensitive response to Pto:avr Pto and plants showed increased susceptibility to fungal pathogen *Phytophthora infestans* (Restrepo et al. 2005; Slaymaker et al. 2002). SABP3 exhibited antioxidant activity in yeast (Clark et al. 2004).

NPR1 (Non-expressor of Pathogenesis-Related genes 1) is a key regulator of SA-mediated plant immune signaling (Dong 2004; Shah et al. 1997; Pieterse and Van Loon 2004; Volko et al. 1998). Besides immune signaling, NPR1 is also implicated in growth and development of plant (Dong 2004). NPR1 has an ankyrin repeat and BTB/POX domain, which shares similarity with IκB-like mammalian immune signaling protein (Aravind and Koonin 1999; Oeckinghaus and Ghosh 2009). Plants lacking NPR1 fail to express PR1 proteins and are defective in mounting robust SAR response. NPR1 is transcriptionally regulated by WRKY transcription factors (Yu et al. 2001; Eulgem and Somssich 2007). Changes in concentration and oligomerization of NPR1 in cytoplasm and nucleus are key to an effective SAR response (Durrant and Dong 2004). Increased SA levels in cytoplasm releases NPR1 monomers from its oligomeric state (Mou et al. 2003). The NPR1 monomers travel to the nucleus with the help of its nuclear localization signal (Kinkema et al. 2000). Further studies show that nuclear accumulation of NPR1 is not sufficient for the induction of *PR1* gene expression (Kinkema et al. 2000). The BTB/POZ domain of NPR1 has an autoinhibitory effect, which could only be relieved by the binding of SA to NPR1 (Wu et al. 2012). Activation of *PR1* genes by NPR1 is mediated by TGA family of transcription factors (Gatz 2013). Interaction of an epigenetic regulator, ELP2 with NPR1 indicates its role in pathogen-induced chromatin remodeling (Wang et al. 2013). To suppress spurious untimely activation of immune response, NPR1 are regularly degraded in unchallenged cells (Spoel et al. 2009). Both unphosphoryated and phosphorylated NPR1 are subjected to degradation but degradation of phosphorylated NPR1 is required for the activation of immune response (Mukhtar et al. 2009; Spoel et al. 2009). The interaction between NPR1 and CUL3-mediated degradation machinery is mediated by adapter proteins, NPR3 and NPR4 (Fu et al. 2012). The NPR3 and NPR4 are now suggested to be SA receptors (Fu et al. 2012). The interaction between NPR1 and its adapter protein NPR3 is stimulated by the SA while the interaction between NPR1 and NPR4 is disrupted by the SA (Fu et al. 2012). It is suggested that NPR4 interacts with NPR1 in uninfected cells to limit the untimely activation of immune response (Fu et al. 2012). In the event of infection and subsequent accumulation of high levels of SA in the infected cells, NPR4 interacts with NPR1 and causes cell death. In systemic tissues, intermediate level of SA causes the NPR1 to accumulate and subsequent turnover resulting in SAR response (Fu et al. 2012). In another study, the SA was shown to bind to NPR1, which was not detected in any of the earlier attempts made so far (Fu et al. 2012; Wu et al. 2012). Involvement of copper was shown in SA binding to NPR1 (Wu et al. 2012).

Several new SA-binding proteins (glutathione *S*-transferases, GSTF2, GSTF8, GSTF10, and GSTF11; E2 subunit of α-ketoglutarate dehydrogenase) were recently identified using a combination of highly sensitive photoaffinity labeling coupled with surface plasmon resonance technology (Tian et al. 2012).

12.2.3 Transgenerational SAR

There has been great interest in the duration of SAR in plants in response to primary infections. Various studies provided some insights into SAR lasting beyond one generation but no systemic study was conducted to understand the mechanisms until recently (Luna and Ton 2012; Holeski et al. 2012). Using *Arabidopsis* plants, it was shown that effects of memory of induced resistance lasted at least two generation (Luna et al. 2012). Involvement of defense pathways mediated by SA was also implicated. Plants exhibiting resistance due to infection with *Pseudomonas syringae* when analyzed showed increased transcription of SA-responsive genes. One explanation for increased transcription was attributed to the enhanced H3 lysine 9 acetylation at the SA-responsive promoters (Luna et al. 2012). Progeny of *Arabidopsis* plants exhibiting transgenerational resistance show overall hypomethylation of DNA (Luna et al. 2012; Kathiria et al. 2010).

12.2.4 Mitogen-Activated Protein Kinases

Mitogen-activated protein kinases (MAPK) cascades transduce both biotic and abiotic signals for downstream processing through transcription factors (TF). The FLS2, an LRR receptor like kinase with transmembrane domain interacts with biotic stress signal by its interaction with PAMPs (e.g., bacterial flagellin) to trigger an MAPK cascade

(Tena et al. 2011). MAPK cascades are also activated in response to abiotic stress signals (Danquah et al. 2014). Interestingly, some MAPK cascades can transduce both biotic and abiotic signals and therefore mediate cross talk between various signals (Atkinson and Urwin 2012). MPK6, an *Arabidopsis* MAPK is known to mediate pathogen and abiotic signals, e.g., salt stress, cold stress, and stomatal control (Wang et al. 2007; Bethke et al. 2009a, b; Takahashi et al. 2007). Overexpression of rice MAPK (OsMAPK5) makes plants more tolerant to abiotic stresses while its RNAi-mediated silencing results in increased resistance against bacterial and fungal pathogens (Xiong and Yang 2003). The MAPK cascades are also activated by H_2O_2 (ROS), a factor in both biotic and abiotic stress signaling (Takahashi et al. 2011; Apel and Hirt 2004).

12.3 SA in Abiotic Stress

SA is a signal molecule that is predominantly known to be a key player in defense against biotic or pathogen infection, but lately it has been implicated to play a role in the defense against abiotic stressors. SA is now known to play a functional role in defense against water stress such as drought and salinity. Plants have developed drought tolerance through signaling cascades that allow them to withstand decreased water conditions until more is available. Understanding the role of SA in mediating these abiotic stress signaling require more attention and need to be explored.

SA is synthesized via the PAL pathway and the ICS pathway from chorismate. Both *ICS1* and *ICS2* genes are localized in the chloroplasts suggesting major site of SA synthesis is in plastids (Garcion et al. 2008). Plastids are also the site of biosynthesis of other phytohormones such as auxin, jasmonic acid, abscisic acid, gibberellins, brassinosteroids, or their precursors. The expression of salicylate hydroxylase (*NahG*) in nucleus is well known to significantly affect accumulation of SA, in response to biotic or abiotic stress stimuli (Vlot et al. 2009; Gaffney et al. 1993). Similarly, the overexpression of *NahG* in chloroplasts also resulted in compromise in both biotic and abiotic responses (Fragniere et al. 2011). The *Arabidopsis* plants expressing *NahG* gene in the chloroplasts were unable to accumulate significant levels of SA in response to inoculations with bacterial pathogen *Pseudomonas syringae* pv. *maculicola* or upon exposure to UVc treatment for 24 h. The mutant *sid2 Arabidopsis* plants accumulated only minimal levels of SA in response to various abiotic signals such as UV and ozone (O_3) (Nawrath and Metraux 1999). Interestingly, treatment of *Arabidopsis* plants with O_3 increased ICS activity, but it did not affect PAL activity (Ogawa et al. 2007). Treatment of the SA signaling mutants (*npr1*, *eds5* mutants, and *NahG* transgenic) with O_3, resulted in enhanced *ICS1* expression and ICS enzymatic activity (Ogawa et al. 2007).

12.3.1 SA Levels in Plants

The basal levels of SA vary greatly from species to species, e.g., in Solanaceae, *Nicotiana tabacum* contains lower levels approximately <100 ng g^{-1} fresh weight (FW) while *Solanum tuberosum* contains higher levels reaching 10 μg g^{-1} FW

(Rivas-San Vicente and Plasencia 2011). Monocots, rice plants contain 5–30 μg g^{-1} FW (Yang et al. 2004). Transgenic rice plants, which fail to accumulate SA, exhibit elevated levels of ROS. In addition, it has been found that both low levels and very high levels of SA attribute to susceptibility to abiotic stresses in plants. Given that different plants are accustomed to a typical gradient of SA, treatment with the same amount of SA in different plants can have opposite affects across species. Depending on the amount of SA, resistance to one stressor may increase while another becomes susceptible (Yuan and Lin 2008).

12.3.2 Effect of SA on Stomatal Closure

Stomata opening and closing is key event in mediating water stress, gas exchange, pathogen entry etc. (Blatt et al. 1999; Melotto et al. 2006, 2008; Liang et al. 2005; Hetherington and Woodward 2003). Stomatal pore is controlled by the osmotic changes in the guard cells (Liu and Luan 1998). The plant hormones also play key role in modulating stomatal opening and closing (Miura and Tada 2014). The ABA plays important role in stomatal response in plants under water stress (Tardieu and Davies 1992; Leckie et al. 1998). Other hormones known to influence stomatal aperture opening and closing include IAA, CK, ethylene (ET), BR, JA, and SA (Tanaka et al. 2006; Suhita et al. 2004; Desikan et al. 2006; Rajasekaran and Blake 1999; Jewer and Incoll 1980; Irving et al. 1992; Gehring et al. 1998; Melotto et al. 2006). Exogenous application of SA induces stomatal closure. The plants (*NahG*, *eds16-2*), which fail to accumulate SA are compromised in stomatal closure (Melotto et al. 2006). While the SA accumulating plants (*cpr5*, acd6) exhibit stomatal closure even in the absence of SA treatment (Bowling et al. 1997; Rate et al. 1999). Interestingly, stomata in *Arabidopsis* close within 1 h of pathogen inoculations (Melotto et al. 2006). Outward movement of K^+ mediated by guard cell outward rectifying channel (GORK) has central role in closure of stomata. Pretreatment with SA ameliorates the toxic effects in plants exposed to salinity stress (Hayat et al. 2010; Horvath et al. 2007). This could be most likely through the SA-mediated reduction in K^+ efflux through GORK channels (Jayakannan et al. 2013).

12.3.3 SA and Seed Germination

In *Arabidopsis*, exogenous treatments with SA (>1.0 mM) can either slow down or inhibit seed germination but treatments with lower levels of SA can improve germination (Rajjou et al. 2006). The addition of 0.05–0.5 mM SA improves seed germination under salt stress. The *Arabidopsis* seeds exposed to salt stress result in only 50 % germination, while the addition of low level of SA increases the germination rate to 80 % (Rivas-San Vicente and Plasencia 2011). Interestingly, seeds of various plants respond differently to SA, which they could tolerate during germination. Corn seed germination is completely inhibited by the treatments with 3–5 mM SA, while the barley seeds only tolerates <0.25 mM SA in germinating media (Xie et al.

2007; Guan and Scandalios 1995). These observations clearly show a positive role for SA in seed germination especially under salt stress. Role of SA in seed germination was highlighted by the delay in germination of SA deficient *Arabidopsis sid2* mutant seeds under salt stress (Alonso-Ramirez et al. 2009). Interestingly, *NahG* transgenic defective in SA accumulations did not have any effect on the seed germination under high salinity conditions (Borsani et al. 2001). Major difference between the *sid2* mutant and *NahG* transgenic is that while both fail to accumulate SA, *sid2* is defective in SA biosynthesis while in the NahG transgenic plants, SA is quickly degraded to catechol. Accumulation of catechol in *NahG* transgenic is likely cause of its protective effects under salinity stress (Lee et al. 2010).

12.3.4 Effects of SA on Drought Tolerance

Drought is a serious issue in crop productivity. Plants respond to water deficit by extending their roots and by controlling water loss through transpiration (Miura and Tada 2014). During water stress, the translocation of carbohydrates is reduced and at the same time nutrient uptake through roots also slows down (Miura and Tada 2014). SA plays a role in plants response to water deficit. SA was shown to accumulate differentially across species following water deficit as well as the induction of *PR1* and *PR2*, which are SA-inducible genes were found to be affected during drought stress (Miura et al. 2013). Levels of SA increased by twofold in barley seedlings exposed to water stress (Bandurska and Stroinski 2005). Treatments with higher levels of SA (2–3 mM) suppressed drought tolerance, while the addition of lower levels (<0.5 mM) increased tolerance to drought (Miura and Tada 2014; Senaratna et al. 2000; Korkmaz et al. 2007; Hamada 2001; Kang et al. 2012). SA accumulating mutants, *cpr5*, *acd6*, *siz1*, *adr1*, and *myb96-1d* were all drought tolerant (Miura et al. 2013; Bowling et al. 1997; Grant et al. 2003; Chini et al. 2004; Seo and Park 2010). Pretreatment of wheat seedlings with low levels (0.5 mM) of SA influences ascorbate-glutathione cycle to enhance drought tolerance (Miura and Tada 2014).

12.3.5 Effects of SA on Salinity-Induced Stress

Higher level of salt in soil/water is a major limiting factor for plant growth, which adversely affects agricultural productivity worldwide. Many studies have shown that exogenous treatment with SA reduces some of the harmful effects of salt stress on plant growth. One mechanism, which provides possible explanation to the protective action of SA is through its effect on antioxidant system. SA affects activity of many of the antioxidant enzymes, e.g., superoxide dismutase, catalases, peroxidases (Vlot et al. 2009). Treatment with 100 mM NaCl caused less damage to the *NahG Arabidopsis* plants, which do not accumulate SA compared to wild type plants (Borsani et al. 2001). A homolog of tobacco *SABP2* gene was found to accumulate at higher levels in response to salt treatment (200 mM NaCl) in a halophytic wild tomato line PI365967 (Sun et al. 2010). This raises the possibility of involvement of SA synthesized via the catalytic activity of SABP2 in salt stress. Several genes encoding glutathione *S*-transferases (GST) were also found to be upregulated in salt-treated PI365967 (Sun et al. 2010; Csiszar et al. 2014). Interestingly, several GST were found to be SA-binding proteins (Tian et al. 2012). Treatments with millimolar levels of SA provided protection against salinity stress and resulted in the accumulation of ABA in tomato plants (Szepesi et al. 2009). The other mechanism by which SA reduces the harmful effects of salinity is by decreasing K^+ efflux via GORK (guard cell outward rectifying) channel and by counteracting NaCl-induced membrane depolarization (Jayakannan et al. 2013).

12.3.6 SA in Chilling Tolerance

Cold temperatures are major limiting factors in growing crop and can cause either reduced growth or could be lethal depending on plants tolerance (Scott et al. 2004; Gray et al. 1997). High levels of SA applied to tomato plants resulted in no effect on chilling tolerance while low levels resulted in increased tolerance to chilling (Miura and Tada 2014). There are two prominent factors that may be significant in the role of SA-induced defense against these abiotic stressors: (1) balance of SA concentration, and (2) relationship between SA and ROS. Low levels, 0.01–0.05 mM, does not effectively induce ROS accumulation but still maintains the ability to increase tolerance to cold stress. Alternative oxidase (AOX) and heat shock protein (HSP) can be induced to scavenge ROS and protect the membrane structure. Moderate levels, 0.1–0.5 mM, allow SA to inhibit antioxidant enzymes such as peroxidases and catalases increasing the concentration of ROS. As ROS levels increase, they can act as a secondary messenger to enhance cell protective enzymes. Moderate SA concentration is predominantly ROS dependent. Higher levels, i.e., above 0.5 mM, is predominantly associated with attack by biotic pathogens. Increase in levels of SA leads to the increase in ROS, by inhibiting the protective antioxidant enzymes, e.g., catalases, peroxidases. Higher levels of SA and ROS induce *PR* genes, which are not induced at lower levels. At this level, H_2O_2 and SA create a positive feedback loop simultaneously overproducing ROS and inhibiting scavenging abilities. This leads to the programmed cell death (PCD) usually associated with a biotic attack (Yuan and Lin 2008). SA defense role can be attributed to the moderate and lower level concentrations. This gradient allows for defense-specific modulation of the SA pathway. SA also activates MAP kinases that are known to be activated by abiotic stressors leading to downstream regulation of genes. It is possible that the MEKK1-MPK4 cascade may be a regulatory element in ROS-SA accumulation (Miura and Tada 2014).

12.3.7 Stomatal Closure Mediated by SA

Abiotic stressors such as salinity and drought use stomatal closure as a defense strategy. Plants use this mechanism to regulate osmotic pressure via the opening and closing of guard cells that flank stomatal opening. It has long been thought that

ABA played the dominant role in regulating stomatal movement, but the role of SA in stomatal closure has recently come to light (Montillet and Hirt 2013). SA-induced stomatal closure has been attributed mainly to defense against biotic pathogen resistance, but it has also been implicated to cross talk with ABA for abiotic stress resistance. Stomatal closure can keep biotic pathogens from entering the cell as well as regulate osmotic stress during abiotic conditions. Both occur due to the generation of ROS by either one of two mechanisms: NADPH oxidase mediated or cell wall peroxidase mediated (Miura and Tada 2014). ABA is predominantly represented in relation to the NADPH oxidase-mediated ROS while SA is represented by inducing the peroxidases. Contrary to this, a recent study demonstrated that SA does make use of the NADPH oxidase. After treatment of exogenous SA, stomatal closure showed 200 % decrease compared to a 10 % decrease in stomatal aperture by the NADPH oxidase mutant *rbohD* (Kalachova et al. 2013).

SA-mediated defense against abiotic stress still remains unclear but the advances made thus far suggests its prominence in regulation and increasing tolerance to various abiotic stress conditions. It is important to understand the varying SA levels among species and its implication between abiotic and biotic conditions. Concurrent biotic and abiotic stresses may allow the visualization of the convergence between separate hormone signaling pathways that interact and converge for overall defense or inhibit one another.

12.4 Cross Talk Between SA and Other Hormones to Regulate Biotic or Abiotic Stress

Recent studies have suggested cross talk between SA and abiotic stress hormone ABA to mediate abiotic/biotic stress signaling (Fan et al. 2009). T-DNA insertion mutants adjacent to ABA biosynthetic genes caused enhanced accumulation of ABA resulting in enhanced susceptibility to *Pseudomonas syringae* (Fan et al. 2009). In contrast, ABA biosynthetic mutant *aba3-1* with reduced levels of ABA showed reduced susceptibility to *P. syringae*. Rice pathogen, *Xanthomonas oryzae* uses ABA produced by plant to suppress SA-mediated defense mechanisms to promote susceptibility (Xu et al. 2013). This study shows a clear link between ABA and SA signaling pathways. Consistent with this observation it was shown that exogenous treatment with ABA made rice plants more susceptible to *X. oryzae* (Xu et al. 2013). Targeted suppression of defense pathways by successful pathogens is in general believed to be mediated by cross talk between various plant hormones (Pieterse et al. 2009; 2012). This is a well-studied phenomenon in case of *Pseudomonas syringae*, which secreted coronatine, a phytotoxin that mimics JA and help in suppression of SA pathways (Brooks et al. 2005). ABA accumulation also caused reduced resistance against *Hyaloperonospora arabidopsis*.

A cross talk between SA, ABA, and GA has been described during seed germination. Treatment with GA_3, increases the levels of SA synthesis in *Arabidopsis* seeds and growing the *Arabidopsis* seedlings in the presence of GA_3 showed enhanced expression of *ICS* and *NPR1* genes. Higher doses of SA could lead to antagonistic effects on GA as seen in barley seed germination and subsequent inhibitory effect on seedling growth. The expression of a barley *WRKY38* (inhibitor of α-amylases) is downregulated by GA treatments, while it is upregulated by SA and ABA treatments (Xie et al. 2007). Expression of GA and ABA-regulated α-amylases is important for seed germination (Sun and Gubler 2004). Another evidence of ABA and SA cross talk comes from the observation that expression of ABA-regulated proteins such as LEA (late embryogenesis abundant), dehydrins, and HSP (heat shock proteins) is increased in *Arabidopsis* seeds germinated in the presence of SA (Rajjou et al. 2006).

Auxin, a plant growth hormone is now implicated in abiotic stress signaling. Auxin signaling is transduced through its receptors, TIR1 and AFB (Dharmasiri et al. 2005). Mutant *tir1afb2* plants were more tolerant to salt stress and expressed higher levels of *PR1* (Iglesias et al. 2011). SA inhibits auxin signaling through repression of TIR1/AFB auxin receptors during stress condition (Iglesias et al. 2011).

Conclusions

Increasing body of literature suggests role of signaling pathways mediated by various phytohormones in biotic and abiotic stress signaling (Pieterse et al. 2012). SA has been widely studied for its direct role in mediating plant responses to microbial pathogens. Many virulent pathogens actively seek to suppress SA-mediated signaling pathway. It will be important to study these signaling processes in plants experiencing both abiotic and biotic stress signaling. Many of the SA signaling components of biotic stress response now appears to be involved in mediating abiotic stress responses in plants. It is important to understand the abiotic stress responses in plants with differential basal SA levels. How the local concentration of SA is regulated is also important to understand, given the fact that certain levels of SA could be lethal for the cell? Is new SA synthesized during abiotic stress or the inactive forms of SA in the cell are converted into active forms? These are several questions, which still needs to be addressed to make significant progress in understanding the role of SA and its cross talk with other plant hormones in mediating abiotic stress signaling.

Acknowledgement Authors acknowledge the financial support from National Science Foundation Grant (MCB# 1022077), RDC grants (E82023, 10-018M, and 14-021M) from East Tennessee State University.

References

Alonso-Ramirez A, Rodriguez D, Reyes D, Jimenez JA, Nicolas G, Lopez-Climent M, Gomez-Cadenas A, Nicolas C (2009) Evidence for a role of gibberellins in salicylic acid-modulated early plant responses to abiotic stress in Arabidopsis seeds. Plant Physiol 150:1335–44

An C, Mou Z (2011) Salicylic acid and its function in plant immunity. J Integr Plant Biol 53:412–28

Apel K, Hirt H (2004) Reactive oxygen species: metabolism, oxidative stress, and signal transduction. Annu Rev Plant Biol 55:373–99

Aravind L, Koonin EV (1999) Fold prediction and evolutionary analysis of the POZ domain: structural and evolutionary relationship with the potassium channel tetramerization domain. J Mol Biol 285:1353–61

Atkinson NJ, Urwin PE (2012) The interaction of plant biotic and abiotic stresses: from genes to the field. J Exp Bot 63:3523–43

Bandurska H, Stroinski A (2005) The effect of Salicylic acid on barley response to water deficit. Acta Physiol Plant 27:379–86

Bethke G, Scheel D, Lee J (2009a) Sometimes new results raise new questions: the question marks between mitogen-activated protein kinase and ethylene signaling. Plant Signal Behav 4:672–4

Bethke G, Unthan T, Uhrig JF, Poschl Y, Gust AA, Scheel D, Lee J (2009b) Flg22 regulates the release of an ethylene response factor substrate from MAP kinase 6 in Arabidopsis thaliana via ethylene signaling. Proc Natl Acad Sci U S A 106:8067–72

Blatt MR, Grabov A, Brearley J, Hammond-Kosack K, Jones JD (1999) K+ channels of Cf-9 transgenic tobacco guard cells as targets for Cladosporium fulvum Avr9 elicitor-dependent signal transduction. Plant J 19:453–62

Boatwright JL, Pajerowska-Mukhtar K (2013) Salicylic acid: an old hormone up to new tricks. Mol Plant Pathol 14:623–34

Borsani O, Valpuesta V, Botella MA (2001) Evidence for a role of salicylic acid in the oxidative damage generated by NaCl and osmotic stress in Arabidopsis seedlings. Plant Physiol 126:1024–30

Bowling SA, Clarke JD, Liu Y, Klessig DF, Dong X (1997) The cpr5 mutant of Arabidopsis expresses both NPR1-dependent and NPR1-independent resistance. Plant Cell 9:1573–84

Brooks DM, Bender CL, Kunkel BN (2005) The Pseudomonas syringae phytotoxin coronatine promotes virulence by overcoming salicylic acid-dependent defences in Arabidopsis thaliana. Mol Plant Pathol 6:629–39

Catinot J, Buchala A, Abou-Mansour E, Metraux JP (2008) Salicylic acid production in response to biotic and abiotic stress depends on isochorismate in Nicotiana benthamiana. FEBS Lett 582:473–8

Chadha KC, Brown SA (1974) Biosynthesis of phenolic acids in tomato plants infected with Agrobacterium tumefaciens. Can J Bot 52:2041–47

Chen Z, Ricigliano JW, Klessig DF (1993a) Purification and characterization of a soluble salicylic acid-binding protein from tobacco. Proc Natl Acad Sci U S A 90:9533–7

Chen Z, Silva H, Klessig DF (1993b) Active oxygen species in the induction of plant systemic acquired resistance by salicylic acid. Science 262:1883–6

Chini A, Grant JJ, Seki M, Shinozaki K, Loake GJ (2004) Drought tolerance established by enhanced expression of the CC-NBS-LRR gene, ADR1, requires salicylic acid, EDS1 and ABI1. Plant J 38:810–22

Clark D, Rowlett RS, Coleman JR, Klessig DF (2004) Complementation of the yeast deletion mutant DeltaNCE103 by members of the beta class of carbonic anhydrases is dependent on carbonic anhydrase activity rather than on antioxidant activity. Biochem J 379:609–15

Conrath U, Chen Z, Ricigliano JR, Klessig DF (1995) Two inducers of plant defense responses, 2,6-dichloroisonicotinec acid and salicylic acid, inhibit catalase activity in tobacco. Proc Natl Acad Sci U S A 92:7143–7

Coquoz JL, Buchala A, Metraux JP (1998) The biosynthesis of salicylic acid in potato plants. Plant Physiol 117:1095–101

Csiszar J, Horvath E, Vary Z, Galle A, Bela K, Brunner S, Tari I (2014) Glutathione transferase supergene family in tomato: Salt stress-regulated expression of representative genes from distinct GST classes in plants primed with salicylic acid. Plant Physiol Biochem 78:15–26

Danquah A, de Zelicourt A, Colcombet J, Hirt H (2014) The role of ABA and MAPK signaling pathways in plant abiotic stress responses. Biotechnol Adv 32:40–52

Delaney TP, Uknes S, Vernooij B, Friedrich L, Weymann K, Negrotto D, Gaffney T, Gut-Rella M, Kessmann H, Ward E, Ryals J (1994) A central role of salicylic acid in plant disease resistance. Science 266:1247–1250

Desikan R, Last K, Harrett-Williams R, Tagliavia C, Harter K, Hooley R, Hancock JT, Neill SJ (2006) Ethylene-induced stomatal closure in Arabidopsis occurs via AtrbohF-mediated hydrogen peroxide synthesis. Plant J 47:907–16
Dharmasiri N, Dharmasiri S, Estelle M (2005) The F-box protein TIR1 is an auxin receptor. Nature 435:441–5
Dong X (2004) NPR1, all things considered. Curr Opin Plant Biol 7:547–52
Du H, Klessig DF (1997) Identification of a soluble, high-affinity salicylic acid-binding protein in tobacco. Plant Physiol 113:1319–1327
Durner J, Klessig DF (1996) Salicylic acid is a modulator of tobacco and mammalian catalases. J Biol Chem 271:28492–501
Durrant WE, Dong X (2004) Systemic acquired resistance. Annu Rev Phytopathol 42:185–209
El-Basyouni SZ, Chen D, Ibrahim RK, Neish AC, Towers GH (1964) The biosynthesis of hydroxybenzoic acids in higher plants. Phytochemistry 3:485–92
Enyedi AJ, Yalpani N, Silverman P, Raskin I (1992a) Localization, conjugation, and function of salicylic acid in tobacco during the hypersensitive reaction to tobacco mosaic virus. Proc Natl Acad Sci U S A 89:2480–4
Enyedi AJ, Yalpani N, Silverman P, Raskin I (1992b) Signal molecules in systemic plant resistance to pathogens and pests. Cell 70:879–86
Eulgem T, Somssich IE (2007) Networks of WRKY transcription factors in defense signaling. Curr Opin Plant Biol 10:366–71
Fan J, Hill L, Crooks C, Doerner P, Lamb C (2009) Abscisic acid has a key role in modulating diverse plant-pathogen interactions. Plant Physiol 150:1750–61
Forouhar F, Yang Y, Kumar D, Chen Y, Fridman E, Park SW, Chiang Y, Acton TB, Montelione GT, Pichersky E, Klessig DF, Tong L (2005) Structural and biochemical studies identify tobacco SABP2 as a methyl salicylate esterase and implicate it in plant innate immunity. Proc Natl Acad Sci U S A 102:1773–8
Fragniere C, Serrano M, Abou-Mansour E, Metraux JP, L'Haridon F (2011) Salicylic acid and its location in response to biotic and abiotic stress. FEBS Lett 585:1847–52
Fu ZQ, Yan S, Saleh A, Wang W, Ruble J, Oka N, Mohan R, Spoel SH, Tada Y, Zheng N, Dong X (2012) NPR3 and NPR4 are receptors for the immune signal salicylic acid in plants. Nature 486:228–32
Gaffney T, Friedrich L, Vernooij B, Negrotto D, Nye G, Uknes S, Ward E, Kessmann H, Ryals J (1993) Requirement of salicylic acid for the induction of systemic acquired resistance. Science 261:754–756
Garcion C, Lohmann A, Lamodiere E, Catinot J, Buchala A, Doermann P, Metraux JP (2008) Characterization and biological function of the ISOCHORISMATE SYNTHASE2 gene of Arabidopsis. Plant Physiol 147:1279–87
Gatz C (2013) From pioneers to team players: TGA transcription factors provide a molecular link between different stress pathways. Mol Plant Microbe Interact 26:151–9
Gehring CA, McConchie RM, Venis MA, Parish RW (1998) Auxin-binding-protein antibodies and peptides influence stomatal opening and alter cytoplasmic pH. Planta 205:581–6
Grant JJ, Chini A, Basu D, Loake GJ (2003) Targeted activation tagging of the Arabidopsis NBS-LRR gene, ADR1, conveys resistance to virulent pathogens. Mol Plant Microbe Interact 16:669–80
Gray GR, Chauvin LP, Sarhan F, Huner N (1997) Cold acclimation and freezing tolerance (a complex interaction of light and temperature). Plant Physiol 114:467–474
Guan L, Scandalios JG (1995) Developmentally related responses of maize catalase genes to salicylic acid. Proc Natl Acad Sci U S A 92:5930–4
Hamada AM (2001) Salicylic acid versus salinity-drought-induced stress on wheat seedlings. Rostl Vyr 47:444–50
Hayat Q, Hayat S, Irfan M, Ahmad A (2010) Effect of exogenous salicylic acid under changing environment: a review. Environmental and Experimental Botany 68:14–25
Hetherington AM, Woodward FI (2003) The role of stomata in sensing and driving environmental change. Nature 424:901–8

Holeski LM, Jander G, Agrawal AA (2012) Transgenerational defense induction and epigenetic inheritance in plants. Trends Ecol Evol 27:618–26

Horvath E, Szalai G, Janda T (2007) Induction of abiotic stress tolerance by salicylic acid signaling. J Plant Growth Regul 26:290–300

Huang J, Gu M, Lai Z, Fan B, Shi K, Zhou YH, Yu JQ, Chen Z (2010) Functional analysis of the Arabidopsis PAL gene family in plant growth, development, and response to environmental stress. Plant Physiol 153:1526–38

Iglesias MJ, Terrile MC, Casalongue CA (2011) Auxin and salicylic acid signalings counteract the regulation of adaptive responses to stress. Plant Signal Behav 6:452–4

Irving HR, Gehring CA, Parish RW (1992) Changes in cytosolic pH and calcium of guard cells precede stomatal movements. Proc Natl Acad Sci U S A 89:1790–4

Jayakannan M, Bose J, Babourina O, Rengel Z, Shabala S (2013) Salicylic acid improves salinity tolerance in Arabidopsis by restoring membrane potential and preventing salt-induced K+ loss via a GORK channel. J Exp Bot 64:2255–68

Jewer PC, Incoll LD (1980) Promotion of stomatal opening in the grass Anthephora pubescens nees by a range of natural and synthetic cytokinins. Planta 150:218–21

Jin Q, Yao Y, Cai Y, Lin Y (2013) Molecular cloning and sequence analysis of a phenylalanine ammonia-lyase gene from Dendrobium. PLoS One 8:e62352

Kalachova T, Iakovenko O, Kretinin S, Kravets V (2013) Involvement of phospholipase D and NADPH-oxidase in salicylic acid signaling cascade. Plant Physiol Biochem 66:127–33

Kang G, Li G, Xu W, Peng X, Han Q, Zhu Y, Guo T (2012) Proteomics reveals the effects of salicylic acid on growth and tolerance to subsequent drought stress in wheat. J Proteome Res 11:6066–79

Kathiria P, Sidler C, Golubov A, Kalischuk M, Kawchuk LM, Kovalchuk I (2010) Tobacco mosaic virus infection results in an increase in recombination frequency and resistance to viral, bacterial, and fungal pathogens in the progeny of infected tobacco plants. Plant Physiol 153:1859–70

Kim DS, Hwang BK (2014) An important role of the pepper phenylalanine ammonia-lyase gene (PAL1) in salicylic acid-dependent signalling of the defence response to microbial pathogens. J Exp Bot 65(9):2295–306

Kinkema M, Fan W, Dong X (2000) Nuclear localization of NPR1 is required for activation of PR gene expression. Plant Cell 12:2339–2350

Korkmaz A, Uzunlu M, Demirkiran A (2007) Treatments with acetyl salicylic acid protects muskmelon seedlings against drought stress. Acta Physiol Plant 29:503–508

Kuc J (1982) Induced immunity to plant disease. BioScience 32:854–860

Kumar D (2014) Salicylic acid signaling in disease resistance. Plant Sci 228:127–34

Kumar D, Klessig DF (2003) High-affinity salicylic acid-binding protein 2 is required for plant innate immunity and has salicylic acid-stimulated lipase activity. Proc Natl Acad Sci U S A 100:16101–6

Kumar D, Gustafsson C, Klessig DF (2006) Validation of RNAi silencing specificity using synthetic genes: salicylic acid-binding protein 2 is required for innate immunity in plants. Plant J 45:863–8

Kumar D, Hotz T, Hossain MA, Chigurupati P, Mayakoti A, Binda N, Zhao B, Tripathi D (2013) Methyl salicylate esterases in plant immunity. Plant stress 7:47–51

Leckie CP, McAinsh MR, Allen GJ, Sanders D, Hetherington AM (1998) Abscisic acid-induced stomatal closure mediated by cyclic ADP-ribose. Proc Natl Acad Sci U S A 95:15837–42

Lee S, Kim SG, Park CM (2010) Salicylic acid promotes seed germination under high salinity by modulating antioxidant activity in Arabidopsis. New Phytol 188:626–37

Leon J, Shulaev V, Yalpani N, Lawton MA, Raskin I (1995) Benzoic acid 2-hydroxylase, a soluble oxygenase from tobacco, catalyzes salicylic acid biosynthesis. Proc Natl Acad Sci U S A 92:10413–7

Liang XW, Dron M, Cramer CL, Dixon RA, Lamb CJ (1989a) Differential regulation of phenylalanine ammonia-lyase genes during plant development and by environmental cues. J Biol Chem 264:14486–92

Liang XW, Dron M, Schmid J, Dixon RA, Lamb CJ (1989b) Developmental and environmental regulation of a phenylalanine ammonia-lyase-beta-glucuronidase gene fusion in transgenic tobacco plants. Proc Natl Acad Sci U S A 86:9284–8

Liang YK, Dubos C, Dodd IC, Holroyd GH, Hetherington AM, Campbell MM (2005) AtMYB61, an R2R3-MYB transcription factor controlling stomatal aperture in Arabidopsis thaliana. Curr Biol 15:1201–6

Liu K, Luan S (1998) Voltage-dependent K+ channels as targets of osmosensing in guard cells. Plant Cell 10:1957–70

Luna E, Ton J (2012) The epigenetic machinery controlling transgenerational systemic acquired resistance. Plant Signal Behav 7:615–8

Luna E, Bruce TJ, Roberts MR, Flors V, Ton J (2012) Next-generation systemic acquired resistance. Plant Physiol 158:844–53

Mauch-Mani B, Slusarenko AJ (1996) Production of salicylic acid precursors is a major function of phenylalanine ammonia-lyase in the resistance of Arabidopsis to Peronospora parasitica. Plant Cell 8:203–212

Melotto M, Underwood W, Koczan J, Nomura K, He SY (2006) Plant stomata function in innate immunity against bacterial invasion. Cell 126:969–80

Melotto M, Underwood W, He S (2008) Y. Role of stomata in plant innate immunity and foliar bacterial diseases. Annu Rev Phytopathol 46:101–22

Meuwly P, Molders W, Buchala A, Metraux JP (1995) Local and Systemic Biosynthesis of Salicylic Acid in Infected Cucumber Plants. Plant Physiol 109:1107–1114

Miura K, Tada Y (2014) Regulation of water, salinity, and cold stress responses by salicylic acid. Front Plant Sci 5:4

Miura K, Okamoto H, Okuma E, Shiba H, Kamada H, Hasegawa PM, Murata Y (2013) SIZ1 deficiency causes reduced stomatal aperture and enhanced drought tolerance via controlling salicylic acid-induced accumulation of reactive oxygen species in Arabidopsis. Plant J

Montillet JL, Hirt H (2013) New checkpoints in stomatal defense. Trends Plant Sci 18:295–7

Mou Z, Fan W, Dong X (2003) Inducers of plant systemic acquired resistance regulate NPR1 function through redox changes. Cell 113:935–44

Mukhtar MS, Nishimura MT, Dangl J (2009) NPR1 in plant defense: it's not over 'til it's turned over. Cell 137:804–6

Nawrath C, Metraux JP (1999) Salicylic acid induction-deficient mutants of Arabidopsis express PR-2 and PR-5 and accumulate high levels of camalexin after pathogen inoculation. Plant Cell 11:1393–404

Oeckinghaus A, Ghosh S (2009) The NF-kappaB family of transcription factors and its regulation. Cold Spring Harb Perspect Biol 1:a000034

Ogawa D, Nakajima N, Tamaoki M, Aono M, Kubo A, Kamada H, Saji H (2007) The isochorismate pathway is negatively regulated by salicylic acid signaling in O_3-exposed Arabidopsis. Planta 226:1277–85

Pallas JA, Paiva NL, Lamb C, Dixon RA (1996) Tobacco plants epigenetically suppressed in phenylalanine ammonia lyase expression do not develop systemic acquired resistance in response to infection by tobacco mosaic virus. Plant J 10:281–293

Park SW, Kaimoyo E, Kumar D, Mosher S, Klessig DF (2007) Methyl salicylate is a critical mobile signal for plant systemic acquired resistance. Science 318:113–6

Payyavula RS, Navarre DA, Kuhl JC, Pantoja A, Pillai SS (2012) Differential effects of environment on potato phenylpropanoid and carotenoid expression. BMC Plant Biol 12:39

Pieterse CM, van Loon LC (2004) NPR1: the spider in the web of induced resistance signaling pathways. Curr Opin Plant Biol 7:456–64

Pieterse CM, Leon-Reyes A, van der Ent S, van Wees SC (2009) Networking by small-molecule hormones in plant immunity. Nat Chem Biol 5:308–16

Pieterse CM, van der Does D, Zamioudis C, Leon-Reyes A, van Wees SC (2012) Hormonal modulation of plant immunity. Annu Rev Cell Dev Biol 28:489–521

Rajasekaran LR, Blake TJ (1999) New Plant Growth Regulators Protect Photosynthesis and Enhance Growth Under Drought of Jack Pine Seedlings. J Plant Growth Regul 18:175–181

Rajjou L, Belghazi M, Huguet R, Robin C, Moreau A, Job C, Job D (2006) Proteomic investigation of the effect of salicylic acid on Arabidopsis seed germination and establishment of early defense mechanisms. Plant Physiol 141:910–23

Rate DN, Cuenca JV, Bowman GR, Guttman DS, Greenberg JT (1999) The gain-of-function Arabidopsis acd6 mutant reveals novel regulation and function of the salicylic acid signaling pathway in controlling cell death, defenses, and cell growth. Plant Cell 11:1695–708

Restrepo S, Myers KL, del Pozo O, Martin GB, Hart AL, Buell CR, Fry WE, Smart CD (2005) Gene profiling of a compatible interaction between Phytophthora infestans and Solanum tuberosum suggests a role for carbonic anhydrase. Mol Plant Microbe Interact 18:913–22

Rivas-San Vicente M, Plasencia J (2011) Salicylic acid beyond defence: its role in plant growth and development. J Exp Bot 62:3321–38

Ross AF (1961) Systemic acquired resistance induced by localized virus infections in plants. Virology 14:340–58

Ryals J, Uknes S, Ward E (1994) Systemic acquired resistance. Plant Physiol 104:1109–1112

Scott IM, Clarke SM, Wood JE, Mur LA (2004) Salicylate accumulation inhibits growth at chilling temperature in Arabidopsis. Plant Physiol 135:1040–9

Senaratna T, Touchell D, Bunn E, Dixon K (2000) Acetyl salicylic acid (aspirin) and salicylic acid induce multiple stress tolerance in bean and tomato plants. Plant Growth Regul 30:157–61

Seo PJ, Park CM (2010) MYB96-mediated abscisic acid signals induce pathogen resistance response by promoting salicylic acid biosynthesis in Arabidopsis. New Phytol 186:471–83

Seskar M, Shulaev V, Raskin I (1998) Endogenous methyl salicylate in pathogen-inoculated tobacco plants. Plant Physiol 116:387–392

Shah J (2003) The salicylic acid loop in plant defense. Curr Opin Plant Biol 6:365–71

Shah J, Tsui F, Klessig DF (1997) Characterization of a salicylic acid-insensitive mutant (sai1) of Arabidopsis thaliana, identified in a selective screen utilizing the SA-inducible expression of the tms2 gene. Mol Plant Microbe Interact 10:69–78

Shulaev V, Silverman P, Raskin I (1997) Airborne signalling by methyl salicylate in plant pathogen resistance. Nature 385:718–721

Slaymaker DH, Navarre DA, Clark D, del Pozo O, Martin GB, Klessig DF (2002) The tobacco salicylic acid-binding protein 3 (SABP3) is the chloroplast carbonic anhydrase, which exhibits antioxidant activity and plays a role in the hypersensitive defense response. Proc Natl Acad Sci U S A 99:11640–5

Spoel SH, Mou Z, Tada Y, Spivey NW, Genschik P, Dong X (2009) Proteasome-mediated turnover of the transcription coactivator NPR1 plays dual roles in regulating plant immunity. Cell 137:860–72

Suhita D, Raghavendra AS, Kwak JM, Vavasseur A (2004) Cytoplasmic alkalization precedes reactive oxygen species production during methyl jasmonate- and abscisic acid-induced stomatal closure. Plant Physiol 134:1536–45

Sun TP, Gubler F (2004) Molecular mechanism of gibberellin signaling in plants. Annu Rev Plant Biol 55:197–223

Sun W, Xu X, Zhu H, Liu A, Liu L, Li J, Hua X (2010) Comparative transcriptomic profiling of a salt-tolerant wild tomato species and a salt-sensitive tomato cultivar. Plant Cell Physiol 51:997–1006

Szepesi A, Csiszar J, Gemes K, Horvath E, Horvath F, Simon ML, Tari I (2009) Salicylic acid improves acclimation to salt stress by stimulating abscisic aldehyde oxidase activity and abscisic acid accumulation, and increases Na+ content in leaves without toxicity symptoms in Solanum lycopersicum L. J Plant Physiol 166:914–25

Takahashi F, Yoshida R, Ichimura K, Mizoguchi T, Seo S, Yonezawa M, Maruyama K, Yamaguchi-Shinozaki K, Shinozaki K (2007) The mitogen-activated protein kinase cascade MKK3-MPK6 is an important part of the jasmonate signal transduction pathway in Arabidopsis. Plant Cell 19:805–18

Takahashi F, Mizoguchi T, Yoshida R, Ichimura K, Shinozaki K (2011) Calmodulin-dependent activation of MAP kinase for ROS homeostasis in Arabidopsis. Mol Cell 41:649–60

Tanaka Y, Sano T, Tamaoki M, Nakajima N, Kondo N, Hasezawa S (2006) Cytokinin and auxin inhibit abscisic acid-induced stomatal closure by enhancing ethylene production in Arabidopsis. J Exp Bot 57:2259–66

Tardieu F, Davies WJ (1992) Stomatal response to abscisic Acid is a function of current plant water status. Plant Physiol 98:540–5

Tena G, Boudsocq M, Sheen J (2011) Protein kinase signaling networks in plant innate immunity. Curr Opin Plant Biol 14:519–29

Tenhaken R, Rubel C (1997) Salicylic acid is needed in hypersensitive cell death in soybean but does not act as a catalase inhibitor. Plant Physiol 115:291–298

Tian M, von Dahl CC, Liu PP, Friso G, van Wijk KJ, Klessig DF (2012) The combined use of photoaffinity labeling and surface plasmon resonance-based technology identifies multiple salicylic acid-binding proteins. Plant J

Uppalapati SR, Ishiga Y, Wangdi T, Kunkel BN, Anand A, Mysore KS, Bender CL (2007) The phytotoxin coronatine contributes to pathogen fitness and is required for suppression of salicylic acid accumulation in tomato inoculated with Pseudomonas syringae pv. tomato DC3000. Mol Plant Microbe Interact 20:955–65

van Tegelen LJ, Moreno PR, Croes AF, Verpoorte R, Wullems GJ (1999) Purification and cDNA cloning of isochorismate synthase from elicited cell cultures of Catharanthus roseus. Plant Physiol 119:705–12

Vlot AC, Dempsey DA, Klessig DF (2009) Salicylic Acid, a multifaceted hormone to combat disease. Annu Rev Phytopathol 47:177–206

Volko SM, Boller T, Ausubel FM (1998) Isolation of new Arabidopsis mutants with enhanced disease susceptibility to Pseudomonas syringae by direct screening. Genetics 149:537–48

Wang H, Ngwenyama N, Liu Y, Walker JC, Zhang S (2007) Stomatal development and patterning are regulated by environmentally responsive mitogen-activated protein kinases in Arabidopsis. Plant Cell 19:63–73

Wang Y, An C, Zhang X, Yao J, Zhang Y, Sun Y, Yu F, Amador DM, Mou Z (2013) The Arabidopsis elongator complex subunit2 epigenetically regulates plant immune responses. Plant Cell 25:762–76

Ward ER, Uknes SJ, Williams SC, Dincher SS, Wiederhold DL, Alexander DC, Ahl-Goy P, Metraux JP, Ryals JA (1991) Coordinate gene activity in response to agents that induce systemic acquired resistance. Plant Cell 3:1085–1094

White RF (1979) Acetylsalicylic acid (aspirin) induces resistance to tobacco mosaic virus in tobacco. Virology 99:410–2

Wildermuth MC, Dewdney J, Wu G, Ausubel FM (2001) Isochorismate synthase is required to synthesize salicylic acid for plant defence. Nature 414:562–5

Wu Y, Zhang D, Chu JY, Boyle P, Wang Y, Brindle ID, de Luca V, Despres C (2012) The Arabidopsis NPR1 protein is a receptor for the plant defense hormone salicylic acid. Cell Rep 1:639–47

Xie Z, Zhang ZL, Hanzlik S, Cook E, Shen QJ (2007) Salicylic acid inhibits gibberellin-induced alpha-amylase expression and seed germination via a pathway involving an abscisic-acid-inducible WRKY gene. Plant Mol Biol 64:293–303

Xiong L, Yang Y (2003) Disease resistance and abiotic stress tolerance in rice are inversely modulated by an abscisic acid-inducible mitogen-activated protein kinase. Plant Cell 15:745–59

Xu J, Audenaert K, Hofte M, de Vleesschauwer D (2013) Abscisic acid promotes susceptibility to the rice leaf blight pathogen *Xanthomonas oryzae* pv *oryzae* by suppressing salicylic acid-mediated defenses. PLoS One 8:e67413

Yang Y, Qi M, Mei C (2004) Endogenous salicylic acid protects rice plants from oxidative damage caused by aging as well as biotic and abiotic stress. Plant J 40:909–19

Yu D, Chen C, Chen Z (2001) Evidence for an important role of WRKY DNA binding proteins in the regulation of NPR1 gene expression. Plant Cell 13:1527–40

Yuan S, Lin H (2008) Role of salicylic acid in plant abiotic stress. Z Naturforsch C 63:313–20

Yuan Y, Chung JD, Fu X, Johnson VE, Ranjan P, Booth SL, Harding SA, Tsai CJ (2009) Alternative splicing and gene duplication differentially shaped the regulation of isochorismate synthase in Populus and Arabidopsis. Proc Natl Acad Sci U S A 106:22020–5

Chapter 13
Methylglyoxal, Triose Phosphate Isomerase, and Glyoxalase Pathway: Implications in Abiotic Stress and Signaling in Plants

Charanpreet Kaur, Shweta Sharma, Sneh Lata Singla-Pareek, and Sudhir Kumar Sopory

Abstract Methylglyoxal (MG) is a cytotoxic metabolite inevitably produced as a side product of primary metabolic pathways via both enzymatic and non-enzymatic reactions. In plants, spontaneous generation of MG through breakdown of triose sugars (dihydroxyacetone phosphate and glyceraldehyde 3-phosphate) is believed to be the major route for MG formation. MG is maintained at basal levels in plants under normal conditions that accumulate to higher concentrations under various stresses, probably as a general consequence of all abiotic stresses. The toxic effects of MG is due to its ability to induce oxidative stress in cells, either directly through increased generation of reactive oxygen species (ROS) or indirectly via advanced glycation end product (AGE) formation. Thus, elevated MG levels have implications in inhibition of growth and development in plants. To keep MG levels in check, the two-step glyoxalase pathway comprising glyoxalase I (GLYI) and glyoxalase II (GLYII) enzymes has evolved as the major MG-scavenging detoxification system that converts MG to D-lactate using glutathione as a cofactor in this process. Over-expression of glyoxalase pathway has been shown to confer tolerance to multiple stresses that works by controlling MG levels and maintaining glutathione homeostasis in plants. Moreover, increased activity of triose phosphate isomerase under different stresses that use up triose sugars via glycolysis further prevents MG levels from accumulating in the system along with increasing the energy status of plants. Considering the fact that MG levels are maintained at a threshold concentration in plants even under physiological conditions and also observed MG-dependent induction in expression

C. Kaur (✉) • S. Sharma • S.L. Singla-Pareek
International Centre for Genetic Engineering and Biotechnology,
Aruna Asaf Ali Marg, New Delhi, India
e-mail: charanpreet06@gmail.com

S.K. Sopory
International Centre for Genetic Engineering and Biotechnology,
Aruna Asaf Ali Marg, New Delhi, India

Jawaharlal Nehru University, Aruna Asaf Ali Marg, New Delhi, India

G.K. Pandey (ed.), *Elucidation of Abiotic Stress Signaling in Plants*,
DOI 10.1007/978-1-4939-2211-6_13

of triose phosphate isomerase, a role for MG in signaling pathways is suggested. Here, we provide an insight to the role of MG and glyoxalases in plant stress response with special mention about the possible involvement of MG in signaling pathway.

Keywords Glyoxalase I • Glyoxalase II • Glutathione • Methylglyoxal • Triose phosphate isomerase • Abiotic stress response • Heavy metal stress • Salinity • Stress tolerance

13.1 Introduction

Abiotic stresses negatively impact plant growth and development resulting in extensive losses to agricultural production worldwide (Boyer 1982). Perception of stress followed by signal transmission to turn on the adaptive responses is the key step leading to plant stress tolerance. However, tolerance or sensitivity to stresses is a very complex event, with more than one stress simultaneously affecting the plant and that too at multiple stages of development. Collectively as a common consequence of stress, various metabolic pathways are affected in living systems and the intricate nature of these pathways poses a challenge for the identification of key regulatory components involved in abiotic stress response (Fraire-Velázquez and Balderas-Hernández 2013). Also associated with stress, is the increased generation of deleterious chemical entities, namely, reactive oxygen species (ROS) and methylglyoxal (MG), that are otherwise constantly produced as by-products of metabolic processes and scavenged by plant antioxidative defense system to maintain at certain steady-state levels. Accumulation of these toxic molecules disrupts the delicate balance leading to oxidative stress, which is ultimately responsible for alteration in metabolic behavior observed during stress. Though implications of ROS toxicity and signaling in plants have been well investigated by various research groups but the role of MG in abiotic stress response and signaling is not fully known.

The generation of MG in living systems is from the triose sugars via dissociable intermediate of the reaction catalyzed by triose phosphate isomerase (TPI) in glycolysis (Phillips and Thornalley 1993; Richard 1993). MG levels have been reported to elevate during various abiotic stress conditions and are responsible, to some extent, for the damage that occurs to cellular machinery under stress (Yadav et al. 2005a; Hoque et al. 2012a). The toxicity of MG is believed to be due to its ability to interact with and modify protein and nucleotide moieties leading to loss of cell viability culminating in cell death (Thornalley 1998, 2008). In order to counter the toxic effects of MG, the ubiquitous glyoxalase pathway has evolved that converts MG to D-lactate, thereby alleviating ill-effects of MG from the system (Thornalley 1993). In this regard, transgenic plants over-expressing glyoxalase genes exhibited improved tolerance to various stresses by resisting an increase in the levels of MG and maintaining redox homeostasis (Singla-Pareek et al. 2003, 2006). However, the relevance of indirect route of MG generation via the glycolytic enzyme TPI in stress response is largely undiscovered. We have reported that MG can induce TPI expression and activity, suggesting an involvement of this enzyme in stress response (Sharma et al. 2012). This, in corroboration to previously reported steady-state levels of MG maintained at some threshold concentration even in glyoxalase over-expressing plants, indicates towards involvement of MG as a signaling molecule even in the plant system. In this chapter, the implications of MG generating and scavenging pathways in abiotic stress response and the possible role of MG as a signal molecule are presented.

13.2 How Methylglyoxal Brings About Toxicity in Cells?

Methylglyoxal was first prepared by von Pechmann in 1887 on warming isonitrosoacetone with dilute sulfuric acid (von Pechmann 1887). It is a well-known reactive α-oxoaldehyde that is both a mutagen and genotoxic agent. The toxic effects of MG arise due to the ability of this compound to modify both proteins and nucleotides through the two functional groups, aldehyde and ketone, present in MG (Thornalley 2008). It is, however, the aldehydic group that is more prone to attack by any other functional groups than the ketonic group, elucidating the mode of action of MG in biological systems (Leoncini 1979). MG along with glyoxal is the most potent glycating agent, modifying amino groups of proteins. Generally, guanidine groups of arginine are susceptible for modifications, resulting in the formation of advanced glycation end products (AGEs) that are the mediators of MG-induced toxicity in biological systems (Thornalley 2008). MG reacts with arginine residues to form hydroimidazolone derivate (MG-H, with three related structural isomers), argpyrimidine, and THP (tetrahydropyrimidine) (Thornalley et al. 2003; Ahmed et al. 2002, 2005). In addition, the cross-linking between lysine residues and MG leads to the formation of CEL [Nε-(carboxyethyl)lysine] and MOLDs (methylglyoxal–lysine dimers) (Ahmed and Thornalley 2002; Ahmed et al. 2002). However, quantitative analysis of AGE revealed hydroimidazolone MG-H1 to be the major glycation adduct formed with argpyrimidine. In this context, a "dicarbonyl proteome" has been defined, which consists of proteins susceptible to modification by MG and undergo functional impairment as a consequence of these modifications (Rabbani and Thornalley 2012). The component proteins are linked to mitochondrial dysfunction in diabetes and ageing, oxidative stress, dyslipidemia, cell detachment, and anoikis and apoptosis and include albumin, hemoglobin, transcription factors, mitochondrial proteins, extracellular matrix proteins, and lens crystallins. Further, activity of several NADPH-generating enzymes is also reported to be reduced on exposure to MG (Morgan et al. 2013). This is due to the irreversible modification of arginine residues, which otherwise form an essential component of active sites and are required for $NADP^+$ binding.

Besides proteins, nucleic acids and basic phospholipids can also be irreversibly modified by MG-mediated glycation reactions at the amino groups (Brown et al. 2005; Thornalley et al. 2010). Deoxyguanosine is the most reactive nucleotide susceptible to MG modification in physiological conditions. In vivo, the major MG-derived nucleotide AGEs are the imidazopurinone derivatives, which have been shown to be responsible for loss in genomic integrity associated with genotoxic effects (Thornalley et al. 2010).

In plants, adverse effect of MG on barley seed germination was demonstrated where growth inhibition was shown to be proportional to the concentration of MG (Mankikar and Rangekar 1974). Inhibition in response to low concentration of MG was recoverable to some extent but at higher than 1 μM MG concentration, the damage was extensive and irreversible. Addition of cysteine or methionine counteracted the detrimental effects of MG at 0.1 μM concentrations and less, thereby implicating the involvement of sulfhydryl groups of key enzymes (Mankikar and Rangekar 1974). Other growth processes, apart from seed germination, such as root elongation and chlorosis are also affected by MG (Hoque et al. 2012a). Hoque and coworkers have shown that 0.1 mM MG delays root elongation in *Arabidopsis*, whereas a week-long exposure to 1 mM MG inhibited germination by 21 % and also repressed root elongation along with inducing chlorosis. Further, MG is also shown to inhibit activities of various enzymes involved in antioxidant defense such as glutathione *S*-transferase (GST) activity, which can be reversed by the exogenous application of glutathione (Hoque et al. 2010). Also, MG can inhibit activity of cytosolic ascorbate peroxidase (APX), an enzyme playing a key role in the protection of cells from oxidative damage by scavenging ROS (Hoque et al. 2012b). The authors suggested that inhibition of GST and APX activities was mainly due to their modification by MG, which thereby lowers the affinity of the enzymes for their respective substrates. In spinach, addition of MG to chloroplasts is known to stimulate photosynthetic electron transport in thylakoid membranes (Saito et al. 2011). MG has been identified as a Hill oxidant that catalyses the photoreduction of O_2 at photosystem I, leading to the production of O_2^- thereby enhancing oxidative stress, which ultimately disrupts photosynthesis (Saito et al. 2011).

13.3 Synthesis and Turnover of Methylglyoxal

Methylglyoxal being a ubiquitous product of cellular metabolism is produced as a result of both enzymatic and non-enzymatic reactions. Its production is inevitable being tightly coupled to glycolysis. However, the rate of production varies depending upon the organism, tissue, cell metabolism, and physiological niche.

Undoubtedly, the major pathway for MG synthesis in biological systems is via its spontaneous generation in a glycolytic bypass, as a result of decomposition of triose sugars, GAP and DHAP (glyceraldehydes 3-phosphate and dihydroxyacetone phosphate, respectively) (Richard 1991; Phillips and Thornalley 1993). This reaction was initially reported in mid-1930s by Meyerhof and Lohmann (1934), but was ignored by referring to it as a mere experimental artifact. Later, Richard (1993) who investigated the mechanism of formation of MG from triose phosphates showed the physiological significance of this reaction. At the physiological pH, there is a high tendency for loss of α-carbonyl proton from the triose phosphates, producing an enediolate phosphate intermediate possessing low energy barrier for phosphate group elimination (Richard 1984). It is thus the deprotonation followed by spontaneous β-elimination of phosphate group of triose phosphates that leads to the formation of MG as the by-product of glycolysis (Richard 1993). However, TPI catalyzing the reversible interconversion of triose phosphates, DHAP and GAP, avoids the spontaneous degradation of the transition state intermediate into MG by stabilizing the enzyme-bound enediolate phosphate intermediate. In fact, the enzyme-bound enediolate phosphate intermediate is protonated 10^6-fold faster compared to the rate at which phosphate group is expelled (Richard 1991). Even then, the reaction catalyzed by TPI is not perfect and the enediolate intermediate may leak from the active site, producing MG in a side reaction. Thus, MG formation from triose sugars is believed to be the major route of its production under physiological conditions. Other minor routes of non-catalyzed MG generation include metabolism of acetone and aminoacetone (Kalapos 1999). MG metabolism through various pathways in plants is depicted in Fig. 13.1.

The enzyme-catalyzed production of MG has been reported only in animals and bacteria and not in plants. These include oxidation of aminoacetone in the catabolism of L-threonine, catalyzed by the enzyme semicarbazide-sensitive amine oxidase (Lyles and Chalmers 1992), the oxidation of ketone bodies by myeloperoxidase (Aleksandrovskii 1992), and the oxidation of acetone by cytochrome P450 (Casazza et al. 1984; Koop and Casazza 1985). In addition, under pathological conditions such as ketosis or diabetic ketoacidosis, the oxidation of ketone bodies also seems to be an important source of MG (Turk et al. 2006). Unlike the above-described pathways where MG is generated as a side product, MG synthase is the only known enzyme, which specifically catalyzes MG synthesis and for this it uses the triose sugar DHAP as the substrate (Hopper and Cooper 1971, 1972). Interestingly, MG synthase is co-operatively inhibited by inorganic phosphate (P_i), and this regulation controls the glycolytic flux depending on the availability of P_i (Cooper 1984). It has

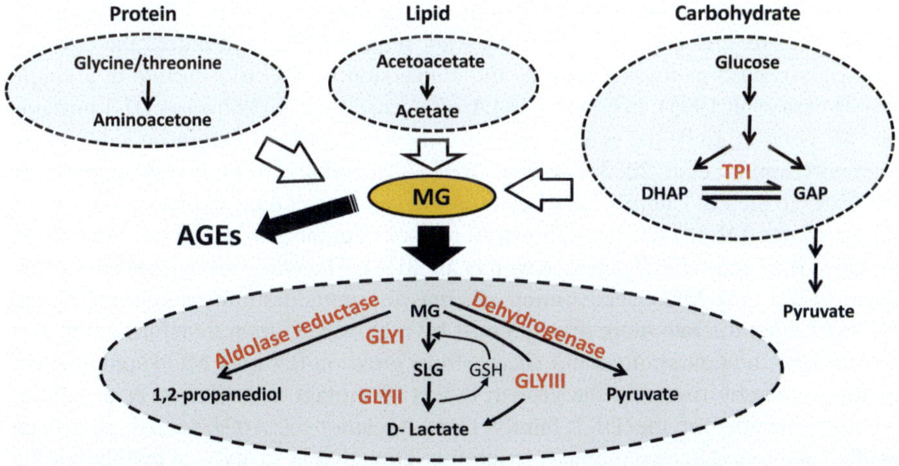

Fig. 13.1 Metabolism of MG in plants. MG is generated spontaneously as a side product of glucose, lipid, and protein metabolic pathways. Detoxification of MG occurs majorly via glyoxalase pathway comprising GLYI and GLYII enzymes. However, GSH-independent GLYIII, aldose reductase, and dehydrogenase are also capable of reducing MG

been reported in many organisms such as bacteria (Hopper and Cooper 1971, 1972), yeast (Babel and Hofmann 1981; Murata et al. 1985), goat liver (Ray and Ray 1981), but no such activity has been yet reported from plants.

The breakdown of MG is regarded as an important detoxification mechanism, which protects the system from its detrimental effects. The primary route for MG detoxification is through the two-step glyoxalase pathway comprising glyoxalase I (GLYI; S-D lactoylglutathione lyase) and glyoxalase II (GLYII; hydroxyacylglutathione hydrolase) enzymes, which act sequentially to convert MG into D-lactate (Fig. 13.1). The first enzyme, GLYI isomerizes hemithioacetal formed from the spontaneous combination of MG and GSH to S-D lactoylglutathione, which is then hydrolyzed by the GLYII enzyme to yield D-lactate, regenerating GSH in the process (Racker 1951; Crook and Law 1952). The result of glyoxalase system was earlier believed to be a dead-end product until the determination of D-lactate dehydrogenase activity in the biological systems through which D-lactate is converted to pyruvate (Long and Kaplan 1968; Pratt et al. 1979). Like MG, glyoxalase pathway is also ubiquitously present in the biological systems, highlighting the fundamental importance of this MG-scavenging mechanism in plants.

Apart from glyoxalases, other enzymes such as aldo-keto reductases can also consume MG thereby reducing it to the corresponding alcohol (Ko et al. 2005; Simpson et al. 2009; Narawongsanont et al. 2012). Over-expression of MG detoxifying enzymes, both glyoxalases (Veena and Sopory 1999; Singla-Pareek et al. 2003, 2006, 2008; Saxena et al. 2011; Alvarez Viveros et al. 2013) and aldo-keto reductases (Hegedüs et al. 2004; Simpson et al. 2009; Turóczy et al. 2011) has resulted in transgenic plants with improved tolerance to abiotic stresses, which keep MG levels in check and thereby contribute to proper growth and sustenance of plants under stress conditions.

A novel type of MG utilizing glyoxalase has been reported in bacteria and recently in *Arabidopsis* and humans, which does not require GSH, and unlike the conventional glyoxalase pathway catalyzes the conversion of MG to D-lactate in a single step (Misra et al. 1995). In humans and *Arabidopsis*, they are known as DJ-1 proteins and are believed to be the genetic cause for the early onset of Parkinson's disease in humans (Bonifati et al. 2003; Lee et al. 2012; Kwon et al. 2013). In *Arabidopsis*, six members in the DJ-1 family have been reported and the ectopic expression of one of the members (DJ-1d) has been shown to restore tolerance to MG in bacterial strain lacking *GLYI* and *GLYIII* genes (Kwon et al. 2013). The wild-type bacteria are resistance to 0.7 mM MG concentration whereas the mutant strain lacking *GLYI* and *GLYIII* cannot tolerate more than 0.5 mM MG. However, upon transformation with *DJ-1d* gene, mutant strain could successfully grow on 0.7 mM MG-supplemented medium, thereby reverting the growth defect of mutant cells (Kwon et al. 2013). Another member of the DJ-1 family from *Arabidopsis*, *AtDJ1C*, has also been studied and found to be required for viability. Though it is an atypical member of the DJ-1 family since it lacks a conserved cysteine residue required for enzymatic activity in other superfamily members, it is essential for proper chloroplast development and homozygous disruption of the *AtDJ1C* gene results in nonviable albino seedlings (Lin et al. 2011).

13.4 Methylglyoxal Levels Under Stress Conditions

Endogenous MG generation has been reported in all biological systems that rapidly increases under stress and disease conditions in bacteria, animals, mammals, and yeast (Thornalley 1990; Kalapos et al. 1992; Wu and Juurlink 2002). Similarly in plants, MG levels have been reported to increase in response to abiotic, biotic stresses as well as other stimuli such as white light, 2,4-D and ABA (Chen et al. 2004; Yadav et al. 2005a; Hossain et al. 2009). Under physiological conditions, MG concentration lies within 30–75 μM range in various plant species such as rice, *Pennisetum*, tobacco, and *Brassica* seedlings and is same in both leaves and roots except rice, where MG levels are lower in roots compared to leaves. However, a 2–6-fold increase in MG levels is observed in response to salinity, drought, and cold stress conditions (Yadav et al. 2005a). Noticeably under salt stress, MG levels increased to as high as 75–200 μM in these plants, suggesting accumulation under salinity stress being a universal response. These values are much higher than what has been reported in yeast and animals (Chaplen et al. 1998; Martins et al. 2001). It is possible that increased rate of glycolysis under stress leads to higher levels of MG in plants. However, another possibility which cannot be undermined is the spontaneous degradation of triosephosphates to MG during pre-analytical processing of samples. Further studies using newer and more sensitive detection methods are likely to resolve these inconsistencies in measurement.

Methylglyoxal levels are also reported to increase significantly in response to white light (2.21-fold) in pumpkin seedlings (Hossain et al. 2009), where treatment with exogenous MG, ABA, and salinity and drought stress also increases MG production manifold (which is in accordance with increased GLYI activity under these stresses). Also, infection with *Aspergillus flavus* increases MG production by 2.5-fold in a maize genotype, consequently resulting in the production of aflatoxin, a carcinogenic secondary metabolite (Chen et al. 2004). Banu et al. (2010) have also demonstrated a significant ~2-fold increase in MG levels in tobacco BY-2 cells in response to 200 mM NaCl stress. Taken together, it can be inferred that accumulation in MG under different stress conditions is unavoidable as it is probably a common consequence of both abiotic and biotic stresses in plants (Chen et al. 2004; Yadav et al. 2005a; Hossain et al. 2009).

13.5 Role of Glyoxalase Pathway in Plant Stress Physiology

Since generation of MG and increase in its concentration under stress conditions is an unavoidable phenomenon in biological systems, manipulation of glyoxalase pathway holds the potential for protection of plants against toxic effects of MG. Initial studies describing an upregulation in transcript levels of glyoxalase genes in response to various stress cues had opened the vistas for future of glyoxalase research in plant stress response. Espartero and coworkers (1995) showed a 2–3-fold increase in GLYI transcripts in roots, stems, and leaves of tomato plants treated with NaC1, mannitol, and ABA. Thereafter, a GLYI gene from *Brassica juncea* was transformed in tobacco in order to investigate the functional significance of glyoxalases *in planta*. Veena and Sopory (1999) reported that over-expression of GLYI gene is capable of inducing tolerance towards salinity stress and exogenous MG application in plants. This is achieved through controlling MG levels and maintaining reduced GSH pool. Similar results were also obtained when GLYI was transformed in *Vigna mungo* (Bhomkar et al. 2008). Further, over-expression of GLYII gene from rice has been carried out in different plants such as tobacco, rice, and recently in *Brassica juncea,* which imparts significant tolerance to high MG and salt treatments similar to GLYI gene (Singla-Pareek et al. 2003, 2008; Wani and Gosal 2011; Saxena et al. 2011). A balance was maintained in Na^+/K^+ ratio in the transgenic rice plants compared to wild-type (WT) in both shoot and root that correlated well with normal growth of these plants and formed the basis of minimizing Na^+ toxicity under salt stress (Singla-Pareek et al. 2008). Also, the transgenic tobacco plants could grow, flower, and set normal viable seeds under continuous salinity stress conditions (Singla-Pareek et al. 2003). Likewise, the transgenic *Brassica* plants over-expressing *OsGLYII* gene also showed significant levels of salinity stress tolerance by delaying senescence (Saxena et al. 2011). Interestingly, the double-transgenic tobacco plants expressing the entire pathway (GLYI+GLYII) outperformed the single transgenic lines, expressing either GLYI or GLYII genes or also non-transformed WT plants under salinity and heavy metal stresses (Singla-Pareek et al. 2003, 2006). The transgenic plants could grow well in the presence of 5 mM $ZnCl_2$ without any yield penalty and could tolerate toxic concentrations of other heavy metals as cadmium and lead. A reduction in MG levels along with maintaining higher levels of reduced GSH under salinity stress and increased phytochelatin production after zinc treatment is believed to confer stress tolerance in the transgenic plants (Yadav et al. 2005b; Singla-Pareek et al. 2006). Similarly, transgenic tomato plants expressing similar construct showed improved salinity stress tolerance by decreasing oxidative stress (Alvarez Viveros et al. 2013). In corroboration to observed heavy metal tolerance of plants expressing glyoxalase pathway, transformation of a wheat GLYI gene in tobacco also leads to increased tolerance to zinc when compared to untransformed control (Lin et al. 2010). Over-expression of sugar beet GLYI in tobacco also confers enhanced tolerance to MG, salt, mannitol, and H_2O_2 treatments (Wu et al. 2013).

Transcriptome and proteome analysis has also indicated the role of glyoxalase pathway in stress response. A recent genome-wide study in rice and *Arabidopsis* has indicated the multiple stress inducible nature of glyoxalases in these species, which also undergo developmental and tissue-specific variations (Mustafiz et al. 2011; Kaur et al. 2013). In *Brassica juncea*, GLYI transcript is upregulated in response to salinity, mannitol, and heavy metal stresses (Veena and Sopory 1999). Additionally, GLYI has been identified as dehydration-induced gene in foliage grass *Sporobolus stapfianus* (Blomstedt et al. 1998). In wheat, GLYI expression is induced in response to *F. graminearium* infection and NaCl and $ZnCl_2$ treatments (Lin et al. 2010). The assessment of stress transcriptome and proteomes in stress-sensitive and stress-tolerant varieties has clearly indicated the role of glyoxalase genes in stress response with stress-tolerant species exhibiting higher expression of glyoxalases even under nonstress conditions (Chao et al. 2005; Witzel et al. 2009; Sun et al.

2010). A comparative transcriptome profiling of salt-tolerant wild tomato and a salt-sensitive tomato cultivar revealed two GLYI genes being salt stress inducible only in the wild tomato suggesting a more effective detoxification system in the tolerant species (Sun et al. 2010). The inspection of root proteome of barley genotypes with contrasting response towards salinity stress also revealed glyoxalase proteins to be either downregulated or present at low levels in sensitive varieties compared to constitutive or increased expression in tolerant varieties (Witzel et al. 2009). Moreover, glyoxalase activity is also induced under stress conditions as reported in onion bulbs and pumpkin seedlings, providing further conformity to the role of glyoxalases in stress response (Hossain et al. 2009; Hossain and Fujita 2009).

13.6 Triose Phosphate Isomerase: Regulation Under Stress and Role in Maintaining Methylglyoxal Homeostasis

TPI, being an important component of glycolysis, has been well studied and is found to be highly conserved in nature, exhibiting roughly 50 % sequence conservation from bacteria to humans (Joseph-McCarthy et al. 1994). TPI adjusts the rapid equilibrium between DHAP and GAP, produced via aldolase during glycolysis and thereby serves an important physiological role which is also reflected through studies manifesting effects of TPI deficiency or loss in its activity in humans. TPI deficiency is a rare autosomal recessive multisystem genetic disease, characterized by reduced enzyme activity in all tissues leading to the elevation of DHAP levels in erythrocytes (Schneider 2000). It was initially described in humans in 1965 and is associated with a progressive and severe neurological disorder, characterized by chronic hemolytic anemia frequently leading to death in early childhood (Schneider et al. 1965). Although there is no indication that DHAP accumulation is toxic, but it is the spontaneous decomposition of DHAP to MG that results in extensive damage to the system (Phillips and Thornalley 1993) as it can readily modify both proteins and DNA molecules (Thornalley 2008).

In plants, different organelle-specific forms of TPI viz. cytosol and chloroplast localized forms have been reported, but much is still to be known about the role of these different isozymes (Pichersky and Gottlieb 1984). However, a plastid-localized TPI has been shown to be crucial for the transition from heterotrophic to autotrophic growth during post-germinative seedling establishment in *Arabidopsis* (Chen and Thelen 2010). It was found that the reduction in activity of the plastid-localized TPI leads to stunted growth and abnormal chloroplast development, probably due to MG toxicity developed as a result of DHAP and MG accumulation in the developing plastids. Further, the role of cytosolic TPI has also been investigated, where reduction in cTPI activity in roots of potato leads to significant changes in several pathways of carbon metabolism such as modifications in glycolysis, pentose phosphate pathway, amino acid pool, and lipid metabolism (Dorion et al. 2010).

TPI has also been reported to be involved in plant stress response and alteration in expression levels is observed under abiotic or biotic stress conditions (Riccardi et al. 1998; Morris and Djordjevic 2001). Transcript levels of cytosolic *TPI* from rice have been shown to rise gradually under submergence stress, reaching maxima at 24 h and are maintained thereafter till 48 h (Umeda and Uchimiya 1994). In fact after 20 h of submergence, a marked increase in *TPI* levels occurs in roots and culms in rice seedlings but no such change is observed in leaf tissues (Xu et al. 1994). Likewise, *cTPI* expression is also induced in response to water deficit conditions in maize (Riccardi et al. 1998) and iron deficiency in *Arabidopsis* (Thimm et al. 2001). Selective alterations with respect to *TPI* induction have also been reported in response to desiccation, salt, oxygen deprivation, and high temperature stress (Minhas and Grover 1991). Transcript levels increase under desiccation, salt, oxygen deprivation, and high temperature stresses in shoots and oxygen deprivation and high temperature treatment in roots (Minhas and Grover 1991). In addition, we have recently shown that there is an increase in *OscTPI* transcript, protein, and enzyme activity in rice in response to various abiotic stresses and MG treatment (Sharma et al. 2012). Interestingly, MG treatment led to a ~2-fold increase in *OscTPI* transcript levels and also induced corresponding activity in a concentration-dependent manner (Sharma et al. 2012). This may probably help in restoring the balance in the glycolytic cycle towards ATP generation and ultimately rescuing the plant from stress. Also, we checked the stress response of recombinant *OscTPI* in *E. coli* (Fig. 13.2).

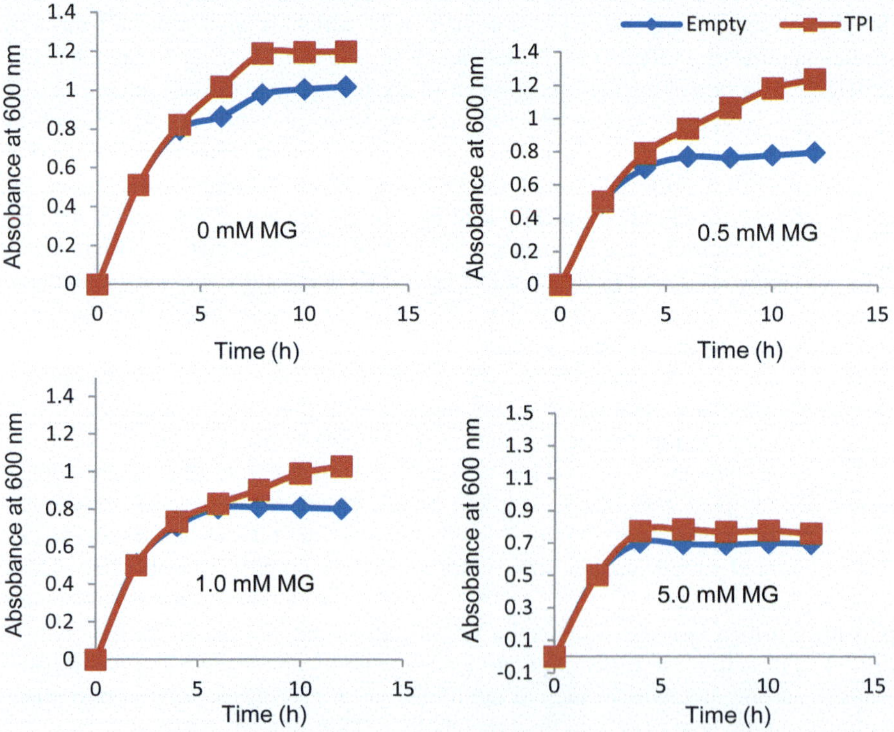

Fig. 13.2 Over-expression of *OsTPI* in *E. coli* confers tolerance towards methylglyoxal. A comparison of growth curves of untransformed *E. coli* (BL21) cells and *E. coli* cells transformed with pET28aOscTPI at different MG concentrations

The bacterial cells over-expressing *OscTPI* displayed better survival ability in response to low MG concentration (0.5 mM) whereas at 5 mM MG, no significant differences in growth pattern of untransformed and *OscTPI* transformed cells could be noticed. This is because higher MG concentration may be lethal to the bacteria.

Taken together, it can be said that environmental stresses affect mitochondrial respiration and photosynthetic processes thereby impairing oxygen uptake, energy levels, and ATP content (Botha et al. 1984; Purvis and Shewfelt 1993; Grass and Burris 1995; Tezara et al. 1999). As a result, glycolysis becomes the primary mode of energy production in plant tissues under low oxygen conditions, leading to increased activity of most of the glycolytic enzymes under such hypoxic and anoxic conditions (Mustroph and Albrecht 2003).

13.7 Methylglyoxal as a Signaling Molecule in Biological Systems

Despite the ubiquitous presence of glyoxalase pathway in biological systems and existence of multiple forms of these enzymes in plants, MG is still maintained at threshold levels (30–75 μM) even in glyoxalase over-expressing plants (Yadav et al. 2005a). These findings suggest a role for MG, which is much beyond a mere toxic compound. MG-mediated protein modifications besides having implications in glycation have also been investigated in the context of stress-induced signaling in different species. Role of MG as a signaling molecule has been demonstrated in yeast and animals but not much work has been done in plants.

MG is known to initiate signal transduction through the high osmolarity glycerol-mitogen-activated protein kinase (HOG-MAPK) cascade in yeast (Maeta et al. 2005). It activates the Yap1 transcription factor, which is important for the oxidative-stress response in *Saccharomyces cerevisiae*, thereby facilitating its translocation to nucleus (Maeta et al. 2004). Further, osmotic stress or MG stimulation has been shown to phosphorylate Hog1 via pbs2 leading to its translocation to nucleus, where transcription factors Msn2/4 are then recruited to the promoter region of stress-responsive genes possessing stress response elements (STRE). Also, MG can activate the uptake of Ca^{2+} in yeast cells, thereby stimulating the calcineurin/Crz1-mediated Ca^{2+} signaling (Maeta et al. 2005). In *Schizosaccharomyces pombe*, Pap1 along with the Sty1/Spc1 stress-activated protein kinase (SAPK) pathway has been identified as respective homologues of Yap1 transcription factor and HOG-MAPK pathway of *S. cerevisiae,* which are involved in response to MG toxicity (Zuin et al. 2005; Takatsume et al. 2006). MG is also reported to attenuate the rate of overall protein synthesis in *S. cerevisiae* by activating the protein kinase Gcn2 to phosphorylate the alpha subunit of translation initiation factor 2 (eIF2α) (Nomura et al. 2008).

In animals, MG is shown to induce signals, which trigger processes such as apoptosis, and also have implications in vascular complications of diabetes, through two distinct signal cascades, protein tyrosine kinase (PTK)-dependent control of

ERK1 and ERK2 (extracellular signal-regulated kinase) and PTK-independent redox-linked activation of JNK (c-Jun N-terminal kinase)/p38 MAPK and caspases (Akhand et al. 2001). It has been shown to induce oxidative stress-mediated apoptosis by facilitating the phosphorylation of p38 MAPK in nerve-derived Schwann cells of rat (Fukunaga et al. 2005). Recently, the phosphorylation of Akt1 (protein kinase B) by MG has been reported in adipose tissues, which thereby stimulates adipogenesis in obese Zucker rats (Jia et al. 2012).

In plants, until now very preliminary information is available regarding the role of MG in stress-induced signaling. The involvement of MG in inducing stomatal closure, ROS production, and cytosolic-free calcium concentration has been investigated in order to clarify its role in *Arabidopsis* guard cells (Hoque et al. 2012c). MG was found to reduce stomatal apertures in a dose-dependent manner with the process being reversible at only low MG concentration (1 mM). Further, it induced O_2^- production in whole leaves and ROS accumulation in the guard cells, which was completely abolished by 1 mM salicylhydroxamic acid (SHAM). MG at 1 mM concentration could increase cytosolic Ca^{2+} oscillations in the guard cells that were suppressed by pretreatment with 1 mM SHAM. Collectively, this data indicated that MG-induced stomatal closure involves an extracellular oxidative burst, which diffuses into the intracellular space leading to intracellular ROS accumulation in guard cells, and this result in stomatal closure via Ca^{2+}-dependent pathway (Hoque et al. 2012c). In addition, MG can also inhibit light-induced stomatal opening in *Arabidopsis* by inhibiting K^+ influx into the guard cells, which most likely occurs via the modification of C-terminal region of the inward-rectifying potassium channel (Hoque et al. 2012d). The role of MG in closure of stomatal aperture is believed to be important to withstand extreme environmental conditions.

The stress-responsive gene expression in *Arabidopsis* has been investigated using an ABA-deficient mutant, *aba2-2* in response to MG (Hoque et al. 2012a). For this purpose, the transcription of ABA-independent gene, *RD29A*- and ABA-dependent genes, *RD29B* and *RAB18,* was monitored in the presence and absence of MG in 2-week-old *Arabidopsis* wild type and *aba2*-2 mutant seedlings. The MG treatment did not affect transcript level of *RD29A* in either plant type, but significantly increased transcription levels of ABA-dependent genes, *RD29B* (fivefold at 1 mM MG) and *RAB18* (threefold at 1 mM MG) in concentration-dependent manner in the wild-type seedlings. However in the *aba2-2* mutant, MG did not induce *RD29B* or *RAB18* transcription even at 1.0 mM concentration. This suggested that MG can act through an ABA-dependent pathway to affect transcription and also plant developmental processes in *Arabidopsis*.

Our studies have shown that MG can induce *OscTPI* expression, which in turn, increases protein and enzyme activity (Sharma et al. 2012). This shifts the reaction towards GAP formation, thereby decreasing DHAP levels and consequently lowering MG concentration by feedback mechanism (Fig. 13.3). MG-dependent regulation of *OscTPI* seems advantageous to the system as increased *OscTPI* expression, on the one hand, is likely to reduce MG-mediated toxicity and on the other hand, will increase the energy status of the cell by adjusting the equilibrium towards GAP formation. Moreover, we have also observed an upregulation in transcript levels of several glyoxalase genes in rice (Mustafiz et al. 2011) and also non-MG metabolism-related

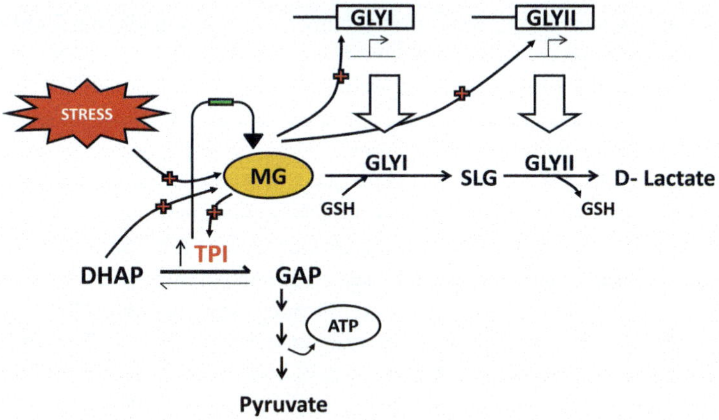

Fig. 13.3 MG, glyoxalase, and TPI interplay during stress. During stress, MG levels increase as a consequence of imbalance in metabolic pathways. Elevated MG levels increase TPI expression and activity, shifting the equilibrium towards completion of glycolytic cycle. Also, MG upregulates glyoxalase pathway genes thereby accelerating its detoxification through feedback mechanism and hence, keeping MG levels in check. "+" indicates activation and "–" indicates inhibition of the particular gene/pathway

gene, *OsETHE1*, in response to exogenous MG application (Kaur et al. 2014). We believe that MG has profound effect on transcriptome and proteome of plant species, which is indeed quite possible since MG is perceived as a stress by the plant, like various abiotic stresses. This is even supported by our recent transcriptome studies in rice in response to MG treatment, which also indicates towards a massive alteration in cellular functioning of a cell after MG application (unpublished data). Collectively, this data gives strong indication towards an important role of MG as signal molecule in plants as well, being involved in diverse stress adaptation pathways.

13.8 Functional Genomics Perspective in Triose Phosphate Isomerase and Glyoxalase Research

Functional genomics approaches, including transcriptome, proteome, and metabolome-based studies, are being increasingly used for simultaneous analysis of large number of genes. Using these techniques, it has become easier to analyze the effect of hormonal, chemical, or environmental responses on various aspects of plant development at a large scale. In this context, stress transcriptomes and proteomes have been assessed for alteration in gene and protein expression profiles in order to identify the underlying mechanism of plant defense and stress adaptation, which can help in raising stress-tolerant plants.

A number of such studies have led to the identification of glyoxalases as stress-responsive genes with increased expression in response to stress. For instance, transcriptome studies describing the response of *Arabidopsis* seedlings to usual contaminants present in soil, such as heavy metals and xenobiotic compounds, have indicated an involvement of glyoxalases under such conditions. A threefold induction in GLYII transcript in response to the xenobiotic, 2,4,6-trinitrotoluene (TNT); increased expression of GLYI on application of herbicide atrazine; or upregulation in response to arsenite; suggest a role of fundamental importance for glyoxalases in stress and adaptive responses (Ekman et al. 2003; Ramel et al. 2007; Chakrabarty et al. 2009). Further, ABA treatment in shoots of rice seedlings, or application of high hydrostatic pressure in germinating rice seeds, has also been shown to increase abundance of GLYI transcripts (Lin et al. 2003; Liu et al. 2008). Even comparative transcriptome studies in stress-sensitive and stress-tolerant plant varieties highlight the significance of glyoxalases. Highly salt-tolerant rice variety, Nona Bokra, showed increase in transcript levels of two glyoxalase genes under salt stress. In contrast, the salt-sensitive rice variety IR28, exhibited a significantly different regulation pattern for these genes (Chao et al. 2005).

Similarly, several proteome studies have reported an upregulation in levels of TPI protein in response to stress. Though the direct role of metabolism-related proteins in defense is less known but their role in maintaining metabolite pool in order to drive the metabolic processes for countering stress cannot be undermined. For instance, levels of TPI protein along with other enzymes of glycolysis were induced in response to oxidative stress, in the green alga *Haematococcus pluvialis* (Wang et al. 2004); and likewise exposure to cadmium in poplar also enhanced levels of TPI (Kieffer et al. 2009). Moreover, proteomic analysis of salt stress-responsive proteins in rice roots led to identification of TPI as one of the upregulated proteins (Yan et al. 2005). Likewise, inspection of salt-stressed proteome of the halophyte C_4 plant, *Aeluropus lagopoides,* revealed an induction in levels of both TPI and GLYI proteins (Sobhanian et al. 2010). Measuring metabolite content in shoots after salt treatment revealed a decrease in DHAP and corresponding increase in GAP levels, in correlation with the observed increase in TPI expression under salt stress. Also, reduced glutathione levels were found to be relatively lower under stress, which is logical since GSH serves an important role in maintaining redox homeostasis in cells, being used in several antioxidant reactions, such as one catalyzed by GLYI (Sobhanian et al. 2010). In addition, developmental events such as grain filling and seed maturation in barley using two-dimensional gel electrophoresis also revealed the involvement of glyoxalases in developmental processes. The glyoxalase protein was found to be present throughout the development process, but showed maximum accumulation at the desiccation stage, suggestive of its involvement in seed development. On the other hand, TPI was found to be expressed throughout the development and as multiple spots on the gel, probably serving a housekeeping role as a component of energy cycle (Finnie et al. 2002).

Thus, with the advent of high-throughput techniques, our knowledge regarding plant stress response has considerably increased, and in fact, these techniques have helped us better understand the underlying mechanisms of plant responses to various stimuli.

13.9 Conclusions

Methylglyoxal is a potent cytotoxin even in plants, which needs to be removed from the system to protect it from the undesirable effects of MG. In response to stress conditions, MG levels increase in plants thereby conferring toxicity in the system mainly through the modifications of protein and nucleotide moieties. MG produced as a result of both enzymatic and non-enzymatic reactions is an unavoidable consequence of stress, and thus detoxification mechanisms play an important role in reducing the concentration of MG from the system, which is largely generated from the triose sugars as a dissociable intermediate of the reaction catalyzed by TPI. This probably indicates an involvement of this glycolytic enzyme in stress response where expression and activity of TPI in the physiological systems and the alteration in stress due to imbalance in metabolism, may in part determine the rate of MG generation. The glyoxalase pathway by scavenging MG has been shown to confer tolerance to multiple stresses mainly by reducing MG levels and maintaining redox homeostasis, thus playing a role of immense significance. Moreover, MG initiates apoptosis in animals, controls stomatal movements in plants, and also induces gene expression. Thus, it is possible that similar to ROS, MG might also be acting as an important messenger in the intricate signaling cascade.

References

Ahmed N, Thornalley PJ (2002) Chromatographic assay of glycation adducts in human serum albumin glycated *in vitro* by derivatisation with aminoquinolyl-N-hydroxysuccimidyl-carbamate and intrinsic fluorescence. Biochem J 364:15–24

Ahmed N, Argirov OK, Minhas HS, Cordeiro CA, Thornalley PJ (2002) Assay of advanced glycation endproducts (AGEs): surveying AGEs by chromatographic assay with derivatisation by 6-aminoquinolyl-N-hydroxysuccimidyl-carbamate and application to *N*ε-carboxymethyl-lysine- and *N*ε-(1-carboxyethyl)lysine-modified albumin. Biochem J 364:1–14

Ahmed N, Babaei-Jadidi R, Howell SK, Beisswenger PJ, Thornalley PJ (2005) Degradation products of proteins damaged by glycation, oxidation and nitration in clinical type 1 diabetes. Diabetologia 48:1590–1603

Akhand AA, Hossain K, Mitsui H, Kato M, Miyata T, Inagi R, Du J, Takeda K, Kawamoto Y, Suzuki H, Kurokawa K, Nakashima I (2001) Glyoxal and methylglyoxal trigger distinct signals for map family kinases and caspase activation in human endothelial cells. Free Radic Biol Med 31:20–30

Aleksandrovskii YA (1992) Antithrombin III, C1 inhibitor, methylglyoxal, and polymorphonuclear leukocytes in the development of vascular complications in diabetes mellitus. Thromb Res 67:179–189

Alvarez Viveros MF, Inostroza-Blancheteau C, Timmermann T, González M, Arce-Johnson P (2013) Overexpression of GlyI and GlyII genes in transgenic tomato (*Solanum lycopersicum* Mill.) plants confers salt tolerance by decreasing oxidative stress. Mol Biol Rep 40:3281–3290

Babel W, Hofmann KH (1981) The conversion of triosephosphate via methylglyoxal, a bypass to the glycolytic sequence in methylotrophic yeasts? FEMS Microbiol Lett 10:133–136

Banu MN, Hoque MA, Watanabe-Sugimoto M, Islam MM, Uraji M, Matsuoka K, Nakamura Y, Murata Y (2010) Proline and glycinebetaine ameliorated NaCl stress via scavenging of hydrogen peroxide and methylglyoxal but not superoxide or nitric oxide in tobacco cultured cells. Biosci Biotechnol Biochem 74:2043–2049

Bhomkar P, Upadhyay CP, Saxena M, Muthusamy A, Prakash NS, Poggin K, Hohn T, Sarin NB (2008) Salt stress alleviation in transgenic *Vigna mungo* L. Hepper (blackgram) by overexpression of the glyoxalase I gene using a novel *Cestrum* yellow leaf curling virus (CmYLCV) promoter. Mol Breed 22:169–181

Blomstedt CK, Gianello RD, Hamill JD, Neale AD, Gaff DF (1998) Drought-stimulated genes correlated with desiccation tolerance of the resurrection grass *Sporobolus stapfianus*. Plant Growth Regul 24:153–161

Bonifati V, Rizzu P, van Baren MJ, Schaap O, Breedveld GJ, Krieger E, Dekker MC, Squitieri F, Ibanez P, Joosse M, van Dongen JW, Vanacore N, van Swieten JC, Brice A, Meco G, van Duijn

CM, Oostra BA, Heutink P (2003) Mutations in the DJ-1 gene associated with autosomal recessive early-onset parkinsonism. Science 299:256–259
Botha FC, Small JGC, Grobbelaar N (1984) The effect of water stress on the respiration and some aspects of respiratory metabolism of *Citrullus lanatus* seeds. Seed Sci Technol 12:585–595
Boyer JS (1982) Plant productivity and environment. Science 218:443–448
Brown BE, Dean RT, Davies MJ (2005) Glycation of low-density lipoproteins by methylglyoxal and glycolaldehyde gives rise to the in vitro formation of lipid-laden cells. Diabetologia 48:361–369
Casazza JP, Felver ME, Veech RL (1984) The metabolism of acetone in rat. J Biol Chem 259:231–236
Chakrabarty D, Trivedi PK, Misra P, Tiwari M, Shri M, Shukla D, Kumar S, Rai A, Pandey A, Nigam D, Tripathi RD, Tuli R (2009) Comparative transcriptome analysis of arsenate and arsenite stresses in rice seedlings. Chemosphere 74:688–702
Chao DY, Luo YH, Shi M, Luo D, Lin HX (2005) Salt-responsive genes in rice revealed by cDNA microarray analysis. Cell Res 15:796–810
Chaplen FW, Fahl WE, Cameron DC (1998) Evidence of high levels of methylglyoxal in cultured Chinese hamster ovary cells. Proc Natl Acad Sci U S A 95:5533–5538
Chen M, Thelen JJ (2010) The plastid isoform of triose phosphate isomerase is required for the postgerminative transition from heterotrophic to autotrophic growth in *Arabidopsis*. Plant Cell 22:77–90
Chen ZY, Brown RL, Damann KE, Cleveland TE (2004) Identification of a maize kernel stress-related protein and its effect on aflatoxin accumulation. Phytopathology 94:938–945
Cooper RA (1984) Metabolism of methylglyoxal in microorganisms. Annu Rev Microbiol 38:49–68
Crook EM, Law K (1952) Glyoxalase: the role of the components. Biochem J 52:492–499
Dorion S, Clendenning A, Jeukens J, Salas JJ, Parveen N, Haner AA, Law RD, Force EM, Rivoal J (2010) A large decrease of cytosolic triosephosphate isomerase in transgenic potato roots affects the distribution of carbon in primary metabolism. Planta 236:1177–1190
Ekman DR, Lorenz WW, Przybyla AE, Wolfe NL, Dean JF (2003) SAGE analysis of transcriptome responses in *Arabidopsis* roots exposed to 2,4,6-trinitrotoluene. Plant Physiol 133:1397–1406
Espartero J, Sanchez-Aguayo I, Pardo JM (1995) Molecular characterization of glyoxalase-I from a higher plant: upregulation by stress. Plant Mol Biol 29:1223–1233
Finnie C, Melchior S, Roepstorff P, Svensson B (2002) Proteome analysis of grain filling and seed maturation in barley. Plant Physiol 129:1308–1319
Fraire-Velázquez S, Balderas-Hernández VE (2013) Abiotic stress in plants and metabolic responses. In: Vahdati K, Leslie C (eds) Abiotic stress—plant responses and applications in agriculture. Intechopen, Rijeka. ISBN 978-953-51-1024-8
Fukunaga M, Miyata S, Higo S, Hamada Y, Ueyama S, Kasuga M (2005) Methylglyoxal induces apoptosis through oxidative stress-mediated activation of p38 mitogen-activated protein kinase in rat Schwann cells. Ann N Y Acad Sci 1043:151–157
Grass L, Burris JS (1995) Effect of heat stress during seed development and maturation on wheat (*Triticum durum*) seed quality. II. Mitochondrial respiration and nucleotide pools during early germination. Can J Plant Sci 75:831–839
Hegedüs A, Erdei S, Janda T, Tóth E, Horváth G, Dudits D (2004) Transgenic tobacco plants overproducing alfalfa aldose/aldehydes reductase show higher tolerance to low temperature and cadmium stress. Plant Sci 166:1329–1333
Hopper DJ, Cooper RA (1971) The regulation of Escherichia coli methylglyoxal synthase: a new control site in glycolysis? FEBS Lett 13:213–216
Hopper DJ, Cooper RA (1972) The purification and properties of Escherichia coli methylglyoxal synthase. Biochem J 128:321–329
Hoque MA, Uraji M, Banu MNA, Mori IC, Nakamura Y, Murata Y (2010) The effect of methylglyoxal on glutathione S-transferase from *Nicotiana tabacum*. Biosci Biotechnol Biochem 74:2124–2126

Hoque MA, Uraji M, Torii A, Banu MN, Mori IC, Nakamura Y, Murata Y (2012a) Methylglyoxal inhibition of cytosolic ascorbate peroxidase from *Nicotiana tabacum*. J Biochem Mol Toxicol 26:315–321

Hoque TS, Okuma E, Uraji M, Furuichi T, Sasaki T, Hoque MA, Nakamura Y, Murata Y (2012b) Inhibitory effects of methylglyoxal on light-induced stomatal opening and inward K+ channel activity in *Arabidopsis*. Biosci Biotechnol Biochem 76:617–619

Hoque TS, Uraji M, Tuya A, Nakamura Y, Murata Y (2012c) Methylglyoxal inhibits seed germination and root elongation and up-regulates transcription of stress-responsive genes in ABA-dependent pathway in *Arabidopsis*. Plant Biol (Stuttg) 14:854–858

Hoque TS, Uraji M, Ye W, Hossain MA, Nakamura Y, Murata Y (2012d) Methylglyoxal-induced stomatal closure accompanied by peroxidase-mediated ROS production in *Arabidopsis*. J Plant Physiol 169:979–986

Hossain MA, Fujita M (2009) Purification of glyoxalase I from onion bulbs and molecular cloning of its cDNA. Biosci Biotechnol Biochem 73:2007–2013

Hossain MA, Hossain MZ, Fujita M (2009) Stress-induced changes of methylglyoxal level and glyoxalase I activity in pumpkin seedlings and cDNA cloning of glyoxalase I gene. Aus J Crop Sci 3:53–64

Jia X, Chang T, Wilson TW, Wu L (2012) Methylglyoxal mediates adipocyte proliferation by increasing phosphorylation of Akt1. PLoS One 7:e36610

Joseph-McCarthy D, Lolis E, Komives EA, Petsko GA (1994) Crystal structure of the K12M/G15A triosephosphate isomerase double mutant and electrostatic analysis of the active site. Biochemistry 33:2815–2823

Kalapos MP (1999) Methylglyoxal in living organisms: chemistry, biochemistry, toxicology and biological implications. Toxicol Lett 110:145–175

Kalapos MP, Garzo T, Antoni F, Mandal J (1992) Accumulation of S-D lactoylglutathione and transient decrease of glutathione level caused by methylglyoxal load in isolated hepatocytes. Biochim Biophys Acta 113:159–64

Kaur C, Vishnoi A, Ariyadasa TU, Bhattacharya A, Singla-Pareek SL, Sopory SK (2013) Episodes of horizontal gene-transfer and gene-fusion led to co-existence of different metal-ion specific glyoxalase I. Sci Rep 3:3076

Kaur C, Mustafiz A, Sarkar A, Ariyadasa TU, Singla-Pareek SL, Sopory SK (2014) Expression of abiotic stress inducible ETHE1-like protein from rice is higher in roots and is regulated by calcium. Physiol Plant 152:1–16

Kieffer P, Schröder P, Dommes J, Hoffmann L, Renaut J, Hausman J-F (2009) Proteomic and enzymatic response of poplar to cadmium stress. J Proteomics 72:379–396

Ko J, Kim I, Yoo S, Min B, Kim K, Park C (2005) Conversion of methylglyoxal to acetol by *Escherichia coli* aldo-keto reductases. J Bacteriol 187:5782–5789

Koop DR, Casazza JP (1985) Identification of ethanol-inducible P-450 isozyme 3a as the acetone and acetol monooxygenase of rabbit microsomes. J Biol Chem 260:13607–13612

Kwon K, Choi D, Hyun JK, Jung HS, Baek K, Park C (2013) Novel glyoxalases from *Arabidopsis thaliana*. FEBS J 280:3328–3339

Lee JY, Song J, Kwon K, Jang S, Kim C, Baek K, Kim J, Park C (2012) Human DJ-1 and its homologs are novel glyoxalases. Hum Mol Genet 21:3215–3225

Leoncini G (1979) The role of alpha-oxoaldehydes in biological systems. Ital J Biochem 28:285–294

Lin F, Xu SL, Ni WM, Chu ZQ, Xu ZH, Xue HW (2003) Identification of ABA-responsive genes in rice shoots via cDNA macroarray. Cell Res 13:59–68

Lin F, Xu J, Shi J, Li H, Li B (2010) Molecular cloning and characterization of a novel glyoxalase I gene TaGly I in wheat (*Triticum aestivum* L.). Mol Biol Rep 37:729–735

Lin J, Nazarenus TJ, Frey JL, Liang X, Wilson MA, Stone JM (2011) A plant DJ-1 homolog is essential for *Arabidopsis thaliana* chloroplast development. PLoS One 6:e23731

Liu X, Zhang M, Duan J, Wu K (2008) Gene expression analysis of germinating rice seeds responding to high hydrostatic pressure. J Plant Physiol 165:1855–1864

Long GL, Kaplan NO (1968) D-lactate specific pyridine nucleotide lactate dehydrogenase in animals. Science 162:685–686

Lyles GA, Chalmers J (1992) The metabolism of aminoacetone to methylglyoxal by semicarbazide-sensitive amine oxidase in human umbilical artery. Biochem Pharmacol 43:1409–1414

Maeta K, Izawa S, Okazaki S, Kuge S, Inoue Y (2004) Activity of the Yap1 transcription factor in *Saccharomyces cerevisiae* is modulated by methylglyoxal, a metabolite derived from glycolysis. Mol Cell Biol 24:8753–8764

Maeta K, Izawa S, Inoue Y (2005) Methylglyoxal, a metabolite derived from glycolysis, functions as a signal initiator of the high osmolarity glycerol-mitogen-activated protein kinase cascade and calcineurin/Crz1-mediated pathway in *Saccharomyces cerevisiae*. J Biol Chem 280:253–260

Mankikar S, Rangekar P (1974) Effects of methylglyoxal on germination of barley. Fyton 32:9–16

Martins AM, Cordeiro CA, Ponces Freire AM (2001) *In situ* analysis of methylglyoxal metabolism in *Saccharomyces cerevisiae*. FEBS Lett 499:41–44

Meyerhof O, Lohmann K (1934) Über die enzymatische Gleichgewichtsreaktion zwischen Hexosediphosphorsäure und Dioxyacetonphosphorsäure. Biochem Z 271:89–110

Minhas D, Grover A (1991) Transcript levels of genes encoding various glycolytic and fermentation enzymes change in response to abiotic stresses. Plant Sci 146:41–51

Misra K, Banerjee AB, Ray S, Ray M (1995) Glyoxalase III from *Escherichia coli*: a single novel enzyme for the conversion of methylglyoxal into D-lactate without reduced glutathione. Biochem J 305:999–1003

Morgan PE, Sheahan PJ, Pattison DI, Davies MJ (2013) Methylglyoxal-induced modification of arginine residues decreases the activity of NADPH-generating enzymes. Free Radic Biol Med 61C:229–242

Morris AC, Djordjevic MA (2001) Proteome analysis of cultivar-specific interactions between *Rhizobium leguminosarum* biovar trifolii and subterranean clover cultivar Woogenellup. Electrophoresis 22:586–598

Murata K, Fukuda Y, Watanabe K, Saikusa T, Shimosaka M, Kimura A (1985) Characterization of methylglyoxal synthase in *Saccharomyces cerevisiae*. Biochem Biophys Res Commun 131:190–198

Mustafiz A, Singh AK, Pareek A, Sopory SK, Singla-Pareek SL (2011) Genome-wide analysis of rice and *Arabidopsis* identifies two glyoxalase genes that are highly expressed in abiotic stresses. Funct Integr Genomics 11:293–305

Mustroph A, Albrecht G (2003) Tolerance of crop plants to oxygen deficiency stress: fermentative activity and photosynthetic capacity of entire seedlings under hypoxia and anoxia. Physiol Plant 117:508–520

Narawongsanont R, Kabinpong S, Auiyawong B, Tantitadapitak C (2012) Cloning and characterization of AKR4C14, a rice aldo-keto reductase, from Thai Jasmine rice. Protein J 31:35–42

Nomura W, Maeta K, Kita K, Izawa S, Inoue Y (2008) Role of Gcn4 for adaptation to methylglyoxal in *Saccharomyces cerevisiae*: Methylglyoxal attenuates protein synthesis through phosphorylation of eIF2alpha. Biochem Biophys Res Commun 376:738–742

Phillips SA, Thornalley PJ (1993) The formation of methylglyoxal from triose phosphates. Investigation using a specific assay for methylglyoxal. Eur J Biochem 212:101–105

Pichersky E, Gottlieb LD (1984) Plant triose phosphate isomerase isozymes: purification, immunological and structural characterization, and partial amino Acid sequences. Plant Physiol 74:340–347

Pratt EA, Fung LW, Flowers JA, Ho C (1979) Membrane D-lactate dehydrogenase from Escherichia coli. Purification and properties. Biochemistry 18:312–316

Purvis AC, Shewfelt RL (1993) Does the alternative pathway ameliorate chilling injury in sensitive plant tissues? Physiol Plant 88:712–718

Rabbani N, Thornalley PJ (2012) Methylglyoxal, glyoxalase 1 and the dicarbonyl proteome. Amino Acids 42:1133–1142

Racker E (1951) The mechanism of action of glyoxalase. J Biol Chem 190:685–696

Ramel F, Sulmon C, Cabello-Hurtado F, Taconnat L, Martin-Magniette ML, Renou JP, El Amrani A, Couée I, Gouesbet G (2007) Genome-wide interacting effects of sucrose and herbicide-mediated stress in *Arabidopsis thaliana*: novel insights into atrazine toxicity and sucrose-induced tolerance. BMC Genomics 8:450

Ray S, Ray M (1981) Isolation of methylglyoxal synthase from goat liver. J Biol Chem 256:6230–6233

Riccardi F, Gazeau P, de Vienne D, Zivy M (1998) Protein changes in response to progressive water deficit in maize. Quantitative variation and polypeptide identification. Plant Physiol 117:1253–1263

Richard JP (1984) Acid–base catalysis of the elimination and isomerization-reactions of triose phosphates. J Am Chem Soc 106:4926–4936

Richard JP (1991) Kinetic parameters for the elimination reaction catalyzed by triosephosphate isomerase and an estimation of the reaction's physiological significance. Biochemistry 30:4581–4585

Richard JP (1993) Mechanism for the formation of methylglyoxal from triosephosphates. Biochem Soc Trans 21:549–553

Saito R, Yamamoto H, Makino A, Sugimoto T, Miyake C (2011) Methylglyoxal functions as Hill oxidant and stimulates the photoreduction of O(2) at photosystem I: a symptom of plant diabetes. Plant Cell Environ 34:1454–1464

Saxena M, Roy SB, Singla-Pareek SL, Sopory SK, Bhalla-Sarin N (2011) Overexpression of the Glyoxalase II gene leads to enhanced salinity tolerance in *Brassica juncea*. Open Plant Sci J 5:23–28

Schneider AS (2000) Triosephosphate isomerase deficiency: historical perspectives and molecular aspects. Baillieres Best Pract Res Clin Haematol 13:119–140

Schneider AS, Valentine WN, Hattori M, Heins HL Jr (1965) Hereditary hemolytic anemia with triosephosphate isomerase deficiency. N Engl J Med 272:229–235

Sharma S, Mustafiz A, Singla-Pareek SL, Shankar Srivastava P, Sopory SK (2012) Characterization of stress and methylglyoxal inducible triose phosphate isomerase (*OscTPI*) from rice. Plant Signal Behav 7:1337–1345

Simpson PJ, Tantitadapitak C, Reed AM, Mather OC, Bunce CM, White SA, Ride JP (2009) Characterization of two novel aldo-keto reductases from *Arabidopsis*: expression patterns, broad substrate specificity, and an open active-site structure suggest a role in toxicant metabolism following stress. J Mol Biol 392:465–480

Singla-Pareek SL, Reddy MK, Sopory SK (2003) Genetic engineering of the glyoxalase pathway in tobacco leads to enhanced salinity tolerance. Proc Natl Acad Sci U S A 100: 14672–14677

Singla-Pareek SL, Yadav SK, Pareek A, Reddy MK, Sopory SK (2006) Transgenic tobacco overexpressing glyoxalase pathway enzymes grow and set viable seeds in zinc-spiked soils. Plant Physiol 140:613–623

Singla-Pareek SL, Yadav SK, Pareek A, Reddy MK, Sopory SK (2008) Enhancing salt tolerance in a crop plant by overexpression of glyoxalase II. Transgenic Res 17:171–180

Sobhanian H, Motamed N, Rastgar Jazii F, Nakamura T, Komatsu S (2010) Salt stress induced differential proteome and metabolome response in the shoots of *Aeluropus lagopoides* (Poaceae), a Halophyte C4 Plant. J Proteome Res 9:2882–2897

Sun W, Xu X, Zhu H, Liu A, Liu L, Li J, Hua X (2010) Comparative transcriptomic profiling of a salt-tolerant wild tomato species and a salt-sensitive tomato cultivar. Plant Cell Physiol 51:997–1006

Takatsume Y, Izawa S, Inoue Y (2006) Methylglyoxal as a signal initiator for activation of the stress-activated protein kinase cascade in the fission yeast *Schizosaccharomyces pombe*. J Biol Chem 281:9086–9092

Tezara W, Mitchell VJ, Driscoll SD, Lawlor DW (1999) Water stress inhibits plant photosynthesis by decreasing coupling factor and ATP. Nature 401:914–917

Thimm O, Essigmann B, Kloska S, Altmann T, Buckhout TJ (2001) Response of *Arabidopsis* to iron deficiency stress as revealed by microarray analysis. Plant Physiol 127:1030–1043

Thornalley PJ (1990) The glyoxalase system: new developments towards functional characterization of a metabolic pathway fundamental to biological life. Biochem J 269:1–11

Thornalley PJ (1993) The glyoxalase system in health and disease. Mol Aspects Med 14:287–371

Thornalley PJ (1998) Glutathione-dependent detoxification of α-oxoaldehyde by the glyoxalase system: involvement in disease mechanisms and antiproliferative activity of glyoxalase I inhibitors. Chem Biol Interact 111–112:137–151

Thornalley PJ (2008) Protein and nucleotide damage by glyoxal and methylglyoxal in physiological systems - role in ageing and disease. Drug Metabol Drug Interact 23:125–150

Thornalley PJ, Battah S, Ahmed N, Karachalias N, Agalou S, Babaei-Jadidi R, Dawnay A (2003) Quantitative screening of advanced glycation end products in cellular and extracellular proteins by tandem mass spectrometry. Biochem J 375:581–592

Thornalley PJ, Waris S, Fleming T, Santarius T, Larkin SJ, Winklhofer-Roob BM, Stratton MR, Rabbani N (2010) Imidazopurinones are markers of physiological genomic damage linked to DNA instability and glyoxalase 1-associated tumour multidrug resistance. Nucleic Acids Res 38:5432–5442

Turk Z, Nemet I, Varga-Defteardarovic L, Car N (2006) Elevated level of methylglyoxal during diabetic ketoacidosis and its recovery phase. Diabetes Metab 32:176–180

Turóczy Z, Kis P, Török K, Cserháti M, Lendvai A, Dudits D, Horváth GV (2011) Overproduction of a rice aldo-keto reductase increases oxidative and heat stress tolerance by malondialdehyde and methylglyoxal detoxification. Plant Mol Biol 75:399–412

Umeda M, Uchimiya H (1994) Differential transcript levels of genes associated with glycolysis and alcohol fermentation in rice plants (*Oryza sativa* L.) under submergence stress. Plant Physiol 106:1015–1022

Veena RVS, Sopory SK (1999) Glyoxalase I from *Brassica juncea*: molecular cloning, regulation and its over-expression confer tolerance in transgenic tobacco under stress. Plant J 17:385–395

von Pechmann H (1887) Über die Spaltung der Nitrosoketone. Ber Dtsch Chem Gesells 20:3213–3214

Wang SB, Chen F, Sommerfeld M, Hu Q (2004) Proteomic analysis of molecular response to oxidative stress by the green alga *Haematococcus pluvialis* (Chlorophyceae). Planta 220:17–29

Wani SH, Gosal SS (2011) Introduction of OsglyII gene into *Oryza sativa* for increasing salinity tolerance. Biol Plant 55:536–540

Witzel K, Weidner A, Surabhi GK, Börner A, Mock HP (2009) Salt stress-induced alterations in the root proteome of barley genotypes with contrasting response towards salinity. J Exp Bot 60:3545–3557

Wu L, Juurlink BH (2002) Increased methylglyoxal and oxidative stress in hypertensive rat vascular smooth muscle cells. Hypertension 39:809–814

Wu C, Ma C, Pan Y, Gong S, Zhao C, Chen S, Li H (2013) Sugar beet M14 glyoxalase I gene can enhance plant tolerance to abiotic stresses. J Plant Res 126:415–425

Xu Y, Yu H, Hall TC (1994) Rice triosephosphate isomerase gene 5′ sequence directs [beta]-glucuronidase activity in transgenic tobacco but requires an intron for expression in rice. Plant Physiol 106:459–467

Yadav SK, Singla-Pareek SL, Ray M, Reddy MK, Sopory SK (2005a) Methylglyoxal levels in plants under salinity stress are dependent on glyoxalase I and glutathione. Biochem Biophys Res Commun 337:61–67

Yadav SK, Singla-Pareek SL, Ray M, Reddy MK, Sopory SK (2005b) Transgenic tobacco plants overexpressing glyoxalase enzymes resist an increase in methylglyoxal and maintain higher reduced glutathione levels under salinity stress. FEBS Lett 579:6265–6271

Yan S, Tang Z, Su W, Sun W (2005) Proteomic analysis of salt stress-responsive proteins in rice root. Proteomics 5:235–244

Zuin A, Vivancos AP, Sansó M, Takatsume Y, Ayté J, Inoue Y, Hidalgo E (2005) The glycolytic metabolite methylglyoxal activates Pap1 and Sty1 stress responses in *Schizosaccharomyces pombe*. J Biol Chem 280:36708–36713

Chapter 14
Plant Immunophilins: A Protein Family with Diverse Functions Beyond Protein Folding Activity

Lujing Shi and Aigen Fu

Abstract Immunophilins were discovered as cellular receptors for immunosuppressive drugs: cyclosporine A and FK506. Cyclophilins (CYPs, receptors for cyclosporine A) and FK506-binding proteins (FKBPs, receptors for FK506) do not share sequence homology, but have a common feature of peptidyl-prolyl *cis-trans* isomerase (PPIase) activity that catalyzes the *cis-trans* conversion of X-Pro peptide bonds, a rate-limiting step in the process of protein folding. Immunophilins are widely present in organisms from bacteria and fungi, to animals and plants. Genomics studies revealed that plants possess the largest immunophilin family. However, the physiological function of plant immunophilins is poorly understood. In this review, starting with a brief introduction of the immunophilin family and the current knowledge about their physiological roles in diverse organisms, the recent advances in the elucidation of plant immunophilin's physiological roles, largely via functional genomics tools, are summarized here. Notably, a striking feature of plant immunophilins is that a large fraction is localized in the chloroplast. Recent studies reveal that chloroplast immunophilins play a central role in the assembly and maintenance of photosynthetic complexes.

Keywords Plant immunophilins • Photosynthesis • Chloroplast • Stress response • Peptidyl-prolyl *cis-trans* isomerase • Plant development

L. Shi • A. Fu (✉)
College of Life Sciences, Northwest University,
229 North Taibai Road, Xi'an, Shaanxi 710069, China
e-mail: aigenfu@nwu.edu.cn

G.K. Pandey (ed.), *Elucidation of Abiotic Stress Signaling in Plants*,
DOI 10.1007/978-1-4939-2211-6_14

14.1 Introduction

Organ rejection due to immune response is a huge risk for patients receiving organ transplantation. During the course to screen immunosuppressive drugs, cyclosporin A (CsA), a nonpolar cyclic oligopeptide isolated from the fungus *Tolypocladium inflatum*, had been shown to reduce the immune response with minor side effects in the early 1970s (Calne et al. 1978; Starzl et al. 1981). After long time of clinical trials, CsA was approved for use in 1983, and it has been the primary tool to prevent immune rejection following kidney, heart, liver, and bone marrow transplantation (Svarstad et al. 2000). In an effort to understand the molecular mechanisms of CsA in immune suppression, the cellular receptor for CsA had been successfully purified from animal cells, and therefore named as cyclophilin (CYP) (Handschumacher et al. 1984). Almost the same time, two macrolide lactone metabolite: FK506 (also called tacrolimus or fujimycin) and rapamycin (also known as sirolimus) were discovered from the bacteria *Streptomyces tsukubaensis*, and shown to suppress the immune rejection of organ graft, similar to the pharmacological properties of CsA (Kino et al. 1987). FK-506 and rapamycin bind to a family of cellular receptors, which is named as FK-506-binding protein (FKBP) (Harding et al. 1989; Siekierka et al. 1989; Standaert et al. 1990). The two protein families of receptors for the immunosuppressive drugs, CYPs and FKBPs, are collectively referred as immunophilins (Schreiber 1991). CYPs and FKBPs do not share primary sequence similarity, but they do suppress immune response in a similar manner (Schreiber 1991; Schreiber and Crabtree 1992; Rao et al. 1997). Immunophilins recognize their respective cognate ligands to form receptor–ligand complexes, and then the complexes bind calcineurin, a calmodulin-dependent protein phosphatase, and further inhibit the phosphatase activity of calcineurin. The activity of calcineurin is required for activating the nuclear factor of activated T cells (NF-AT), which translocate from the cytosol to the nucleus and initiates the transcription of early T-cell activation genes related to immune response. The inhibition of calcineurin activity by immunophilin-drug complexes leads to the immunosuppressive effect of CsA and FK506 (Romano et al. 2004).

14.2 PPIase Activity and Protein Folding

Soon after cyclophilin was identified as the receptor of CsA, a peptidyl-prolyl *cis-trans* isomerase (PPIase) was purified and found to be identical to cyclophilin in two separate experiments (Fischer et al. 1989; Takahashi et al. 1989). The other groups of immunophilins, FKBPs, were also found to possess the PPIase activity like CYPs (Harding et al. 1989; Siekierka et al. 1989; Schreiber 1991). PPIases catalyze the *cis-trans* isomerization of peptide bonds between proline residue and its preceding residue in a protein sequence.

Right after the process of translation, newly synthesized linear primary peptide chains are rapidly folded into functional tertiary structures. Most peptide bonds in proteins are in the *trans* conformation; but peptidyl-prolyl bonds occur in both *cis* and *trans* conformations. The *cis* form of X-prolyl bonds are uncommon compared to the *trans* isomer, due to unfavorable contacts between prolyl residue and its adjacent amino acid residue (MacArthur and Thornton 1991). The folding of polypeptides happens on a millisecond timescale, but the isomerization of the X–prolyl bond is a much slower step, making it a rate-limiting step (Brandts et al. 1975; Kiefhaber et al. 1990). Since PPIases catalyze the rapid isomerization of the rate-limiting step in the folding process, it is also referred to as protein foldase or rotamase (He et al. 2004).

Discovery of PPIase activity of immunophilins led to the hypothesis that the inhibition of the PPIase function by CsA or FK-506 may be required for the suppression of the immune response. A synthetic FK-506 analogue, 506BD, binds to FKBP very efficiently and inhibits its PPIase activity, but 506BD does not suppress the immune response via the T-cell activation (Bierer et al. 1990). It revealed that PPIase activity of immunophilins is not related to their immunosuppressive function, and that the mechanism of the immune suppression was more complex than expected. Amino acid mutagenesis analysis revealed that key residues of immonuphilins essential for calcineurin binding differ from residues required for PPIase activity or drug-binding sites. And immunosuppressive drugs can bind with immunophilins lacking PPIase activity and still be able to inhibit the phosphatase activity of calcineurin (Zydowsky et al. 1992; Etzkorn et al. 1994; Futer et al. 1995). It demonstrated that there is no correlation between PPIase activity of immunophilins and their calcineurin inhibition activity.

The immunosuppressive activity is a gain-of-function for immunophilins in the presence of their ligands. However, it does not address the physiological functions of immunophilins in cells without the presence of their drug ligands. Although the PPIase activity of immunophilins has no physiological relevance in the process of immunosuppression, it still led to speculation that immunophilins may promote protein folding in vivo; and evidence for this role is accumulating (Luan 1998). For instance, CsA is able to inhibit collagen folding process in chick embryo fibroblast cells via inhibiting the PPIase activity of CYPs (Steimann et al. 1991). CPR3, a CYP member found in yeast mitochondria, has been shown to accelerate the folding of dihydrofoalte reductase (Matouschek et al. 1995). CYPs had been found to be able to accelerate the early folding of carbonic anhydrase and prevent its aggregation in a manner of chaperones (Freskgard et al. 1992). In a study of steroid-receptor folding, CYPs were also shown to assist protein folding as molecular chaperone (Freeman et al. 1996). NinaA, a cyclophilin in fruit flies, is found to be required for maintaining the normal configuration of rhodopsin in the endoplasmic reticulum (Stamnes et al. 1991).

Evidence also suggested that each member of immunophilins might interact specifically with their partner proteins. And those protein interactions may or may not be related to protein folding (Buchanan and Luan 2005). During the past two decades, a rapid growing number of immunophilins have been characterized from a wide range of organisms, including bacteria, fungi, algae, animals, and higher plants. The ubiquitous presence of immunophilins among wild-ranging organisms and their distributions in almost all the cellular compartments indicated that these proteins may play essential roles in important cellular processes.

14.3 Genomics Analysis of Immunophilin Family in Representative Organisms

Immunophilins are initially identified as the receptors of immunosuppressive drugs and the PPIases. Along with fast-accumulating data of genomics, immunophilins are identified by sequence comparison based on the PPIase domain of previously identified CYPs and FKBPs (He et al. 2004; Vallon 2005). Therefore, the definition of immunophilins is extended to proteins containing PPIase domain of CYP-type or FKBP-type immunophilins, whether they are active PPIase or not. Immunophilins may contain a single-PPIase domain, so called as single-domain immunophilins; whereas some immunophilins may comprise one or more additional functional domains, often referred to as multidomain immunophilins.

The presence of immunophilins from all tested organisms suggests that the PPIase catalytic domain represents an important portion of proteomes during evolution. Though the focus of this chapter is the plant immunophilins and their physiological roles, in order to demonstrate their functional and structural diversity of this protein family, a brief description of immunophilins found in representative organisms, besides plant species, is presented here.

14.3.1 Bacterial Immunophilins

Escherichia coli possesses at least five members of immunophilin family, including two cyclophilins (PpiA, and PpiB), and three FKBP proteins (FkpA, FkpC or SlyD, SlpA) (Justice et al. 2005; Romano et al. 2004).

PpiA is a homolog of human cyclophilin A, but it showed a very limited affinity to the immunosuppressive drug CsA. The protein is located in the periplasm, where it is believed to play a role in refolding of secreted proteins (Liu and Walsh 1990). The secondly identified cyclophilin in *E. coli* is named PpiB, which is a protein located in cytoplasm. However, the cytoplasmic PpiB is more similar to human cyclophilin A, which is also a cytoplasmic protein (Edwards et al. 1997). PpiB binds with CsA very efficiently, just as human cyclophilin A.

FkpA is a FKBP-type immunophilin located in periplasm, which is found to be required for the toxicity to *E. coli* by a toxin (colicin M). Active colicin M kills *E. coli* wild-type cells efficiently at a normal dosage, where mutant cells lack of FkpA are highly resistant to the toxin at a tenfold concentration. It is generally accepted that FkpA function as chaperone to facilitate to fold colicin M into an active form in the periplasm and the PPIase activity of FkpA is essential for the toxicity of colicin M (Hullmann et al. 2008). FkpA also plays a vital role in assisting the biogenesis of outer membrane proteins under heat shock stress conditions. Under normal condition, the function of FkpA is less significant than the other two periplasmic chaperones: SurA and Skp, where FkpA takes a leading role at heat shock response (Ge et al. 2014).

SlyD (sensitive to lysis D), also known as FkpC, is another FKBP-type immunophilin. SlyD is found to function in the biosynthesis of the metal cluster in the [NiFe]-

hydrogenase enzymes (Kaluarachchi et al. 2011; Chung and Zamble 2011). Site-directed mutagenesis analysis revealed that there is no connection between the PPIase activity of SlyD and the physiological roles of SlyD playing in hydrogenase activity (Zhang et al. 2007). Besides the PPIase domain, SlyD contains a metal-binding region and a small insert-in-flap (IF) domain. The IF domain, rather than the PPIase domain, is responsible for the chaperone activity of SlyD. SlyD also acts as a chaperone in the twin-arginine translocation (Tat) pathway, which translocates proteins through the bacterial cytoplasmic membrane to periplasm space (Graubner et al. 2007).

SlpA is a SlyD-like protein without the metal-binding regions. It consists of an FKBP-type PPIase domain and a small insert-in-flap (IF) domain which are responsible for its chaperone activity. It is suggested that SlpA is involved in ribosome assembly process (Quistgaard et al. 2012).

14.3.2 Yeast Immunophilins

At least eight cyclophilins and four FKBPs were identified in yeast *Saccharomyces cerevisiae*. Yeast cyclophilins are also designated as CPH (cyclophilin) or CPR (CsA-binding proline rotamase); and FKBP named as FPR (FK506-binding proline rotamase) or RBP (rapamycin-binding protein) (Dhillon 1996; Arevalo-Rodriguez et al. 2004). Usually, eight cyclophilins and four FKBPs in yeast are described as CPR1 to CPR8, and FPR1 to FPR4, respectively. Surprisingly, yeast mutants lack individual 12 immunophilins or all 12 immunophilins were viable, suggesting that none of 12 immunophilins is essential (Dolinski et al. 1997).

FPR1 is a cytosolic protein, sharing a high similarity to human FKBP12. Null mutations in the *FPR1* gene do not cause detectable defects under various growth conditions. However, the null mutant cells are totally insensitive to the growth inhibition by rapamycin, suggesting that FPR1 is the only receptor for rapamycin in yeast, and it is responsible for mediating the toxic effects of the drug (Heitman et al. 1991).

FPR2, a homolog of human FKBP13, is localized to both the endoplasmic reticulum and the microsomal fraction (Partaledis et al. 1992). It is believed that FPR2 plays a role in stress response since the level of FPR2 mRNA is upregulated when yeast cells are subjected to certain stress conditions (Partaledis and Berlin 1993).

FPR3 is multidomain FKBP-type immunophilin with a molecular weight of 70 kDa. FPR3 is localized to the nucleus, and it is found to be a physiological substrate of the tyrosine-specific phosphoprotein phosphatase Ptp1 (Wilson et al. 1995). FPR3 is also found to play a significant role in controlling meiosis checkpoint. FPR3 is able to interact with protein phosphatase 1 (PP1) through its PPIase domain, and further regulates PP1 localization to functions in cell division (Hochwagen et al. 2005).

FPR4 is another nucleus-localized FKBP, which shares similarity to FPR3. It contains an acidic domain in the N-terminus, a basic domain in the middle, and the conserved PPIase domain at the C-terminus. FPR4 possesses a histone chaperone activity; it is able to deposit histones onto DNA for nucleosome assembly. The chaperone activity of FPR4 resides in the N-terminal acidic domain, and the C-terminal PPIase domain inhibits the chaperone activity (Xiao et al. 2006).

CPR1 is a cytosolic cyclophilin with a molecular weight of 18 kDa. Since the absence of CPR1 causes a complete resistance to CsA, it is believed to be the primary receptor of CsA in yeast (Dhillon 1996). The physiological role of CPR1 in yeast is likely to mediate the defense response against oxidative stress conditions (Arevalo-Rodrıguez and Heitman 2005; Kim et al. 2007).

CPR2, CPR4, CPR5, and Cpr8 are cyclophilins localized to the endoplasmic reticulum (ER). It is believed that they play roles in assisting the folding process of proteins in the ER lumen or in the secretory pathway (Dhillon 1996; Dolinski et al. 1997; Arevalo-Rodriguez et al. 2004).

CPR3 is a 20 kDa cyclophilin residing in the mitochondrial matrix. It is likely to play a role in maintaining the proper conformation of proteins residing in the electron transfer chain at high temperatures (Davis et al. 1992).

CPR6 and CPR7, homologs of human cyclophilin-40, were shown to be able to form complexes with heat shock protein 90 (Hsp90), a well-known chaperone functioning in several signal transduction pathways (Duina et al. 1996). CPR6 and CPR7 can modulate the conformational changes of Hsp90 and stabilize Hsp90 structure (Zuehlke and Johnson 2012). However, CPR6 is much more stable than CPR7 under thermal denaturation condition, and CPR6 displays up to 100-fold higher PPIase activity than CPR7. However, the chaperone activity detected in CPR6 protein is much lower than that of CPR7 (Mayr et al. 2000).

14.3.3 Immunophilins in C. elegans

The *C. elegans* genome encodes 18 cyclophilins (CYP1–CYP18) and eight FKBPs (FKB1–FKB8). According to their cellular localization and functions, 24 immunophilins can be classified into four major classes; type-A, conserved cytosolic forms; type-B, secreted forms; type C, mitochondrial forms, and type-D, multidomain or divergent forms (Bell et al. 2006).

CYP-2, -3, and -7 can be classified into type-A cyclophilins, which are homologs of human CYPA. Global RNAi experiments showed that *C. elegans* exhibited wild-type phenotypes with reduced expression of the genes encoding these three isoforms. However, a severe inhibition of CYP-7 gene expression resulted in a phenotype of embryonic lethality (Kamath et al. 2003).

CYP5 and CYP6 are the type-B-secreted cyclophilins. Inhibition of gene expression of CYP5 and CYP6 by the RNAi method did not lead to any visible phenotypical changes, either individually or in combination (Picken et al. 2002). The CYP-1 is characterized as a type-C cyclophilin, since it contains a putative N-terminal mitochondrial signal peptide, but the function of CYP1 remains unclear (Page et al. 1996).

CYP4 and CYP8-18 belong to the type-D cyclophilins, which are divergent or multidomain cyclophilins. They are the largest class of CYPs in *C. elegans*, since this group includes 12 members. They have divergent CYP domains and/or contain additional noncyclophilin domains. They broadly function at diverse physiological process, including: muscle protein folding, sexual differentiation, protein folding, and RNA splicing (Bell et al. 2006).

The FKBP family comprises type-A (cytosolic forms), type-B (secreted forms), and type-D (divergent, multidomain forms). Interestingly, none of the FKBP family members was predicted to be localized into the mitochondria. This same situation could also be found in yeast: no FKBP present in yeast mitochondria.

FKB2 and FKB8 are two type-A cytosolic FKBPs in *C. elegans*. FKB2 is a single-domain FKBP, where FKB8 has two FKBP domains. These genes for FKB2 and FKB8 are closely arranged on chromosome I, suggesting they probably arose from a recent gene duplication event (Bell et al. 2006). There are 4 type-B secreted FKBPs (FKB-1, -3, -4, and -5) in *C. elegans*. They all contain both a secretory signal peptide and a conserved ER retention signal. FKB1 is a small single-domain FKBP, where the remaining FKB-3, -4, and -5 all have two FKBP domains.

The divergent type-D) isoforms consist of two members: FKB6 and FKB7. FKB6 contains TPR domains in addition to the FKBP domain. It shares homology to human FKBP51/52 structurally and functionally, which are steroid-receptor-associated Hsp-90-binding co-chaperones (Riggs et al. 2003). The other type-D FKBP, FKB-7, has a single FKBP domain and two EF-hand domains.

14.3.4 Human (Homo sapiens) Immunophilins

At least 18 FKBP isoforms are encoded in the human genome. Human FKBPs possess one to four PPIase domains with molecular weight varying from 13 to 135 kDa. Twenty four genes encoding cyclophilins were identified in the human genome with the homolog comparison to human CYPA. The molecular weight of human CYPs varies from 17 to 324 kDa (He et al. 2004). Genes for immunophilins are generally not linked to each other in the genome. Although the gain-of-functions of human immunophilins in the presence of immunosuppressive drugs are well studied, little is known about cellular functions of these 42 human immunophilins (Kumari et al. 2013).

14.3.5 Cyanobacterial Immunophilins

There are a total of four immunophilins in the cyanobacteria *Synochocystis* sp. PCC6803 genome, including one FKBP-type immunophilin: FKBP, and three cyclophilin-type immunophilns: PPIase A, PPIase B, and PPIase C. The functions of immunophilins in *Synochocystis* remain to be determined.

The cyanobacterial FKBP is a 201 AA protein, encoded by the gene slr1761. It contains a secretary pathway signal peptide in the N-terminus, which could direct it to localize in the periplasm or the thylakoid lumen. PPIase A (gene sll0408) is a protein with a sequence of 402 AA. It contains a secretary signal at the N-terminal, a PsbQ domain in the middle, and a PPIase domain at the C-terminal. The PPIase B is encoded by gene sll0227 with a length of 246 AA, it also contains a secretary signal peptide at N-terminal, and a PPIase domain at C-terminal. PPIase C is a 171 AA protein, which is encoded by the gene slr1251. It does not contain any signal peptide; it is likely a cytosolic protein.

14.3.6 Chlamydomonas Immunophilins

The eukaryote unicellular algae *Chlamydomonas reinhardtii* is an excellent system to study molecular processes and regulatory mechanisms of photosynthesis. Twenty three FKBPs and 28 cyclophilins were found in the *Chlamydomonas* proteome (Vallon 2005). Very interestingly, every single immunophilin protein could be found as one-to-one ortholog of *Arabidopsis* immunophilin protein family (Vallon 2005).

The physiological roles of all immunophilins were not experimentally characterized, except the smallest immunophilin, FKPB12. *Chlamydomonas* lacking FKBP12 are fully resistant to rapamycin, whereas the growth of wild-type cells is sensitive to the drug, indicating that this protein is responsible for the inhibition of cell growth by rapamycin. However, the cellular function of FKBP12 in *Chlamydomonas* remains unclear (Crespo et al. 2005).

14.4 Plant Immunophilins

14.4.1 Genomics

Along with the development of genomics, a full repertoire of immunophilin family was discovered in several higher plant species, including *Arabidopsis*, rice, wheat, and grapevine (Ahn et al. 2010; Gollan and Bhave 2010; Shangguan et al. 2013; Kumari et al. 2012; Wang et al. 2012). Usually immunophilin family in higher plants contains 52–60 protein members. A clear relation of orthology could be observed when comparing immunophilin family among vascular plants. However, some immunophilin proteins were novel; they could not match with any immunophilin protein in proteome of *Arabidopsis* and *Chlamydomonas*. Several imunnophilin genes are results from recent genetic duplication event.

The high similarity of the FKBP and cyclophilin family among eukaryotic photosynthetic organisms indicates that the diversity of immunophilin family occurred in their last common ancestor, probably an event close to the root of the "green" lineage of plants.

14.4.2 Arabidopsis thaliana Immunophilins

Arabidopsis thaliana is the best-studied vascular plant so far, most of the significant progress on plant biology was achieved on this model plant. The *Arabidopsis* genome sequence was released first in the year of 2000. Using previously identified FKBPs and CYPs as reference, a total of 52 immunophilins could be identified by sequence comparison, including 23 FKBPs and 29 CYPs (He et al. 2004). As for the immunophilin nomenclature, the proteins are generally named with prefix letters to indicate species (At for *Arabidopsis*) and a suffix number to show the hypothetic molecular weight of its mature form. For example, AtFKBP13 is a 13 kDa FKBP protein in *Arabidopsis*. The physiological roles of plant immunophilins are studied most extensively in *Arabidopsis* via functional genomics approaches. In this chapter, I will be focusing on the functional genomics studies on *Arabidopsis* immunophilins, while mentioning some processes on other higher plant species.

14.4.3 *Arabidopsis* FKBPs

Sixteen of the 23 identified AtFKBPs can be classified as single-domain FKBPs, which contain a single-FKBP domain, or additionally harbor a targeting peptide defining the subcellular localization (Fig. 14.1; He et al. 2004). Among these, 11

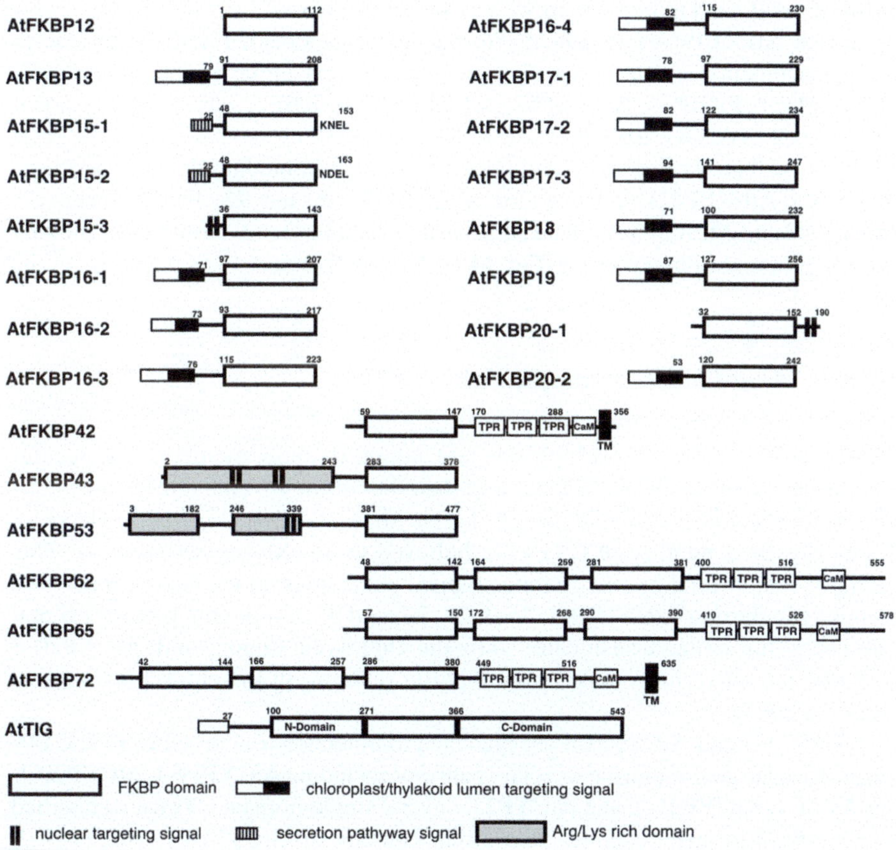

Fig. 14.1 Domain architecture of the *Arabidopsis* FKBPs (He et al 2004)

AtFKBPs contain chloroplast targeting sequences and two AtFKBPs possess ER targeting sequences. Interestingly, all the 11 chloroplast-localized FKBPs contain thylakoid lumen targeting signal peptides, suggesting they are all thylakoid lumenal proteins, which is confirmed by several proteomics studies (Peltier et al. 2002; Schubert et al. 2002). The lumenal AtFKBPs are quite small, with a molecular weight of 13–20 kDa.

Besides the 11 chloroplast FKBPs, the other 12 FKBPs were predicted to be located in the ER, nucleus, and cytosol. However, none of *Arabidopsis* FKBPs is predicted to be present in the mitochondria, just like the situation occurs in other organisms (Fig. 14.1). AtFKBP15-1 and AtFKBP15-2 are two putative ER-located proteins, which possess a secretory signal peptide at the N-terminus and an ER retention signal at the C-terminus. The nucleus-localized FKBP group includes AtFKBP15-3, AtFKBP20-1, AtFKBP43, AtFKBP53, and AtFKBP72, according to the presence of putative nuclear localization signal(s) in their protein sequences. The other five FKBPs may be located in the cytosol due to lack of any identifiable targeting signal (He et al. 2004; Romano et al. 2004). However, the precise locations of FKBPs still need to be verified with experimental works.

Several AtFKBPs are classified into the multiple-domain FKBP group due to additional functional domains in their sequences. This group includes AtFKB42, AtFKBP43, AtFKBP53, AtFKBP62, AtFKBP65, AtFKBP72, and a protein previously named as AtTIG. AtFKB42, AtFKBP62, AtFKBP65, and AtFKBP72 are characterized by their TPR domain and putative calmodulin-binding domain, besides a single (for AtFKBP42) or triple (for AtFKBP62, 65, and 72) FKBP domain(s). AtFKBP43 and AtFKBP53 contain Arg/Lys rich domain(s) at N-terminus, in addition to the C-terminal PPIase domain (Fig. 14.1; He et al. 2004).

14.4.4 Arabidopsis Cyclophilins

Twenty one out of 29 cyclophilins are single-domain members (molecular weight ranging from 18 to 37 kDa) and the other eight are characterized as multidomain proteins (molecular weight higher than 38 kDa) (Fig. 14.2; He et al. 2004). Again, a number of CYPs are believed to be chloroplast proteins. This group of CYPs (total of six) contains typical chloroplast at the N-terminus. Five members including AtCYP20-2, 26-3, 28, 37, and 38 contain the bipartite signal sequences for being translocated into the thylakoid lumen, and AtCYP20-3 could be the only chloroplast immunophilin residing in the chloroplast stroma (He et al 2004).

AtCYP21-3 and AtCYP21-4 are predicted to mitochondrial proteins. Five CYPs are predicted to go through the secretary pathway, including AtCYP19-4, AtCYP20-1, AtCYP21-1, AtCYP21-2, and AtCYP23. Several single-domain CYPs are classified to be cytosol-located proteins. This group consists of AtCYP18-1, AtCYP18-2, AtCYP18-3, AtCYP18-4, AtCYP19-1, At- CYP19-2, and AtCYP19-3. They are highly conserved and may have similar functions. AtCYP26-1 could be another

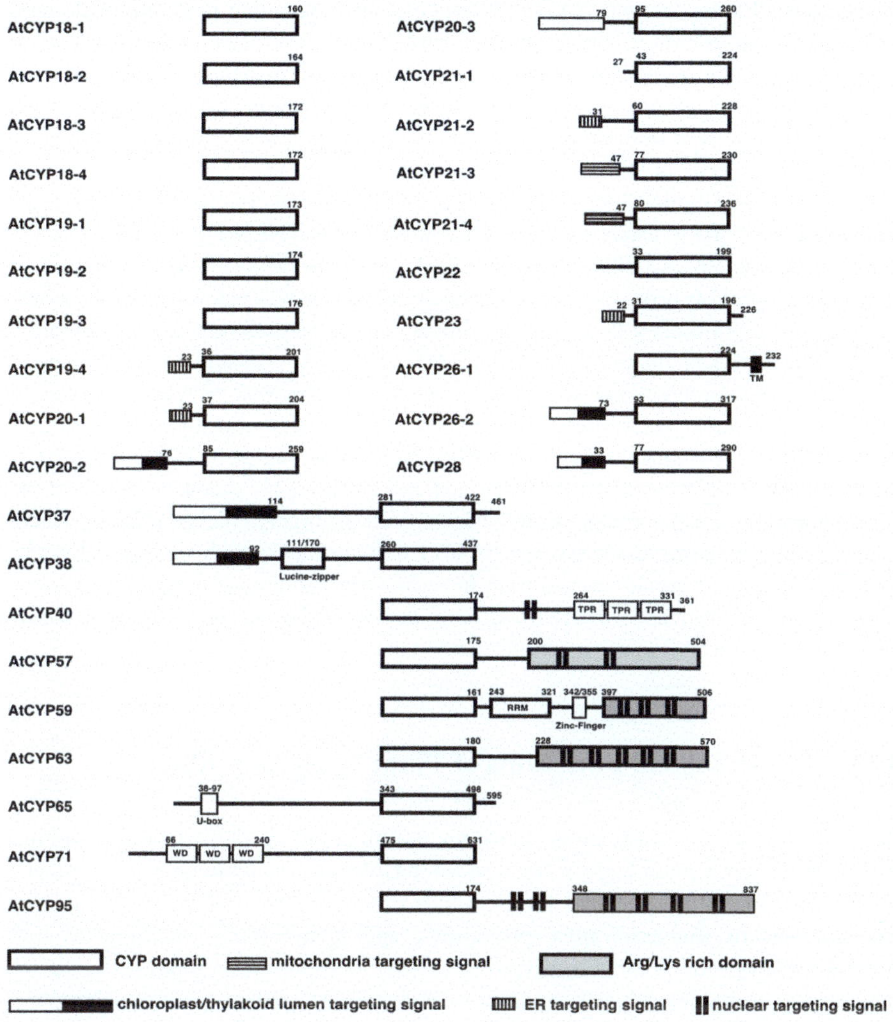

Fig. 14.2 Domain architecture of the *Arabidopsis* cyclophilins (He et al 2004)

cytosolic protein; however it is a possible membrane protein because of the potential transmembrane domain at its C-terminus.

Eight *Arabidopsis* cyclophilins are characterized as multiple-domain CYPs because of the presence of other functional domains. The chloroplast thylakoid lumenal protein AtCYP38 was regarded to contain a leucine-zipper domain at N-terminus of its mature form (He et al. 2004; Sirpiö et al. 2008), but X-ray analysis of crystal structure revealed that the N-terminus of AtCYP38 contains a PsbQ-like domain, rather than a leucine-zipper domain (Vasudevan et al. 2012). AtCYP40 contains an N-terminus PPIase domain, a nuclear targeting signal in the middle and three TPR domains at the C-terminus. AtCYP57 contains a domain with two NLS within it at the C-terminus, in addition to the

N-terminus PPIase domain. AtCYP59 contains an N-terminal cyclophilin domain followed by an RRM region, a putative zinc-finger motif, and a highly charged C-terminal domain with three putative nuclear localization signals (NLSs). The 90 amino acid RRM domain was found in a number of RNA-binding proteins (Birney et al. 1993). The zinc-finger domain is regarded to be involved in RNA binding, DNA binding, and protein–protein interaction (Matthews and Sunde 2002). AtCYP63 contains a N-terminal PPIase domain and a highly charged, Ser/Lys-rich domain at the C-terminus interrupted with five putative nuclear localization signals (NLSs). AtCYP65 is a 595 amino acid protein with a PPIase domain at the C-terminus and a U-box domain near the N-terminus. The U-box is a highly conserved domain present in ubiquitin ligases, which are responsible for adding polyubiquitin chains to proteins undergoing degradation by proteasome (Hatakeyama and Nakayama 2003). AtCYP71 contains a CYP domain at the C-terminus and three WD-40 repeats at the N-terminus. WD-40 repeat is about 40 amino acid domains involved in a protein–protein interaction. AtCYP95 is the largest immunophilin in *Arabidopsis* with a peptide chain consisting of 837 amino acid residues. It has a CYP domain at the N-terminal end and a highly charged domain at the C-terminus. Four NLS domains could be detected in the highly charged domain, where other two NLS domains reside between the CYP domain and the highly charged domain. The highly charged domain is also rich in serine and arginine residues (SR repeats), which is a common feature for RNA splicing factors (Zahler et al. 1992; Neugebauer et al. 1995).

14.5 Physiological Functions of Plant Immunophilins

Functional genomics tools open doors for the elucidation of plant immunophilin functions. Generation of large collections of T-DNA insertion mutants and the RNA interference methods allow the knock out or downregulation of genes of interest. The analysis of transcriptomes, proteomes, and metabolomes facilitate to link immunophilin proteins to their in vivo roles. By far, plant immunophilins are demonstrated to be involved in several important processes, such as plant development, stress response, and chloroplast functions.

14.5.1 The Roles of Immunophilins in Plant Development

14.5.1.1 AtFKBP12

AtFKBP12 is the smallest immunophilin, which is a cytosolic protein. It is demonstrated that AtFKBP12 is able to interact with a DNA-binding protein, AtFIP37. AtFKBP12 is believed to play important roles in mRNA splicing, cell-cycle regulation, and embryo development (Faure et al. 1998; Vespa et al. 2004). FKBP12 in the conifer species *Picea wilsonii* was shown to play a role in pollen tube development through interaction with HAP5, a subunit of a histone-associated transcription factor (Yu et al. 2011). These results suggest that plant FKBP12 functions in regulating gene expression via an epigenetic manner.

14.5.1.2 AtFKBP42 (TWD1)

Arabidopsis twisted dwarf 1 (*TWD*) mutants showed developmental abnormalities with stunted growth and helical rotation of roots and shoots (Geisler et al. 2003), and it has been found to be caused by the disruptions of the *AtFKBP42* gene. AtFKBP42 consists of an inactive PPIase domain at N-terminus, followed by three TPR motifs and a calmodulin-binding domain, and a transmembrane domain at its C-terminus (Granzin et al. 2006).

AtFKBP42 has been shown to interact with the C-terminal domain of both AtPGP1 and AtPGP19, homologues of *Arabidopsis* p-glycoprotein ABC transporter (Geisler et al. 2003). Although AtFKBP42 is shown to be an inactive PPIase, the PPIase domain is required for the interaction between the AtFKBP42 and ABC transporters. AtFKBP42 also interacts with the multidrug resistance-associated MRP/ABCC-like ABC transporters AtMRP1, and its homologue AtMRP2 (Geisler et al. 2004). ABC transporters are responsible for cellular auxin efflux, lack of AtFKBP42 leads to a malfunction of assembly of auxin transport machinery in the plasma membrane, consequently the interrupted transport of auxin results in the dramatic phenotype of the *twd1* mutant (Bouchard et al. 2006; Wu et al. 2010).

14.5.1.3 AtFKBP53

The AtFKBP53 is a multidomain FKBP with a PPIase domain and several highly charged domains. It is reported that AtFKBP53 is able to interact with histone H3 through its highly charged acidic domains, acting as a histone chaperone. Interestingly the highly charged acidic domains are sufficient for the histone chaperone- or histone-binding activity, whereas the PPIase domain is dispensable. AtFKBP53 is found to localize to chromatin harboring 18S ribosomal RNA genes. The reduction or elimination of AtFKBP53 activity leads to overexpression of 18S rRNA gene in *Arabidopsis*. It indicated that AtCYP53 plays a role to repress the transcription of 18S rRNA genes at the chromatin level (Li and Luan 2010).

The highly charged domain at N-terminal end of AtFKBP53 is also present in its duplicate AtFKBP43 and in their paralogues in rice, OsFKBP53a and OsFKBP53b. Single-knockout mutants of AtFKBP43 or AtFKBP53 do not show abnormal growth phenotypes, indicating a functional overlap between nuclear duplicates (Li and Luan 2010).

14.5.1.4 AtFKBP72

AtFKBP72 dysfunction causes a block in the biosynthesis of VLCFAs (very long chain fatty acids), which are essential for normal cell division and differentiation (Roudier et al 2010). The *Arabidopsis* mutant plants exhibited a phenotype known as "PASTICCINO" caused by unregulated cell division (Vittorioso et al. 1998). AtFKBP72, also known as PAS1, is regarded as a scaffold for the assembly of the VLCFA elongation complex via interacting with VLCFA synthesis enzymes (Vittorioso et al. 1998). AtFKBP72 is found to regulate cell proliferation by assisting translocation of the FAN transcription factor to the nucleus (Carol et al. 2001). The C-terminal CaM domain of AtFKBP72 is required for interactions with the VLCFA elongase and with FAN transcription factor (Vittorioso et al. 1998; Smyczynski et al. 2006).

The C-terminal domain of AtFKBP72, containing a membrane anchor and a CaM module, regulates its cellular distribution (Smyczynski et al. 2006). It is hypothesized that the membrane anchor tethers AtFKBP72 in the cytosol, then a signal for cell differentiation causes CaM bind to its partners, which is disrupting the membrane anchor and allowing the complex translocate into the nucleus.

14.5.1.5 AtCYP18-3 (ROC1)

AtCYP18-3 is a cytosolic immunophilin, which is also identified as ROC1 (Rotamase cyclophilin 1). *Arabidopsis* mutants with altered levels of AtCYP18-3 have defects in the plant de-etiolation process, which are mediated by plant light signal receptors, such as phytochrome A, phytochrome B, and cryptochrome 1. These mutants also show higher sensitivity to plant hormones, brassinosteroids, in the light but not in the dark. The AtCYP18-3 mutants displayed different phosphorylation patterns of BES1, the transcription factor for controlling plant sensitivity to brassinosteroids. It is likely that perception of light by phyA, cry1, or phyB activates AtCYP18-3; in turn AtCYP18-3 reduces the intensity of brassinosteroid signaling and finetunes seedling de-etiolation (Trupkin et al. 2012).

A gain-of-function mutant AtCYP18-3 with serine173 converted to phenylalanine in *Arabidopsis* mutant *roc1* showed a phenotype of reduced stem elongation and increased shoot branching. And such phenotypes displayed in mutant plants are strongly affected by temperature and photoperiod. Downregulation of AtCYP18-3 level through an RNAi method did not cause obvious growth alteration. It was found that exogenous plant hormone gibberellic acid (GA) inhibited stem elongation even more in the gain-of-function mutant. GA treatment was found to induce the accumulation of the mutated AtCYP18-3 protein; however the level of the wild-type protein was not affected. The AtCYP18-3 mutation does interfere with GA biosynthesis or the GA signaling pathway (Ma et al. 2013). Under normal growth conditions, the wild-type AtCYP18-3 proteins do not play a role in the process of stem elongation. It is believed the point mutation in the *AtCYP18*-3 gene resulted in the mutated protein a new function, in which the mutated AtCYP18-3 could interact with new target proteins. And the new target proteins enable the mutated AtCYP18-3 to function in regulating stem elongation (Ma et al. 2013)

AtCYP18-3 was shown to interact with *Agrobacterium* VirD2, an endonuclease covalently bound to the T-DNA. Because T-DNA transfer is inhibited by CsA, CYP interaction with VirD2 may be important for Agrobacterial infection (Deng et al. 1998).

14.5.1.6 AtCYP19-2 (CYP2, ROC6)

Auxin is a vital plant hormone functioning in plant development. It is illustrated that auxin can bind auxin/IAA receptor proteins, which are suppressors for the expression of auxin/IAA responsive genes. The binding with auxin induces the interaction between auxin/IAA proteins and the ubiquitin protein ligases, which further leads to the degradation of the AUX/IAA proteins. A mutation in rice CYP19-2 gene was found to show defects in auxin responses. In the rice *cyp19-2* mutants, lateral root initiation was arrested before the first anticlinal division of the pericycle cell. An interaction between OsCYP19-2 and the co-chaperone OsSGT1, a suppressor functioning in the auxin signal pathway, was detected via yeast two-hybrid and in vitro pull-down assays. It is likely that OsCYP19-2 plays a role in auxin signal transduction via interacting with OsSGT1 (Kang et al. 2013).

Deficiency of CYP19-2 in rice causes developmental defects in rice, but no visible altered growth phenotype was observed in *Arabidopsis cyp19-2* mutants. Sequence comparison showed no close homolog of CYP19-2 was found in rice, and 3 CYPs are found to be highly similar to CYP19-2 in *Arabidopsis* (Tridedi et al. 2012). The redundancy of *Arabidopsis* CYP19-2 homologues might explain the phenotypical discrepancy of *cyp19-2* mutants in rice and in *Arabidopsis*.

14.5.1.7 AtCYP40

AtCYP40 is an ancient, highly conserved protein from yeast, animal to plants. It contains an N-terminal PPIase domain, a nuclear targeting signal in the middle and three TPR domains at the C-terminus. *Arabidopsis* mutants lack of AtCYP40 showed reduced number of juvenile leaves but had no effect on reproductive growth (Berardini et al. 2001). Further analysis revealed that AtCYP40 is able to activate the AGO1 protein, which is primarily responsible for miRNA-directed posttranscriptional silencing. It is suggested that AtCYP40 regulates vegetative phase change via the miRNA-directed posttranscriptional silencing mechanism in *Arabidopsis* (Smith et al. 2009). AtCYP40 is also shown to interact with cytoplasmic Hsp90 proteins. The interaction between AtCYP40 and Hsp90 requires the TPR domain in the N-terminal of AtCYP40 (Earley and Poethig 2011). The interaction of AtCYP40 and Hsp90 could be an upstream event to affect AGO1 by AtCYP40.

14.5.1.8 AtCYP71

AtCYP71 is a multidomain immunophilin containing a WD40 domain. It plays a role in organogenesis, probably via regulating gene repression in *Arabidopsis*. Disruption of AtCYP71 gene resulted in activation of genes that are regulating meristem development (Li et al. 2007). The *atcyp71* mutant plants displayed dramatic developmental defects, including abnormal lateral organ formation, and arrested root growth. AtCYP71 was found to be associated with the chromatin and can physically interact with histone H3. The *atcyp71* mutant showed reduced level of methylation of H3K27 of genes involved in meristem development, which is correlated with the de-repression of these genes in the mutant. It is proposed that AtCYP71 serves as a histone remodeling factor playing key roles in chromatin-based gene silencing in plants (Li et al. 2007).

AtCYP71 was also shown to physically interact with AtFKBP72 (a chromatin assembly factor) and LHP1 (a heterochromatin protein), respectively, to control their distribution on chromatin. The *lhp1cyp71* double mutant was observed in more severe phenotypes than that of each single mutant, indicating that AtCYP71 and LHP1 proteins synergistically control plant development. It is illustrated that the association between LHP1 with its target loci requires AtCYP71 function. Again AtCYP71 is required for the function of AtFKBP72 on its target locus. AtCYP71 is essential for chromatin assembly and histone modification in *Arabidopsis* (Li and Luan 2011).

14.5.2 Roles in Defense Against Biotic Stress

14.5.2.1 AtCYP18-3 (ROC1)

AtCYP18-3, also known as ROC1 was identified as is an activator of AvrRpt2, an effector secreted by *P. syringae* triggering plant resistance. AvrRpts is a cysteine protease, which is responsible for the cleavage of RIN4, a negative regulator of plant immunity. RIN4 cleavage serves as a signal to activate plant resistance. The PPIase enzymatic activity of AtCYP18-3 is responsible for the conversion of AvrRpt2 from an unfolded to a folded state (Coaker et al. 2005, 2006).

AtCYP18-3 is also found to inhibit the replication of tomato bushy stunt tombusvirus (TBSV), via binding to the viral replication proteins and the viral RNA. The binding effect of AtCYP18-3 to the virus causes malfunction in several important processes in viral replication, such as RNA recruitment, the assembly of the replicase complex, and RNA synthesis (Kovalev and Nagy 2013).

14.5.2.2 At19-3, AtCYP57, and AtFKBP65

In a study of plant and microbe interaction, three immunophilins, AtCYP19-1 (ROC3), AtCYP57, and AtFKBP65 (ROF2), were found to be involved in plant response to microbial infection. AtCYP19-1 is a cytosolic protein, AtCYP57 and AtFKBP65 could be localized into the nucleus. At first, the expression of *AtCYP19-1*, *AtCYP57*, and *AtFKBP65* genes were found to be locally induced in *Arabidopsis* after inoculated with *P. syringae*. *Arabidopsis* triple mutants that lack these three genes showed an increased susceptibility to *P. syringae* compared to wild type, whereas overexpression of these genes in *Arabidopsis* led to an enhanced resistance to the pathogen. The mechanism of these immunophilins function in this aspect is still unclear, though there is evidence showing that AtCYP19 could be involved in reactive oxygen species production, and both AtCYP57 and AtFKBP65 play a role in the cell (Pogorelko et al. 2014).

14.5.3 Roles in Defense Against Abiotic Stress

14.5.3.1 AtFKBP62 (ROF1) and AtFKBP65 (ROF2)

The multidomain isoforms AtFKBP62 (ROF1) and AtFKBP65 (ROF2) share 85 % identity and it is generally believed that they are homologous duplicates (He et al. 2004). ROF1 and ROF2 work antagonistically in response to high temperature via controlling the expression of some small heat shock proteins (sHSP) (Aviezer-Hagai et al. 2007; Meiri and Breiman 2009; Meiri et al. 2010). AtFKBP62 binds HSP90 via its TPR domain, and further binds the heat shock transcription factor HsfA2 to form the AtFKBP62–HSP90–HsfA2 complex. The resulting complex relocates from the cytosol to the nucleus, and induced expression of sHSPs and AtFKBP65 (Meiri and Breiman 2009). On the other hand, AtFKBP65 binds to AtFKBP62 at its FKBP domain; and the interaction between AtFKBP65 and AtFKBP62 interrupts the nuclear AtFKBP62–HSP90–HsfA2 complex. As a result, sHSP expression will be downregulated after heat shock (Meiri et al. 2010). *Arabidopsis* mutants lacking AtFKBP62 showed severely compromised capacity to deal with high temperature stress, similar to the effect of HsfA2 deficiency. Mutant plants lack AtFKBP65 and plants overexpressing AtFKBP62 showed better long-term heat tolerance than the wild-type plants (Meiri et al. 2010).

AtFKBP62 and AtFKBP65 were recently shown to be involved in maintaining the intracellular pH and membrane polarity. The *rof1rof2* double-knockout mutant displayed a higher sensitivity to acid inhibition than single *rof1* or *rof2* knockouts, indicating a collaborative functionality of this pair (Bissoli et al. 2012). AtFKBP62/AtFKBP65 pairs could also be found in other higher plants, such as wheat (Blecher et al. 1996), rice (Magiri et al. 2006), and maize (Yu et al. 2012). Their domain structures and expression profiles are homologous to their counterparts in *Arabidopsis*, suggesting that the antagonistic collaboration of the AtFKBP62/AtFKBP65 pair is a well-conserved mechanism in higher plants.

14.5.3.2 FKBP20-1a and FKBP20-1b

OsFKBP20-1a and OsFKBP20-1b, two FKBPs found in rice, share 85 % identity. The expression level of these two genes is locally induced under treatment of heat and desiccation stress, although expression profiles of each gene are different. Right after high temperature and drought stress, the expression of OsFKBP20-1a is induced to a very high level in all tissues, while elevated expression levels of OsFKBP20-1b gene are observed after 24 h of stress treatment. Both OsFKBP20-1a and OsFKBP20-1b are located in the nucleus, but OsFKBP20-1b could be found in the cytosol as well (Ahn et al. 2010). A physical interaction was found between OsFKBP20-1a and the SUMO-conjugating enzyme (Sce), which adds ubiquitin-like modifier (SUMO) proteins to their target proteins. It is suggested that the function of OsFKBP20-1a in heat stress response is distributed to its interaction with Sce (Nigam et al. 2008).

14.5.3.3 FKBP15-1 and FKBP15-2

Besides the AtFKBP62/AtFKBP65 and the OsFKBP20-1a/OsFKBP20-1b pairs, a third set of FKBP duplicates involved in plant stress response is FKBP15-1 and FKBP15-2, which share 70 % identity. Both of them are believed to be localized to the ER, although their function in this cellular compartment remains unknown. It is found that the expression of AtFKBP15-1 and AtFKBP15-2 are upregulated under heat stress (He et al. 2004). The expression level of ZmFKBP15-3, a third duplicate of FKBP15-1/FKBP15-2 identified in maize, is strongly upregulated under high temperature, while the expression level of ZmFKBP15-1 and ZmFKPB15-2 displayed a stable pattern during heat stress (Yu et al. 2012).

14.5.4 The Chloroplast: A Focal Point of Immunophilin Functions

14.5.4.1 Lumenal PPIase

There are 17 immunophilins located in the chloroplast. A striking feature of chloroplast immunophilin proteins is that 16 are predicted to be present in the thylakoid lumen, except the chloroplast stroma-localized AtCYP20-3 (ROC4). The compact space of thylakoid lumen contains about 100 proteins. The immunophilin family proteins contribute 15 % of thylakoid lumen proteome, and most of them are abundant proteins, which could be easily detected by proteomics studies. The vast number of immunophilins in the lumen and their PPIase activity lead to an assumption that the protein-folding process catalyzed by PPIase would be essential for chloroplast function (He et al 2004).

Sequence analyses demonstrated that only AtCYP20-2 and AtFKBP13 possess all of the amino acid residues essential for PPIase activity, which are characterized by site-directed mutagenesis studies performed on the human CYPA and FKBP12. The PPIase enzyme assay confirmed that AtCYP20-2 and AtFKBP13 are robust PPIase enzymes. The other 14 immunophilins, lacking most of the key amino acid residues for PPIase activity, were found to be inactive PPIase in an in vitro assay. The thylakoid isolated from *Arabidopsis* mutant plants lack of both AtFKBP13 and AtCYP20-3 cannot catalyze the cis/trans isomerization of L-prolyl peptide bonds, indicating that chloroplast immunophilin proteins, except AtFKBP13 and AtCYP20-3, are not active PPIase (Shapiguzov et al 2006, Edvardsson et al 2007).

Surprisingly, *Arabidopsis* mutants lacking each or both of AtFKBP13 and AtCYP20-2 did not show any visible altered phenotype compared to the wild-type plants under normal, cold stress or high light conditions. Photosynthetic capacity and protein composition were not affected by the deficiency of two PPIase enzymes. The dispensability of PPIase activity of lumen immunophilins suggested that the functions of these proteins are not related to their PPIase capacity and should be investigated beyond this enzymatic activity (Ingelsson et al 2009).

14.5.4.2 Cyclophilin 20-3 (ROC 4)

AtCYP20-3, also designated as ROC4, is the only chloroplast immunophilin found in the stroma of chloroplasts. It contains all key amino acid residues required for PPIase activity, and it is confirmed that AtCYP20-3 is an active PPIase (Lippuner et al 1994). An *Arabidopsis* T-DNA-knockout mutant lack of AtCYP20-3 did not display inefficient growth or development, or deficiency of photosynthetic capacity or protein composition under normal growth condition. However, a very fine photoinhibition assay found an increased sensitivity to high light stress in the mutant, suggesting that AtCYP20-3 could play a role in the repair of photodamaged PSII (Cai et al. 2008).

AtCYP20-3 contains two pairs of conserved cysteines, which could form intramolecular disulfide bonds. A site-directed mutagenesis assay showed the PPIase activity of AtCYP20-3 is regulated by a mechanism of dithiol-disulfide transitions. The activation of AtCYP20-3 activity is regulated by chloroplast peroxiredoxins (Prx) under light in a response to the redox state of chloroplast (Laxa et al. 2007). AtCYP20-3 function is controlled by 2-Cys peroxiredoxin (2-CysPrx). 2-CysPrx is a multifunctional regulator that changes different conformational states depending on its redox state.

During oxidative stress conditions in the presence of high light condition and with the application of rose bengal, *Arabidopsis* mutant plants lacking AtCYP20-3 displayed defective growth and development. In a yeast two-hybrid screening assay, chloroplast serine acetyltransferase (SAT1) was identified as an interacting partner for AtCYP20-3. The activity of SAT1, a key enzyme in cysteine biosynthesis, increased significantly under stress conditions in wild-type plants, but not in *cyp20-3* mutant plants, indicating that AtCYP20-3 is required for the induced upregulation of SATl activity under stress condition. Together with that AtCYP20-3 foldase activity is enhanced by thioredoxin-mediated reduction, it is suggested that AtCYP20-3 is a protein factor governing the stress-responsive cellular redox homeostasis. It links photosynthetic electron transport and the cysteine-based thiol biosynthesis pathway, and allows plants to adjust to light and stress conditions (Dominguez-Solis et al. 2008).

Recently, AtCYP20-3 is identified as a chloroplast acceptor for jasmonates (JA), well-known plant hormones functioning in plant development and stress acclimation. After binding with (+)-12-oxo-phytodienoic acid (OPDA), a jasmonate molecule, AtCYP20-3 forms a complex with SAT 1, which leads to the formation of a hetero-oligomeric cysteine synthase complex (CSC) in chloroplasts. The CSC formation promotes sulfur assimilation, consequently resulting in increased levels of thiol metabolites and cellular reduction potential. The enhanced redox capacity in chloroplasts regulates the expression of OPDA-responsive genes. It is concluded that CYP20-3 is a key effecter linking the JA signaling pathway to the cellular redox homeostasis in stress responses (Park et al. 2013)

It is proposed that CYP20-3 binds to 2-CysPrx and form a CYP20-3/2-CysPrx complex under normal conditions; but during stress, an increased level of OPDA is generated, and it binds CYP20-3 and releases CYP20-3 from the CYP20-3/2-CysPrx complex, then the OPDA-binding CYP20-3 recruits SAT1 to form CSC. Finally, buildup of cellular reduction potential triggered by OPDA/CYP20-3 binding reverses the conformational states of 2-CysPrx and/or CYP20-3, which leads to the reformation of CYP20-3/2-CysPrx complex (Park et al. 2013; Kopriva 2013).

In plants, the previously identified major receptor for JA is a nucleus-localized protein, named as COI1, which is key subunit of ubiquitin E3 ligase complex. Upon binding of JA, COI1 targets a number of repressor proteins and promotes the degradation of repressor proteins. As a result, removal of repressors relieves inhibition of a specific set of JA-responsive transcription factors and activates JA-responsive gene expression. The mechanism of JA signaling seems well understood, but there are some certain JA-regulated genes that are not COI1-dependent, leading to the speculation that there is an alternative JA signaling pathway, in addition to the COI1-dependent cascade (Kopriva 2013). Identification of chloroplast AtCYP20-3 as a JA receptor not only confirms the existence of the alternative JA signaling pathway, also connects JA signaling pathway with chloroplast retrograde signaling, sulfur metabolism, and redox regulation (Park et al. 2013).

14.5.5 The Function of Lumen Immunophilins

14.5.5.1 AtFKBP13

AtFKBP13 is an active PPIase with a very high enzymatic activity. There are two proximal pairs of Cys residues in its primary structure, which are able to form two disulfide bonds. The two pairs of Cys residues are found tightly related to the PPIase activity of AtFKBP13 protein. AtFKBP13 is in an oxidized form under native conditions. Disruption of either of the two-disulfide bonds, or reduction of the protein by thioredoxin, leads to a reduction in its PPIase activity (Gopalan et al. 2004). For most enzymes involved biosynthesis, the reduction (conversion of S–S to 2SH) usually activates their activity. Interestingly, the PPIase activity of AtFKBP13 is activated by oxidation (conversion of 2SH to S–S) (Gopalan et al. 2004).

A yeast two-hybrid screening identified the Rieske protein, Fe–S subunit of the cytochrome *b6f* complex, as an interacting partner of AtFKBP13 protein. It is interesting that interaction between the two proteins can only occur when AtFKBP13 is in its precursor form, which contains the chloroplast, the transit peptide (Gupta et al. 2002). AtFKBP13 may be involved in the regulation or/and assembly of cytochrome *b6f* levels in a noncanonical mechanism.

14.5.5.2 AtFKBP16-1

AtFKBP16-1 is a divergent immunophilin, containing only conserved amino acid residues essential for PPIase activity. It is very likely that it is not an active PPIase. AtFKBP16-1 mRNA and protein levels are upregulated by high light stress, hinting that this might play a role under stress condition. Indeed, transgenic plants with a high level of AtFKBP16-1 expression exhibited an increased tolerance against high light stress. Plants overexpressing AtFKBP16-1 showed increased accumulation of PsaL, a subunit of PSI, under various abiotic stresses. The role of AtFKBP16-1 playing in the response to photosynthetic stress conditions could be related to its function in regulating PsaL stability (Seok et al. 2014).

14.5.5.3 AtFKBP 16-2

In higher plants, the chloroplast thylakoid membrane harbors the NAD(P)H dehydrogenase (NDH) complex, which functions in the photosynthetic cyclic electron transfer (CET) chain and in the chlororespiratory electron transfer chain. The NDH complex usually interacts with the photosystem I (PSI) complex to form the NDH-PSI supercomplex. The NDH complex consists of the stroma subcomplex, the membrane subcomplex, and the lumen subcomplex. The lumen-localized AtFKBP16-2 is detected in the NDH-PSI supercomplex by mass spectrometry, making AtFKBP16-2 a lumen subunit of the NDH complex. It is hypothesized that the FKBP domain of AtFKBP16-2 plays a role in the assembly or/and maintaining of the NDH complex (Peng et al 2009).

14.5.5.4 AtFKBP20-2

Two of the five amino acids required for PPIase activity was found in the AtFKBP20-2 sequence. A PPIase assay with recombinant AtFKBP20-2 revealed that it possessed a very limited PPIase activity (500 times lower than that of AtFKBP13), however, the activity is too low to detect in biological samples isolated from chloroplasts with native levels of the protein (Lima et al. 2006; Edvardsson et al. 2007).

A T-DNA insertion in the AtFKBP20-2 gene leads to the mutant plants with a reduced growth, a lower rate of PSII activity, and a reduced oxygen evolution rate. Analysis of isolated thylakoid membranes showed a reduced amount of the photosystem II (PSII) supercomplex in mutant plants. This suggested a role of AtFKBP20-2 in the assembly of the PSII supercomplex, but the mechanism remains unclear (Lima et al. 2006).

14.5.5.5 AtCYP38

Arabidopsis cyclophilin38 (AtCYP38) is a highly divergent cyclophilin. The X-ray analysis of its crystal structure revealed that AtCYP38 contains two distinct domains: an α-helix bundle structure at the N-terminus, an acidic loop in the middle, and a typical CYP β-barrel structure at its C-terminus. The N-terminal helical domain is closely packed with the C-terminal CYP domain, forming a strong intramolecular interaction. The unique structure of AtCYP38 suggested there is an autoinhibition mechanism of its function through an intramolecular interaction (Vasudevan et al. 2012). It is hypothesized that under a lower energy condition, the interaction between the N-terminal domain and the C-terminal domain prevent other proteins getting access into the functional CYP domain. Under a higher energy condition, AtCYP38 changes its confirmation and disrupts the interaction of its N- and C-termini, and then the CYP domain is free to its target proteins.

An in vitro enzyme activity assay showed that AtCYP38 does not possess PPIase activity. AtCYP38 could interact with the E-loop of chlorophyll protein47 (CP47), a component of PSII, through its CYP domain. It suggested that the CYP domain plays a vital role in the function of AtCYP38 via an unidentified mechanism, rather than PPIase activity.

AtCYP38 was found to be essential for the assembly and maintenance of PSII. Mutant plants with a T-DNA insertion in AtCYP38 gene showed a phenotype of stunted growth and a very high hypersensitivity to high light. The mutant plants had a decreased level of in PSII. Deficiency of AtCYP38 did not impair the biosynthesis of individual thylakoid proteins in chloroplast, but did lead to a malfunction in the assembly of the PSII supercomplex. D1 and D2 proteins, two key core subunits of PSII, showed an elevated degradation rate in the mutant plants lacking AtCYP38 protein. The severe phenotype displayed by the mutant plants entitled the significance of AtCYP38 in PSII function (Fu et al 2007; Sirpiö et al. 2008).

The *Arabidopsis* AtCYP38 is an inactive PPIase, while its spinach orthologue TLP40 is known to be an active PPIase, which is observed with an in vitro with enzymatic activity assay (Fulgosi et al., 1998). The reason resulting in the difference of these two proteins is to be determined, since they share a high overall sequence identity (82 % sequence identity). Spinach TLP40 was proposed to be involved in the phosphorylation process of thylakoid proteins (Fulgosi et al. 1998).

14.5.5.6 AtCYP20-2

AtCYP20-2 is one of two immunophilins with an active PPIase activity in thylakoid lumen (Edvardsson et al. 2007). A blue native gel analysis revealed that AtCYP20-2 co-migrates with the NAD(P)H dehydrogenase (NDH) complex. Like the AtFKBP16-2, AtCYP20-2 is another lumen subunit of NDH complex. However, mutant plants lacking AtCYP20-2 did not show severe malfunction of NDH, suggesting redundancy in the function of lumen immunophilins (Sirpiö et al. 2009).

It seems that AtCYP20-2 is not essential. It is reported that the mRNA and protein level of rice CYP20-2 (OsCYP20-2) is upregulated under abiotic stress. Overexpression of OsCYP20-2 in both tobacco and *Arabidopsis* leads to transgenic plants with enhanced tolerance against osmotic stress and high light condition. Based on these results, CYP20-2 is believed to take a vital role in helping plants deal with environmental stresses (Kim et al. 2012).

Both yeast two-hybrid and pull-down assays identified AtCYP20-2 as an interacting partner of BZR1, a transcription factor responding to the plant hormone brassinosteroids (BRs). It is well established that BRs regulate plant development in many aspects. It is found that BZR1 can change its confirmation upon interacting with AtCYP20-2. It is thought that AtCYP20-2 catalyzed the conformational change of BZR1 via its PPIase activity, further regulated flowering timing through modulation of FLD expression (Zhang et al. 2013).

However, AtCYP20-2 is a protein located in the chloroplast thylakoid lumen, which is reported in several studies (Romano et al. 2004; Schubert et al. 2002; Peltier et al. 2002). The BZR1 is a transcription factor located in the nucleus. How the lumen-localized AtCYP20-2 regulates the nucleus-localized BZR1 is a very interesting question to be determined further.

14.6 Concluding Remarks

Comparing to animals, plants possess a much larger immunophilin family. It is intriguing and perplexing to answer why plants need such a big family of immunophilins. After 20 years of studying the physiological roles of higher plant immunophilins, plant immunophilins have established their involvement in plant development, stress response, and photosynthesis. Their structure and functions have reiterated their diversity and complexity. A large number of immunophilins are abundantly present in the chloroplast, indicating they are important for chloroplast function. However, only a few chloroplast immunophilins are found to be essential for photosynthesis. Chloroplast immunophilins probably are evolved to protect photosynthesis in response to environmental stress. Future investigations into the influence of immunophilins to interacting proteins' modification during stress conditions could gain insight into their roles in the chloroplasts.

Acknowledgements Research in AF's lab is supported by funding from National Natural Science Foundation of China (No. 31270284) and BRJH of Shaanxi Province, China.

References

Ahn J, Kim DW, You Y, Seok M, Park J, Hwang H, Kim BG, Luan S, Park HS, Cho HS (2010) Classification of rice (Oryza sativa L. japonica Nipponbare) immunophilins (FKBPs, CYPs) and expression patterns under water stress. BMC Plant Biol 10:253–275

Arevalo-Rodrıguez M, Heitman J (2005) Cyclophilin A is localized to the nucleus and controls meiosis in *Saccharomyces cerevisiae*. Eukarot Cell 4(1):17–29

Arevalo-Rodriguez M, Wu X, Hanes SD, Heitman J (2004) Prolyl isomerases in yeast. Front Biosci 9:2420–46

Aviezer-Hagai K, Skovorodnikova J, Galigniana M, Farchi-Pisanty O, Maayan E, Bocovza S, Efrat Y, Koskull-Doring P, Ohad N, Breiman A (2007) Arabidopsis immunophilins ROF1 (AtFKBP62) and ROF2 (AtFKBP65) exhibit tissue specificity, are heat-stress induced, and bind Hsp90. Plant Mol Biol 63:237–255

Bell A, Monaghan P, Page AP (2006) Peptidyl-prolyl cis–trans isomerases (immunophilins) and their roles in parasite biochemistry, host–parasite interaction and antiparasitic drug action. Inter J Parasit 36:261–276

Berardini TZ, Bollman K, Sun H, Poethig RS (2001) Regulation of vegetative phase change in Arabidopsis thaliana by cyclophilin 40. Science 291:2405–2407

Bierer BE, Somers PK, Wandless TJ, Burakoff SJ, Schreiber SL (1990) Probing immunosuppressant action with a non-natural immunophilin ligand. Science 250:556–559

Bissoli G, Ninoles R, Fresquet S, Palombieri S, Bueso E, Rubio L, Garcia-Sanchez M, Fernandez JA, Mulet JM, Serrano R (2012) Peptidyl-prolyl cis-trans isomerase ROF2 modulates intracellular pH homeostasis in Arabidopsis. Plant J 70:704–716

Blecher O, Erel N, Callebaut I, Aviezer K, Breiman A (1996) A novel plant peptidyl-prolyl cis-trans-isomerase (PPIase): cDNA cloning, structural analysis, enzymatic activity and expression. Plant Mol Biol 32:493–504

Bouchard R, Bailly A, Blakeslee JJ, Oehring SC, Vincenzetti V, Lee OR, Paponov I, Palme K, Mancuso S, Murphy AS, Schulz B, Geisler M (2006) Immunophilin-like TWISTED DWARF1 modulates auxin efflux activities of Arabidopsis P-glycoproteins. J Biol Chem 281(41):30603–12

Brandts JF, Halvorson HR, Brennan M (1975) Consideration of the possibility that the slow step in protein denaturation reactions is due to cis-trans isomerism of proline residues. Biochemistry 14:4953–4963

Buchanan BB, Luan S (2005) Redox regulation in the chloroplast thylakoid lumen: a new frontier in photosynthesis research. J Exp Bot 56(416):1439–1447

Cai W, Ma J, Guo J, Zhang L (2008) Function of ROC4 in the efficient repair of photodamaged photosystem II in Arabidopsis. Photochem Photobiol 84(6):1343–8

Calne RY, White DJG, Thiru S, Evans DB, McMaster P, Dunn DC (1978) Cyclosporin A in patients receiving renal allografts from cadaver donors. The Lancet II:1323–1327

Carol RJ, Breiman A, Erel N, Vittorioso P, Bellini C (2001) Pasticcino1 (AtFKBP70) is a nuclear-localised immunophilin required during Arabidopsis thaliana embryogenesis. Plant Sci 161:527–535

Chung KC, Zamble DB (2011) The Escherichia Coli metal-binding chaperone SlyD interacts with the large subunit of [NiFe]-hydrogenase 3. FEBS Lett 585(2):291–4

Coaker G, Falick A, Staskawicz B (2005) Activation of a phytopathogenic bacterial effector protein by a eukaryotic cyclophilin. Science 308(5721):548–50

Coaker G, Zhu G, Ding Z, Van Doren SR, Staskawicz B (2006) Eukaryotic cyclophilin as a molecular switch for effector activation. Mol Microbiol 61(6):1485–96

Crespo JL, Díaz-Troya S, Florencio FJ (2005) Inhibition of target of rapamycin signaling by rapamycin in the unicellular green alga Chlamydomonas reinhardtii. Plant Physiol 139(4):1736–49

Davis ES, Becker A, Heitman J, Hall MN, Brennan MB (1992) Proc Natl Acad Sci U S A 89:11169–11172

Deng W, Chen L, Wood DW, Metcalfe T, Liang X, Gordon MP, Comai L, Nester EW (1998) Agrobacterium VirD2 protein interacts with plant host cyclophilins. Proc Natl Acad Sci U S A 95:7040–7045

Dhillon N, Jeremy TJ (1996) Immunophilins in the yeast Saccharomyces cerevisiae: a different spin on proline rotamases. Methods 9:165–176

Dolinski K, Muir S, Cardenas M, Heitman J (1997) All cyclophilins and FK506 binding proteins are, individually and collectively, dispensable for viability in Saccharomyces cerevisiae. Proc Natl Acad Sci U S A 94(24):13093–8

Dominguez-Solis JR, He Z, Lima A, Ting J, Buchanan BB, Luan S (2008) A cyclophilin links redox and light signals to cysteine biosynthesis and stress responses in chloroplasts. Proc Natl Acad Sci U S A 105(42):16386–16391

Duina AA, Chang HC, Marsh JA, Lindquist S, Gaber RF (1996) A cyclophilin function in Hsp90-dependent signal transduction. Science 274(5293):1713–5

Earley KW, Poethig RS (2011) Binding of the cyclophilin 40 ortholog SQUINT to Hsp90 protein is required for SQUINT function in Arabidopsis. J Biol Chem 286(44):38184–9

Edvardsson A, Shapiguzov A, Petersson UA, Schröder WP, Vener AV (2007) Immunophilin AtFKBP13 sustains all peptidyl-prolyl isomerase activity in the thylakoid lumen from Arabidopsis thaliana deficient in AtCYP20-2. Biochemistry 46(33):9432–42

Edwards KJ, Ollis DL, Dixon NE (1997) Crystal structure of cytoplasmic Escherichia coli peptidyl-prolyl isomerase: evidence for decreased mobility of loops upon complexation. J Mol Biol 271(2):258–65

Etzkorn FA, Chang ZY, Stolz LA, Walsh CT (1994) Cyclophilin residues that affect noncompetitive inhibition of the protein serine phosphatase activity of calcineurin by the cyclophilin-Cyclosporin A complex. Biochemistry 33:2380–2388

Faure JD, Gingerich D, Howell SH (1998) An Arabidopsis immunophilin, AtFKBP12, binds to AtFIP37 (FKBP interacting protein) in an interaction that is disrupted by FK506. Plant J 15:783–789

Fischer G, Wittmann-Liebold B, Lang K, Kiefhaber T, Schmid FX (1989) Cyclophilin and peptidyl-prolyl *cis-trans* isomerase are probably identical proteins. Nature 337:476–478

Freeman BC, Toft DO, Morimoto RI (1996) Molecular chaperone machines: chaperone activities of the cyclophilin Cyp-40 and the steroid aporeceptor-associated protein p23. Science 274:1718–1720

Freskgard PO, Bergenhem N, Jonsson BH, Svensson M, Carlsson U (1992) Isomerase and chaperone activity of prolyl isomerase in the folding of carbonic anhydrase. Science 258:466–468

Fu A, He Z, Cho HS, Lima A, Buchanan BB, Luan S (2007) A chloroplast cyclophilin functions in the assembly and maintenance of photosystem II in Arabidopsis thaliana. PNAS USA 104(40):15947–15952

Fulgosi H, Vener AV, Altschmied L, Herrmann RG, Andersson B (1998) A novel multi-functional chloroplast protein: identification of a 40 kDa immunophilin-like protein located in the thylakoid lumen. EMBO J 17:1577–1587

Futer O, DeCenzo MT, Aldape RA, Livingston DJ (1995) FK506 binding protein mutational analysis. J Biol Chem 270:18935–18940

Ge X, Lyu ZX, Liu Y, Wang R, Zhao XS, Fu X, Chang Z (2014) Identification of FkpA as a key quality control factor for the biogenesis of outer membrane proteins under heat shock conditions. J Bacteriol 196(3):672–80

Geisler M, Kolukisaoglu HU, Bouchard R, Billion K, Berger J, Saal B, Frangne N, Koncz-Kalman Z, Koncz C, Dudler R, Blakeslee JJ, Murphy AS, Martinoia E, Schulz B (2003) TWISTED DWARF1, a unique plasma membrane-anchored immunophilin-like protein, interacts with Arabidopsis multidrug resistance-like transporters AtPGP1 and AtPGP19. Mol Biol Cell 14(10):4238–49

Geisler M, Girin M, Brandt S, Vincenzetti V, Plaza S, Paris N, Kobae Y, Maeshima M, Billion K, Kolukisaoglu UH, Schulz B, Martinoia E (2004) Arabidopsis immunophilin-like TWD1 functionally interacts with vacuolar ABC transporters. Mol Biol Cell 15(7):3393–405

Gollan PJ, Bhave M (2010) Genome-wide analysis of genes encoding FK506-binding proteins in rice.Plant. Mol Biol 72(1–2):1–16

Granzin J, Eckhoff A, Weiergräber OH (2006) Crystal structure of a multi-domain immunophilin from Arabidopsis thaliana: a paradigm for regulation of plant ABC transporters. J Mol Biol 364(4):799–809

Graubner W, Schierhorn A, Bruser T (2007) DnaK plays a pivotal role in Tat targeting of CueO and functions beside SlyD as a general Tat signal binding chaperone. J Biol Chem 282: 7116–7124

Handschumacher RE, Harding MW, Rice J, Drugge RJ, Speicher DW (1984) Cyclophilin: a specific cytosolic binding protein for cyclosporin A. Science 226:544–547

Harding MW, Galat A, Uehling DE, Schreiber SL (1989) A receptor for the immunosuppressant FK-506 is a *cis-trans* peptidyl-prolyl isomerase. Nature 341:758–760

He Z, Li L, Luan S (2004) Immunophilins and parvulins. Superfamily of peptidyl prolyl isomerases in Arabidopsis. Plant Physiol 134:1248–1267

Heitman J, Movva NR, Hiestand PC, Hall MN (1991) FK 506-binding protein proline rotamase is a target for the immunosuppressive agent FK 506 in Saccharomyces cerevisiae. Proc Natl Acad Sci U S A 88(5):1948–52

Hochwagen A, Tham WH, Brar GA, Amon A (2005) The FK506 binding protein Fpr3 counteracts protein phosphatase 1 to maintain meiotic recombination checkpoint activity. Cell 122(6): 861–73

Hullmann J, Patzer SI, Römer C, Hantke K, Braun V (2008) Periplasmic chaperone FkpA is essential for imported colicin M toxicity. Mol Microbiol 69(4):926–37

Ingelsson B, Shapiguzov A, Kieselbach T, Vener AV (2009) Peptidyl-prolyl isomerase activity in chloroplast thylakoid lumen is a dispensable function of immunophilins in Arabidopsis thaliana. Plant Cell Physiol 50(10):1801–14

Justice SS, Hunstad DA, Harper JR, Amy R, Duguay AR, Pinkner JS, Bann J, Frieden C, Silhavy TJ, Hultgren SJ (2005) Periplasmic peptidyl prolyl cis-trans isomerases are not essential for viability, but SurA is required for pilus biogenesis in Escherichia coli. J Bacteriol 187(22):7680–7686

Kaluarachchi H, Zhang JW, Zamble DB (2011) Escherichia coli SlyD, more than a Ni(II) reservoir. Biochemistry 50(50):10761–3

Kamath RS, Fraser AG, Dong Y, Poulin G, Durbin R, Gotta M, Kanapin A, Le Bot N, Moreno S, Sohrmann M, Welchman DP, Zipperlen P, Ahringer J (2003) Systematic functional analysis of the Caenorhabditis elegans genome using RNAi. Nature 421(6920):231–7

Kang B, Zhang Z, Wang L, Zheng L, Mao W, Li M, Wu Y, Wu P, Mo X (2013) OsCYP2, a chaperone involved in degradation of auxin-responsive proteins, plays crucial roles in rice lateral root initiation. Plant J 74(1):86–97

Kiefhaber T, Quaas R, Hahn U, Schmid FX (1990) Folding of ribonuclease T1. 1. Existence of multiple unfolded states created by proline isomerization. Biochemistry 29:3053–3061

Kim IS, Yun HS, Kwak SH, Jin IN (2007) The physiological role of CPR1 in Saccharomyces cerevisiae KNU5377 against menadione stress by proteomics. J Microbiol 45(4):326–32

Kim SK, You YN, Park JC, Joung Y, Kim BG, Ahn JC, Cho HS (2012) The rice thylakoid lumenal cyclophilin OsCYP20-2 confers enhanced environmental stress tolerance in tobacco and Arabidopsis. Plant Cell Rep 31(2):417–26

Kino T, Hatanaka H, Miyata S, Inamura N, Nishiyama M, Yajima T, Goto T, Okuhara M, KohsakaM AH (1987) FK-506, a novel immunosuppressant isolated from Streptomyces. II. Immunosuppressive effect of FK-506 in vitro. J Antibiot 40:256–1265

Kopriva S (2013) 12-Oxo-phytodienoic acid interaction with cyclophilin CYP20-3 is a benchmark for understanding retrograde signaling in plants. Proc Natl Acad Sci U S A 110(23): 9197–8

Kovalev N, Nagy PD (2013) Cyclophilin A binds to the viral RNA and replication proteins, resulting in inhibition of tombusviral replicase assembly. J Virol 87(24):13330–42

Kumari S, Roy S, Singh P, Singla-Pareek SL, Pareek A (2013) Cyclophilins: proteins in search of function. Plant Signal Behav 8(1):e22734

Laxa M, König J, Dietz KJ, Kandlbinder A (2007) Role of the cysteine residues in Arabidopsis thaliana cyclophilin CYP20-3 in peptidyl-prolyl cis-trans isomerase and redox-related functions. Biochem J 401(1):287–97

Li H, Luan S (2010) AtFKBP53 is a histone chaperone required for repression of ribosomal RNA gene expression in Arabidopsis. Cell Res 20(3):357–66

Li H, Luan S (2011) The cyclophilin AtCYP71 interacts with CAF-1 and LHP1 and functions in multiple chromatin remodeling processes. Mol Plant 4(4):748–58

Li H, He Z, Lu G, Lee SC, Alonso J, Ecker JR, Luan S (2007) A WD40 domain cyclophilin interacts with histone H3 and functions in gene repression and organogenesis in Arabidopsis. Plant Cell 19(8):2403–16

Lima A, Lima S, Wong JH, Phillips RS, Buchanan BB, Luan S (2006) A redox-active FKBP-type immunophilin functions in accumulation of the photosystem II supercomplex in Arabidopsis thaliana. Proc Natl Acad Sci U S A 103(33):12631–6

Lippuner V, Chou IT, Scott SV, Ettinger WF, Theg SM, Gasser CS (1994) Cloning and characterization of chloroplast and cytosolic forms of cyclophilin from Arabidopsis thaliana. J Biol Chem 269(11):7863–8

Liu J, Walsh CT (1990) Peptidyl-prolyl cis-trans-isomerase from Escherichia coli: a periplasmic homolog of cyclophilin that is not inhibited by cyclosporin A. Proc Natl Acad Sci U S A 87(11):4028–32

Luan S (1998) Immunophilins in animals and higher plants. Bot Bullet Acad Sini 39:217–223

Ma X, Song L, Yang Y, Liu D (2013) A gain-of-function mutation in the ROC1 gene alters plant architecture in Arabidopsis. New Phytol 197(3):751–62

MacArthur MW, Thornton JM (1991) Influence of proline residues on protein conformation. J Mol Biol 218:397–412

Magiri EN, Farchi-Pisanty O, Avni A, Breiman A (2006) The expression of the large rice FK506 binding proteins (FKBPs) demonstrate tissue specificity and heat stress responsiveness. Plant Sci 170:695–704

Matouschek A, Rospert S, Schmid K, Glick BS, Schatz G (1995) Cyclophilin catalyzes protein folding in yeast mitochondria. Proc Natl Acad Sci U S A 92:6319–6323

Mayr C, Richter K, Lilie H, Buchner J (2000) Cpr6 and Cpr7, two closely related Hsp90-associated immunophilins from Saccharomyces cerevisiae, differ in their functional properties. J Biol Chem 275(44):34140–6

Meiri D, Breiman A (2009) Arabidopsis ROF1 (FKBP62) modulates thermotolerance by interacting with HSP90.1 and affecting the accumulation of HsfA2-regulated sHSPs. Plant J 59:387–399

Meiri D, Tazat K, Cohen-Peer R, Farchi-Pisanty O, Aviezer-Hagai K, Avni A, Breiman A (2010) Involvement of Arabidopsis ROF2 (FKBP65) in thermotolerance. Plant Mol Biol 72:191–203

Nigam N, Singh A, Sahi C, Chandramouli A, Grover A (2008) Sumoconjugating enzyme (Sce) and FK506-binding protein (FKBP) encoding rice (Oryza sativa L.) genes: genome-wide analysis, expression studies and evidence for their involvement in abiotic stress response. Mol Genet Genomics 279:371–383

Page AP, MacNiven K, Hengartner MO (1996) Cloning and biochemical characterisation of the cyclophilin homologues from the free-living nematode Caenorhabditis elegans. Biochem J 317:179–185

Park SW, Li W, Viehhauser A, He B, Kim S, Nilsson AK, Andersson MX, Kittle JD, Ambavaram MM, Luan S, Esker AR, Tholl D, Cimini D, Ellerström M, Coaker G, Mitchell TK, Pereira A, Dietz KJ, Lawrence CB (2013) Cyclophilin 20-3 relays a 12-oxo-phytodienoic acid signal during stress responsive regulation of cellular redox homeostasis. Proc Natl Acad Sci U S A 110(23):9559–64

Partaledis JA, Berlin V (1993) The FKB2 gene of Saccharomyces cerevisiae, encoding the immunosuppressant-binding protein FKBP-13, is regulated in response to accumulation of unfolded proteins in the endoplasmic reticulum. Proc Natl Acad Sci U S A 90(12):5450–4

Partaledis JA, Fleming MA, Harding MW, Berlin V (1992) Saccharomyces cerevisiae contains a homolog of human FKBP-13, a membrane-associated FK506/rapamycin binding protein. Yeast 8(9):815

Peltier JB, Emanuelsson O, Kalume DE, Ytterberg J, Friso G, Rudella A, Liberles DA, Söderberg L, Roepstorff P, von Heijne G, van Wijk KJ (2002) Central functions of the lumenal and peripheral thylakoid proteome of Arabidopsis determined by experimentation and genome-wide prediction. Plant Cell 14:211–236

Peng L, Fukao Y, Fujiwara M, Takami T, Shikanai T (2009) Efficient operation of NAD(P)H dehydrogenase requires supercomplex formation with photosystem I via minor LHCI in Arabidopsis. Plant Cell 21(11):3623–40

Picken NC, Eschenlauer S, Taylor P, Page AP, Walkinshaw MD (2002) Structural and biological characterisation of the gut-associated cyclophilin B isoforms from Caenorhabditis elegans. J Mol Biol 322:15–25

Pogorelko GV, Mokryakova M, Fursova OV, Abdeeva I, Piruzian ES, Bruskin SA (2014) Characterization of three Arabidopsis thaliana immunophilin genes involved in the plant defense response against Pseudomonas syringae. Gene 538(1):12–22

Quistgaard EM, Nordlund P, Löw C (2012) High-resolution insights into binding of unfolded polypeptides by the PPIase chaperone SlpA. FASEB J 26(10):4003–13

Rao A, Luo C, Hogan PG (1997) Transcription factors of the NFAT family: regulation and function. Annu Rev Immuno 15:707–747

Riggs DL, Roberts PJ, Chirillo SC, Cheung-Flynn J, Prapapanich V, Ratajczak T, Gaber R, Picard D, Smith D (2003) The Hsp90-binding peptidylprolyl isomerase FKBP52 potentiates glucocorticoid signaling in vivo. Eur Mol Biol Org J 22:1158–1167

Romano PGN, Edvardsson A, Ruban AV, Andersson B, Vener AVV, Gray JE, Horton P (2004) Arabidopsis AtCYP20-2 is a light-regulated cyclophilintype peptidyl-prolyl cis-trans isomerase associated with the photosynthetic membranes. Plant Physiol 134:1244–1247

Roudier F, Gissot L, Beaudoin F, Haslam R, Michaelson L, Marion J, Molino D, Lima A, Bach L, Morin H, Tellier F, Palauqui JC, Bellec Y, Renne C, Miquel M, DaCosta M, Vignard J, Rochat C, Markham JE, Moreau P, Napier J, Faure JD (2010) Very-long-chain fatty acids are involved in polar auxin transport and developmental patterning in Arabidopsis. Plant Cell 22:364–375

Schreiber SL (1991) Chemistry and biology of immunophilins and their immunosuppressive ligands. Science 251:283–287

Schreiber SL, Crabtree GR (1992) The mechanism of action of cyclosporine A and FK-506. Immunp Tod 13:136–142

Schubert M, Petersson UA, Haas BJ, Funk C, Schröder WP, Kieselbach T (2002) Proteome map of the chloroplast lumen of Arabidopsis thaliana. J Biol Chem 277:8354–8365

Seok MS, You YN, Park HJ, Lee SS, Aigen F, Luan S, Ahn JC, Cho HS (2014) AtFKBP16-1, a chloroplast lumenal immunophilin, mediates response to photosynthetic stress by regulating PsaL stability. Physiol Plant 150(4):620–31

Shangguan L, Kayesh E, Leng X, Sun X, Korir NK, Mu Q, Fang J (2013) Whole genome identification and analysis of FK506-binding protein family genes in grapevine (Vitis vinifera L.). Mol Biol Rep 40(6):4015–31

Shapiguzov A, Edvardsson A, Vener AV (2006) Profound redox sensitivity of peptidyl-prolyl isomerase activity in Arabidopsis thylakoid lumen. FEBS Lett 580:3671–3676

Siekierka JJ, Hung SH, Poe M, Lin CS, Sigal NH (1989) A cytosolic binding protein for the immunosuppressant FK506 has peptidyl-prolyl isomerase activity but is distinct from cyclophilin. Nature 341:755–757

Sirpiö S, Khrouchtchova A, Allahverdiyeva Y, Hansson M, Fristedt R, Vener AV, Scheller HV, Jensen PE, Haldrup A, Aro EM, 4 (2008) AtCYP38 ensures early biogenesis, correct assembly and sustenance of photosystem II. Plant J 55:639–51

Sirpiö S, Holmström M, Battchikova N, Aro EM (2009) AtCYP20-2 is an auxiliary protein of the chloroplast NAD(P)H dehydrogenase complex. FEBS Lett 583(14):2355–8

Smith MR, Willmann MR, Wu G, Berardini TZ, Möller B, Weijers D, Poethig RS (2009) Cyclophilin 40 is required for microRNA activity in Arabidopsis. Proc Natl Acad Sci U S A 106(13):5424–9

Smyczynski C, Roudier F, Gissot L, Vaillant E, Grandjean O, Morin H, Masson T, Bellec Y, Geelen D, Faure JD (2006) The C terminus of the immunophilin PASTICCINO1 is required for plant development and for interaction with a NAC-like transcription factor. J Biol Chem 281:25475–25484

Stamnes MA, Shieh BH, Chuman L, Harris GL, Zuker CS (1991) The cyclophilin homolog ninaA is a tissue-specific integral membrane protein required for the proper synthesis of a subset of Drosophila rhodopsins. Cell 65:219–227

Standaert RF, Galat A, Verdine GL, Schreiber SL (1990) Molecular cloning and overexpression of the human FK506-binding protein FKBP. Nature 346:671–674

Starzl TE, Klintmalm GB, Porter KA, Iwatsuki S, Schröter GP (1981) Liver transplantation with use of cyclosporin a and prednisone. N Engl J Med 305(5):266–9

Steimann B, Bruckner P, Superti-Furga A (1991) Cyclosporin A slows collagen triple-helix formation in vivo: indirect evidence for a physiologic role of peptidyl-prolyl cis-trans isomerase. J Biol Chem 266:1299–1303

Svarstad H, Bugge HC, Dhillion SS (2000) From Norway to Novartis: Cyclosporin from Tolypocladium inflatum in an open access bioprospecting regime. Biodiver Conser 9(11):1521–1541

Takahashi N, Hayano T, Suzuki M (1989) Peptidyl-prolyl *cis-trans* isomerase is the cyclosporin A-binding protein cyclophilin. Nature 337:473–475

Trivedi DK, Yadav S, Vaid N, Tuteja N (2012) Genome wide analysis of Cyclophilin gene family from rice and Arabidopsis and its comparison with yeast. Plant Signal Behav 7(12):1653–66

Trupkin SA, Mora-García S, Casal JJ (2012) The cyclophilin ROC1 links phytochrome and cryptochrome to brassinosteroid sensitivity. Plant J 71(5):712–23

Vallon O (2005) *Chlamydomonas* immunophilins and parvulins: survey and critical assessment of gene models. Eukarot Cell 4(2):230–241

Vasudevan D, Fu A, Luan S, Swaminathan K (2012) Crystal structure of Arabidopsis cyclophilin38 reveals a previously uncharacterized immunophilin fold and a possible autoinhibitory mechanism. Plant Cell 24(6):2666–74

Vespa L, Vachon G, Berger FDR, Perazza D, Faure JD, Herzog M (2004) The immunophilin-interacting protein AtFIP37 from Arabidopsis is essential for plant development and is involved in trichome endoreduplication. Plant Physiol 134:1283–1292

Vittorioso P, Cowling R, Faure JD, Caboche M, Bellini C (1998) Mutation in the Arabidopsis PASTICCINO1 gene, which encodes a new FK506- binding protein-like protein, has a dramatic effect on plant development. Mol Cell Biol 18:3034–3043

Wang WW, Ma Q, Xiang Y, Zhu SW, Cheng BJ (2012) Genome-wide analysis of immunophilin FKBP genes and expression patterns in Zea mays. Genet Mol Res 11(2):1690–700

Wang B, Bailly A, Zwiewka M, Henrichs S, Azzarello E, Mancuso S, Maeshima M, Friml J, Schulz A, Geisler M (2013) Arabidopsis TWISTED DWARF1 functionally interacts with auxin exporter ABCB1 on the root plasma membrane. Plant Cell 25(1):202–14

Wilson LK, Benton BM, Zhou S, Thorner J, Martin GS (1995) The yeast immunophilin Fpr3 is a physiological substrate of the tyrosine-specific phosphoprotein phosphatase Ptp1. J Biol Chem 270(42):25185–93

Wu G, Otegui MS, Spalding EP (2010) The ER-localized TWD1 immunophilin is necessary for localization of multidrug resistance-like proteins required for polar auxin transport in Arabidopsis roots. Plant Cell 22(10):3295–304

Xiao H, Jackson V, Lei M (2006) The FK506-binding protein, Fpr4, is an acidic histone chaperone. FEBS Lett 580(18):4357–64

Yu Y, Li Y, Huang G, Meng Z, Zhang D, Wei J, Yan K, Zheng C, Zhang L (2011) PwHAP5, a CCAAT-binding transcription factor, interacts with PwFKBP12 and plays a role in pollen tube growth orientation in Picea wilsonii. J Exp Bot 62:805–4817

Yu Y, Zhang H, Li W, Mu C, Zhang F, Wang L, Meng Z (2012) Genome-wide analysis and environmental response profiling of the FK506-binding protein gene family in maize (Zea mays L.). Gene 498:212–222

Zhang JW, Leach MR, Zamble DB (2007) The peptidyl-prolyl isomerase activity of SlyD is not required for maturation of Escherichia coli hydrogenase. J Bacteriol 189(21):7942–4

Zhang Y, Li B, Xu Y, Li H, Li S, Zhang D, Mao Z, Guo S, Yang C, Weng Y, Chong K (2013) The cyclophilin CYP20-2 modulates the conformation of BRASSINAZOLE-RESISTANT1, which binds the promoter of FLOWERING LOCUS D to regulate flowering in Arabidopsis. Plant Cell 25(7):2504–21

Zuehlke AD, Johnson JL (2012) Chaperoning the chaperone: a role for the co-chaperone Cpr7 in modulating Hsp90 function in Saccharomyces cerevisiae. Genetics 191(3):805–14

Zydowsky LD, Etzkorn FA, Chang HY, Ferguson SB, Stolz LA, Ho SI, Walsh CT (1992) Active site mutants of human cyclophilin A separate peptidyl-prolyl isomerase activity from cyclosporine A binding and calcineurin inhibition. Prot Sci 1:1092–1099

Index

G.K. Pandey (ed.), *Elucidation of Abiotic Stress Signaling in Plants*,
DOI 10.1007/978-1-4939-2211-6

B

The manufacturer's authorised representative in the EU is Springer
Nature Customer Service Centre GmbH, Europaplatz 3, 69115 Heidelberg,
Germany. If you have any concerns regarding our products, please
contact ProductSafety@springernature.com

Printed and bound by CPI Group (UK) Ltd, Croydon, CR0 4YY
14/10/2026
02233163-0006